A BIBLIOGRAPHY OF
CONIFERS

SECOND EDITION

Majestic trees of *Abies magnifica* in the Sierra Nevada of California. Photograph Aljos Farjon, September 1992

A BIBLIOGRAPHY OF

CONIFERS

SECOND EDITION

SELECTED LITERATURE ON TAXONOMY AND
RELATED DISCIPLINES OF THE

CONIFERALES

COMPILED AND ANNOTATED BY

ALJOS FARJON

ROYAL BOTANIC GARDENS, KEW

PLANTS PEOPLE
POSSIBILITIES

First published in 2005 by
Royal Botanic Gardens, Kew
Richmond, Surrey,
TW9 3AB, UK
www.kew.org

ISBN 1 84246 120 6

Production Editor: Ruth Linklater
Typesetting and page layout: Margaret Newman
Cover design: Jeff Eden, Media Resources

Information Services Department,
Royal Botanic Gardens, Kew

British Library Cataloguing in Publication Data
A catalogue record for this book is available from the British Library.

Printed in Great Britain
by Cambridge Printing

For information or to purchase all Kew titles please visit
www.kewbooks.com or email publishing@kew.org

All proceeds go to support Kew's work in saving the world's plants for life

TABLE OF CONTENTS

INTRODUCTION

The first edition of A BIBLIOGRAPHY OF CONIFERS was published in 1990 by the International Association for Plant Taxonomy (IAPT) as volume 122 in the series Regnum Vegetabile. It is a companion volume to my book PINACEAE (*Regnum Vegetabile* Vol. 121), which is an illustrated taxonomic treatment of all species in that family, but excluding those in the genus *Pinus*, which had been the subject of an earlier book PINES (Farjon, 1984, second edition 2005). The first edition of A BIBLIOGRAPHY OF CONIFERS treated taxonomic literature of conifers, with emphasis on the families Cupressaceae, Pinaceae and the former Taxodiaceae, and contained 2130 references. The present, second edition of that work has been greatly expanded, both in scope and number of references, which now total 3787. The other families have now been treated more comprehensively especially to include the references wherein new taxa were described and named, while the literature of the originally included families has been fully updated. Another field of expansion has been the literature on fossil conifers, from the Permian conifers to the Recent taxa. Since the publication of the first edition, research in molecular phylogenetics using DNA data has become a major field in systematic botany and numerous papers have been added on this subject. In the following paragraphs I intend to explain in more detail how the second edition of A BIBLIOGRAPHY OF CONIFERS is structured, what its contents are, and how it can be used in book form ('hard copy'). It is envisaged that eventually an electronic database searchable by computer manipulation will be made available. A 'primitive' version of this in DOS format exists at present as a private database; it has been kept up-to-date simultaneously with the presently published second edition of A BIBLIOGRAPHY OF CONIFERS.

Selection of literature on conifers

As has been explained in the first edition, the literature, popular as well as scientific, on conifers is very large and ever expanding. Since the first edition was published in 1990, the Internet has caused a virtual explosion of information not available, or only partially available, in traditional printed form. All such electronically disseminated information, in the form of 'websites' or 'databases' or any other format, has been excluded from A BIBLIOGRAPHY OF CONIFERS, which deals exclusively with printed issue, either published in periodicals or as 'independent publications' *sensu* TL-2 (Stafleu & Cowan, 1976–88). An exception, of course, has been made for all those papers that have been published 'online' *as well as* in printed journals, but reference is made to the printed version only. In this author's opinion, much of the information on the Internet is as yet too unconsolidated to be included in a bibliography of permanent literature. A website that exists today can disappear tomorrow, or be altered substantially without notice. There are no public library catalogues to this source of information (again with the exception of those printed journals that now also publish an 'online' version of their papers) although powerful 'search engines' now freely available make it possible to find many specified items and subjects on the Internet. I am aware of the probability that valuable information, in particular database information, has been omitted in A BIBLIOGRAPHY OF CONIFERS as a result of this exclusion. Less valuable information may have been included merely because it appeared in print. Unless a method or protocol can be found and agreed to guarantee greater permanence of Internet information and professional bibliographic reference to its contents e.g. via institutional or public libraries, I am reluctant to include it. These shortcomings have played a decisive rôle in the ruling against electronic publishing so far of new names of taxa under the current codes of biological nomenclature, a situation that is not expected to be resolved in favour of exclusive electronic publishing of taxonomic information in the near future, even though proposals to admit electronic publication have recently been formulated for the International Code of Botanical Nomenclature (Zander, 2004).

As in the first edition, I have followed Clive Stace's maxim that taxonomy is a fundamental discipline and a synthesis of all biological knowledge (Stace, 1989) to guide my choice of literature that is not strictly taxonomic for inclusion. The experienced taxonomist makes use of a wide range of biological information and this bibliography should be an aid to find it. I have therefore continued to expand not only on descriptions and classifications of taxa, but on a wide range of 'subsidiary' disciplines. The coverage of palaeobotanical literature, started in the first edition, has been greatly increased and many more entries are now dealing with fossil conifers. Another subject of study that sees many more entries in this bibliography is phylogenetics, increasingly based on molecular (DNA) data analysis. This is, of course, due to a spectacular growth in this field since the first edition was published 14 years ago. In my own research, and in comparing the results of these molecular studies, I have increasingly been impressed with the importance of palaeobotany for a true understanding of the phylogeny "that probably happened" in conifers, as opposed to the hypotheses inferred from the molecules of extant taxa. The inclusion of palaeobotany to a greater extent than before is in part an attempt to redress the balance between more traditional disciplines and the modern approaches to conifer systematics, again with Stace's maxim in mind. Additions since the first edition include literature published prior to 1990 not only because of the addition of other families and palaeobotany, but also simply as a result of further searches through references in consulted publications. Completeness will never be achieved, but the second edition has come closer to that elusive objective than the first.

As before, I have largely excluded forestry and horticulture, fields of knowledge with less relevance to the science of taxonomy. However, there is a degree of overlap in certain types of books on conifers often used by these two disciplines. As outlined below, the second edition has been restructured to make these more accessible by grouping them as a particular class of publications. Most of these works are compilations of useful information on a large number of species (and often cultivars) and many include descriptions or other information that can be used for identification, which is again relevant to taxonomists. More specifically focused horticultural references and forestry research papers that lack such information are definitely excluded from this bibliography.

Since 1990 I have completed a monograph of Cupressaceae, the fruit of 13 years of study and undoubtedly my magnum opus on conifers. This family includes the former Taxodiaceae (but not the sister group Sciadopityaceae) which is therefore no longer recognised as a distinct family. This merger, first proposed almost a century ago by Saxton (1913) and reiterated by Eckenwalder (1976) and many others since, is supported by evidence from morphology, DNA, phylogenetic analysis and palaeobotany. However, for a bibliography that serves as a tool to find literature, a heading 'Taxodiaceae' is maintained under which references are listed that deal with all or most of the taxa traditionally included. It does not mean that I endorse a classification that recognises this family. Palaeobotanical literature largely continues to refer to Taxodiaceae as a family of conifers and of course there are different concepts here, such as the notion that all fossil names are names for organs or parts of plants, not for whole plants and hence not for species or higher taxa in the same sense as for living plants. However, these ideas are not shared by all palaeobotanists, and some will use names as equivalent to extant taxa. Other taxonomic developments ('progress') in conifers have occurred since the first edition, as a result of revisions, phylogenetic studies and other research. Current nomenclature as indicated in the comments to entries has been updated as much as possible. The main source for this has been my WORLD CHECKLIST AND BIBLIOGRAPHY OF CONIFERS (Farjon, 1998, second edition 2001) which lists all names and synonyms of conifers from family down to variety. The comments or annotations added to many entries are now clearly separated by being placed between brackets [...] and include new names and combinations published in the entry as before, with some exceptions. Indication of synonymy has not been pursued extensively as this information is now more fully available and easier to find in my checklist of conifers cited above.

Structure of the bibliography

As in the first edition, entries are numbered in an alphabetical order by author(s). Original numbers have been retained wherever possible, but problems have arisen with the intercalation of entries, which required additional numbering. In the original setup, terminal numbers 1–9 could be inserted between entries to admit additional literature. This space has not been sufficient in many cases and letters a, b, c, etc. had to be added to new numbers. The purpose of the numbers is to find literature about taxa in the taxonomic index and this system could be maintained in this way. The continuous alphabetical listing of authors from A–Z throughout the bibliography in the first edition has now been broken up into groups or classes. Within each, the order is alphabetical by author and also numerical, but the numbers are not sequential as they are based on the entire list of entries. I have classified the entries as follows:

- Bibliographies and Plant Dictionaries
- Floras
- Manuals
- Gymnosperms (general titles)
- Coniferales (general titles on conifers, including 'taxads')
- Araucariaceae (general titles)
- Cupressaceae (general titles)
- Phyllocladaceae (general titles)
- Pinaceae (general titles)
- Podocarpaceae (general titles)
- Taxaceae (general titles)
- Taxodiaceae (general titles)
- Taxa below family rank

Only entries in the bibliography that have as their subjects taxa with ranks from family down to variety are listed in the taxonomic index. The other more general literature has not been indexed but has now been grouped into useful classes and is presented in alphabetical order by author(s). A BIBLIOGRAPHY OF CONIFERS starts therefore with a list of other bibliographies and dictionaries relevant to conifers. Then follow works classified as Floras, Manuals and those whose contents deal with gymnosperms in general, including conifers, or with conifers only. To distinguish between publications that consider families in a general sense or more specifically certain taxa within these was not always unambiguous, but when in doubt, I have listed the work in the last class, Taxa below family rank. This is still by far the largest part of the bibliography, but further division is difficult as it would introduce increasingly arbitrary pigeon-holing and make literature more difficult to find. I hope that with this new structure the bibliography has been made more user-friendly than before.

Each entry has the following information:

1. A five-digit reference number, where necessary suffixed with a lower case letter to allow new entries without altering the entire numbering system. With this number the entries are indicated under taxon names in the taxonomic index. Protologues and places of publication of new combinations are indicated in the index with bold numbers.
2. The name(s) of the author(s) or editor(s) of the works cited, or of the authors of chapters or taxa published in the works of other authors. If the latter circumstance is the case, there can be two

separate entries: one of the author or editor of the entire work, another of the author of the chapter or taxon (taxa). Annotations are with the second, while reference to this entry follows the first after a taxon name in the index. Treatments of conifers in Floras by separate authors are dealt with in the same manner.

3. The year(s) of actual publication. These dates may be at variance with those indicated on title pages etc. An important source of information on the actual dates of publication of older literature has been 'TL-2' (Stafleu & Cowan, 1976–88) and its Supplements (Stafleu & Mennega, 1992–2000).

4. The title of the publication. Exceptionally lengthy titles, common in eighteenth and nineteenth century publications, have been shortened to the first phrase, followed by ... and an indication of the volumes and/or editions concerned. In some cases, a single title or volume has been split into two or more numbered references when this seemed appropriate in the light of the annotations to be added. Titles of Chinese, Japanese, Korean and often Russian publications are given in translation, mostly taken from the abstracts provided, or from subtitles, but sometimes from other bibliographies.

5. The place of publication. For papers in journals, the abbreviations follow *Botanico-Periodicum-Huntianum* (B-P-H) (Lawrence *et al.*, 1968) and its supplement 'B-P-H/S' (Bridson & Smith, 1991). For independent publications usually only the municipality is given, not the name of the publisher. Pagination is usually indicated in full for papers in periodicals; for books dealing with many other plant groups only the section for gymnosperms or conifers is indicated by page numbers.

6. Comments, including a mention of taxa treated in the publication, are now placed within brackets [...] to separate them clearly from titles and bibliographic references; the latter may include indication of language or the presence of illustrations, placed in parentheses (...). For new taxa or combinations page numbers are usually indicated in the comments, but for a full and complete reference to this detail I can now refer the reader to my WORLD CHECKLIST AND BIBLIOGRAPHY OF CONIFERS (Farjon, 1998; 2ⁿᵈ ed. 2001).

This second edition of A BIBLIOGRAPHY OF CONIFERS includes literature published up to the end of the year 2003 and a few into the first half of 2004. Several authors have expressed their appreciation of this work by sending me unsolicited reprints of their papers, or even in some instances their books. If relevant, they have been rewarded with inclusion and are here thanked for their courtesy. I have also had considerable help, especially from Dr. Thomas A. Zanoni at the New York Botanical Garden, who for years scanned new library acquisitions for periodicals that would often be rare in European libraries and sent me the information. The selection for inclusion was mine, but his efforts have yielded numerous entries and I am in debt of gratitude to him. Assistance of a different kind came from Dr. David G. Frodin at the Royal Botanic Gardens, Kew and Chelsea Physic Garden, London, who advised me to restructure the contents which led to the division in classes made in the second edition. Lastly, I apologize to all those authors who may find (some of) their publications missing in this bibliography. As before, this work has been a spare-time task and time was often at a premium. Perhaps they will be so kind as to send me their reprints for a third edition.

Aljos Farjon

August 2004, at the Royal Botanic Gardens, Kew

References

Bridson, G. D. R. & E. R. Smith (1991). Botanico-Periodicum-Huntianum. Supplement (B-P-H/S). Hunt Institute for Botanical Documentation, Pittsburg, Pa.

Eckenwalder, J. E. (1976). Re-evaluation of Cupressaceae and Taxodiaceae: a proposed merger. Madroño 23 (5): 237–256.

Farjon, A. (1984, 2005). PINES. Drawings and descriptions of the genus *Pinus*. E. J. Brill/Dr. W. Backhuys, Leiden.

Farjon, A. (1990). PINACEAE. Drawings and descriptions of the genera *Abies, Cedrus, Pseudolarix, Keteleeria, Nothotsuga, Tsuga, Cathaya, Pseudotsuga, Larix* and *Picea*. Regnum Vegetabile Vol. 121. Koeltz Scientific Books, Königstein.

Farjon, A. (1998, 2001). World Checklist and Bibliography of Conifers. Royal Botanic Gardens, Kew, Richmond, U.K.

Lawrence, G. H. M., A. F. Günther-Buchheim, G. S. Daniels & H. Dolezal, eds. (1968). Botanico-Periodicum-Huntianum. (B-P-H). Hunt Botanical Library, Pittsburg, Pa.

Saxton, W. T. (1913). The classification of conifers. New Phytologist 12: 242–262.

Stace, C. A. (1989). Plant taxonomy and biosystematics; second edition. Edward Arnold, London – Melbourne – Auckland.

Stafleu, F. A. & R. S. Cowan (1976–88). Taxonomic literature. A selective guide to botanical publications and collections with dates, commentaries and types; second edition (TL-2). Regnum Vegetabile Vols. 94, 98, 105, 110, 112, 115, 116. Bohn, Scheltema & Holkema, Utrecht – Antwerpen/W. Junk, The Hague – Boston.

Stafleu, F. A. & E. A. Mennega (1992–2000). Taxonomic literature. (TL-2). Supplements 1–6. Regnum Vegetabile Vols. 125, 130, 132, 134, 135, 137. Koeltz Scientific Books, Königstein.

Zander, R. H. (2004). (180–181) Report of the Special Committee on Electronic Publishing with two proposals to amend the Code. Taxon 53 (2): 592–594.

NUMBERED ALPHABETICAL LIST OF REFERENCES

BIBLIOGRAPHIES & PLANT DICTIONARIES

03720 Farr, E. R., J. A. Leussink & F. A. Stafleu (eds.) (1979). Index nominum genericorum (plantarum). (ING). 3 Vols.: Regnum Veg. 100–102; Utrecht. [genus names, cit., type spp. etc.].

03720a Farr, E. R., J. A. Leussink & G. Zijlstra (eds.) (1986). Index nominum genericorum (plantarum): Supplementum I. (ING-S). Regnum Veg. 113; The Hague. [genus names, cit., type spp. etc.].

03825 Ferguson, E. R. (1970). Eastern redcedar: an annotated bibliography. U.S. Dep. Agr. Forest Serv. Res. Pap. SO-64, 21 pp. [*Juniperus virginiana*].

04580 Frodin, D. G. (1984). Guide to standard floras of the world. An annotated, geographically arranged systematic bibliography of the principal floras, enumerations, checklists, and chorological atlases of different areas. Cambridge Univ. Press, Cambridge.

04581 Frodin, D. G. (2001). Guide to standard floras of the world. An annotated, geographically arranged systematic bibliography of the principal floras, enumerations, checklists, and chorological atlases of different areas. Second edition, Cambridge Univ. Press, Cambridge. [greatly expanded and updated commentary and coverage, many 'tree floras' included].

04830 Gaussen, H. (1979). Bibliographie. Trav. Lab. Forest. Toulouse, T. 2, vol. 1, part. 3: 1–180. [bibliography for Les Gymnospermes actuelles et fossiles].

05460 Harris, A. S. & R. H. Ruth (1970). Sitka spruce, a bibliography with abstracts. U.S. Forest Serv. Res. Paper PNW-105. Pacific Northwest Forest and Range Exp. Stat., Portland, Oregon. [*Picea sitchensis*, bibliogr.].

05715 Hennon, P. E. & A. S. Harris (1997). Annotated bibliography of *Chamaecyparis nootkatensis*. U.S.D.A. Forest Service, Gen. Tech. Rep. PNW-GTR-413, Portland, Oregon. [= *Xanthocyparis nootkatensis* (Farjon & al., 2002); 680 citations covering a wide range of subjects in relation to 'yellow cedar', including taxonomy].

06610 Jackson, B. D. (1895). Index Kewensis plantarum Phanerogamarum... an enumeration of the genera and species of flowering plants from the time of Linnaeus to the year 1885 inclusive. (2 vol.) Oxford. [see also Supplementa 1–18, 1901–1985, Royal Botanic Gardens, Kew, Richmond].

07890 Langman, I. K. (1964). A Selected Guide to the Literature on the Flowering Plants of Mexico. Univ. Pennsylvania Press, Philadelphia. [gymnosperms, angiosperms, bibliography, 1015 pp.].

08440 Little, E. L., Jr. (1949). Lambert's "Description of the genus Pinus", 1832 edition. Madroño 14: 33–47. [Pinaceae, (dates of) nomenclature].

08670 Little, E. L., Jr. & B. H. Honkala (1976). Trees and shrubs of the United States: a bibliography for identification. U.S. Forest Serv. Misc. Publ. 1336. Washington, D.C. (56 pp.).

08873 Mabberley, D. J. (1987). The Plant Book – A portable dictionary of the higher plants. Cambridge University Press, Cambridge, U.K. (ed. 2, 1997). [gymnosperms, angiosperms, vascular cryptogams; families, genera, example spp., descr., numbers of spp., distrib., uses etc. (ed. 1: xii + 707 pp., ed. 2: xvi + 858 pp.)].

10020 Merrill, E. D. & E. H. Walker (1938). A Bibliography of Eastern Asiatic Botany. Jamaica Plain, Mass. [gymnosperms, angiosperms, cryptogams; xlii + 719 pp., see for suppl.: E. H. Walker, 1960].

11890a Rehder, A. (1911–18). The Bradley Bibliography. A guide to the literature of the woody plants of the world published before the beginning of the twentieth century. Vol. 1. Dendrology, part 1; Vol. 2. Dendrology, part 2; Vol. 3. Arboriculture; Vol. 4. Forestry; Vol. 5. Index. Publ. Arnold Arbor. 3. Cambridge, Mass. [Coniferales, Ginkgoales, angiosperms].

11950 Rehder, A. (1949). Bibliography of cultivated trees and shrubs hardy in the cooler temperate regions of the Northern Hemisphere. (xl + 825 pp.). Jamaica Plain, Mass. [Coniferales, Ginkgoales, angiosperms].

12000 Renkema, H. W. & J. Ardagh (1930). Aylmer Bourke Lambert and his "Description of the Genus Pinus". J. Linn. Soc., Bot. 48: 439–466.

13640 Stafleu, F. A. & R. S. Cowan (1976–88). Taxonomic literature. A selective guide to botanical publications and collections with dates, commentaries and types; second edition (TL-2). Vol. I–VII; Regnum Veg. 94, 98, 105, 110, 112, 115, 116. Utrecht/Antwerpen – The Hague/Boston. [A comprehensive standard bibliography of independent taxonomic publications between 1753–1940].

13641 Stafleu, F. A. & E. A. Mennega (1992–2000). Taxonomic literature. A selective guide to botanical publications and collections with dates, commentaries and types. Supplement I: A–Ba; Supplement II: Be–Bo; Supplement II: Br–Ca; Supplement IV: Ce–Cz; Supplement V: Da–Di; Supplement VI: Do–E; Regnum Veg. 125, 130, 132, 134, 135, 137. Königstein, Germany. [supplements updating the first volume of TL-2 (A–G) publ. 1976 to balance subsequent volumes in scope, but in fact exceeding these, including much non-taxonomic literature].

13730 Stapf, O. (1929–41). Index Londinensis to illustrations of flowering plants, ferns and fern allies. Vol. I–VI, Supplement Vol. I–II. Oxford. [includes gymnosperms].

14462 Tralau, H., ed. (1969). Index Holmensis I. (a world phytogeographic index); Equisetales, Isoetales, Lycopodiales, Psilotales, Filicales, Gymnospermae. Zürich. [phytogeographic bibliography].

14880 Walker, E. H. (1960). A Bibliography of Eastern Asiatic Botany, supplement 1. Amer. Inst. Biol. Sci. Publ., Washington, D.C. [gymnosperms, angiosperms, cryptogams; see also E. D. Merrill & E. H. Walker, 1938].

FLORAS

00570a Bao, G. Z., ed. (1985). Flora Intramongolica. T. 1. (Chin.). [Gymnospermae pp. 125–157, ill., Pinaceae pp. 125–141, *Picea wilsonii, P. meyeri, P. crassifolia, P. koraiensis, Larix gmelinii, L. principis-rupprechtii, L. olgensis, Pinus tabuliformis, P. sylvestris* var. *mongolica, P. bungeana, P. pumila, P. armandii,* Cupressaceae pp. 141–149, *Platycladus orientalis, Sabina* (= *Juniperus*) *chinensis, S. przewalskii S. davurica, S. vulgaris* (= *J. sabina*), *Juniperus rigida, J. sibirica,* Ephedraceae].

00572 Barkley, T. M., ed. (1977). Atlas of the Flora of the Great Plains. Iowa State Univ. Press, Ames, Iowa. [dot maps of Pinaceae and of *Juniperus* spp., pp. 11–13].

00573 Barkley, T. M., ed. (1986). Flora of the Great Plains. Univ. Press of Kansas, Lawrence, Kansas. [Cupressaceae, *Juniperus*, Pinaceae, *Picea, Pinus*; contrib. by R. E. Brooks, pp. 71–76].

00597 Bartel, J. A. (1993). Cupressaceae – Cypress family. J. Arizona-Nevada Acad. Sci. 27 (2): 195–200. [*Cupressus arizonica, Juniperus* spp.].

00597a Bartel, J. A. (1993). Cupressaceae – Cypress family. In: J. C. Hickman (ed.). The Jepson Manual, higher plants of California. (pp. 111–115). Berkeley, California. [treats fam. in strict sense, incl. *Calocedrus, Chamaecyparis, Cupressus, Juniperus*, questions status of *Chamaecyparis*].

00646 Beaman, J. H. & R. S. Beaman (1993). The gymnosperms of Mount Kinabalu. Contrib. Univ. Michigan Herb. 19: 307–340. [Coniferales, Gnetales, *Agathis borneensis* (= *A. dammara*), *A. kinabaluensis, A. lenticulata, Phyllocladus, Dacrycarpus, Dacrydium, Falcatifolium, Nageia, Podocarpus, Sundacarpus, Gnetum,* ill. of herb. specimens].

00647 Beaman, J. H. & R. S. Beaman (1998). The plants of Mount Kinabalu 3. Gymnosperms and non-orchid monocotyledons. Natural History Publications (Borneo), Kota Kinabalu, in assoc. with Royal Botanic Gardens, Kew. [gymnosperms, Coniferales, pp. 9–31; see also Beaman & Beaman 1993].

00821 Bentham, G. (1873). Flora Australiensis: a description of the plants of the Australian Territory. Vol. VI. Order 116. Coniferae. (pp. 232–248). London. [Cupressaceae, *Frenela* = *Callitris, Actinostrobus, Diselma, Athrotaxis,* Araucariaceae, Podocarpaceae].

00955 Blume, C. L. von (1827–28). Enumeratio plantarum Javae et insularum adjacentium minus cognitarum vel novarum ex herbariis Reinwardtii, Kuhlii, Hasseltii et Blumii. (in 2 fasc.). Leiden. [*Podocarpus bracteatus* sp. nov., fasc. 1, p. 88, *P. amarus* sp. nov. = *Sundacarpus amarus,*fasc. 1, p. 88, *P. imbricatus* sp. nov. = *Dacrycarpus imbricatus,* fasc. 1, p. 89].

01020 Bobrov, E. G., B. A. Fedchenko, A. V. Fomin, M. M. Iljin, A. N. Krishtofovich, V. L. Komarov & S. V. Yuzepchuk (1968). Flora of the U.S.S.R.; Vol. I, Archegoniatae and Embryophyta. Chief ed. V. L. Komarov (Leningrad 1934, transl. from Russ.). [gymnosperms].

01050 Boissier, P. E. (1884). Flora orientalis sive enumeratio plantarum in Oriente a Graecia et Aegypto ad Indiae fines hucusque observatarum. Vol. 5. Basel Genève. (Ord. CXLII. Coniferae pp. 693–713). [Pinaceae, Cupressaceae, Taxaceae, *Picea smithiana* (Wallich) comb. nov., p. 700, *Abies pectinata* var. *equi-trojani* Aschers. et Sint. ex Boiss. = *A. nordmanniana* ssp. *equi-trojani*, p. 701, *Juniperus macropoda* sp. nov., p. 709].

01193 Brandis, D. (1874). The forest flora of north-west and central India: a handbook of the indigenous trees and shrubs of those countries. London. [Coniferales pp. 502–539; *Pinus longifolia* Roxb. = *P. roxburghii, P. gerardiana, P. excelsa* = *P. wallichiana, Cedrus deodara, Abies smithiana* = *Picea smithiana, A. dumosa* = *Tsuga dumosa, A. webbiana* = *A. spectabilis, Larix griffithii* = *L. griffithiana, Cupressus sempervirens, C. torulosa, Juniperus communis, J. recurva, J. wallichiana* Hook. f. ex Brandis = *J. indica* Bertol., p. 537, *J. excelsa*].

01194 Brandis, D. (1907). Indian Trees. London. [Coniferae pp. 688–697, Pinaceae (incl. Cupressaceae), Taxaceae (incl. Cephalotaxaceae, Podocarpaceae].

01220a   Breitenbach, F. von & J. von Breitenbach (1992). Tree atlas of southern Africa [Boomatlas van suider-Afrika] Section 1. Dendrological Foundation, Pretoria. [gymnosperms (*Welwitschia*), Cupressaceae, *Widdringtonia nodiflora* (incl. *W. whytei*), *W. cedarbergensis*, *W. schwarzii*, Podocarpaceae, ill., maps].

01271   Browicz, K. & J. Zielinski (1982). Chorology of Trees and Shrubs in South-West Asia and Adjacent Regions. Vol. 1. Warszawa – Poznan. [Cupressaceae, Pinaceae, Taxaceae, pp. 7–17, maps 1–18].

01648   Carabia, J. P. (1941). Contribuciones al estudio de la flora Cubana Gymnospermae. Caribbean Forester 2: 83–99. [Cycadaceae, Taxaceae (= Podocarpaceae), *Podocarpus*, Pinaceae, *Pinus*, Cupressaceae, *Juniperus*, Flora].

01958   Chapman, A. W. (1860). Flora of the southern United States. New York. [*Taxus floridana* sp. nov., p. 436].

02051   Cheng, W. C. & L. K. Fu, eds. (1978). Flora Reipublicae Popularis Sinicae. Delectis florae reipublicae popularis sinicae agendae academiae sinicae edita. Tomus 7: Gymnospermae. Acad. Sinica, Beijing. (xiv + 542 pp.). [many authors, new comb. mostly by W. C. Cheng & L. K. Fu; Chin. + Lat. ref., ill.; Cycadales, Coniferales, Ginkgoales, Taxales, Ephedrales, Gnetales; *Abies ernestii* var. *salouenensis* (Bord.-Rey et Gaussen) comb. et stat. nov. = *A. chensiensis* ssp., p. 93, *Calocedrus macrolepis* var. *formosana* (Florin) comb. et stat. nov., p. 327].

02295   Coker, W. C. & H. R. Totten (1945). Trees of the Southeastern States. ed. 3, Chapel Hill. [Cupressaceae, *Taxodium*, *Pinus*].

02360   Coode, M. J. E. & J. Cullen (1965). Gymnospermae. in: P. H. Davis (ed.). Flora of Turkey and the East Aegean Islands. Vol. I. (pp. 67–85). Edinburgh Univ. Press, Edinburgh. [*Abies*, *Picea*, *Cedrus*, *Pinus*, *Taxus*, *Cupressus*, *Juniperus*, *Ephedra*, maps].

02381   Correll, D. S. (1966). Gymnospermae. in: C. L. Lundell *et al.* Flora of Texas. Vol. 1, pp. 322–368, pl. 40–47. Texas Res. Found., Renner, Texas. [*Pinus*, *Pseudotsuga*, *Taxodium*, *Cupressus*, *Juniperus*, Ephedraceae].

02395   +++Costermans, L. (1986). Native trees and shrubs of southeastern Australia. Dee Why West, New South Wales. [*Callitris*, maps, ill. of cones, phot. of habit].

02140a   Christensen, K. I. (1986). Gymnospermae. pp. 38–50 in: A. Strid (ed.). Mountain Flora of Greece. Vol. 1. Cambridge. [Cupressaceae, Pinaceae, Taxaceae].

02600   Critchfield, W. B. & E. L. Little, Jr. (1966). Geographic distribution of the pines of the world. U.S. Forest Serv. Misc. Publ. 991. Washington, D.C. [*Pinus* spp., *P.* subsect. *Krempfianae* (as "Krempfiani"), p. 5, subsect. *Contortae*, p. 19, subsect. *Oocarpae*, p. 19, subsect. nov., bibliography, detailed maps].

02610   Cronquist, A., A. H. Holmgren, N. H. Holmgren & J. L. Reveal (1972). Intermountain Flora, Vascular Plants of the Intermountain West, U.S.A. Vol. 1. New York – London. [bot. explor. of the region, plant geography, ecology, habitus photographs of conifers; Pinaceae, Cupressaceae, Ephedraceae, *Juniperus occidentalis* var. *australis* (Vasek) A. & N. Holmgren, comb. nov., p. 239, ill., keys, pp. 222–248].

02790   Davis, P. H., ed. (1965). Flora of Turkey and the East Aegean Islands. Vol. I. Edinburgh Univ. Press, Edinburgh. [see M. J. E. Coode & J. Cullen, 1965].

03216   Duncan, W. H. & M. B. Duncan (1988). Trees of the southeastern United States. Athens, Georgia – London. [Cupressaceae, *Taxodium*, Pinaceae, *Pinus*, colour pl., maps].

03595   Entwisle, T. J. (1994). Conifers (Pinophyta). In: N. G. Walsh & T. J. Entwisle (eds.). Flora of Victoria 2. Ferns and allied plants, conifers and monocotyledons. (pp. 113–121). [*Callitris endlicheri*, *C. glaucophylla*, *C. gracilis*, *C. rhomboidea*, *C. verrucosa*, *Podocarpus lawrencei*, ill. of *Callitris*, grid maps].

03651   Exell, A. W. & H. Wild (1960). Flora Zambesiaca. Vol. 1, Part 1. Gymnospermae, pp. 79–88. London. [Cupressaceae, *Widdringtonia whytei*, *W. nodiflora*, *Juniperus procera*, ill.].

03760   Fedtschenko, B. A. (1929). Flora Transbaicalica. Leningrad. [see V. N. Sukachev, 1929].

03761   Fedtschenko, B. A., M. G. Popov & B. K. Shishkin, eds. (1932). Flora Turkmenii. Vol. 1. Leningrad. [*Juniperus turcomanica* B.A. Fedtsch., sp. nov., p. 14].

04010a   Flora of North America Editorial Committee (ed.) (1993). Flora of North America, North of Mexico. Vol. 2. Pteridophytes and gymnosperms. Oxford Univ. Press, New York, Oxford. [Coniferophyta (conifers) pp. 352–422, Pinaceae pp. 352–398, Cupressaceae s.l. pp. 399–422, treatment of taxa down to rank of var., small distr. maps, ill., keys; lit. ref. for entire vol. pp. 435–458].

04266   Fonseca, R. M. (1994). Flora de Guerrero: No. 2. Cupressaceae y Taxodiaceae. UNAM, Mexico, D.F. (16 pp., ill., maps, keys). [*Cupressus lusitanica*, *Juniperus flaccida*, *J. monticola*, *Taxodium mucronatum*].

04298   Forster, J. G. A. (1786). Florulae insularum australicum prodromus. Göttingen. [*Cupressus columnaris* J. R. Forst. sp. nov., p. 67 = *Araucaria columnaris*, *Dacrydium* Sol. ex G. Forst., p. 92]

04881 Gleason, H. A. (1968). The new Britton and Brown Illustrated Flora of the Northeastern United States and adjacent Canada. vol. 1. The Pteridophyta, Gymnospermae and Monocotyledonae. New York – London. [Coniferales, Pinaceae, Taxodiaceae, Cupressaceae, ill., keys, pp. 56–68].

04980 Graça Silva, M. da (1983). Flora de Moçambique 3: Cupressaceae. Lisboa, I. I. G. T., pp. 41–44. [*Widdringtonia nodiflora*].

05021 Grenier, J. C. M. & D. A. Godron (1855). Flore de France,... Vol. 3. [*Pinus laricio* var. *cebennensis* Gren. et Godron, var. nov., p. 153; see A. Rehder, 1922, for further synonymy].

05030 Greuter, W., H. M. Burdet & G. Long (eds.) (1984). Med-Checklist – A critical inventory of vascular plants of the circum-mediterranean countries. Vol. 1. Pteridophyta (ed. 2), Gymnospermae, Dicotyledones (Acanthaceae-Cneoraceae). Genève. [gymnosperms pp. 26–35; *Cupressus, Juniperus, Tetraclinis, Abies, Cedrus, Larix, Picea, Pinus*, nomenclature, phytogeography in tab. form].

05049 Grierson, A. J. C. & D. G. Long (1983). Flora of Bhutan. Vol. 1, Part 1. Royal Botanic Garden, Edinburgh. [Gymnospermae: 44–57, *Abies, Cupressus, Juniperus, Larix, Picea, Pinus, Tsuga*].

05070 Griffin, J. R. & W. B. Critchfield (1972). The distribution of forest trees in California. USDA Forest Res. Paper No. PSW-82: 1–118. [Coniferales, Cupressaceae, Pinaceae, Taxodiaceae, angiosperms, maps].

05171b Güner, A., N. Özhatay, T. Ekim & K. H. C. Baser (eds.). (2000). Flora of Turkey. Vol. 11 (Supplement 2). Edinburgh. [Pinaceae pp. 5–7, Cupressaceae pp. 8–10].

05380 Hara, H. (1971). Flora of Eastern Himalaya. Univ. of Tokyo Press, Tokyo. [see Y. Satake, 1971].

05408 Harden, G. J. (ed.) (1990). Flora of New South Wales. Vol. 1. Sydney. [Cupressaceae by G. J. Harden & J. Thompson, pp. 83–86, ill. *Callitris* spp.].

05680 Hegi, G. (1981). Illustrierte Flora von Mitteleuropa. Ed. 3, Bd. 1, Teil 2. Berlin – Hamburg. [see H. Zoller, 1981].

05830 Heywood, V. H., ed. (1964). Pinaceae. in: T. G. Tutin *et al.* (eds.). Flora Europaea Vol. I, chapt. 26: 29–40. Cambridge Univ. Press, Cambridge.

05855 Hickman, J. C., ed. (1993). The Jepson Manual: higher plants of California. University of California Press, Berkeley. [Coniferales pp. 111–123, Cupressaceae 111–114, Pinaceae pp. 115–121, Taxaceae p. 121, Taxodiaceae p 122, ill., keys].

05950 Hitchcock, C. L., A. Cronquist & M. Ownbey (1969). Vascular Plants of the Pacific Northwest – Part 1: vascular cryptogams, gymnosperms, and monocotyledons. Univ. of Washington Press, Seattle – London. [Cupressaceae, Pinaceae, ill., keys, pp. 103–133].

05958 Ho, P. H. (1991). Câyco Viêtnam. An Illustrated Flora of Vietnam. Vol. 1, part 1. Sterculiaceae to Fabaceae. Vietnamese; publ. unknown. [Pinaceae, pp. 269–272, Araucariaceae, pp. 272–273, Taxodiaceae, pp. 273–274, Cupressaceae, pp. 275–277, Podocarpaceae, pp. 277–279, Taxaceae, pp. 279–280, Cephalotaxaceae, pp. 280–281, Amentotaxaceae, pp. 281–282, incl. introd. taxa, ill.].

06005 Hoffmann, A. (1982). Flora silvestre de Chile, zona austral. Una guia illustrada para la identificación de las especies de plantas leñosas del sur de Chile. Santiago de Chile. [*Araucaria araucana* pp. 48–49, *Austrocedrus chilensis* pp. 50–51, *Fitzroya cupressoides* pp. 52–53, *Pilgerodendron uviferum* pp. 54–55, Podocarpaceae].

06050b Hooker, J. D. (1852–55). The Botany of the Antarctic Voyage of H. M. discovery ships Erebus and Terror... II. Flora Novae-Zelandiae. London. [*Phyllocladus alpinus* sp. nov., part 1: 235, t. 53 (1853)].

06052 Hooker, J. D. (1857). The Botany of the Antarctic Voyage of H. M. discovery ships Erebus and Terror... III. Flora Tasmaniae Vol. 1 (Dicotyledones). Coniferae: pp. 349–359, pl. 97–100. London. (1860 on title p., but see F. A. Stafleu & R. S. Cowan, 1979: TL-2, Vol. II). [Cupressaceae, Podocarpaceae, *Frenela* spp. = *Callitris* spp., *Diselma archeri* gen. et sp. nov., p. 353, pl. 98, *Athrotaxis* spp.].

06080 Hooker, J. D. (1888). The Flora of British India 5 (15). London. [Coniferales: pp. 640–658, *Cephalotaxus griffithii* sp. nov., p. 648].

06100 Hooker, W. J. (1840). Flora boreali-americana; or, the botany of the northern parts of British America: ... Vol. 2, part 10. London. [Coniferales, pp. 161–167; *Pinus* (*Abies*) *lasiocarpa* sp. nov. = *Abies lasiocarpa* (Hook.) Nutt., p. 163, *Juniperus occidentalis* sp. nov., p. 166].

06131 Hosie, R. C. (1969). Native trees of Canada. Ed. 7. Canad. Forest Serv., Dept. Fish. & Forest., Ottawa. [Coniferales, Cupressaceae, Pinaceae, ill., maps].

06132 Hough, R. B. (1957). Handbook of the Trees of the Northern States and Canada. New York. [Coniferales, maps].

06350 Hultén, E. (1927). Flora of Kamtchatka and the adjacent islands. I. Pteridophyta, Gymnospermae and Monocotyledonae. Kongl. Svenska Vetenskapsakad. Handl., ser. 3, 5 (1): 1–346.

06360   Hultén, E. (1968). Flora of Alaska and Neighboring Territories – a manual of the vascular plants. Stanford University Press, Stanford, California. [Pinaceae, Cupressaceae, pp. 59–66].

06370   Hultén, E. (†) & M. Fries (1986). Atlas of North European vascular plants north of the tropic of cancer. Vols. I–III. Koenigstein, Fed. Rep. Germany. [maps of *Abies sibirica*, *Picea abies*, *P. obovata* (as *P. abies* ssp. *obovata*), *Larix sibirica* = *L. russica*, *L. gmelinii*, *Pinus sylvestris* (+ var. *hamata*), *P. sibirica*, *P. cembra*, *P. pumila*, *Juniperus communis* et var., *Taxus* spp., Vol. I maps 77–83, text Vol. III, pp. 977–978].

06530a   Iwatsuki, K., T. Yamazaki, D. E. Boufford & H. Ohba (1995). Flora of Japan. Vol. I, Pteridophyta and Gymnospermae. Tokyo. [Coniferales, pp. 264–287].

06700   Jalas, J. & J. Suominen (eds.) (1973). Atlas Florae Europaeae 2. Gymnospermae (Pinaceae to Ephedraceae). Helsinki.

06771   Jepson, W. L. (1909). A Flora of California. Vol. 1, part 1: 33–64. San Francisco. [*Cupressus bakeri*, *C. sargentii* spp. nov., p. 61].

06800   Jepson, W. L. (1923). The trees of California. Ed. 2, Berkeley, Calif. [Coniferales, Pinaceae, Cupressaceae, Taxodiaceae, *Cupressus sargentii* var. *duttonii* var. nov., p. 200].

06811   Jessop, J. P. & H. R. Toelken, eds. (1986). Flora of South Australia. Part I. Lycopodiaceae–Rosaceae. Adelaide. [see J. Venning, 1986].

06930   Kanehira, R. (1936). Formosan trees indigenous to the island. (revised). Dept. of Forest., Govt. Res. Inst., Taihoku, Taiwan. [*Pseudotsuga taitoensis* sp. nov., without Latin descr. or diagn., p. 51; Ed. 1 publ. in 1917].

07021   Kearney, T. H. & R. H. Peebles (1942). Flowering plants and ferns of Arizona. U.S.D.A. Misc. Publ. 423. Washington, D.C. [gymnosperms: pp. 58–71; Pinaceae, Cupressaceae, habitus photographs, e.g. *Cupressus arizonica*, pl. 11].

07200   Kingdon-Ward, F. (1935). A sketch of the Botany and Geography of Tibet. J. Linn. Soc., Bot. 50: 239–265.

07210   Kingdon-Ward, F. (1946). Botanical explorations in North Burma. J. Roy. Hort. Soc. (London) 71: 318–325.

07231d   Kirk, T. (1889). The forest flora of New Zealand... (ill.). Wellington. [*Podocarpus hallii* sp. nov. = *P. cunninghamii*, p. 14].

07311   Koch, K. H. E. (1849). Beiträge zu einer Flora des Orientes. Gymnospermae, Nacktsämler. (pp. 291–307). Linnaea 22: 177–464. [*Pinus heterophylla* sp. nov., p. 295, *P. kochiana* Klotsch ex K. Koch, p. 296, *P. armena* sp. nov., p. 297, *Juniperus pygmaea*, *J. polycarpos* (see also A. L. Takhtajan & A. A. Fedorov, 1972), *J. isophyllos* spp. nov., pp. 302–304].

07480   Komarov, V. L. (1927). Flora peninsulae Kamtschatka I, cum summario lingua britannica expresso. Leningrad. (Russ., Eng. summ., 339 pp., 13 pl., 1 map). [Pinaceae, *Haploxylon* (Koehne) Komarov, stat. nov., *H. pumilum* (Regel) comb. nov. = *Pinus pumila*, p. 103].

07500   Komarov, V. L. (ed.) (1934). Flora SSSR (Flora of the U.S.S.R.). Vol. I. Izd. Akad. Nauk S.S.S.R., Leningrad. [gymnosperms; rep. of diagn. of *Juniperus schugnanica* and *J. seravschanica*, p. 190 and p. 187, first publ. in Bot. Žurn. SSSR 27: 481–482, 1932].

07617   Krylov, P. (1927). Flora Zapadnoy Sibiri. Vol. 1. Pteridophyta – Hydrocharitaceae. Tomsk. (Russian). [Pinaceae, Cupressaceae, *Pinus cembra* ssp. *sibirica* (Du Tour) stat. nov., p. 77, *P. cembra* ssp. *sibirica* f. *coronans* (Litv.) comb. nov., p. 79].

07640   Kuan, C. T., (Guan Zhong-tian) ed. (1983). Flora Sitchuanica. Tomus 2 (Gymnospermae). (Chin., Lat. diagn.). [*Abies recurvata* var. *ernestii* (Rehder) C. T. Kuan, comb. et stat. nov., p. 46, et var. *salouenensis* (Bord.-Rey et Gaussen) C. T. Kuan, comb. et stat. nov., p. 48, *Pseudotsuga xichangensis* C. T. Kuan et L. J. Zhou, sp. nov. = *P. sinensis* var. *sinensis*, p. 54, *Picea aurantiaca* var. *retroflexa* (Masters) C. T. Kuan et L. J. Zhou, comb. nov., p. 71, *Pinus tabuliformis* var. *henryi* (Masters) C. T. Kuan, comb. nov., p. 113, *Cupressus chengiana* var. *jiangeensis* (N. Zhao) C. T. Kuan, comb. nov., p. 167, but see Silba, 1981, *Taxus wallichiana* var. *yunnanensis* (W. C. Cheng & L. K. Fu) C. T. Kuan, comb. nov., p. 215].

07640a   Kuan, C. T. (Guan Zhong-tian) (1990). The geography of Conifers in Sichuan. Chengdu, Sichuan. (Chin., Eng. summ., maps, tables). [Pinaceae, Taxodiaceae, Cupressaceae, Podocarpaceae, Cephalotaxaceae, Taxaceae].

07740   Laínz, M., S. Castroviejo, G. López Gonzáles *et al.*, eds. (1986). Flora Iberica. Plantas vasculares de la Peninsula Ibérica e Islas Baleares. Vol. 1. Lycopodiaceae-Papaveraceae. Madrid. [see J. do A. Franco, 1986].

07931a   Lanner, R. M. (1999). Conifers of California. Los Olivos, California, U.S.A. [Cupressaceae, Pinaceae, Taxaceae, Taxodiaceae, all species with many colour ill., maps].

07990   Laubenfels, D. J. de (1972). Flore de la Nouvelle Calédonie et dependances no. 4, Gymnospermes. Mus. Natl. Hist. Nat., Paris. [Coniferales: Araucariaceae, Podocarpaceae, Cupressaceae, *Parasitaxus* gen. nov., p. 44, *P. usta* (Vieill.) comb. nov. ("*ustus*"), p. 45, *Prumnopitys ferruginoides* (R. H. Compton) comb. nov., p. 56, *Podocarpus polyspermus* sp. nov., p. 60, *Araucaria luxurians* (Brongn. & Gris) comb. et stat. nov., p. 92, *Callitris neocaledonica*, *C. sulcata*, *Libocedrus austrocaledonica*, *L. yateensis*, *L. chevalieri*, *Neocallitropsis pancheri* (Carrière) comb. nov., p. 161, ill.].

08006   Laubenfels, D. J. de (1982). Podocarpaceae. In: Z. Luces de Febres (ed.). Flora de Venezuala 11 (2): 7–41 (ill.). Ediciones Fundación Educación Ambiental. [*Podocarpus brasiliensis* sp. nov., p. 31, *P. buchholzii* sp. nov., p. 31, *P. celatus* sp. nov., p. 35].

08020   Laubenfels, D. J. de (1988). Coniferales. in: Flora Malesiana ser. I, 10 (3): 337–453. [Cupressaceae, Pinaceae, Araucariaceae, Podocarpaceae, *Dacrydium cornwallianum* sp. nov., p. 366, *D. gracile* sp. nov., p. 367, *D. ericoides* sp. nov., p. 371, *Agathis spathulata* sp. nov., p. 435, *Libocedrus papuana* var. *arfakensis* (Gibbs) comb. et stat. nov. = *Papuacedrus papuana* var. *arfakensis*, p. 446].

08080   Ledebour, C. F. von (1833). Flora Altaica, Vol. IV. Berlin. [*Pinus sylvestris* var. *sibirica* var. nov., p. 199, *Picea obovata* sp. nov., p. 201, *P. orientalis* var. *longifolia* var. nov. = *P. schrenkiana*, p. 201, *Abies sibirica* sp. nov., p. 202, *Larix sibirica* sp. nov. = *L. russica*, p. 204].

08090   Ledebour, C. F. von (1843). Flora rossica sive enumeratio plantarum... Vol. 2, part 4: 1–204. Stuttgart. [Coniferales].

08120   Lemmon, J. G. (1895). Handbook of West-American Cone-Bearers. Ed. 3, Oakland, California. [ed. 1 publ. Mar 1892; ed. 2 Apr 1892 with *Pinus muricata* var. *anthonyi*, var. nov., p. 10; Coniferales, Pinaceae, Cupressaceae, e.g. *Pinus* subgen. *Strobus* subgen. nov., p. 20 (but see also Lemmon, 1888), *P. radiata* var. *binata* (Engelm.) comb. nov., p. 42, *Cupressus arizonica* var. *bonita* var. nov., p. 76, *C. goveniana* var. *pygmaea* var. nov. (as "pigma"), p. 77, *Juniperus occidentalis* var. *gymnocarpa* var. nov., p. 80].

08190   Li, H. L. (1963). Woody Flora of Taiwan. (Morris Arbor. Monogr.) Narberth, Pa. [*Taxus celebica* (Warb.) comb. nov. = *T. sumatrana*, p. 34].

08195   Li, H. L. (1975). Gymnospermae. in: C. DeVol *et al.* (eds.). Flora of Taiwan. Vol. 1. Taipei, Taiwan. [Pinaceae: pp. 514–529, Cupressaceae and Taxodiaceae: pp. 530–544, keys, ill.].

08255   Lin, X. & S. Y. Zhang (1993). Gymnospermae. in: C. F. Zhang & S. Y. Zhang (eds.). Flora of Zhejiang 1. Pteridophyta – Gymnospermae. (Chin., pp. 338–391, ill.). [e.g. *Abies beshanzuensis, Pseudolarix amabilis, Taiwania flousiana* (not indigenous? = cultivated *T. cryptomerioides*), *Glyptostrobus pensilis, Platycladus orientalis, Fokienia hodginsii, Nageia nagi, Cephalotaxus fortunei, C. sinensis, Pseudotaxus chienii, Amentotaxus argotaenia, Torreya jackii*].

08460   Little, E. L., Jr. (1950). Southwestern trees: A guide to the native species of New Mexico and Arizona. U.S.D.A. Forest Service Handb. 9. Washington, D.C. (109 pp.). [Coniferales].

08500   Little, E. L., Jr. (1953). Check List of native and naturalized trees of the United States (including Alaska). U.S.D.A. Agric. Handb. 41. Washington, D.C.

08600   Little, E. L., Jr. (1979). Checklist of United States trees (native and naturalized). U.S.D.A. Agric. Handb. 541. Washington, D.C.

08610   Little, E. L., Jr. (1980). The Audubon Society field guide to North American trees. Eastern region. New York.

08871   Lundell, C. L. *et al.* (1966). Flora of Texas. Vol. 1. Texas Res. Found., Renner, Texas. [see D. S. Correll, 1966].

08871c   Maas, P. J. M. & L. Y. Th. Westra (1993). Neotropical Plant Families. A concise guide to families of vascular plants in the Neotropics. Koenigstein, Germany. [see A. Farjon, 1993 for the gymnosperms].

08871d   Maas, P. J. M. & L. Y. Th. Westra (1998). Familias de plantas neotropicales. Una guía concisa a las familias de plantas vasculares en la región neotropical. Vaduz – Liechtenstein. [see also Maas & Westra, 1993 for an earlier Eng. ed. and Farjon, 1993, 1998 for the gymnosperms].

08962b   Maiden, J. H. (1907). The forest flora of New South Wales. Vol. 2, part 12, No. 44–51: The Cypress Pinus of New South Wales. Sydney. [*Callitris* spp., pp. 30–62, pl. 46–48].

08962c   Maire, R. (1952). Flore de l'Afrique du Nord. Vol. 1. Pteridophyta – Gymnospermae – Monocotyledonae [in part]. Paris. [Coniferales, Cupressaceae pp. 107–126, *Tetraclinis articulata* (sub *Callitris*), *Juniperus* spp., *Cupressus sempervirens, C. dupreziana*, Araucariaceae, Pinaceae pp. 129–147, *Abies, Cedrus, Pinus*, Taxodiaceae].

08981   Makino, T. (1940). An illustrated Flora of Nippon, with cultivated and naturalized plants. Nippon shokubutsu dzukan. (Japan., ill.). [*Pinus pentaphylla* var. *himekomatsu* (Miyabe et Kudo) comb. nov., p. 903, f. 2709].

08982   Makino, T. & K. Nemoto (1931). Flora of Japan. Ed. 2. Tokyo. (Japan., ill.). [*Pinus hakkodensis* sp. nov., p. 148].

09797   McCarthy, P. M. (vol. ed.) (1998). Flora of Australia, Volume 48. Ferns, Gymnosperms and Allied Groups. CSIRO, Melbourne, Australia. (ill., maps). [(gymnosperms colour pl. insert pp. 497–504), gymnosperms pp. 505–661, Introduction by K. D. Hill, fossil record of conifers pp. 527–537 and cycads by R. S. Hill & L. J. Scriven and R. S. Hill resp., Podocarpaceae, Araucariaceae, Cupressaceae, Pinaceae (introduced) and Cycadophyta by K. D. Hill, distr. maps conifers pp. 694–698, *Callitris gracilis* ssp. *murrayensis* (J. Garden) K. D. Hill, comb. nov., p. 716, *C. oblonga* ssp. *corangensis* K. D. Hill, ssp. nov., p. 716, *C. oblonga* ssp. *parva* K. D. Hill, ssp. nov., p. 717].

09815   McVaugh, R. (1992). Flora Novo-Galiciana. A descriptive account of the vascular plants of western Mexico. Vol. 17: gymnosperms and Pteridophytes. Univ. Michigan Herbarium, Ann Arbor. (gen ed. W. R. Anderson). [gymnosperms pp. 4–119, Cupressaceae, Pinaceae, Taxodiaceae, Podocarpaceae, Ephedraceae, Zamiaceae, *Cupressus lusitanica* pp. 7–11, *Pinus ayacahuite* var. *novogaliciana* Carvajal, var. nov., p. 48].

09843   Medwedew, Y. S. (1905). Derev'ya i Kustarniki Kavkasa, ... (Trees and Shrubs of the Caukasus) I. Gymnospermae. Tiflis. (Russ.).

09870   Meikle, R. D. (1977). Flora of Cyprus. Vol. 1. Spermatophyta Gymnospermae. (pp. 21–35). Roy. Bot. Gard., Kew (Richmond). [Pinaceae, Cupressaceae, *Cedrus libani* ssp. *brevifolia* (Hook. f.) comb. nov., p. 22].

09955   Melville, R. (1958). Gymnospermae. In: W. B. Turrill & E. Milne-Redhead (eds.). Flora of Tropical East Africa. Crown Agents Oversea Govt. & Admin., London. [*Cycas, Encephalartos, Podocarpus, Juniperus procera,* ill.].

10054   Métro, A. & C. Sauvage (1955). Flore des végétaux ligneux de la Mamora. Rabat. [region in N. Algeria E of Rabat; gymnosperms pp. 81–123, ill., *Juniperus oxycedrus, J. phoenicea, Tetraclinis articulata, Cupressus* spp., *Pinus* spp.].

10080   Michaux, A. (1803). Flora boreali-americana, ... Vol. 2; pp. 1–340, pl. 30–51. Paris – Strasbourg. [Coniferales pp. 203–209, Taxales p. 245, *Larix americana* sp. nov. = *L. laricina*, p. 203, *Pinus serotina* sp. nov., p. 205, *Abies balsamifera* sp. nov. = *A. balsamea*, p. 207, *Taxus baccata* var. *minor* var. nov. = *T. canadensis*, p. 245].

10165   Miller, A. G. (1996). Gymnospermae, Family 18. Cupressaceae. In: A. G. Miller & T. A. Cope. Flora of the Arabian Peninsula and Socotra, Vol. 1: 71–75, pl. 1. Edinburgh Univ. Press. [*Cupressus sempervirens, Juniperus excelsa* ssp. *polycarpos, J. phoenicea, J. procera*].

10250   Minnich, R. A. (1987). The distribution of forest trees in northern Baja California, Mexico. Madroño 34 (2): 98–127. [Cupressaceae, Pinaceae, angiosperms].

10251   Minnich, R. A. & R. G. Everett (2001). Conifer tree distributions in southern California. Madroño 48 (3): 177–197. (maps). [Cupressaceae, Pinaceae].

10261a   Miquel, F. A. G. (1855–59). Flora van Nederlands Indië [Flora Indiae Batavae]. (3 volumes). Amsterdam. [*Podocarpus teysmannii* sp. nov., vol. 2 (7): 1072 (1859), *P. eurhynchus* sp. nov. = *Sundacarpus amarus*, vol. 2 (7): 1074 (1859), *Cephalotaxus sumatrana* sp. nov. = *Taxus sumatrana*, vol. 2 (7): 1076 (1859)].

10489   Moore, D. M. (ed.) (1993). Coniferopsida, Taxopsida. Pp. 37–48 in: T. G. Tutin *et al.* (ed.). Flora Europaea 1, 2nd. ed. Cambridge Univ. Press, Cambridge. [Pinaceae, Taxodiaceae, Cupressaceae, Taxaceae; incl. introduced taxa].

10701   Munz, P. A. & D. D. Keck (1959). A California Flora. Univ. California Press, Berkeley – Los Angeles. [Coniferales, Pinaceae, Taxodiaceae, Cupressaceae, other families, ill., keys].

10911a   Nasir, E. & Y. J. Nasir (1987). Gymnospermae. Flora of Pakistan Fam. Nos. 178–186 (ed. E. Nasir & S. I. Ali). Islamabad. [Araucariaceae pp. 4–5, Pinaceae pp. 6–15, Taxodiaceae pp. 15, Cupressaceae pp. 16–25, ill.].

10911b   Nasir, E., M. A. Siddiqi & Z. Ali (1968). Gymnosperms of West Pakistan. Botany Department, Gordon College, Rawalpindi. [Coniferales, Cupressaceae, *Cupressus, Juniperus,* Pinaceae, *Abies, Cedrus, Pinus,* Gnetales, *Ephedra,* native and introduced species].

10970e   Nguyen Tien Hiep & J. E. Vidal (1996). Gymnospermae: Cycadaceae, Pinaceae, Taxodiaceae, Cupressaceae, Podocarpaceae, Taxaceae, Gnetaceae. Flore du Cambodge, du Laos et du Vietnam 28: 1–166. (ill.). [First author also known as Hiep, N. T.; *Amentotaxus hatuyenensis* T. H. Nguyen, sp. nov., p. 126].

11024   Ohwi, J. (1965). Flora of Japan. (Rev. ed.: F. G. Meyer & E. H. Walker). Washington, D.C. [gymnosperms pp. 109–118, photographs of *Pinus pumila, P. thunbergii, Picea jezoensis, Sciadopytis verticillata, Cryptomeria japonica* betw. pp. 108–109, keys].

11166   Ozenda, P. (1977). Flore du Sahara septentrional et central. 2nd. ed. ("revue et completée") C. N. R. S. Paris. (1st. ed. 1958). [*Cupressus dupreziana*, p. 121, ill. p. 123].

11250   Palmer, E. & N. Pitman (1972). Trees of Southern Africa, covering all known indigenous species in the Republic of South Africa, South-West Africa, Botswana, Lesotho & Swaziland. Vol. 1. Cape Town. [gymnosperms on pp. 304–345, many ill., Zamiaceae, Podocarpaceae, Cupressaceae, *Widdringtonia,* Welwitschiaceae].

11631   Polunin, O. & A. Stainton (1988). Flowers of the Himalaya. (Drawings by Ann Farrer). Oxford Univ. Press, Delhi. [Pinaceae, Cupressaceae, pp. 384–391, ill. pp. 510–512; see also A. Stainton, 1988; descr. + ill. under *J. macropoda* in both publ. pertain to *J. semiglobosa*].

11636   Poole, A. L. & N. M. Adams (1990). Trees and shrubs of New Zealand. Field Guide Series. (Repr. 1994) Lincoln, Canterbury, New Zealand. [Coniferales pp. 18–29, ill., *Agathis australis, Libocedrus bidwillii, L. plumosa,* Phyllocladus, Podocarpaceae].

11753a Quezel, P. & S. Santa (1962). Nouvelle flore de l'Algérie et des régions désertiques méridionales. Vol. 1. C. N. R. S. Paris. [gymnosperms pp. 33–39, Cupressaceae: *Tetraclinis articulata* (sub *Callitris*), *Cupressus dupreziana, Juniperus* spp., Pinaceae: *Abies numidica, Cedrus, Pinus* spp., ill.].

11831 Rechinger, K. H., ed. (1965–68). Flora Iranica. Flora des iranischen Hochlandes und der umrahmenden Gebirge. Graz. [see H. Riedl, 1965–68].

12025 Richter, K. (1890). Plantae Europeae. Leipzig. [Coniferae pp. 1–7, *Pinus laricio* ssp. *salzmanii* (Dunal) comb. nov., p. 2; Flora].

12045 Ridley, H. N. (1922–25). The flora of the Malay Peninsula...Vol. 1–5. London. [*Podocarpus deflexus* sp. nov., vol. 5: 283 (1925)].

12042 Riedl, H. (1965). Pinaceae. in: K. H. Rechinger (ed.). Flora Iranica. Lfg./Cont. No. 14: 1–9. Graz. [*Abies spectabilis, A. pindrow, Picea smithiana, Cedrus deodara, Pinus wallichiana, P. gerardiana, P. roxburghii, P. armena, P. halepensis, P. brutia, P. eldarica*, ill. with phot.].

12043 Riedl, H. (1968). Cupressaceae. in: K. H. Rechinger (ed.). Flora Iranica. Lfg./Cont. No. 50: 1–10. Graz. [*Cupressus sempervirens, Platycladus orientalis* (feral?), *Juniperus semiglobosa, J. sabina, J. turcomanica, J. excelsa, J. foetidissima, J. squamata, J. oxycedrus, J. communis*].

12201 Rouy, G. (1913). Flore de France ou descriptions des plantes qui croissent spontanément en France, en Corse et en Alsace-Lorraine. Vol. 14. Gymnospermes – Conifères, pp. 352–375. Paris. [*Juniperus gallica* (Coincy) comb. et stat. nov. = *J. thurifera* var. *gallica*, p. 374].

12211a Roxburgh, W. (1832). Flora indica; or, descriptions of Indian plants. Calcutta, London. (4 volumes edited by W. Carey). [*Juniperus elata* sp. nov. = *Dacrydium elatum*, vol. 3: 838, *J. chinensis* sp. nov. (non L.) = *Podocarpus chinensis*, vol. 3: 840].

12211e Royen, P. van (1979). The Alpine Flora of New Guinea. Vol. 2: Taxonomic Part – Cupressaceae to Poaceae. Vaduz. [Cupressaceae, *Papuacedrus papuana*, Podocarpaceae (incl. *Phyllocladus*) pp. 1–35, ill.].

12435 Sahni, K. C. (1990). Gymnosperms of India and adjacent countries. Dehra Dun. [Cycadales, Ginkgoales, Taxales, Coniferales, Gnetales, Cupressaceae, Pinaceae, Taxodiaceae, ill., keys].

12600 Satake, Y. (1971). Gymnospermae. in: H. Hara. Flora of Eastern Himalaya. Univ. of Tokyo Press, Tokyo.

13071a Seemann, B. (1865–73). Flora vitiensis: a description of the plants of the Viti or Fiji Islands with an account of their history, uses and properties. (in 10 parts, ill.). London. [*Dammara vitiensis* sp. nov. = *Agathis macrophylla*, p. 256, t. 76 (1868), *Podocarpus affinis* sp. nov., p. 266 (1868)].

13080c Seubert, M. (1844). Flora Azorica, quam ex collectionibus schedisque Hochstetteri patris et filii elaboravit et tabulis xv... Bonn. [*Juniperus oxycedrus* var. *brevifolia* var. nov., p. 26].

13150 Sherif, A. S. & A. El-Taife (1986). Flora of Libya – gymnosperms. Al Faateh Univ., Tripoli. (30 pp., ill.). [Ephedraceae, *Pinus halepensis, Cupressus sempervirens, Juniperus oxycedrus* ssp. *macrocarpa, J. phoenicea* + cult. conifers].

13250 Siebold, P. F. von & J. G. Zuccarini (1842–44, –70). Flora Japonica sive plantae, ... Vol. II. Leiden. (descr. + tab. £ 115 publ. 1842; descr. on pp. 45–89 by F. A. W. Miquel, 1870). [Coniferales, *Sciadopitys* gen. nov., *S. verticillata* comb. nov., p. 1, t. 101–102, *Abies tsuga* sp. nov., p. 14, t. 106 = *Tsuga sieboldii, A. firma* sp. nov., p. 15, t. 107, *A. homolepis* sp. nov., p. 17, *A. bifida* sp. nov., p. 18 = *A. firma, A. jezoensis* sp. nov., p. 19, t. 110 = *Picea jezoensis, A. polita* sp. nov., p. 20, t. 111 = *Picea torano, Pinus densiflora* sp. nov., p. 22, t. 112, *P. parviflora* sp. nov., p. 27, t. 115, *P. koraiensis* sp. nov., p. 28, t. 116, *Thujopsis* gen. nov., p. 32 (but see Endlicher, 1842), *T. dolabrata* comb. nov., p. 34, t. 119–120, *Retinispora* gen. nov. = *Chamaecyparis*, p. 36, *R. obtusa* sp. nov., p. 38, *R. pisifera* sp. nov., p. 39, + cultivar, t. 121–123, *Cryptomeria japonica* var. *sinensis* var. nov., p. 52, *Juniperus rigida* sp. nov., p. 56, t. 125, *J. procumbens* sp. nov., p. 59, t. 127, f. 3; but see Lindley & Gordon, 1850].

13430 Small, J. K. (1903). Flora of the southeastern United States. New York. [Coniferales, Cupressaceae, Pinaceae, Taxodiaceae, *Caryopitys* gen. nov., *C. edulis* (Engelm.) comb. nov. = *Pinus edulis*, p. 29, *Picea australis* sp. nov. = *P. rubens*, p. 30].

13431 Small, J. K. & A. M. Vail (1893). List of Pteridophyta and Spermatophyta growing without cultivation in northeastern North America. Mem. Torrey Bot. Club 5: 5–377. [*Taxus minor* (Michx.) Britton ex Small & Vail = *T. canadensis*, p. 19].

13651 Stainton, J. D. A. (1988). Flowers of the Himalaya, a supplement. Oxford Univ. Press, Delhi. [colour photogr. of *Abies spectabilis, A. densa, Tsuga dumosa, Pinus roxburghii, P. wallichiana, Cupressus torulosa, Larix himalaica* = *L. potaninii* var. *himalaica, L. griffithii, Picea smithiana, Cedrus deodara, Juniperus macropoda* auct., non Boiss. = *J. semiglobosa, J. communis, J. recurva, J. squamata, J. indica*, pl. 110–113; see also O. Polunin & A. Stainton, 1988].

13654 Standley, P. C. & J. A. Steyermark (1958). Flora of Guatemala, Part I. Fieldiana, Bot. 24: 1–476. [Cycadaceae – Taxaceae, pp. 11–63, Araucariaceae, pp. 23–26, Cupressaceae, pp. 26–36, *Cupressus lusitanica*, (ill.), *Juniperus comitana*, *J. standleyi* (ill.), Pinaceae, pp. 36–56, *Abies guatemalensis* (ill.), *Pinus* spp. (ill.), Taxodiaceae, pp. 56–60, *Taxodium mucronatum* (ill.)].

13659 Stapf, O. (1894). On the flora of Mount Kinabalu, in North Borneo. Trans. Linn. Soc. London, Bot., ser. 2, 4: 69–263. (ill.). [*Podocarpus neriifolius* var. *brevifolius* var. nov. = *P. brevifolius*, p. 249].

13665 Stapf, O. (1914). Gymnospermae. In: L. S. Gibbs. A contribution to the Flora and plant formations of Mount Kinabalu and the highlands of British North Borneo. J. Linn. Soc. Bot. 42: 191–195. [*Dacrydium gibbsiae* sp. nov., p. 192].

13668 Stapf, O. (1917). Order CXXIX Pinaceae; Order CXXIXA Taxaceae. In: D. Prain (ed.). Flora of Tropical Africa 6 (2): 333–344. London. [Cupressaceae, *Juniperus*, *Widdringtonia*, Podocarpaceae, *Afrocarpus*, *Podocarpus*, *P. dawei* sp. nov. = *A. dawei*, p. 342].

13972 Sudworth, G. B. (1908). Forest Trees of the Pacific Slope. U.S. Forest Serv., Washington, D.C. (ill.). [Coniferales, angiosperms].

14030 Sudworth, G. B. (1927). Check list of the forest trees of the United States, their names and ranges. Rev. ed. 2. U.S.D.A. Misc. Circ. 92. (Ed. 1 publ. 1898).

14040 Sudworth, G. B. (1967). Forest trees of the Pacific Slope. Ed. 2, New York. (ill.). [Coniferales, angiosperms].

14060 Sukachev, V. N. (1929). Pinaceae. in: B. A. Fedtschenko. Flora Transbaicalica; pp. 26–31, f. 18–25. Leningrad. (Russ.).

14150 Takhtajan, A. L. (1986). Floristic regions of the world. (T. Crovello, translator). Univ. California Press, Berkeley.

14151 Takhtajan, A. L. & A. A. Fedorov (1972). Flora Yerevana: opredelitel' dikorastuscikh rastenij Araratskoj kotloviny. Nauka, Leningrad. (Russ.). [*Juniperus excelsa* ssp. *polycarpos* (K. Koch) Takhtajan, comb. nov., p. 53].

14370 Thunberg, C. P. (1784). Flora Japonica, sistens plantas insularum japonicarum ... Leipzig. (publ. Aug. 1784). [Coniferales].

14371 Thunberg, C. P. (1794–1800). Prodromus plantarum Capensium, quas in promontorio Bonae Spei Africes, annis 1772–1775 collegit Carol. Pet. Thunberg. (in 2 parts, publ. 1794 and 1800, bound in one volume) Uppsala. [*Taxus falcata* sp. nov. = *Afrocarpus falcatus* (Podocarpaceae), p. 117 (part 2), *T. latifolia* sp. nov. = *Podocarpus latifolius*, p. 117 (part 2)].

14550 Tutin, T. G. *et al.*, eds. (1964). Flora Europaea Vol. I. Cambridge Univ. Press, Cambridge. [see V. H. Heywood, ed. 1964].

14551 Tutin, T. G. *et al.*, eds. (1993). Flora Europaea Vol. I. Ed. 2, Cambridge Univ. Press, Cambridge. [see D. M. Moore, ed. 1993].

14713 Venning, J. (1983). Cupressaceae. in: B. D. Morley & H. R. Toelken (eds.). Flowering plants in Australia. (pp. 31–32). Adelaide.

14714 Venning, J. (1986). Cupressaceae. in: J. P. Jessop & H. R. Toelken (eds.). Flora of South Australia. Part I. Lycopodiaceae–Rosaceae. Ed. 4. Adelaide. [pp. 105–108, *Callitris*].

14740 Viereck, L. A. & E. L. Little, Jr. (1972). Alaska Trees and Shrubs. U.S.D.A. Agric. Handb. 410. Washington, D.C. [Taxaceae, Pinaceae, Cupressaceae, pp. 43–70, ill., maps].

14750 Viereck, L. A. & E. L. Little, Jr. (1975). Atlas of United States Trees; Vol. 2, Alaska Trees and Common Shrubs. U.S. Forest Serv. Misc. Publ. 1293. Washington, D.C. [maps].

14881 Walker, E. H. (1976). Flora of Okinawa and the Southern Ryukyu Islands. Washington, D.C. [gymnosperms pp. 125–136, ill., *Pinus luchuensis* pp. 130–131, *Juniperus taxifolia* (incl. *J. lutchuensis* Koidz., *J. conferta* Parl. sensu E. H. Wilson, 1916 (p. p.), 1920].

15181 Willis, J. H. (1970). A Handbook to Plants in Victoria. Vol. I. Ferns, conifers and Monocotyledons. 2nd Ed., Univ. Press, Melbourne. [Araucariaceae, Cupressaceae, Podocarpaceae].

15190 Willkomm, M. (1887). Forstliche Flora von Deutschland, Österreich und der Schweiz. 2. Aufl. Leipzig. [Coniferales, *Picea* sect. *Omorika* sect. nov. (as "Omorica"); emend. Mayr, 1890].

15402 Wu, Z. Y. & P. H. Raven (ed. comm.) (1999). Flora of China, Vol. 4, Cycadaceae through Fagaceae. Beijing & St. Louis. [contents gymnosperms: Cycadaceae, Ginkgoaceae, Araucariaceae (introd.), Pinaceae, Sciadopityaceae (introd.), Taxodiaceae, Cupressaceae, Podocarpaceae, Cephalotaxaceae, Taxaceae, Ephedraceae, Gnetaceae, and various angiosperm families].

15421 Yamazaki, T. (1995). Gymnospermae. Pp. 261–287 in: K. Iwatsuki, T. Yamazaki, D. E. Boufford & H. Ohba (eds.). Flora of Japan. Vol. 1. Pteridophyta and Gymnospermae. Tokyo. [*Cryptomeria japonica*, *Sciadopitys verticillata* (each in sep. fam.), *Abies*, *Tsuga*, *Pseudotsuga japonica*, *Larix*, *Picea*, *Pinus*, *Thujopsis dolabrata*, *Thuja standishii*, *Chamaecyparis*, *Juniperus*, *Sabina chinensis* = *J. chinensis*].

15690   Zoller, H. (1981). Abteilung Gymnospermae Nacktsamige Pflanzen. In: G. Hegi. Illustrierte Flora von Mitteleuropa. Ed. 3, Bd. 1, Teil 2., pp. 11–148. Berlin – Hamburg. [gymnosperms, Coniferales, Taxales, Pinaceae, Cupressaceae, Taxaceae, other families, richly ill.].

MANUALS

00120   Aiton, W. (1789). Hortus Kewensis; or, a catalogue of the plants cultivated in the Royal Botanic Garden at Kew. Vol. 3. Monoecia monadelphia, pp. 366–373; Dioecia monadelphia, pp. 413–418. London. (publ. date: 7 Aug.–1 Oct.). [*Pinus sylvestris* var. *divaricata* = *P. banksiana*, p. 366, *P. pinaster* sp. nov., p. 367, *P. inops* sp. nov. = *P. virginiana*, p. 367, *P. resinosa* sp. nov., p. 367, *P. pendula* sp. nov. = *Larix laricina*, p. 369, *P. nigra* Ait. (non Arnold) = *Picea mariana*, p. 370, *P. alba* sp. nov. = *Picea glauca*, p. 371, *Cupressus disticha* var. *nutans* = *Taxodium distichum* var. *imbricatum*, p. 372, *Juniperus communis* var. *montana* var. nov., p. 413, *Taxus elongata* sp. nov. = *Podocarpus elongatus*, p. 415].

00121   Aiton, W. (1813). Hortus Kewensis; or, a catalogue of the plants cultivated in the Royal Botanic Garden at Kew. Vol. 5. Monoecia monadelphia, pp. 314–324; Dioecia monadelphia, pp. 412–416. Second edition, London. [*Araucaria excelsa* (Lamb.) R. Br. = *A. columnaris*, p. 412].

00660   Bean, W. J. (1980). Trees and shrubs hardy in the British Isles. ed. 8, 4 vols. (1970–1980), London. [Coniferales, Ginkgoales, Taxales].

00730   Beissner, L. (1891). Handbuch der Nadelholzkunde. Berlin. (2. Aufl. 1909). [gymnosperms, *Biota* "spp. nov." = *Platycladus orientalis* cultivars, *Keteleeria davidiana* (Bertr.) comb. nov., *K. sacra* (Franch.) comb. nov., p. 426].

00791   Beissner, L. (1909). Handbuch der Nadelholzkunde. 2. Aufl. Berlin. [gymnosperms, Coniferales, *Abies magnifica* var. *argentea* et var. *glauca*, var. nov., p. 167].

00800   Beissner, L. & J. Fitschen (1930). Handbuch der Nadelholzkunde. 3. Aufl. Berlin (rev. ed. of Beissner's Handbuch der Nadelholzkunde with new names and descriptions by J. Fitschen). [gymnosperms, Coniferales, *Keteleeria davidiana* var. *sacra* (Franch.) comb. et stat. nov., p. 185, *Picea alcockiana* var. *acicularis* et var. *reflexa* (Shirasawa) comb. nov., p. 258, *Pinus heldreichii* var. *leucodermis* (Ant.) comb. nov., p. 404, *P. sylvestris* var. *armena* (K. Koch) comb. et stat. nov., p. 417].

01080a   Boland, D. J., M. I. H. Brooker, G. M. Chippendale, N. Hall, B. P. M. Hyland, R. D. Johnston, D. A. Kleinig & J. D. Turner (1984). Forest trees of Australia. Ed. 4, Melbourne. [Araucariaceae, *Agathis*, *Araucaria*, Cupressaceae, *Callitris*, Podocarpaceae, *Dacrydium*, *Phyllocladus*, *Podocarpus*, Taxodiaceae, *Athrotaxis*, ill., maps].

01082   Boom, B. K. (1978). Nederlandse Dendrologie. (Flora der cultuurgewassen van Nederland, deel I). 10e herziene druk, Wageningen. [gymnosperms: pp. 84–111, ill.].

01172   Bowers, N. A. (1965). Cone-bearing trees of the pacific coast. (169 pp.). 2nd Ed. Palo Alto, California. [Cupressaceae, Pinaceae, Taxodiaceae].

01495   Burns, R. M. & B. H. Honkala, tech. coords. (1990). Silvics of North America. Vol. 1, Conifers. U.S.D.A. Forest Service Agric. Handb. 654. Washington, D.C. (675 pp., ill., maps; supersedes Agric. Handb. 271, see H. A. Fowells, 1965). [compilation of papers on *Abies* spp., *Chamaecyparis* spp., *Juniperus* spp., *Larix* spp., *Libocedrus* (= *Calocedrus*) *decurrens*, *Picea* spp., *Pinus* spp., *Pseudotsuga* spp., *Sequoia sempervirens*, *Sequiadendron giganteum*, *Taxodium distichum*, *Taxus brevifolia*, *Thuja* spp., *Torreya taxifolia*, *Tsuga* spp.].

01530   Callen, G. (1976–77). Les Conifères Cultivés en Europe. I–II. Paris. [Coniferales, ill.].

01680   Carrière, E. A. (1855). Traité Général des Conifères, ou description de toutes les espèces et variétés,... Paris. [*Cupressus roylei* sp. nov. = *C. sempervirens*, p. 128, *C. corneyana* sp. nov. = cultivated plant of uncertain origin (cf. J. Silba, 1987), p. 128, *C. lambertiana* sp. nov. = *C. macrocarpa*, p. 146, *Taxodium mexicanum* sp. nov. = *T. mucronatum*, p. 147, *Tsuga* gen. nov., p. 185, *T. sieboldii* sp. nov., p. 186, *T. brunoniana* (Wallich) comb. nov. = *T. dumosa*, p. 188, *T. canadensis* (L.) comb. nov., p. 189, *Abies homolepis* var. *tokunaiae* var. nov. = *A. sachalinensis*, p. 216, *A. cilicica* (Ant. et Kotschy) comb. nov., p. 229, *Picea jezoensis* (Sieb. et Zucc.) comb. nov., p. 255, *P. polita* (Sieb. et Zucc.) comb. nov. = *P. torano*, p. 256, *P. sitchensis* (Bong.) comb. nov., p. 260, *Larix griffithiana* (Lindl. et Gord.) comb. nov., p. 278, *L. kamtschatica* (Endl.) comb. nov. = *L. gmelinii*, p. 279, *Pinus torreyana* sp. nov., p. 326, *Podocarpus ensifolius* R. Br. ex Carrière = *P. spinulosus*, p. 456, *P. endlicherianus* sp. nov. = *P. neriifolius*, p. 468, *Phyllocladus glaucus* sp. nov. = *P. aspleniifolius*, p. 502 ("*glauca*")].

01720   Carrière, E. A. (1867). Traité Général des Conifères, ou description de toutes les espèces et variétés,... Ed. 2, Paris. [*Juniperus cinerea* sp. nov. = *J. thurifera*, p. 35, *Biota* "spp. nov." = *Platycladus orientalis* cultivars, p. 94–96, *Thuja standishii* (Gord.) comb. nov., p. 108, *Cupressus lusitanica* var. *benthamii* (Endl.) comb. et stat. nov., p. 155, *C. cashmeriana* sp. nov. (for neotype see A. Farjon, 1994), p. 161, *Tsuga mertensiana* (Bong.) comb. nov., p. 250, *T. hookeriana* (Andr. Murray) comb. nov. = *T. mertensiana* ssp. *grandicona*, p. 252, *Pseudotsuga* gen. nov., p. 256, *P. douglasii* (D. Don) comb. nov. = *P. menziesii*, p. 256, *Abies religiosa* var. nov., pp. 274–275, *A. gordoniana* sp. nov. = *A. grandis*, p. 298, *Picea obovata* var. *schrenkiana* (Fischer et Meyer) comb. et stat. nov. = *P. schrenkiana*, p. 338, *P. alcockiana* (Lindl.) comb. nov., p. 343, *Cedrus atlantica* (Endl.) Manetti ex Carrière, comb. nov., p. 374, *Pinus tabuliformis* sp. nov. ("*tabulaeformis*"), p. 510, *Araucaria cookii* var. *gracilis* var. nov. = *A. columnaris*, p. 613, *Eutacta subulata* (Vieill.) comb. nov. = *Araucaria subulata*, p. 864 (614), *E. pancheri* sp. nov. = *Neocallitropsis pancheri*, p. 615, *E. rulei* (F. Muell.) comb. nov. = *Araucaria rulei*, p. 864 (615), *E. humilis* sp. nov. = *A. columnaris*, p. 864 (616), *E. rulei* var. *compacta* var. nov., p. 864, *E. minor* sp. nov. = *A. columnaris*, p. 864 (616), *E. muelleri* sp. nov. = *Araucaria mueller*, p. 864, *Nageia minor* sp. nov. = *Retrophyllum minus*, p. 641, *Podocarpus gnidioides* sp. nov., p. 656, *P. taxodioides* sp. nov. = *Falcatifolium taxoides*, p. 657, *Taxus baccata* var. *cuspidata* (Siebold & Zucc.) comb. nov. = *T. cuspidata*, p. 733].

02311   Coltman-Rogers, C. (1920). Conifers and their characteristics. London. [Coniferales].

02720   Dallimore, W. & A. B. Jackson (1923). A Handbook of Coniferae. London. [Coniferales, Ginkgoales, Taxales, *Podocarpus henkelii* Stapf ex Dallim. & A. B. Jacks., p. 47, *Juniperus corneyana* hort. ex Dallim. & A. B. Jacks. (in syn. : *Cupressus torulosa* var. *corneyana*; transf. to *Cupressus* in subseq. eds.) = *Cupressus* sp. (cultivar?), p. 225, *Picea likiangensis* var. *purpurea* (Masters) comb. et stat. nov. = *P. purpurea*, p. 334].

02721   Dallimore, W. & A. B. Jackson (1931–66). A Handbook of Coniferae. Ed. 2 (1931), Ed. 3 (1948), Ed. 3, rev. (1961), Ed. 4, rev.: A Handbook of Coniferae and Ginkgoaceae. (1966). London. [*Picea abies* var. *acuminata* var. nov. in Ed. 3, p. 390 (1948), *Abies pindrow* var. *brevifolia* var. nov. in Ed. 3, rev., p. 158 (1961); see for the entirely revised latest edition also S. G. Harrison, 1966].

02730   Dallimore, W. & E. S. Harrar (1948). Textbook of Coniferae. Ed. 3. London. [Coniferales, Ginkgoales, Taxales; see also W. M. Harlow & E. S. Harrar, 1958, for Ed. 4].

02821   Debazac, E. F. (1965). Manuel des Conifères. Nancy. (172 pp.). [Coniferales].

02850   Den Ouden, P. & B. K. Boom (1965). Manual of Cultivated Conifers, hardy in the cold- and warm-temperate zone. The Hague. (526 pp., ill.).

03410   Ellenberg, H. (1988). Vegetation ecology of Central Europe. (Eng. transl.) Ed. 4. Cambridge Univ. Press, Cambridge.

03411   Elwes, H. J. & A. Henry (1906). The trees of Great Britain and Ireland I. Edinburgh. [Coniferales, Ginkgoales, Taxales, *Araucaria* pp. 43–54, *Picea* pp. 75–97, *Taxus* pp. 98–126, *Cryptomeria* pp. 127–140, *Taxodium* pp. 171–181, *Thuja* pp. 182–200, ill.].

03412   Elwes, H. J. & A. Henry (1907). The trees of Great Britain and Ireland II. Edinburgh. [Coniferales, Ginkgoales, Taxales, *Tsuga pattoniana* var. *jeffreyi* Henry, var. nov. = *T. mertensiana* ssp. *mertensiana* var. *jeffreyi*, p. 231].

03413   Elwes, H. J. & A. Henry (1908). The trees of Great Britain and Ireland III. Edinburgh. [Coniferales, Ginkgoales, Taxales, *Cedrus brevifolia* (Hook. f.) Henry, comb. et stat. nov., p. 467].

03420   Elwes, H. J. & A. Henry (1909). The trees of Great Britain and Ireland IV. Edinburgh. [Coniferales, Ginkgoales, Taxales, *Abies pindrow* var. *intermedia* Henry, var. nov., p. 756].

03430   Elwes, H. J. & A. Henry (1910). The trees of Great Britain and Ireland V. Edinburgh. [Coniferales, Ginkgoales, Taxales, *Pinus parviflora* var. *pentaphylla* (Mayr) Henry, comb. nov., p. 1033, *Pinus halepensis* var. *brutia* (Ten.) Henry, comb. nov., p. 1100, *Cupressus formosensis* (Matsum.) Henry, comb. nov., p. 1149].

03440   Elwes, H. J. & A. Henry (1912). The trees of Great Britain and Ireland VI. Edinburgh. [Coniferales, Ginkgoales, Taxales, *Picea obovata* var. *fennica* (Regel) Henry, comb. et stat. nov. = *P.* × *fennica*, p. 1360, *Juniperus chinensis* var. *sargentii* Henry, var. nov., p. 1432].

03450   Emberger, L. (1960). Gymnospermes. in: Traité de botanique systématique., Vol. II (1): 383–459. Masson et Cie, Paris.

03451   Endlicher, S. L. (1837). Genera plantarum secundum ordines naturales disposita. Vindobonae (Wien) 1836–1840. (Classis XXII. Coniferae, pp. 258–264). [Coniferales].

03452   Endlicher, S. L. (1841). Genera plantarum secundum ordines naturales disposita. Supplementum I. Wien. [*Arthrotaxis* "gen. nov." = *Athrotaxis* D. Don, 1839, p. 1372, *Parolinia* Endl. non Webb = *Widdringtonia*, p. 1372].

03453   Endlicher, S. L. (1841). Enchiridion botanicum, exhibens classes et ordines plantarum, accedit nomenclator generum... Leipzig – Wien. [Coniferales: pp. 138–144, *Chamaepeuce* "Zucc.", nom. nud.].

03454   Endlicher, S. L. (1842). Genera plantarum secundum ordines naturales disposita. Supplementum II. Wien. (publ. date Mar–Jun 1842). [*Thujopsis* gen. nov., p. 24, *Widdringtonia* gen. nov., *W. cupressoides* (L.) comb. nov. = *W. nodiflora*, p. 25, *Cephalotaxus* gen. nov., p. 27].

03570   Engler, A., ed. (1954). Syllabus der Pflanzen-familien. Berlin. [see H. Melchior & E. Werdermann, 1954].

03580   Engler, A. & K. A. E. Prantl, eds. (1887). Die natürlichen Pflanzenfamilien. Leipzig. [see A. W. Eichler, 1887].

03590   Engler, A. & K. A. E. Prantl, eds. (1926). Die natürlichen Pflanzenfamilien. 2. Aufl. Leipzig. [see R. Pilger, 1926].

04320   Fowells, H. A. (1965). Silvics of Forest Trees of the United States. U.S.D.A. Agric. Handb. 271. Washington, D.C. (compiled from numerous papers). [Coniferales, Cupressaceae, Taxodiaceae, Pinaceae].

04608   Fu, L. K. & J. M. Jin (eds.) (1992). China Plant Red Data Book. Rare and endangered plants. Vol. 1. Science Press, Beijing – New York. [gymnosperms pp. 24–153; Cupressaceae, Pinaceae, Podocarpaceae, Taxodiaceae, Cephalotaxaceae, Taxaceae, Ginkgoaceae, Cycadaceae, many rare colour photographs, distr. maps].

04950   Gordon, G. (1858). The Pinetum: being a synopsis of all the coniferous plants at present known,... London. [Coniferales, *Juniperus canariensis* sp. nov. = *J. cedrus*, p. 114, *Larix leptolepis* (Sieb. et Zucc.) comb. nov. = *L. kaempferi*, p. 128, *Pinus divaricata* (Ait.) hort. ex Gord. = *P. banksiana*, p. 163, *Pinus pinceana* sp. nov., p. 204, *Pseudolarix kaempferi* gen. et comb. nov., p. 292 = *P. amabilis* (nom. gen. cons., see W. Greuter *et al.*, 1988)].

04960   Gordon, G. (1862). A Supplement to Gordon's Pinetum: containing descriptions and additional synonymes of all the coniferous plants not before enumerated in that work. London. [*Micropeuce*, gen. nov. = *Tsuga*, p. 13, *Biota excelsa* hort. ex Gord., sp. nov. = *Platycladus orientalis*, p. 17, *Juniperus densa* sp. nov. = *J. squamata*, p. 32, *Pinus lawsonii* Roezl ex Gord., sp. nov., p. 64, *Thujopsis standishii* sp. nov. = *Thuja standishii* (Gord.) Carrière, 1867, p. 100].

04961   Gordon, G. (1875). The Pinetum:... Second edition considerably enlarged and including the former supplement,... London. [*Taxodium nuciferum* Brongn. ex Gord. = *Glyptostrobus pensilis* (?), p. 126].

04970   Gordon, G. (1880). The Pinetum: being a synopsis of all the coniferous plants..., Ed. 3. (partly with R. Glendenning), London.

05440   Harlow, W. M. & E. S. Harrar (1958). Textbook of dendrology. Ed. 4, New York. [Coniferales, Ginkgoales, Taxales].

05470   Harrison, S. G. (rev.), W. Dallimore & A. B. Jackson (1966). A Handbook of Coniferae and Ginkgoaceae. Ed. 4, London. [Coniferales, Ginkgoales, Taxales].

06474   Huxley, A., M. Griffiths & M. Levy, eds. (1992). The new Royal Horticultural Society Dictionary of Gardening. Vol. 1–4. London and Basingstoke. (ed. 2; Vol. 1 (A–C), 2 (D–K), 3 (L–Q), 4 (R–Z); ill., incl. short biogr. of botanists and gardeners). [*Abies* pp. 3–7, *Agathis* pp. 82–83, *Callitris* pp. 468–469, *Calocedrus* p. 469, *Cedrus* pp. 554–555, *Chamaecyparis* pp. 586–587, *Cupressus* pp. 781–783; *Juniperus* pp. 722–725, *Keteleeria* pp. 734–735; *Larix* pp. 15–18, *Libocedrus* pp. 64–65, *Picea* pp. 570–573, *Pinus* pp. 582–594, *Pseudotsuga* pp. 747–748; *Taxodium* p. 437, *Widdringtonia* p. 707].

06600   Jackson, A. B. (1946). The identification of conifers. London. (ill.). [Coniferales, Ginkgoales, Taxales].

07100   Kent, A. H. (1900). Veitch's Manual of the Coniferae, containing a general review of the order, ... 2nd. Ed., London. (564 pp.). [Coniferales, *Laricopsis* Kent (non Fontaine) gen. nov., p. 403, *L. kaempferi* (Lamb.) comb. nov. = *Pseudolarix amabilis*, p. 403, *Abietia* gen. nov., p. 474, *A. douglasii* (Lindl.) comb. nov. = *Pseudotsuga menziesii*, p. 476, *A. fortunei* (Andr. Murray) comb. nov. = *Keteleeria fortunei*, p. 485].

07300   Klika, J., K. Siman, F. Novak & B. Kavka (1953). Jehlicnaté. Praha. (Czech.). [Coniferales, Pinaceae, Cupressaceae, other families].

07310   Knight, J. & T. A. Perry (1850). A Synopsis of the Coniferous Plants grown in Great Britain, and sold by Knight and Perry, at the exotic nursery, King's Road, Chelsea. London. [Coniferales, Ginkgoales, Taxales, *Cupressus corneyana* nom. nud., p. 19. *C. majestica* nom. nud., p. 20 = cultivars of uncertain origin, see J. Silba, 1987].

07350   Koch, K. H. E. (1869–73). Dendrologie I–III. Erlangen. [e.g. *Abies venusta* (Dougl.) comb. nov. (= *A. bracteata*), with basion. *Pinus venusta* Dougl. in Companion Bot. Mag. 2, Dec. 1, 1836, Vol. III, p. 210, 1873; *Abies torano* Sieb. ex K. Koch, sp. nov. (= *Picea torano* (K. Koch) Koehne, 1893), Vol. II, p. 233, 1873, with *A. polita* as syn.; *Cupressus obtusa* (Sieb. et Zucc.) comb. nov. (= *Chamaecyparis obtusa* Sieb. et Zucc. in Endl., 1847), Vol. II, p. 168, 1873; *C. pisifera* (Sieb. et Zucc.) comb nov. (= *Chamaecyparis pisifera* Sieb. et Zucc. in Endl., 1847), Vol. II, p. 170, 1873; *Glyptostrobus pensilis* (Staunton) comb. nov., Vol. II, p. 191, 1873, with basion. *Thuja pensilis* Staunton in Account embassy China, 1797; *Heyderia decurrens* gen. et comb. nov. (basion. : *Libocedrus decurrens* Torrey, 1853) = *Calocedrus decurrens*, Vol. II, p. 177, 179, 1873; *Araucaria araucana* (Molina) comb. nov., Vol. II, p. 206; *Larix laricina* (Du Roi) comb. nov. (basion. : *Pinus laricina* Du Roi in Dissert. Inaug. Observ. Bot., p. xlix, 1771), Vol. II, p. 263, 1873;

Cephalotaxus harringtonii (Knight ex J. Forbes) comb. nov. (basion.: *Taxus harringtonii* Knight ex J. Forbes in Pinetum Woburn.: 217, t. 63 (1839), Vol. II, p. 102].

07380   Koehne, B. A. E. (1893). Deutsche Dendrologie. Stuttgart. [e.g. *Taxus baccata* var. *brevifolia* (Nutt.) comb. nov., p. 6, *Picea torano* (K. Koch) Koehne, comb. nov., p. 22, *Pinus contorta* var. *bolanderi* (Parl. in DC.) Koehne, comb. nov., p. 37].

07610   Krüssmann, G. (1983). Handbuch der Nadelgehölze. 2. Aufl. u. Mitw. v. H.-D. Warda. Berlin – Hamburg. (Also issued in Engl. translation as: Manual of cultivated conifers. Portland, Oregon, 1985). [Coniferales, Ginkgoales, Taxales, descr., ill., maps].

08107   Lemmens, R. H. M. J., I. Soerianegara & W. C. Wong (eds.) (1995). Plant Resources of South-East Asia (PROSEA) 5 (2). Timber trees: minor commercial timbers. Leiden. [*Dacrycarpus* pp. 161–166, *Dacrydium* pp. 167–172, *Falcatifolium = Decussocarpus* pp. 242–245, *Libocedrus = Papuacedrus papuana* pp. 281–284, *Nageia* pp. 356–360, *Phyllocladus* pp. 392–395, *Podocarpus* pp. 395–401, *Prumnopitys = Sundacarpus amarus* pp. 402–405, descr., with ill. of wood anatomy for each genus].

08701a   Liu, T. S. (1960). Illustrations of native and introduced ligneous plants of Taiwan. Vol. 1. Dicksoniaceae – Ulmaceae. Taipei. (Chin., Latin nomencl., ill.). [ill. in colour of *Abies kawakamii*, *Cunninghamia konishii*, *Taiwania cryptomerioides*, *Chamaecyparis formosensis*, *C. obtusa* var. *formosana*, Coniferales pp. 15–70, *Taxus mairei* (Lemée & Lév.) comb. nov. = *T. chinensis* var. *mairei*, p. 16].

08829   Loudon, J. C. (1829). An encyclopaedia of plants, … London. (supplement publ. 1840; ed. 2 publ. 1841; ed. 3 publ. 1855). [Coniferales pp. 802–807; 847–849 treated in the Linnaean system, *Agathis australis* (D. Don) Loudon, p. 802, ill.].

08830   Loudon, J. C. (1830). Hortus britannicus. A catalogue of all the plants indigenous, cultivated in, or introduced to Britain. Part 1, ed. 1. London. [see G. Don, 1830].

08840   Loudon, J. C. (1838). Arboretum et fruticetum britannicum. Vols. 3, 4. London. [*Juniperus dealbata* sp. nov. = *J. communis*, Vol. 3, p. 1090; *Pinus* subsect. *Gerardianae* subsect. nov., Vol. 4, p. 2254, *P.* subsect. *Australes* subsect. nov., p. 2255, *P.* subsect. *Canarienses* subsect. nov., p. 2261, *P. insignis* Dougl. ex Loud., sp. nov. = *P. radiata*, p. 2265, *P. llaveana* sp. nov., p. 2267 (1 July, see also D. F. L. von Schlechtendal, 1838, p. 488), *P.* subsect. *Strobi* subsect. nov., p. 2280, *P. contorta* Dougl. ex Loud., sp. nov., p. 2292, *Abies cephalonica* sp. nov., p. 2325 (see also Gard. Mag. (London) 14: 81, 1838), *A. luscombeana* hort. ex Loud.,

in syn. = *A. cephalonica*, p. 2325, *A. taxifolia* hort. ex Loud. (non Desf.) in syn. ? = *A. cephalonica*, p. 2325, *A. dumosa* (D. Don) comb. nov. = *Tsuga dumosa*, p. 2325, *A. obovata* (Ledeb.) comb. nov. = *Picea obovata*, p. 2329, *Picea pindrow* (D. Don) comb. nov. = *Abies pindrow*, p. 2346, *J. communis* var. *oblonga* (M.-Bieb.) comb. nov., p. 2490].

08840a   Loudon, J. C. (1842). An encyclopedia of trees and shrubs; being the Arboretum et fruticetum brittanicum abridged... London. [*Pinus* spp. on pp. 950–951, 990–994, 1000–1015].

09052   Marshall, H. (1785). Arbustrum americanum, or, an alphabetical catalogue of forest trees and shrubs, natives of the American United States, arranged according to the Linnaean System. Philadelphia. [*Taxus canadensis* sp. nov., p. 151].

09780   Mayr, H. (1906). Fremdländische Wald- und Parkbäume für Europa. pp. i–vii, 1–622, pl. 1–20, f. 1–258. Berlin. [Coniferales, Ginkgoales, Taxales, manual; *Larix cajanderii* sp. nov., p. 297, *L. kurilensis* (Mayr 1890), p. 300, *L. principis-rupprechtii* sp. nov., p. 309, (= var. of *L. gmelinii* Rupr.), *Picea mastersii* sp. nov. = *P. wilsonii*, p. 328, *P. tschonoskii* sp. nov. = *P. maximowiczii*, p. 339].

09855   Mehra, P. N. (1988). Indian Conifers, Gnetophytes and phylogeny of gymnosperms. New Delhi. [Cupressaceae, *Cupressus*, *Juniperus*, Pinaceae, *Abies*, *Cedrus*, *Picea*, *Pinus*, cytology (karyotypes), ill.].

09880   Melchior, H. & E. Werdermann, eds. (1954). A. Englers Syllabus der Pflanzenfamilien I. Allg. Teil. Bakterien bis Gymnospermen. 12. Aufl., Berlin. (reissued 1976). (contrib. on conifers by (R. Pilger) H. Melchior & E. Werdermann, pp. 312–344). [gymnosperms: Cycadales, Ginkgoales, Coniferales, Taxales, Gnetales, Ephedrales, Welwitschiales, Laricoideae Melchior et Werdermann, subfam. nov., p. 331].

10070   Meyer, F. (1952). Die Nadelhölzer einschliesslich Ginkgo. Stuttgart. [Coniferales, Ginkgoales, Taxales].

10400   Mitchell, A. F. (1972). Conifers in the British Isles; A Descriptive Handbook. (drawings by C. Darter). London.

10410   Mitchell, A. F. (1974). A Field Guide to the Trees of Britain and Northern Europe. London – Glasgow.

10560   Morgenthal, J. (1964). Die Nadelgehölze. 4. Aufl. Stuttgart. [Coniferales, Ginkgoales, Taxales].

10621   Mottet, S. (1902). Les Conifères et Taxacées. Traité élémentaire et pratique... Paris. (ill.).

10940 Neger, F. W. (1907). Die Nadelhölzer (Koniferen) und übrigen Gymnospermen. Sammlung Goschen 355: 1–185. [Cupressaceae fam. descr., p. 139; see also F. G. Bartling & H. L. Wendland, 1830 and P. F. Horaninov, 1847].

10941 Neger, F. W. & E. Munch (1952). Die Nadelhölzer (Koniferen) und übrigen Gymnospermen. Sammlung Goschen 355: 1–140. (rev. ed.).

11022 Nuttall, T. (1842–49). The North American sylva: or, a description of the forest trees of the United States, Canada and Nova Scotia, not described in the work of F. Andrew Michaux. Vol. 1–3. Philadelphia. (ill.). [Coniferales, *Taxus brevifolia* sp.nov., Vol. 3: 86, t. 108, 1849, *Abies lasiocarpa* (Hook.) comb. nov., Vol. 3: 138, 1849, *Larix occidentalis* sp. nov., Vol. 3: 143, t. 120, 1849].

11301 Pardé, L. (1937). Les Conifères. (manual, 294 pp.; repr. 1955) Maison Rustique, Paris.

11441 Penhallow, D. P. (1907). A manual of North American gymnosperms. (374 pp., ill.). The Atheneum Press, Boston. [Coniferales (incl. taxads)].

11481 Perry, J. P., Jr. (1991). The Pines of Mexico and Central America. Portland, Oregon. [*Pinus* spp., comprehensive review of all taxa, ill., maps (see index of botanical names)].

11691 Preston, R. J., Jr. (1976). North American trees (exclusive of Mexico and tropical United States). Ed. 3. Iowa State Univ. Press, Ames, Iowa. [Coniferales, 399 pp., ill., maps].

11901 Rehder, A. (1927). Manual of cultivated trees and shrubs hardy in North America exclusive of the subtropical and warmer temperate regions. New York. [Coniferales, *Pinus nigra* var. *caramanica* (Loud.) comb. nov., p. 61].

11941 Rehder, A. (1940). Manual of cultivated trees and shrubs hardy in North America exclusive of the subtropical and warmer temperate regions. Ed. 2, New York. [Coniferales, *Picea asperata* var. *heterolepis* (Rehder et Wilson) comb. et stat. nov., p. 24, *P. brachytyla* var. *complanata* (Masters) comb. et stat. nov., p. 30].

11960 Rehder, A. (1962). Manual of cultivated trees and shrubs hardy in North America exclusive of the subtropical and warmer temperate regions. Ed. 3, New York. [Coniferales].

12370 Rushforth, K. D. (1987). Conifers. London. [Coniferales, Ginkgoales; short descr. of many spp.].

12481 Sargent, C. S. (1884). Report on the forests of North America (exclusive of Mexico). United States 10th Census, 1880, Vol. 9. Washington, D.C. [Coniferales, Cupressaceae, Pinaceae, *Pinus clausa* (Chapm. ex Engelm.) Vasey ex Sarg., comb. et stat. nov. [based on *Pinus clausa* Vasey (1876), publ. without basion. cit. or descr.), p. 199].

12482 Sargent, C. S. (1896). The silva of North America: a description of the trees which grow naturally in North America exclusive of Mexico. Vol. 10. Boston – New York. [Coniferales, Pinaceae, Cupressaceae, *Juniperus monosperma* (Engelm.) comb. et stat. nov., p. 89, pl. 522, *Libocedrus plumosa* (D. Don) comb. nov., p. 134].

12490 Sargent, C. S. (1897). The silva of North America: a description of the trees which grow naturally in North America exclusive of Mexico. Vol. 11. Boston – New York. [Coniferales, Pinaceae, *Pinus roxburghii* nom. nov., syn. : *P. longifolia* Roxb. in Hortus bengalensis, 1814, a later homonym of *P. longifolia* Salisb. in Prodr. stirp. Chap. Allerton, 1796].

12500 Sargent, C. S. (1898). The silva of North America: a description of the trees which grow naturally in North america exclusive of Mexico. Vol. 12. Boston – New York. [Coniferales, *Picea rubens* sp. nov., p. 33, *Tsuga heterophylla* (Raf.) comb. nov., p. 73, pl. 605].

12501 Sargent, C. S. (1905). Manual of the trees of North America (exclusive of Mexico). Gymnospermae: pp. 1–101, ill. Boston – New York.

12521 Sargent, C. S. (1922). Manual of the trees of North America (exclusive of Mexico). Ed. 2, London. [corrected 2nd edition published 1926, Boston & New York].

12530 Sargent, C. S. (1965). Manual of the trees of North America (exclusive of Mexico). Ed. 2 (= 3, with corr.), New York.

12710 Schenck, C. A. (1939). Fremdländische Wald- und Parkbäume; Erster Band: Klimasektionen und Urwaldbilder. Berlin. [Coniferales, numerous photographs of (virgin) old stands esp. in North America and Japan, many tables on climate].

13320 Silba, J. (1986). Encyclopaedia Coniferae. Phytologia Memoirs VIII. Corvallis, Oregon. [Coniferales, Ginkgoales, short descriptions of taxa as accepted in J. Silba, 1984–85].

13350 Silva Tarouca, E. & C. Schneider (1923). Unsere Freiland-Nadelhölzer. Wien – Leipzig. [Coniferales, Taxales].

13510 Sokolov, S. J. & B. K. Shishkin (1949). Trees and Shrubs in USSR. I. Naked-seed plants. Bot. Inst. Komarova, Akad. Nauk SSSR, Leningrad – Moscow. (Russ., distr. maps). [gymnosperms, Coniferales].

14490 Tseng, W. (1961). Dendrology of China. (Chin.). Beijing. [Coniferales].

14520 Tubeuf, C. von (1897). Die Nadelhölzer, mit besonderer Berücksichtigung der in Mitteleuropa winterharten Arten. Stuttgart. [Coniferales, Pinaceae, Cupressaceae, Taxales, Ginkgoales, ill.].

14610   U.S.D.A. Forest Service (1948). Woody plant seed manual. U.S.D.A. Forest Service Misc. Publ. 654. Washington, D.C. [Coniferales, Cupressaceae, Taxodiaceae, Pinaceae; rev. repr. 1974, as U.S.D.A. Agric. Handb. 450, see Anonymus, 1974].

14727   Vidaković, M. (1982). Četinjače. Morfologija i variabilnost. Zagreb. (Serbocroatian, 710 pp. ill. manual; revised ed. based on English ed., publ. Zagreb, 1993). [Araucariaceae, Cupressaceae, Pinaceae, Taxodiaceae, other families].

14728   Vidaković, M. (1991). Conifers. Morphology and variation. 2nd ed. Zagreb. (Engl., 755 pp.). [ill. manual with emphasis on North Temperate (European) taxa and cultivars, revised transl. ed. of no. 14727, 1988 ed.].

15088   Welch, H. J. (1990). The conifer manual. Vol. 1. Dordrecht, The Netherlands. (Forestry series, vol. 34). [genera *Abies* through *Phyllocladus*, alphabetical (41 genera, 182 spp., cultivars), ill.].

15320   Wolf, E. L. (1925). Coniferous trees and shrubs in European and Asiatic sections of U.S.S.R. pp. 1–172, f. 1–59. Leningrad. (Russ.). [Coniferales, manual].

15360a Woodland, D. W. (1991). Contemporary Plant Systematics. Englewood Cliffs, New Jersey. (x, 582 pp., ill.). [Chap. 7, Families of the Pinophyta (gymnosperms) pp. 85–105, Class Pinatae (conifers), with fam. Araucariaceae, Cephalotaxaceae, Cupressaceae, Pinaceae, Podocarpaceae, Taxaceae, Taxodiaceae].

15360b Woodland, D. W. (2000). Contemporary Plant Systematics. Third ed. Andrew University Press, Berrien Springs, Michigan. (xii, 569 pp., ill., CD-ROM Photo Atlas of Vascular Plants). [Chap. 6, Families of the Pinophyta (gymnosperms) pp. 58–79, Class Pinatae (conifers), with fam. Araucariaceae, Cephalotaxaceae, Cupressaceae, Phyllocladaceae, Pinaceae, Podocarpaceae, Sciadopityaceae, Taxaceae].

GYMNOSPERMS
(general titles)

00155a Anderson, J. M. & H. M. Anderson (2003). Heyday of the gymnosperms: systematics and biodiversity of the Late Triassic Molteno fructifications. Strelitzia 15. [Pinopsida (pp. 60–132), Coniferales, Voltziaceae, Cycadopsida, Ginkgoopsida, Class Insertae sedis, Bennetitopsida, Gnetopsida, South Africa, Gondwana, plant and organ reconstruction, Triassic flood plain environment, photographs and drawings of fossils, tables].

00200a Andrews, H. A., ed. (1948). Evolution and classification of gymnosperms. A symposium. Bot. Gaz. (Crawfordsville) 110 (1): 1–103. [see also C. A. Arnold, 1948].

00281   Arnold, C. A. (1948). Classification of gymnosperms from the viewpoint of palaeobotany. Bot. Gaz. (Crawfordsville) 110 (1): 2–12.

00361b Axsmith, B. J., E. L. Taylor & T. N. Taylor (1998). The limitations of molecular systematics: a palaeobotanical perspective. Taxon 47 (1): 105–108. [critique on conifer/gymnosperm phylogenies at higher taxonomic levels based solely on molecular data; see e.g. Chaw & al., 1997].

00550   Banerji, M. L. (1952). Observations on the distribution of gymnosperms in Eastern Nepal. J. Bombay Nat. Hist. Soc. 51: 156–159.

00620   Bauch, J. (1975). Dendrologie der Nadelbäume und übrigen Gymnospermen (Coniferophytina und Cycadophytina). Samml. Göschen 2603, Berlin.

00670   Beauverie, J. (1933). Les gymnospermes vivantes et fossiles. Cours de Botanique professé à la Faculté des Sciences de Lyon. (Text. Vol. + Atlas with 38 plates).

00691   Beck, C. B. (1966). On the origin of gymnosperms. Taxon 15 (9): 337–339. [gymnosperm evolution].

00699   Beck, C. B. (1976). Current status of the Progymnospermopsida. Rev. Paleobot. Palynol. 21: 5–23.

00700   Beck, C. B., ed. (1976). Origin and early evolution of angiosperms. Columbia Univ. Press, New York. [see F. Ehrendorfer, 1976].

00707   Beck, C. B. (1985). Gymnosperm phylogeny – A commentary on the views of S. V. Meyen. Bot. Rev. (Lancaster) 51: 273–294. [see S. V. Meyen, 1984].

00710   Beck, C. B., ed. (1988). Origin and evolution of gymnosperms. Columbia Univ. Press, New York.

00884   Bhatnagar, S. P. & A. Moitra (1996). gymnosperms. New Delhi. [fossil and extant orders, morphology, anatomy, reproductive biology].

01080b Bond, W. J. (1989). The tortoise and the hare: ecology of angiosperm tolerance and gymnosperm persistence. Biol. J. Linn. Soc. 36: 227–249.

01171c Bowe, L. M., G. Coat & C. W. de Pamphilis (2000). Phylogeny of seed plants based on all three genomic compartments: extant gymnosperms are monophyletic and Gnetales' closest relatives are conifers. Proc. Nation. Acad. Sci. U.S.A. 97: 4092–4097. [DNA, cladistic analysis with limited taxon sampling; for problems see Soltis & al., 2002 and Magallón & Sanderson, 2003].

01175 Boyd, A. (1992). Revision of the late Cretaceous Pautût flora from west Greenland: Gymnospermopsida (Cycadales, Cycadeoidales, Caytoniales, Ginkgoales, Coniferales). Palaeontographica Abt. B, Palaeophytol. 225: 105–172. (ill.).

01850 Čelakovsky, L. J. (1890). Die Gymnospermen. Eine morphologisch-phylogenetische Studie. Abh. Königl. Böhm. Ges. Wiss. (Prag) Ser. VII, Band 4, 148 pp.

01851 Čelakovsky, L. J. (1898). Nachtrag zu meiner Schrift über die Gymnospermen. Bot. Jahrb. Syst. 24: 202–231.

01890 Chamberlain, C. J. (1935). gymnosperms. Structure and evolution. Univ. of Chicago Press, Chicago.

01984 Chaw, S. M., C. L. Parkinson, Y. Cheng, T. M. Vincent & J. D. Palmer (2000). Seed plant phylogeny inferred from all three plant genomes: monophyly of extant gymnosperms and origin of Gnetales from conifers. Proc. Nation. Acad. Sci. U.S.A. 97: 4086–4091. [Coniferales, Gnetales, gymnosperms, angiosperms, DNA, cladograms].

01986 Chaw, S. M., A. Zharkikh, H.M. Sung, T.C. Lau & W.H. Li (1997). Molecular phylogeny of extant gymnosperms and seed plant evolution: analysis of nuclear 18S rRNA sequences. Mol. Biol. Evol. 14 (1): 56–68. [phylogeny of conifers and other gymnosperms, cladograms; for critique of phylogenies ignoring fossil evidence in taxa with many extinctions, see Axsmith & al., 1998].

02050 Cheng, W. C. (1961). Chinese Dendrology, Vol. 1. (Chin.). Beijing. [gymnosperms; *Hesperopeuce longibracteata* (Cheng) comb. nov. (in syn., see also W. C. Cheng & L. K. Fu, 1978, p. 108) = *Nothotsuga longibracteata*].

02061 Chesnoy, L. (1987). La réproduction sexuée des gymnospermes. Bull. Soc. Bot. France (Act. Bot.) 134 (1): 63–85.

02400 Coulter, J. M. & C. J. Chamberlain (1917). Morphology of gymnosperms. Ed. 2. Univ. of Chicago Press, Chicago. (465 pp.).

02470e Crane, P. R. (1985). Phylogenetic analysis of seed plants and the origins of angiosperms. Ann. Missouri Bot. Gard. 72: 716–793. (ill., cladograms). [gymnosperms, angiosperms, Coniferales, Gnetales].

02915 Dogra, P. D. (1964). Pollination mechanisms in gymnosperms. In: P. K. K. Nair (ed.). Advances in Palynology, Chapter 7: 142–175. (ill.). National Botanic Gardens, Lucknow.

02920 Dogra, P. D. (1980). Embryogeny of gymnosperms and taxonomy: An assessment. in: P. K. K. Nair (ed.). Glimpses in plant research, Vol. 5: 114–128. New Delhi.

02961 Domin, K. (1938). Gymnospermae. Nakladem ceske Akademie Ved a Umeni v Praze. (Czech., 379 pp.).

03014 Donoghue, M. J. & J. A. Doyle (2000). Seed plant phylogeny: Demise of the anthophyte hypothesis? Current Biol. 10 (3): R106–109. [Discusses implications of inferred phylogeny based on genes from three plant genomes indicating Gnetales to have derived from Coniferales (Pinaceae?), and apparent conflict with analyses of morphological data].

03050c Doyle, J. A. (1996). Seed plant phylogeny and the relationships of Gnetales. Int. J. Plant Sci. 157 (6 Suppl.): S3-S39. [phylogeny of fossil and extant seed plants, gymnosperms, Coniferales, based mainly on morphology; review paper].

03050d Doyle, J. A. & M. J. Donoghue (1987). The importance of fossils in elucidating seed plant phylogeny and macroevolution. Rev. Palaeobot. Palynol. 50: 63–95. (ill.). [gymnosperms, seed habit].

3050e Doyle, J. A. & M. J. Donoghue (1992). Fossils and seed plant phylogeny reanalyzed. Brittonia 44: 89–106. [gymnosperms, palaeobotany, evolution and origin of angiosperms].

03340 Ehrendorfer, F. (1976). Evolutionary significance of chromosomal differentiation patterns in gymnosperms and primitive angiosperms. in: C. B. Beck, ed. Origin and early evolution of angiosperms. (pp. 220–240). Columbia Univ. Press, New York.

03610 Erdtman, G. (1957–65). Pollen and spore morphology/plant taxonomy. Vol. 2. Gymnospermae, Pteridophyta, Bryophyta (Ill.). Vol. 3. Gymnospermae, Bryophyta (Text). Stockholm.

03640 Esau, K. (1960). Anatomy of seed plants. New York – London. [gymnosperm leaves, pp. 288295].

03692 Farjon, A. (1993). Gymnosperms. In: P. J. M. Maas & L. Y. Th. Westra. Neotropical Plant Families. Koenigstein, Germany. [pp. 35–42, Cycadaceae, Zamiaceae, Araucariaceae, Pinaceae, Podocarpaceae, Cupressaceae, Ephedraceae, Gnetaceae].

03696f Farjon, A. (1998). Gimnospermas. In: P. J. M. Maas & L. Y. Th. Westra. Familias de plantas neotropicales. Vaduz – Liechtenstein. [pp. 38–45, Cycadaceae, Zamiaceae, Araucariaceae, Pinaceae, Podocarpaceae, Cupressaceae, Ephedraceae, Gnetaceae].

03953 Feustel, H. (1921). Anatomie und Biologie der Gymnospermen-blätter. Bot. Centralbl., Beih. 38 (2): 177–257. (ill.).

04120 Florin, C. R. (1955). The systematics of the gymnosperms. in: A century of progress in the natural sciences 1853–1953. Calif. Acad. Sci., San Francisco.

04152 Flory, W. S. (1936). Chromosome numbers and phylogeny in the gymnosperms. J. Arnold Arbor. 17 (2): 83–89.

04606 Fu, D. Z., Y. Yong & G. H. Zhu (2004). A new scheme of classification of living Gymnosperms at family level. Kew Bull. 59 (1): 111–116. [Conifers are separated into two groups, one of which includes Gnetales, Ginkgoales, Araucariaceae and Nageiaceae (= Podocarpaceae in part), the other includes the remaining conifer families; Cycadales are basal to 'conifers'].

04751 Gaussen, H. (1944). Les gymnospermes actuelles et fossiles. Fasc. 1–2: Généralités. Introduction. Chapitre I Généralités, II Les Ptéridospermales, III Les Cycadales. Faculté des Sciences, Toulouse (reprinted from diverse articles in Trav. Lab. Forest. Toulouse 1942–44). [evolution, gymnosperms, extinct and extant orders, ill.].

04752 Gaussen, H. (1946). Les gymnospermes actuelles et fossiles. Fasc. 3: Chapitre IV Les Bennetitales, Chapitre V Les autre Cycadophytes, Chapitre VI Les Cordaitales, Chapitre VII Les Ginkgoales. Faculté des Sciences, Toulouse (reprinted from diverse articles in Trav. Lab. Forest. Toulouse 1945–46). [gymnosperms, extinct and extant orders, ill.].

04760 Gaussen, H. (1960). Les gymnospermes actuelles et fossiles. Fasc. 6: Généralités, Genre *Pinus*. Trav. Lab. Forest. Toulouse, T. 2, sect. 1, vol. 1, chap. 11: 1–272. (ill.).

04850f Gerrath, J. M., L. Covington, J. Doubt & D. W. Larson (2002). Occurrence of phi thickenings is correlated with gymnosperm systematics. Canad. J. Bot. 80: 852–860. [Coniferales, Cycadales, Ginkgoales, Gnetales, anatomy of root cortex].

04878 Givnish, T. J. (1980). Ecological constraints on the evolution of breeding systems in seed plants: dioecy and dispersal in gymnosperms. Evolution 34: 959–972. [dioecious spp. zoochorous].

05000 Greguss, P. (1955). Identification of living gymnosperms on the basis of xylotomy. Budapest.

05003 Greguss, P. (1967). Fossil gymnosperm woods in Hungary, from Permian to Pliocene. Akad. Kiádo, Budapest. (136 pp., ill.). [gymnosperms, Coniferales, palaeobotany].

05130 Groff, G. W. (1930). gymnosperms of Kwangtung, China. Lingnan Sci. J. 9: 267–305. [rev. with keys].

05180 Guttenberg, H. von (1961). Grundzüge der Histogenese höherer Pflanzen: II. Die Gymnospermen. in: H. J. Braun, S. Carlquist, P. Ozenda & I. Roth (eds.). Handbuch der Pflanzenanatomie; Band VIII, Teil 4. Berlin – Stuttgart.

05370 Hao, K. S. (1945–51). Gymnospermae sinicae. (Illustrations of Chinese gymno-sperms) pp. 1–152, f. 1–26, 1945; 2nd. ed. (no Lat. title) pp. 1–4, 1–165, f. 1–35, 1951. Beijing.

05410 Hardin, J. W. (1971). Studies of the southeastern United States flora. II. The gymnosperms. J. Elisha Mitchell Sci. Soc. 87: 43–50.

05462 Harris, T. M. (1976). The Mesozoic gymnosperms. Rev. Palaeobot. Palynol. 21 (1): 119–134. [gymnosperms, palaeobotany, ill.].

05662 Hayata, B. (1933). Systematic Botany. Vol. 1. Gymnospermae. (Japan., ill.). Tokyo. [morphology and anatomy, taxon. account with emphasis on Japanese and Chinese taxa].

05685 Hegnauer, R. (1962–86). Chemotaxonomie der Pflanzen. Band 1. Thallophyten, Bryophyten, Pteridophyten und Gymnospermen. Basel – Stuttgart. Band 7. Nachträge zu Band 1 und Band 2. (1986) Basel – Boston – Stuttgart. [Araucariaceae 1: 337–341, 7: 482–487; Cephalotaxaceae 1: 341–342, 7: 487–490; Cupressaceae 1: 342–372, 478–480, 7: 491–504, 801; Pinaceae 1: 373–410, 480–481, 7: 504–523, 801; Podocarpaceae 1: 410–421, 7: 524–535; Taxodiaceae 1: 421–429, 481–482, 7: 535–540; Taxaceae 1: 430–440, 482, 7: 540–546; extensive literature review].

05879 Hill, T. G. & E. de Fraine (1908). On the seedling structure of gymnosperms. I. Ann. Bot. (London) 22: 689–712, pl. 35. [Taxaceae, Podocarpaceae, Cupressaceae (Cupressineae), Taxodiaceae (Taxodiinae)].

05880 Hill, T. G. & E. de Fraine (1909). On the seedling structure of gymnosperms. II. Ann. Bot. (London) 23: 189–227, pl. 15. [Pinaceae (Abietineae), Araucariaceae (Araucarieae)].

05881 Hill, T. G. & E. de Fraine (1909). On the seedling structure of gymnosperms. III. Ann. Bot. (London) 23: 433–458, pl. 30. [Ginkgoaceae, Cycadaceae s.l.].

05882 Hill, T. G. & E. de Fraine (1910). On the seedling structure of gymnosperms. IV. Ann. Bot. (London) 24: 319–333, pl. 22–23. [Gnetales, *Ephedra, Welwitschia, Gnetum*].

06280 Hu, Y. S. & B. J. Yao (1981). Transfusion tissue in gymnosperm leaves. Bot. J. Linn. Soc. 83: 263–272. (ill.).

06471 Hutchinson, J. (1924). Contributions towards a phylogenetic classification of flowering plants: III. The genera of gymnosperms. Bull. Misc. Inform. 1924 (2): 49–66. [Araucariaceae, Cupressaceae, Pinaceae, Taxodiaceae, Podocarpaceae, other families, maps].

06830 Johnson, M. A. (1951). The shoot apex in gymnosperms. Phytomorphology 1: 188–204.

07150 Khoshoo, T. N. (1961). Chromosome numbers in gymnosperms. Silvae Genet. 10: 1–9.

07160 Khoshoo, T. N. (1962). Cytological evolution in the gymnosperms karyotype. in: Proceedings of the Summer School of Botany, Darjeeling. Min. of Sci. Res. and Cult. Affairs, New Delhi.

07642 Kubitzki, K., ed. (1990). The families and genera of vascular plants. Vol. 1. Pteridophytes and gymnosperms. (ed. K. U. Kramer & P. S. Green). Berlin – Heidelberg. [see C. N. Page, 1990].

07690 Kuo, S. R., T. T. Wang & T. C. Huang (1972). Karyotype analysis of some Formosan gymnosperms. Taiwania 17: 66–80.

08050a Lebreton, P. (1982). Les Cupressales: une définition chimiosystématique. (51me communication dans la série: "Recherches chimiotaxonomiques sur les plantes vasculaires"). Candollea 37 (1): 243–256. [classifications of gymnosperms and Cupressaceae].

08050c Lebreton, P. (1990). La chimiotaxonomie des gymnospermes. Bull. Soc. Bot. France 137 (1): 35–46. [summ. traditional classif. of gymnosperms, suggests *Pinus* as sep. fam.; chemistry of *Juniperus thurifera*].

08257 Lindley, J. (1830). An introduction to the Natural System of Botany. Longman & Co., London. ['Gymnospermae' recognized as one of four 'classes' of seed plants, having uncovered ovules].

08405 Liston, A., W. A. Robinson, J. M. Oliphant & E. R. Alvarez-Buylla (1996). Length variation in the nuclear ribosomal DNA Internal Transcribed Spacer region of non-flowering seed plants. Syst. Bot. 21 (2): 109–120. [gymnosperms, Coniferales, Araucariaceae, Cupressaceae, Pinaceae, Podocarpaceae, Taxodiaceae etc. (32 genera), variation range 975–3125 bp (greatest in Pinaceae) and much larger than in angiosperms].

08740 Liu, Y. H. (1947). Gymnosperms of eastern China. Bot. Bull. Acad. Sin. 1: 141–171, f. 1–6. (Chin.). [taxonomic notes].

08936 Magallón, S. & M. J. Sanderson (2002). Relationships among seed plants inferred from highly conserved genes: sorting conflicting phylogenetic signals among ancient lineages. Amer. J. Bot. 89 (12): 1991–2006. [gymnosperms, Coniferales, Gnetales, angiosperms, DNA chloroplast genes *psa*A, *psb*B, codon positions, phylogeny].

08958 Maheshwari, P. & M. Sanwal (1963). The archegonium of gymnosperms. Mem. Indian Bot. Soc. 4: 103–119.

08960 Maheshwari, P. & H. Singh (1967). The female gametophyte of gymnosperms. Biol. Rev. Cambridge Philos. Soc. 42: 88–130. [gymnosperms, *Pinus*, embryology].

10010 Merrill, E. D. (1934). The gymnosperms of Malaysia, the Philippines, and Polynesia. Proc. 5th Pacific Sci. Congr. (Canada 1933) 4: 3267–3271.

10040 Metcalf, F. P. (1935). gymnosperms in China. Lingnan Sci. J. 14: 687–688. [notes suppl. to H. H. Hu, 1934].

10060 Meyen, S. V. (1984). Basic features of gymnosperm systematics and phylogeny as evidenced by the fossil record. Bot. Rev. (Lancaster) 50 (1): 1–112.

10060a Meyen, S. V. (1988). Gymnosperms of the Angaran flora. In: C. B. Beck (ed.). Origin and evolution of gymnosperms. (pp. 338–381, ill.). [gymnosperms, Coniferales, Angaran Permian conifers].

10164 Millay, M. A. & T. N. Taylor (1976). Evolutionary trends in fossil gymnosperm pollen. Rev. Palaeobot. Palynol. 21: 65–91. [loss of sacci is a derived state in conifers].

10209 Miller, C. N., Jr. (1985). A critical review of S. V. Meyen's "Basic features of gymnosperm systematics and phylogeny as evidenced by the fossil record." Bot. Rev. (Lancaster) 51: 295–318.

10480i Mongrand, S., A. Badoc, B. Patouille, C. Lacomblez, M. Chavent, C. Cassagne & J.-J. Bessoule (2001). Taxonomy of gymnospermae: multivariate analyses of leaf fatty acid composition. Phytochemistry 58: 101–115.

10700d Mundry, I. (2000). Morphologische und morphogenetische Untersuchungen zur Evolution der Gymnospermen. Bibl. Bot. 152: 1–90. (ill.). [Coniferales, Pinaceae, *Picea*, *Pinus*, Podocarpaceae, *Podocarpus*, *Lagarostrobos*, Taxaceae, *Torreya*, *Taxus*, Cephalotaxaceae, *Cephalotaxus*].

10797 Mutschler, O. von (1927). Die Gymnospermen des Weissen Jura von Nusplingen. Jahresber. Mitt. Oberrheinischer Geol. Ver. (January): 25–50.

10910 Napp-Zinn, K. (1966). Anatomie des Blattes, 1. Blattanatomie der Gymnospermen. in: W. Zimmermann *et al.* (eds.). Handbuch der Pflanzenanatomie, 2. Aufl. VIII 1. (370 pp.). Berlin-Nikolassee.

11003 Nimsch, H. (1995). A reference guide to the gymnosperms of the world – an introduction to their history, systematics, distribution, and significance. Koeltz Scientific Books, Champlain, Illinois. [gymnosperm families and genera, key, ill.].

11011a Nixon, K. C., W. L. Crepet, D. Stevenson & E. M. Friis (1994). A re-evaluation of seed plant phylogeny. Ann. Missouri Bot. Gard. 81 (3): 484–533. [fossil and extant gymnosperms, conifers, cycads, ginkgoids, anthophytes and selected angiosperms; cladistic analysis and evolution of anatomy and morphology, ill., cladograms].

11780   Raizada, M. B. & K. C. Sahni (1960). Living Indian gymnosperms Part I. (Cycadales, Ginkgoales and Coniferales). Indian Forest Rec., Bot. 5 (2): 73–150.

12191b   Rothwell, G. W. (1985). The role of comparative morphology and anatomy in interpreting the systematics of fossil gymnosperms. Bot. Rev. (Lancaster) 51: 319–326.

12191c   Rothwell, G. W. (1994). Phylogenetic relationships among ferns and gymnosperms; an overview. J. Plant Res. 107: 411–416. [Pteridophytes, gymnosperms, phylogenetic systematics].

12191e   Rothwell, G. W. & R. Serbet (1994). Lignophyte phylogeny and the evolution of spermatophytes: a numerical cladistic analysis. Syst. Bot. 19 (3): 443–482. [gymnosperms, conifers, phylogeny, palaeobotany].

12402   Rydin, C., M. Källersjö & E. M. Friis (2002). Seed plant relationships and the systematic position of Gnetales based on nuclear and chloroplast DNA: conflicting data, rooting problems, and the monophyly of conifers. Int. J. Plant Sci. 163 (2): 197–214. [angiosperms, gymnosperms, Coniferales, Gnetales (both monophyletic), phylogeny].

12845   Schmidt, M. & H. A. W. Schneider-Poetsch (2002). The evolution of gymnosperms redrawn by phytochrome genes: the Gnetaceae appear at the base of the gymnosperms. J. Mol. Evol. 54: 715–724. [DNA, phylogeny (a highly unlikely result)].

12952   Schnarf, K. (1933). Embryologie der Gymnospermen. Handbuch der Pflanzenanatomie, II. Abt., 2. Teil, Band X (2), 304 pp., ill. Berlin.

12953   Schnarf, K. (1937). Anatomie der Gymnospermen-Samen. Handbuch der Pflanzenanatomie, II. Abt., 2. Teil, Band X (1), 156 pp., ill. Berlin.

12955   Schneckenburger, S. (1989). Studien zur Embryogenese und Keimung verschiedener Gymnospermen unter besonderer Berücksichtigung der Suspensorbildung und Keimwurzelgenese. (Diss., 123 pp., ill.). Palmarum Hortus Francofortensis – Wissenschaftliche Berichte 1. Frankfurt am Main, Germany.

12956   Schneckenburger, S. (1993). Embryology and germination in gymnosperms. Acad. Nac. Ciencias, Misc. 91. Córdoba, Argentina.

13080e   Sharp, A. J. (1935). An improvement in method of preparing gymnosperms for the herbarium. Rhodora 37: 257–268.

13380   Singh, H. (1978). Embryology of gymnosperms. in: H. J. Braun, S. Carlquist, P. Ozenda & I. Roth (eds.). Handbuch der Pflanzenanatomie; Band X, Teil 2. Berlin – Stuttgart.

13386   Sitholey, R. V. (1963). Gymnosperms of India – I. Fossil forms. Bull. Nat. Bot Gard. 86 (Lucknow, India). [gymnosperms, conifers pp. 32–56, ill.].

13468   Smoot, E. L. & T. N. Taylor (1986). Evidence of simple polyembryony in Permian seeds from Antarctica. Amer. J. Bot. 73: 1079–1081.

13516   Soltis, D. E., P. S. Soltis & M. J. Zanis (2002). Phylogeny of seed plants based on evidence from eight genes. Amer. J. Bot. 89 (10): 1670–1681. [gymnosperms, angiosperms, Coniferales, Gnetales, Cycadales, *Ginkgo*, DNA sequences, codon positions].

13570   Sporne, K. R. (1965). The morphology of gymnosperms. The structure and evolution of primitive seed plants. Hutchinson Univ. Library, London (216 pp.].

13580   Sporne, K. R. (1974). The morphology of gymnosperms. The structure and evolution of primitive seed plants. Ed. 2, revised. Hutchinson Univ. Library, London.

13820   Sterling, C. (1963). The structure of the male gametophyte in gymnosperms. Biol. Rev. Cambridge Philos. Soc. 38 (2): 167–203.

13826   Stewart, W. N. & G. W. Rothwell (1993). Paleobotany and the evolution of plants. Second edition, Cambridge University Press, Cambridge, UK. (1st. ed. 1983). [chapters 22–29 on gymnosperms, Coniferales, ill.].

13850   Stoffers, A. L., C. Kalkman, F. A. Stafleu & H. C. D. de Wit (1982). Compendium van de Spermatophyta. Utrecht – Antwerpen. [gymnosperms on pp. 81–93].

14112a   Sykes, W. R. (1991). Gymnospermae of Guangxi, South China. Guihaia 11 (4): 339–377. [gymnosperms, Pinaceae pp. 355–366, Taxodiaceae pp. 366–367, Cupressaceae pp. 367–369, map p. 370].

14157   Tarbaeva, V. M. (1990). Comparative seed anatomy and seed coat ultra-sculpture in gymnosperms. Inst. Biol. Komi Science Centre of the Ural Division of the USSR Acad. Sci. (Eng.). [gymnosperms, Cupressaceae, Pinaceae, ill.].

14380   Tieghem, P. van (1869). Anatomie comparée de la fleur femelle et du fruit des Cycadées, des Conifères et des Gnetacées. Ann. Sci. Nat. Bot., sér. 5, 10: 269–304. (ill.). [also publ. in: Compt. Rend. Hebd. Séances Acad. Sci. (Paris) 68: 830–834, 870–874, 1869].

14710b   Velenovsky, J. (1885). Die Gymnospermen der böhmischen Kreideformation. Prag.

14961   Wang, F. H. & Z. K. Chen (1990). An outline of embryological characters of gymnosperms in relation to systematics and phylogeny. Cathaya 2: 1–10.

15046 Watson, J. (1988). The Cheirolepidiaceae. In: C. B. Beck, ed. Origin and evolution of gymnosperms. (pp. 382–447). Columbia Univ. Press, New York. [palaeobotany].

15137 White, M. E. (1994). The greening of Gondwana. 2nd edition, Reed Books, Australia. [palaeobotany, photography of Australian plant fossils by Jim Frazier, Gondwanan palaeoflora, gymnosperms, conifers, e.g. *Agathis jurassica* M. E. White on pp. 160–164].

15361 Worsdell, W. C. (1897). On "transfusion-tissue", its origin and function in the leaves of gymnospermous plants. Trans. Linn. Soc. London., Bot., ser. 2, 5: 301–319. (ill.).

15404 Xi, Y. Z. (1989). Pollen exine ultrastructure of extant Chinese gymnosperms. Cathaya 1: 119–142. (ill.).

15480a Ye, N. G., G. Q. Gou & H. M. Liao (1993). The seedling types of gymnosperms and their evolutionary relationships. Acta Phytotax. Sin. 31 (6): 505–516. (Chin., Eng. summ.) (ill., 3 types and 8 subtypes recognized].

CONIFERALES

(general titles on conifers, incl. 'taxads')

00010 Aase, H. C. (1915). Vascular anatomy of the megasporophylls of conifers. Bot. Gaz. (Crawfordsville) 60 (4): 277–315.

00010d Abbe, L. B. & A. S. Crafts (1939). Phloem of white pine and other coniferous species. Bot. Gaz. (Crawfordsville) 100 (4): 695–722. [Araucariaceae, Cupressaceae, Pinaceae (*Pinus albicaulis*), Taxaceae, Taxodiaceae, anatomy of bark, ill.].

00151b Alvin, K. L. (1982). Cheirolepidiaceae: biology, structure, and paleoecology. Rev. Palaeobot. Palynol. 37: 71–78. [palaeobotany].

00155 Amerom, H. W. J. van (ed.) (2000). Fossilium Catalogus. II. Plantae, pars 102, Gymnospermae (3) (Ginkgophyta et Coniferae) 1. Supplement. *Haborosequoia – Piceoxylon*. Leiden.

00159 André, D. (1956). Contribution à l'étude morphologique du cône femelle de quelques gymnospermes (Cephalotaxacées, Juniperoidées, Taxacées). Nat. Monspel. Bot. 8: 3–35. [Juniperoidées = Cupressaceae s. str.].

00200e Anonymus (1938). Handlist of Coniferae, Cycadaceae & Gnetaceae, grown in the Royal Botanic Gardens, Kew and the National Pinetum Bedgebury. Royal Botanic Gardens Kew & H.M. Stationery Office, London. [principally conifers].

00210 Anonymus (1974). Seeds of woody plants in the United States. U.S.D.A. Agric. Handb. 450. Washington, D.C. [see J.F. Franklin, 1974; R.H. Ruth, 1974; L.O. Safford, 1974].

00230 Antoine, F. (1840–46). Die Coniferen nach Lambert, Loudon und anderen frei bearbeitet. Wien. [Iconography in 11 fasc.; *Pinus firma* (Sieb. et Zucc.) comb. nov., p. 70, t. 27/28, *P. bifida* (Sieb. et Zucc.) comb. nov., p. 79, t. 31, 1846 = *Abies firma*, *Araucaria glauca* sp. nov., p. 105, 1846 = *A. cunninghamii* var. *cunninghamii*, *A. cunninghamii* var. *longifolia* var. nov., p. 102].

00291 Aubréville, A. (1964). Vue d'ensemble sur la géographie et l'écologie des Conifères et Taxacées à propos de l'ouvrage de Rudolf Florin. Adansonia, sér. 2, 4 (1): 8–18.

00300 Aubréville, A. (1965). Les reliques de la flore des Conifères en Australie et en Nouvelle-Calédonie. Adansonia, sér. 2, 5 (4): 481–492.

00310 Aubréville, A. (1973). Déclin des genres des Conifères tropicaux dans le temps et l'espace. Adansonia sér. 2, 13 (1): 5–36.

00311 Aubréville, A. (1973). Distribution des Conifères dans la Pangée; Essais. Adansonia, sér. 2, 13 (2): 125–133.

00320 Axelrod, D. I. (1958). Evolution of the Madro-Tertiary geoflora. Bot. Rev. (Lancaster) 24: 433–509. [Coniferales, angiosperms].

00340 Axelrod, D. I. (1976). History of the coniferous forests, California and Nevada. Univ. Calif. Publ. Bot. 70. [palaeophytogeography].

00370 Bader, F. J. W. (1960). Die Coniferen der Tropen. Decheniana Beih. 113 (1): 71–97. [also publ. in: Nova Acta Leop. 148 (23), Leipzig; phytogeography, tables, maps].

00509 Baillon, H. (1860). Recherches organogéniques sur la fleur femelle des Conifères. Ann. Sci. Nat. Bot. 14: 186–199. (ill.) (also published in Adansonia 1: 1–16, 1861). [Coniferales, ontogeny of ovuliferous cones].

00510 Baillon, H. (1894). Conifères; Histoire des plantes. Vol. 12, pp. 1–45. Paris – London – Leipzig.

00598 Barthelmess, A. (1935). Über den Zusammenhang zwischen Blattstellung und Stelenbau, unter besondere Berücksichtigung der Koniferen. Bot. Archiv 37: 207–260. (ill.). [Coniferales].

00811 Benson, G. T. (1930). The trees and shrubs of western Oregon. Contr. Dudley Herb. 2: 5–170. [Coniferales, Pinaceae, Taxodiaceae, Cupressaceae: pp. 12–34, annot. list. of spp.].

00830 Bentham, G. & J. D. Hooker (1880). Coniferae (Ordo CLXV). in: Genera Plantarum 3 (1): 420–442. London. [Araucariaceae, Cupressaceae, Taxodiaceae, Pinaceae, generic descr., *Callitris* sect. *Pachylepis* (Brongn.) Benth. comb. et stat. nov. = *Widdringtonia*, p. 424, *Libocedrus macrolepis* (S. Kurz) comb. nov. = *Calocedrus macrolepis* S. Kurz, 1873, p. 426, *Agathis vitiensis* (Seem.) Benth. & Hook. f. = *A. macrophylla*, p. 436].

00861 Bernard, C. (1926). Preliminary note on branch fall in the Coniferales. Proc. Linn. Soc. New South Wales 51: 114–128. [physiology].

00880 Bertrand, M. C. E. (1874). Anatomie comparée des tiges et des feuilles chez les Gnetacées et les Conifères. Ann. Sci. Nat. Bot., sér. 5, 20: 51–53. [*Pseudotsuga davidiana* sp. nov. = *Keteleeria davidiana*, pp. 86–87].

00920 Biswas, K. (1933). The distribution of wild conifers in the Indian Empire. J. Indian Bot. Soc. 12: 24–47, pl. 2. (1 map).

01262 Brongniart, A. (1871). Note sur la constitution du cône des Conifères. Bull. Soc. Bot. France 18: 141–143.

01325 Brunsfeld, S. J., P. S. Soltis, D. E. Soltis, P. A. Gadek, C. J. Quinn, D. D. Strenge & T. A. Ranker (1994). Phylogenetic relationships among the genera of Taxodiaceae and Cupressaceae: evidence from rbcL sequences. Syst. Bot. 19: 253–262.

01351 Buchholz, J. T. (1920). Embryo development and polyembryony in relation to the phylogeny of conifers. Amer. J. Bot. 7: 125–145.

01351a Buchholz, J. T. (1926). Origin of cleavage polyembryony in conifers. Bot. Gaz. (Crawfordsville) 81: 55–71.

01361 Buchholz, J. T. (1933). The classification of Coniferales. Trans. Illinois State Acad. Sci. 25: 112–133.

01400 Buchholz, J. T. (1948). Generic and subgeneric distribution of the Coniferales. Bot. Gaz. (Crawfordsville) 110 (1): 80–91.

01415 Buchholz, J. T. (1950). Embryology of gymnosperms. Pp. 374–375 in: H. Osvald & E. Aaberg (eds.). Proceedings of the Seventh International Botanical Congress, Stockholm, 12–20 July 1950. Chronica Botanica Company, Waltham, MA.

01502 Burtt, A. H. (1899). Ueber den Habitus der Coniferen. Inaug. Diss. Doct. Eberhard-Karls-Universität zu Tübingen; pp. [1–7] 8–86, fig., 50 tab., 3 pl. + explanation on p. 88.

01510 Butts, D. & J. T. Buchholz (1940). Cotyledon numbers in conifers. Trans. Illinois State Acad. Sci. 33: 58–62.

01832 Cassie, D. V. (1954). New Zealand conifers. J. Arnold Arbor. 35 (3): 268–272. [Araucariaceae, Podocarpaceae, Cupressaceae, *Libocedrus bidwillii*, *L. plumosa*].

01841 Čelakovsky, L. J. (1879). Zur Gymnospermie der Coniferen. Flora 62: 257–264, 272–283.

01860 Čelakovsky, L. J. (1900). Neue Beiträge zum Verständniss der Fruchtschuppe der Coniferen. Jahrb. Wiss. Bot. 35: 407–448.

01930 Chaney, R. W. & D. I. Axelrod (1959). Miocene floras of the Columbian plateau. Publ. Carnegie Inst. Wash. 617. [Coniferales, angiosperms].

01950 Chang, Y. P. (1954). Bark structure of North American conifers. U.S.D.A. Forest Service Tech. Bull. 1095: 1–86. [anatomy of bark, ill.].

01973 Chase, M. W., D. E. Soltis, R. G. Olmstead, D. Morgan, D. H. Les, B. D. Mishler, M. R. Duvall, R. A. Price, H. G. Hills, Y. L. Qiu, K. A. Kron, J. H. Rettig, E. Conti, J. D. Palmer, J. R. Manhart, K. J. Sytsma, H. J. Michaels, W. J. Kress, K. G. Karol, W. D. Clark, M. Hedrén, B. S. Gaut, R. K. Jansen, K. J. Kim, C. F. Wimpee, J. F. Smith, G. R. Furnier, S. H. Strauss, Q. Y. Xiang, G. M. Plunkett, P. S. Soltis, S. M. Swensen, S. E. Williams, P. A. Gadek, C. J. Quinn, L. E. Eguiarte, E. Golenberg, G. H. Learn, S. W. Graham, S. C. H. Barrett, S. Dayanandan & V. A. Albert (1993). Phylogenetics of seed plants: an analysis of nucleotide sequences from the plastid gene rbcL. Ann. Missouri Bot. Gard. 80: 528–580. [includes conifer families and species in some cladograms].

01975 Chavchavadze, E. S. (1979). Coniferous wood. Nauka, Leningrad.

01976 Chaw, S. M., H. M. Sung, H. Long, A. Zharkikh & W. H. Li (1995). The phylogenetic positions of the conifer genera *Amentotaxus*, *Phyllocladus* and *Nageia* inferred from 18S rRNA sequences. J. Mol. Evol. 41: 224–230.

01978a Chen, Z. K. & F. H. Wang (1990). On the embryology and relationship of the Cephalotaxaceae and Taxaceae. Cathaya 2: 41–52. (ill.). [*Cephalotaxus*, *Amentotaxus*, *Austrotaxus*, *Pseudotaxus*, *Taxus*].

02010 Cheng, W. C. (1933). The studies of Chinese conifers I. Contr. Biol. Lab. Chin. Assoc. Advancem. Sci., Sect. Bot. 9: 18–23. [syst. treatm. + keys].

02020 Cheng, W. C. (1939). Les forêts du Se-tchouan et du Si-kang oriental. Trav. Lab. Forest. Toulouse, T. 5, vol. 1, art. 2: 1–233. [Coniferales, *Juniperus pingii* sp. nov., without Lat. descr., for this see W. C. Cheng in Y. de Ferré in Bull. Soc. Hist. Nat. Toulouse 79: 76, 1944].

02060a Cheng, Y. C., R. G. Nicolson, K. Tripp & S. M. Chaw (2000). Phylogeny of Taxaceae and Cephalotaxaceae genera inferred from chloroplast *mat*K gene and nuclear rDNA ITS region. Mol. Phylogen. Evol. 14 (3): 353–365.

02080 Chevalier, L. (1957). Les conifères actuellement connus en Nouvelle Calédonie. Etudes Mélanésiennes, II (1): 105–118.

02090 Chigira, Y. (1935). An anatomical study of the leaves of the Coniferae in Manchuria. Res. Bull. Agric. Exp. Sta. Kung-chu-ling Manchoukuo 14: 1–9, pl. 1–6. (Japan.).

02120   Chowdhury, C. R. (1962). The embryogeny of conifers: a review. Phytomorphology 12: 313–338.

02210   Church, A. H. (1920). Elementary notes on conifers. Bot. Mem. 8, Humphrey Milford, Oxford Univ. Press.

02220   Church, A. H. (1920). Form-factors in Coniferae. Bot. Mem. 9, Humphrey Milford, Oxford Univ. Press.

02240   Clausen, J. (1965). Population studies of alpine and subalpine races of conifers and willows in the California High Sierra Nevada. Evolution 19: 56–65.

02252   Clifford, H. T. & J. Constantine (1980). Ferns, fern allies and conifers of Australia. (xvii + 150 pp., 24 figs.). Univ. of Queensland Press, Brisbane. [Araucariaceae, Cupressaceae, Podocarpaceae].

02260   Clinton-Baker, H. (1909–13). Illustrations of conifers. 3 vols. (1, i–xxii, 1–75, 69 pl., 1909; 2, 1–79, 91 pl., 1909; 3, 1–89, 70 pl., 1913). Hertford.

02280   Clinton-Baker, H. & A.B. Jackson (1935). Illustrations of New Conifers. Hertford.

02331   Conkle, M. T. (1972). Analyzing genetic diversity in conifers... izozyme resolution by starch gel electrophoresis. U.S. Forest Serv. Note PSW-264. Pacific Southwest Forest and Range Exp. Stat., San Francisco.

02332   Conkle, M. T. (1992). Genetic diversity – seeing the forest through the trees. In: W. T. Adams, S. H. Strauss, D. L. Copes & A. R. Griffin (eds.). Population genetics of forest trees. New Forests 6: 5–22. [Coniferopsida, Pinaceae, *Pinus*, *Abies*, Cupressaceae, *Cupressus*].

02640   Curtis, W., first ed. (1787–hodie). The Botanical Magazine. (cont. as: Curtis's Botanical Magazine, including The Kew Magazine). Vols. 1–184, + 2 companion vols. (1787–1983), Vols. 1–xx. (1984–hodie). London. [iconography and descriptions of various conifer species have been enumerated separately in this bibliography].

02829a   Debreczy, Z. & I. Rácz (2000). Fenyók a föld körül (Conifers round the world). (Hungarian, Eng. summ., ill.). Budapest. [Coniferales, "333 conifers in color"].

02881   Dijkstra, S. J. (1975). Fossilium Catalogus. II. Plantae. (ed. S. J. Dijkstra & F. Schaarschmidt), pars 87, Gymnospermae IX (Ginkgophyta et Coniferae), pp. 937–1094. The Hague.

02881a   Dijkstra, S. J. & H. W. J. van Amerom (1999). Fossilium Catalogus. II. Plantae. (ed. H. J. W. van Amerom), pars 100, Gymnospermae (1) (Ginkgophyta et Coniferae) 1. Supplement *Aachenia* – Czekanowskiales. Leiden.

02881b   Dijkstra, S. J. & H. W. J. van Amerom (1999). Fossilium Catalogus. II. Plantae. (ed. H. W. J. van Amerom), pars 101, Gymnospermae (2) (Gynkgophyta et Coniferae) 1. Supplement *Dacrycarpus* – Gymnospermenzweig. Leiden.

02882   Dluhosch, H. (1937). Entwicklungsgeschichtliche Untersuchungen über die Mikrosporophyllgestaltung der Coniferen. [Die Blüten der Coniferen III (ed. M. Hirmer)] Bibl. Bot. 114 (3): 1–24. (ill.).

02923   Dogra, P. D. (1986). Conifers of India and their natural gene resources in relation to forestry and Himalayan environment. In: P. K. K. Nair (ed.). Glimpses in plant research 7: 129–194. New Delhi. (maps, ill.).

03010   Don, G. (1830). Coniferae. in: J. C. Loudon. Hortus britannicus. A catalogue of all the plants indigenous, cultivated in, or introduced to Britain. Part 1. (ed. 1). [*Pinus cembra* var. *sibirica* (Du Tour) comb. et stat. nov., p. 387, *Cedrus deodara* (Roxb.) comb. nov., p. 388; see also ed. 4 + supplement 1850, Araucariaceae, Pinaceae, Cupressaceae, Taxodiaceae, Podocarpaceae pp. 387–388].

03070   Doyle, J. C. (1945). Developmental lines in pollination mechanisms in the Coniferales. Sci. Proc. Roy. Dublin Soc. 24: 43–62.

03090   Doyle, J. C. & M. Brennan (1971). Cleavage polyembryony in conifers and taxads, a survey 1. podocarps, taxads and taxodioids. Sci. Proc. Roy. Dublin Soc., ser. A, 4 (6): 57–88.

03100   Doyle, J. C. & M. Brennan (1972). Cleavage polyembryony in conifers and taxads, a survey 2. Cupressaceae, Pinaceae and conclusions. Sci. Proc. Roy. Dublin Soc., ser. A, 4 (10): 137–158.

03131a   Doyle, M. F. (1998). Gymnosperms of the SW Pacific -1. Fiji. Endemic and indigenous species: changes in nomenclature, key, annotated checklist, and discussion. Harvard Papers in Botany 3 (1): 101–106. [*Agathis macrophylla*, *Acmopyle sahniana*, *Dacrycarpus imbricatus* var. *patulus*, *Dacrydium nausoriense*, *D. nidulum*, *Podocarpus affinis*, *P. neriifolius*, *Retrophyllum vitiense*, 2 non-coniferous gymnosperms].

03210   Dümmer, R. A. (1913). The conifers of the Lindley Herbarium; Botany School, Cambridge. J. Roy. Hort. Soc. 39: 63–91.

03260   Eckenwalder, J. E. (1976). Comments on "A new classification of the Conifers" (by H. Keng, 1975). Taxon 25 (2–3): 337–339.

03280   Edelin, C. (1977). Images de l'Architecture des Conifères. Thesis, Univ. Sci. Techn. Languedoc, Montpellier. (256 pp., ill.).

03290   Edelin, C. (1981). Quelques aspects de l'architecture végétative des conifères. Bull. Soc. Bot. France (Lettres bot.) 128 (3) 177–188.

03295 Edmunds, G. F. & D. N. Alstad (1978). Coevolution in insect herbivores and conifers. Science 199: 941–945.

03351 Eichler, A. W. (1873). Sind die Coniferen Gymnospermen oder nicht? Flora 54: 241–247, 260–272.

03360 Eichler, A. W. (1881). Ueber die weiblichen Blüthen der Coniferen. Monatsber. Königl. Preuss. Akad. Wiss. Berlin, Nov. 1881: 1–32, 2 pl.

03390 Eichler, A. W. (1889). Coniferae. in: A. Engler & K. A. E. Prantl (eds.). Die natürlichen Pflanzenfamilien II. Teil, 1. Abt., pp. 28–116. Leipzig. [Araucariaceae, Cupressaceae, Pinaceae, Podocarpaceae, Taxodiaceae, other families, *Tsuga dumosa* (D. Don, 1825) comb. nov., p. 80, morphology, anatomy, taxonomy, ontogeny].

03395 El-Kassaby, Y. A. (1991). Genetic variation within and among conifer populations: review and evaluation of methods. Pp. 61–76 in: S. Fineschi *et al.* (eds.). Biochemical markers in the population genetics of forest trees. SPB Academic Publication, The Hague.

03450a Emberger, L. (1968). Les plantes fossiles dans leurs rapports avec les végétaux vivants. 2nd. enlarged ed., Paris. [gymnosperms (sensu Emberger) pp. 505–588, Coniferales pp. 523–543, fossil conifers pp. 544–588, bibliography pp. 653–710].

03460 Endlicher, S. L. (1847). Synopsis Coniferarum. Scheitlin und Zollikofer, Sangalli (Sankt Gallen). [Coniferales, *Juniperus* sect. *Caryocedrus* sect. nov., p. 8, *J. communis* var. *caucasica* = *J. communis* var. *oblonga* (M.-Bieb.) Loud., *J. incurva*, *J. lambertiana* spp. nov. = *J. recurva*, pp. 18, 19, *J. sabinoides* = *J. thurifera*, p. 23, *J. procera* Hochst. ex Endl., sp. nov., p. 26, *J. virginiana* var. *vulgaris* var. nov., p. 28 = *J. virginiana* var. *virginiana*, *J. virginiana* var. *australis* var. nov., p. 28 = *J. bermudiana*, *J. gracilis* sp. nov. = *J. virginiana* L., p. 31, *Widdringtonia juniperoides* (L.) comb. nov. = *W. cedarbergensis*, p. 32, *Frenela fruticosa* sp. nov. = *Callitris oblonga*, p. 36, *F. rhomboidea* (L. C. Rich.) comb. nov. = *C. rhomboidea*, p. 36, *F. roei* sp. nov. = *C. roei*, p. 36, *F. australis* Mirb. ex Endl. sp. nov. = *C. endlicheri* p.p., p. 37, *F. verrucosa* A. Cunn. ex Endl. sp. nov. = *C. verrucosa*, p. 37, *F. robusta* A. Cunn. ex Endl. sp. nov. = *C. preissii*, p. 37, *F. gunnii* (Hook. f.) comb. nov. ("*gunii*") = *C. oblonga*, p. 38, *Libocedrus* gen. nov., p. 42, *L. doniana* sp. nov. = *L. bidwillii*, p. 43, *L. chilensis* (D. Don) comb. nov. = *Austrocedrus chilensis*, p. 44, *L. tetragona* (Hook.) comb. nov. = *Pilgerodendron uviferum*, p. 44, *Biota orientalis* (L.) gen. et comb. nov. = *Platycladus orientalis*, p. 47, *Cupressus funebris* sp. nov. = *Chamaecyparis funebris*, p. 58, *Cupressus benthamii* = *C. lusitanica* var. *benthamii*, p. 59 (but see G. Kunze, 1847), *Chamaecyparis thurifera* (Kunth) comb. nov. = *Juniperus flaccida* var. *poblana* (see M. Martínez, 1946) p. 62, *C. obtusa* (Sieb. et Zucc.) comb. nov., p. 63, *C. pisifera* (Sieb. et Zucc.) comb. nov., p. 64, *Glyptostrobus* gen. nov., p. 69, *G. heterophyllus* (Brongn.) comb. nov. = *G. pensilis*, p. 70, *Schubertia nucifera* sp. nov. = *G. pensilis*, p. 70, *Glyptostrobus pendulus* (J. Forbes) comb. nov. = *Taxodium distichum* var. *distichum*, p. 71, *Pinus larix* var. *russica* var. nov. = *Larix sibirica*, p. 134, *P. atlantica* sp. nov. = *Cedrus atlantica*, p. 137, *P. rudis* sp. nov., p. 151, *P. insularis* sp. nov., p. 157, *P. bungeana* sp. nov., p. 166, *Araucaria cookii* R. Br. "Species indescripta", p. 188 = *A. columnaris*, *Sequoia* gen. nov., p. 197 (nom. cons.), *S. sempervirens* (D. Don) comb. nov., p. 198, *S. gigantea* sp. nov. = *S. sempervirens*, p. 198, Podocarpaceae fam. nov., p. 203, *Podocarpus nageia* R. Br. ex Endl. = *Nageia nagi*, p. 207, *P. salicifolius* Klotzsch & H. Karst. ex Endl., p. 209, *P. sellowii* Klotzsch ex Endl., p. 209, *P. lambertii* sp. nov., p. 211, *P. rigidus* Klotzsch ex Endl. = *P. glomeratus*, p. 211, *P. elatus* R. Br. ex Endl., p. 213, *P. polystachyus* R. Br. ex Endl., p. 215, *P. japonicus* Siebold ex Endl. = *P. chinensis*, p. 217, *P. falcatus* (Thunb.) comb. nov. = *Afrocarpus falcatus*, p. 219, *P. andinus* Poepp. ex Endl. = *Prumnopitys andina*, p. 219, *Cephalotaxus umbraculifera* Siebold ex Endl. = *Taxus cuspidata*, p. 239, *Chamaecyparites* fossil. gen. nov., p. 277].

03465 Engelhardt, H. & F. Kinkelin (1908). I. Oberpliocäne Flora und Fauna des Untermaintales, insbesondere des Frankfurter Klärbeckens. Abh. Senckenbergischen Naturf. Ges. 29 (3): 15–306, pl. 22–35. [Coniferales, Cupressaceae: *Frenelites europaeus*, *Callitris*, *Libocedrus* (*Calocedrus*), Taxodiaceae: *Taxodium*, *Sequoia*, Pinaceae: *Pinus*, *Picea*, *Larix*, *Abies*, *Keteleeria loehrii* (Geyler et Kinkelin) comb. nov. ("*loehri*") = *Pseudotsuga* aff. *sinensis* on pp. 216–217, pl. 26, f. 7a–b].

03594 Enright, N. J. & R. S. Hill (eds.) (1995). Ecology of the Southern Conifers. Melbourne University Press. [Araucariaceae, Cupressaceae, Podocarpaceae, Phyllocladaceae, Taxodiaceae].

03630 Ern, H. (1974) Zur Ökologie und Verbreitung der Koniferen im östlichen Zentralmexiko. Mitt. Deutsch. Dendrol. Ges. 67: 164–198. [Pinaceae, Cupressaceae].

03655 Falcon-Lang, H. J. & D. J. Cantrill (2000). Cretaceous (Late Albian) Coniferales of Alexander Island, Antarctica. 1: Wood taxonomy: a quantitative approach. Rev. Palaeobot. Palynol. 111: 1–17. [fossil wood anatomy, form genera *Araucarioxylon*, *Araucariopitys*, *Podocarpoxylon* and *Taxodioxylon*, ill.].

03696d Farjon, A. (1996). A world list of threatened conifers: How much do we know? In: D. Hunt (ed.). Temperate trees under threat. Proceedings of an IDS symposium on the conservation status of temperate trees, University of Bonn, 30 September – 1 October 1994, pp. 151–160. International Dendrology Society. (ill., tables).

03696g   Farjon, A. (1998). World checklist and bibliography of conifers. Royal Botanic Gardens, Kew. [Coniferales, 8 families, 68 genera, 629 species, 176 infraspecific taxa, 3225 synonyms, 73 *incertae sedis*, taxonomy, nomenclature, distribution, conservation, 30 plates].

03696h   Farjon, A. (2001). World checklist and bibliography of conifers. Second edition. Royal Botanic Gardens, Kew. [Coniferales, 8 families, 69 genera, 630 species, 180 infraspecific taxa, 3368 synonyms, 82 *incertae sedis*, taxonomy, nomenclature, distribution, conservation, 30 plates].

03696l   Farjon, A. (2003). The remaining diversity of conifers. In: R. R. Mill (ed.). Proceedings of the Fourth International Conifer Conference. Acta Horticulturae 615: 75–89. International Society for Horticultural Science (ISHS). [Coniferales, Cupressaceae, Pinaceae, Podocarpaceae, phylogeny, taxonomy, cladograms].

03704a   Farjon, A. & C. N. Page (compilers) (1999). Conifers: Status survey and Conservation Action Plan. IUCN-SSC Conifer Specialist Group. IUCN, Gland, Switzerland and Cambridge, UK. [includes Global Red List of Conifers and regional as well as species accounts by several members of the Conifer Specialist Group].

03704b   Farjon, A. & C. N. Page (2001). An Action Plan for the world's conifers. Plant Talk 22/23 (July/October 2000): 43–47. (ill.).

03705   Farjon, A., C. N. Page & N. Schellevis (1993). A preliminary world list of threatened conifer taxa. Biodiv. & Conserv. 2 (3): 304–326. [416 taxa of conservation concern, 42 endangered, distribution and status indicated].

03790   Ferguson, D. K. (1967). On the phytogeography of Coniferales in the European Cenozoic. Palaeogeogr. Palaeo-climatol. Palaeoecol. (Amsterdam) 3 (1): 73–110.

03841   Ferré, Y. de (1941). Morphologie des graines de Gymnospermes. Trav. Lab. Forest. Toulouse, T. 2, vol. 2, art. 1: 1–14. [Coniferales, ill.].

03842   Ferré, Y. de (1941). Morphologie des plantules de Gymnospermes. Trav. Lab. Forest. Toulouse, T. 2, vol. 2, art. 2: 1–12. [Coniferales, ill.].

03845   Ferré, Y. de (1944). Morphologie des graines de Gymnospermes (suite I). Bull. Soc. Hist. Nat. Toulouse 79: 73–80. [*Juniperus pingii* W. C. Cheng in Ferré, Lat. diagn. in footnote on p. 76, see also W. C. Cheng, 1939].

03901   Ferré, Y. de (1952). Morphologie des plantules de Gymnospermes (suite I). Trav. Lab. Forest. Toulouse, T. 2, sect. 1, vol. 2, art. 2: 5–12. [Coniferales, ill.].

03911   Ferré, Y. de (1958). Problèmes actuels relatifs aux Conifères. Bull. Soc. Bot. France 105 (4): 155–205. [discussion, bibliography].

04000   Fitzpatrick, H. M. (1926, rev. 1965). Conifers: Keys to the Genera and Species, with economic notes. Sci. Proc. Roy. Dublin Soc., ser. A., 2 (7): 67–129, t. 4–10 + index.

04050   Florin, C. R. (1931). Untersuchungen zur Stammesgeschichte der Coniferales und Cordaitales. Erster Teil: Morphologie und Epidermisstruktur der Assimilationsorgane bei den rezenten Koniferen. Kongl. Svenska Vetenskapsakad. Handl. 10 (1): 1–588, t. 1–58.

04060   Florin, C. R. (1934). Die von E. L. Ekman (†) in Westindien gesammelten Koniferen. Ark. Bot., Band 25-A, No. 5: 1–22, pl. 1–3. [*Pinus, Podocarpus, Juniperus, J. ekmanii* sp. nov., p. 14].

04065   Florin, C. R. (1939). The morphology of the female fructifications in cordaites and conifers of Palaeozoic age. Bot. Not. 36: 547–565.

04070   Florin, C. R. (1938–45). Die Koniferen des Oberkarbons und des unteren Perms. Palaeontographica Abt. B, Paläophytol. 85 (1–8): 1–729. [Coniferopsida (palaeobotany), morphology of reproductive organs in recent Coniferales and Taxales (all genera) in No. 7: 526–654, with descr. per genus; *Neocallitropsis* gen. nov., *N. araucarioides* (R. H. Compton) comb. nov. = *N. pancheri*, p. 590, 1944].

04080   Florin, C. R. (1940). The Tertiary fossil conifers of south Chile and their phytogeographical significance. Kongl. Svenska Vetenskapsakad. Handl. Ser. 3, Vol. 19, No. 2.

04090   Florin, C. R. (1948). Enumeration of gymnosperms collected on Swedish Expeditions to Western and North-western China in 1930–34. Acta Horti Berg. 14 (8): 343–384, pl. 1–6. [Coniferales, *Juniperus arenaria* (E. H. Wilson) comb. et stat. nov. (basion. : *J. chinensis* var. *arenaria* E.H. Wilson, 1928), p. 353, pl. 4, f. 2–4, *Taxus wallichiana* var. *chinensis* (Pilg.) comb. nov. = *T. chinensis*, p. 355, *T. speciosa* sp. nov. = *T. chinensis* var. *mairei*, p. 382, pl. 6].

04096   Florin, C. R. (1950). Upper Carboniferous and Lower Permian conifers. Bot. Rev. (Lancaster) 16: 258–282.

04100   Florin, C. R. (1951). Evolution in Cordaites and Conifers. Acta Horti Berg. 15 (2): 285–388, 1 pl. [Coniferopsida; compare also J. A. Clement-Westerhof, 1984 and H. Kerp *et al.*, 1989].

04111   Florin, C. R. (1954). The female reproductive organs of Conifers and Taxads. Biol. Rev. Cambridge Philos. Soc. 29: 367–389.

04135   Florin, C. R. (1958). On Jurassic taxads and conifers from northeastern Europe and eastern Greenland. Acta Horti Bergiani 17: 259–388.

04140   Florin, C. R. (1963). The distribution of conifer and taxad genera in time and space. Acta Horti Berg. 20 (4): 121–312. [Coniferales, Taxales, Ginkgoales, palaeobotany, maps, bibliography].

04141   Florin, C. R. (1966). "The distribution of conifer and taxad genera in time and space"; Additions and Corrections. Acta Horti Berg. 20 (6): 319–326. [Coniferales, maps].

04260   Floyd, A. G. (1977). N. S. W. rainforest trees: part 4. Families: Podocarpaceae, Araucariaceae, Cupressaceae, Fagaceae, Ulmaceae, Moraceae, Urticaceae. Forest. Commiss. Res. Note 34. (63 p.) [Coniferales, ill. of *Callitris*].

04350   Franchet, A. R. (1884). Plantae davidianae ex sinarum imperio. Part I. Plantes de Mongolie du nord et du centre de la Chine. Coniferae. T. VII, pp. 285–293, pl. 12–14. Paris (repr. New York, 1970; *P. armandii* also publ. in Nouv. Arch. Mus. Hist. Nat., ser. 2, 7: 95. Dec 1884). [*Pinus armandii* sp. nov., p. 285, *Abies davidiana*, *A. sacra* sp. nov. = *Keteleeria davidiana*, p. 290, *Juniperus chinensis* var. *pendula* var. nov., p. 291].

04360   Franchet, A. R. (1899). Plantarum sinensium ecloge tertia. Coniferae. J. Bot. (Morot) 13 (8, suite, 9, fin): 253–266. [*Pinus yunnanensis* sp. nov., p. 253, *Abies delavayi* sp. nov., p. 255, *A. fargesii* sp. nov., p. 256, *A. fargesii* var. *sutchuenensis* var. nov., p. 256, *A. likiangensis* sp. nov. = *Picea likiangensis*, p. 257, *A. brachytyla* sp. nov. = *Picea brachytyla*, p. 258, *A. yunnanensis* sp. nov. = *Tsuga dumosa*, p. 258, *A. chinensis* sp. nov. = *Tsuga chinensis*, p. 259, *Larix thibetica* sp. nov. = *L. potaninii*, p. 262, *Thuja sutchuenensis* sp. nov., (9) p. 262, *Torreya fargesii* sp. nov. = *T. grandis* var. *fargesii*, p. 264].

04391   Franco, J. do A. (1942). Carpologia das Coníferas. Anais Inst. Super. Agron. (Lisboa) 13: 1–18.

04460   Franco, J. do A. (1952). Nomenclatura de algumas Coníferas. Anais Inst. Super. Agron. (Lisboa) 19: 5–23. [*Glyptostrobus lineatus* = *G. pensilis*, *Larix kaempferi*, *Pseudolarix amabilis*, *Thuja*, *Araucaria heterophylla* (Salisb.) comb. nov., p. 11, *Callitris hugelii* (Carrière) comb. nov., p. 11, *C. preissii*].

04530   Franklin, J. F., & C. T. Dyrness (1973). Natural vegetation of Oregon and Washington. U.S. Forest Serv. Gen. Techn. Rep. PNW-8. Pacific Northwest Forest and Range Exp. Stat., Portland, Oregon. [Pinaceae, Cupressaceae, *Taxus*].

04570   Frenzel, B. (1968). Grundzüge der pleistozänen Vegetationsgeschichte Nord-Eurasiens. Erdwiss. Forschungen 1. Wiesbaden. [Coniferales, Ginkgoales].

04741   Gaussen, H. (1948). Interprétation de l'écaille du cône chez les Conifères. Compt. Rend. Hebd. Séances Acad. Sci. (Paris) 227: 731–733.

04753   Gaussen, H. (1950–52). Les gymnospermes actuelles et fossiles. Fasc. 4: Annexe au Chapitre IV Les Pentoxylées, Chapitre VIII Les Coniferales. Première partie: les Pinoidines. Faculté des Sciences, Toulouse (reprinted from diverse articles in Trav. Lab. Forest. Toulouse). [gymnosperms, Coniferales, extant families, ill.].

04754   Gaussen, H. (1955). Les gymnospermes actuelles et fossiles. Fasc. 5: Annexe au Chapitre II, Chapitre IX Les Coniferales, Chapitre X Addenda et errata, indexes et tables. Faculté des Sciences, Toulouse (reprinted from diverse articles in Trav. Lab. Forest. Toulouse). [gymnosperms, Coniferales, extant families, ill.].

04820   Gaussen, H. (1976). La répartition des conifères et ses problèmes. Compt. Rend. Sommaire Séances Soc. Biogéogr., nos. 452–460: 38–45.

04887a   Golte, W. (1974). Öko-physiologische und phylogenetische Grundlagen der Verbreitung der Coniferen auf der Erde. Erdkunde 28 (2): 81–101. [maps on conifer distribution].

04987   Grauvogel-Stamm, L. & J. Galtier (1998). Homologies among Coniferophyte cones: further observations. C. R. Acad. Sci. Paris, Earth & Planetary Sci. 326 (Palaeontology): 513–520. [Coniferales, Cordaitales, pollen cones].

05020   Greguss, P. (1972). Xylotomy of the living conifers. Akadémiai Kiadó, Budapest. (172 pp., ill.).

05040   Greuter, W. *et al.* (1988). International Code of Botanical Nomenclature, adopted by the Fourteenth International Botanical Congress, Berlin, July–August 1987. (ICBN). Koenigstein, Germany. [Pinaceae = Pinaceae + Cupressaceae s.l. (incl. Taxodiaceae) + Araucariaceae: nom. gen. cons.: *Agathis* Salisb., *Cedrus* Trew, *Pseudolarix* Gord., *Cunninghamia* R. Br., *Sequoia* Endl., *Metasequoia* Hu et Cheng].

05041   Greuter, W. *et al.* (1994). International Code of Botanical Nomenclature, adopted by the Fifteenth International Botanical Congress, Tokyo–Yokohama, August 1993. (ICBN). Koenigstein, Germany. [Pinaceae, Cupressaceae s.l. (incl. Taxodiaceae), Araucariaceae, Phyllocladaceae, Taxaceae: nom. gen. cons.: *Agathis* Salisb., *Cedrus* Trew, *Pseudolarix* Gord., *Cunninghamia* R. Br., *Sequoia* Endl., *Metasequoia* Hu et Cheng, *Phyllocladus* L. C. et A. Rich., *Torreya* Arn.].

05042   Greuter, W. *et al.* (2000). International Code of Botanical Nomenclature, adopted by the Sixteenth International Botanical Congress, St. Louis, Missouri, August 1999. (ICBN). Koenigstein, Germany. [Pinaceae, Cupressaceae s.l. (incl. Taxodiaceae), Araucariaceae, Phyllocladaceae, Podocarpaceae, Taxaceae: nom. gen. cons.: *Agathis* Salisb., *Cedrus* Trew, *Pseudolarix* Gord., *Cunninghamia* R. Br., *Sequoia* Endl., *Metasequoia* Hu et Cheng, *Phyllocladus* L. C. et A. Rich., *Podocarpus* Pers., *Thujopsis* Endl., *Torreya* Arn.].

05080   Griffith, W. (1847). Journals of Travels in Assam, Burma, Bootan, Afghanistan and the neighbouring countries. Posthumous papers ... arranged by John M'Clelland. Calcutta. [Coniferales].

05090   Griffith, W. (1848). Itinerary Notes of plants collected in the Khasyah and Bootan Mountains, 1837–38, in Afghanistan and neighbouring countries, 1839 to 1841. Posthumous papers ... arranged by John M'Clelland. Calcutta. [Coniferales, *Abies spinulosa* sp. nov. = *Picea spinulosa*, pp. 259, 265, 275, *Taxus contorta* sp. nov. = *T. wallichiana*, p. 351].

05100   Griffith, W. (1854). Notulae ad Plantas Asiaticas. Part 4. Posthumous papers ... arranged by John M'Clelland. Calcutta. [Coniferales, *Pinus khasyana* sp. nov. = *P. insularis*, p. 18, *Abies densa* sp. nov., p. 19].

05110   Griffith, W. (1854). Icones Plantarum Asiaticarum. Part 4. Posthumous papers ... arranged by John M'Clelland. Calcutta. [Coniferales; plates].

05160a   Groves, E. W. & D. T. Moore (1989). A list of the cryptogams and gymnospermous plant specimens in the British Museum (Natural History) gathered by Robert Brown in Australia 1801–5. Proc. Linn. Soc. New South Wales 111 (2): 65–102. [Araucariaceae, Cupressaceae, *Callitris*, Podocarpaceae, pp. 96–98, all Bennett nos.].

05190   Hagerup, O. (1933). Zur Organogenie und Phylogenie der Koniferen-Zapfen. Biol. Meddel. Kongel. Danske Vidensk. Selsk. 10: 1–82. [Pinaceae, Cupressaceae, other families; see also K. Lanfer, 1934].

05310   Handel-Mazzetti, H. (1921). Übersicht über die wichtigsten Vegetationsstufen und formationen von Yunnan und SW.-Setschuan. Bot. Jahrb. Syst. 56: 578597. [Coniferales].

05321   Handel-Mazzetti, H. (1929). Symbolae sinicae. Botanische Ergebnisse der Expedition der Akademie der Wissenschaften in Wien nach Südwest-China 1914/1918. T. VII Anthophyta, Lief. 1, Gymnospermae: 1–18. Wien. [gymnosperms, Coniferales].

05340   Handel-Mazzetti, H. (1932). Hochland und Hochgebirge von Yünnan und Südwest-Setchuan. II. Die temperierte Stufe. in: G. H. H. Karsten & H. Schenck. Vegetationsbilder, Vol. 22 (8). Jena. [Coniferales, Pinaceae, ecology].

05462a   Harris, T. M. (1976). Two neglected aspects of fossil conifers. Amer. J. Bot. 63 (6): 902–910. [palaeobotany].

05480   Hart, J. A. (1987). A cladistic analysis of conifers: Preliminary results. J. Arnold Arbor. 68: 269–307.

05508   Hay, J. A. & U. V. Dellow (1952). New Zealand conifers; a note on their uses and importance. Tuatara 4: 108–117. (J. of the Biol. Soc., Victoria Univ. Coll., Wellington, New Zealand). (ill., key). [*Agathis, Libocedrus, Dacrydium, Phyllocladus, Podocarpus*].

05520   Hayashi, Y. (1951). The natural distribution of important trees, indigenous to Japan, Conifers Rpt. 1.; Bull. Gov. Forest Exp. Sta. (Meguro, Tokyo) No. 48. (Japan., Eng. summ., ill., maps).

05530   Hayashi, Y. (1952). The natural distribution of important trees, indigenous to Japan, Conifers Rpt. 2.; Bull. Gov. Forest Exp. Sta. (Meguro, Tokyo) No. 55. (Japan., Eng. summ., ill., maps).

05540   Hayashi, Y. (1954). The natural distribution of important trees, indigenous to Japan, Conifers Rpt. 3.; Bull. Gov. Forest Exp. Sta. (Meguro, Tokyo) No. 75. (Japan., Eng. summ., ill., maps).

05541   Hayashi, Y. (1960). Taxonomical and phytogeographical study of Japanese Conifers. Tokyo. (Japan., 440 pp., 138 phot., 91 maps).

05550   Hayashi, Y. (1969). Illustrations of useful trees (Part of Forest Trees). Shinkosha. (Japan.). [*Picea maximowiczii* var. *senanensis* var. nov., *P. shirasawae* sp. nov. = *P. alcockiana* var. *acicularis*, text with fig. 43].

05560   Hayata, B. (1905). On the distribution of the Formosan conifers. Bot. Mag. (Tokyo) 19: 43–60.

05590   Hayata, B. (1908). New conifers from Formosa. Gard. Chron., ser. 3, 43: 194. [*Pinus mastersiana* sp. nov., *P. morrisonicola* sp. nov., *Tsuga formosana* sp. nov., *Cunninghamia konishii* sp. nov., *Juniperus morrisonicola* sp. nov., *Keteleeria formosana* sp. nov., *Chamaecyparis obtusa* f. *formosana* f. nov., publ. 28 March 1908; *Chamaecyparis obtusa* Sieb. & Zucc. var. *formosana* [stat. nov. without author cit.; comb. validly publ. by A. Rehder in L. H. Bailey, Stand. Cycl. Hort. 2: 731, 1914] in Repert. Spec. Nov. Regni Veg. 8: 365–367 (1910), where the Latin descr. of the spp. were republished].

05600 Hayata, B. (1908). On some new species of Coniferae from the Island of Formosa. J. Linn. Soc., Bot. 38: 297–300, pl. 22–23. [*Juniperus morrisonicola* sp. nov., p. 298, *Cunninghamia konishii* sp. nov., p. 299, *Pinus formosana* sp. nov., p. 297; the first two also in Gard. Chron., ser. 3, 43: 194 (1908)].

05610 Hayata, B. (1908). Flora Montana Formosae – Gymnospermae. J. Coll. Sci. Imp. Univ. Tokyo 25, Art. 19: 207–224, figs. [Coniferales, *Libocedrus macrolepis* = *Calocedrus formosana*, p. 207, *Chamaecyparis formosensis*, *C. obtusa* f. *formosana* f. nov. (earlier in Gard. Chron. 43), p. 208, *Juniperus formosana* sp. nov., p. 209, *J. morrisonicola*, *Cunninghamia konishii*, *Taiwania cryptomerioides*, *Pinus armandii* var. *mastersiana* (Hayata) comb. et stat. nov., p. 217, f. 8, *P. formosana* sp. nov. = *P. morrisonicola*, p. 217, *Picea morrisonicola* sp. nov., p. 220, *Keteleeria davidiana* var. *formosana* (Hayata) comb. et stat. nov., p. 221, *Tsuga formosana* = *T. chinensis*, *Abies mariesii* var. *kawakamii* var. nov. = *A. kawakamii*, p. 223].

05640 Hayata, B. (1911–1921). Icones plantarum Formosanarum nec non et contributiones ad floram Formosanam. (with Eng. subtitle). Vol. 1–10. Taihoku (Govt. of Formosa). [Coniferales in several volumes, *Cephalotaxus wilsoniana* sp. nov., Vol. 4: 22 (1914), *Pseudotsuga wilsoniana* sp. nov. = *P. sinensis*, Vol. 5: 204, t. 15 (1915), *Cunninghamia kawakamii* sp. nov. = *C. konishii*, Vol. 5: 207 (1915), *Podocarpus nakaii* sp. nov., Vol. 6: 66 (1916)].

05661 Hayata, B. (1934). The phytogeography of Conifers in Japan and successional stages in the conifer forests of Mount Fuji. Proc. 5th Pacific Sci. Congr. (Canada 1933) 4: 3289–3293.

05671 He, S.A. (Shanan) (ed.). Rare and precious plants of China. Shanghai. (Chin., Eng., ill.). [gymnosperms, Coniferales (pp. 7–33) e.g. *Cathaya argyrophylla*, *Keteleeria* spp., *Pseudolarix kaempferi* = *P. amabilis*, *Metasequoia glyptostroboides*, *Taiwania cryptomerioides*].

05710 Henkel, J. B. & W. Hochstetter (1865). Synopsis der Nadel-hölzer, ... Stuttgart. [gymnosperms, Coniferales, *Dammara hypoleuca* sp. nov. = *Agathis ovata*, p. 217, *Athrotaxis doniana* sp. nov. = *A. laxifolia*, p. 221, *Chamaecyparis leptoclada* sp. nov. = *C. pisifera*? cultivar, p. 257, *Caryotaxus* Zucc. ex Henkel & Hochst. = *Torreya*, p. 366, *C. nucifera* (L.) comb. nov. = *T. nucifera*, p. 366, *C. grandis* (Fortune ex Lindl.) comb. nov. = *T. grandis*, p. 367, *C. taxifolia* (Arn.) comb. nov. = *T. taxifolia*, p. 367, *C. myristica* (Hook.) comb. nov. = *T. californica*, p. 368].

05791 Herbin, G. A. & P. A. Robbins (1968). Studies on plant cuticular waxes. III. The leaf wax alkanes and w-hydroxyacids of some members of the Cupressaceae and Pinaceae. Phytochemistry 7 (8): 1325–1337.

05810 Herrmann, Geheimrat (1933). Zapfen-Bestimmungstabellen. Mitt. Deutsch. Dendrol. Ges. 45: 135–143, f. 1–40. [Coniferales, Araucariaceae, Cupressaceae, Taxodiaceae, Pinaceae].

05812 Herzfeld, S. (1914). Die weibliche Koniferenblüte. Österreich. Bot. Zeitschr. 64: 321–358. (ill.).

05850 Hickel, R. (1911). Graines et plantules des Conifères. Bull. Soc. Dendrol. France 19: 13–115.

05876a Hill, R. S. & T. Brodribb (1999). Southern conifers in time and space. Austral. J. Bot. 47: 639–696. [Araucariaceae, Cupressaceae, Podocarpaceae, leaf morphology, conifer diversity, palaeobotany, extant conifers, ill.].

05930 Hirmer, M. (1936). Entwicklungsgeschichte und vergleichende Morphologie des weiblichen Blütenzapfens der Coniferen. [Die Blüten der Coniferen I (ed. M. Hirmer)] Bibl. Bot. 114 (1): 1–100. (ill.).

05930a Hirmer, M. (1937). Über die fossilen Reste der männlichen Coniferen-Blüten. Bibl. Bot. 114 (3, Anhang): 21–24. (ill.).

06170 Hsü, J. (1936). Anatomy of the leaves of conifers of North China. Sci. Rep. Natl. Univ. Peking (Beijing) 1 (2): 39–58, pl. 1, f. 1–39.

06210 Hu, H. H. (1934). Distribution of taxads and conifers in China. Proc. 5th Pacific Sci. Congr. (Canada, 1933) 4: 3273–3288.

06379 Humboldt, F. H. A. von (1849). Ansichten der Natur, mit wissenschaftliche Erläuterungen. Zweiter Band, p. 182–207, Nadelhölzer. Stuttgart und Tübingen. [Coniferales].

06380 Humboldt, F. H. A. von, A. J. Bonpland & C. S. Kunth (1817). Nova genera et species plantarum. Vol. 2 (5), Paris. [Coniferales, *Podocarpus taxifolius* sp. nov. = *Prumnopitys montana*, p. 2, *Cupressus sabinioides* sp. nov. = *Juniperus sabinioides* = *J. monticola*, p. 3, *C. thurifera* sp. nov. = *J. flaccida* var. *poblana* (see M. Martínez, 1946), p. 3, *Pinus occidentalis* Kunth in H. B. K. non Sw., 1788 = *P. montezumae* Lamb., p. 4, *Taxodium distichum* Kunth in H. B. K., non Rich., 1810 = *T. mucronatum*, p. 4, *Pinus religiosa*, *P. hirtella* spp. nov. = *Abies religiosa*, p. 5; all descr. of new taxa are by C. S. Kunth].

06420 Hustich, I. (1953). The Boreal Forest limit of conifers. Arctic 6: 149–160. [Pinaceae, Cupressaceae].

06530 Iwata, T. & M. Kusaka (emend. H. Takeda) (1952–54). Coniferae Japonicae illustratae. pp. 1–228, f. 1–125, 1–83 (ed. 1, 1952); 1–15, 1–247, pl. 1–32, f. 1–89 (ed. 2, 1954). (Japan.). [*Abies veitchii* var. *sikokiana* (Nakai) Kusaka, comb. et stat. nov. (basion. : *A. sikokiana* Nakai in Bot. Mag. (Tokyo) 42: 452, 1928), p. 212, 1954].

06730 Janchen, E. (1950). Das System der Koniferen. Sitzungsber. Österreich. Akad. Wiss., Math.-Naturwiss. Kl., Abt. I, 158 (for 1949, 3): 155–262. Wien.

06810 Jepson, W. L. (1934). Phytogeography of the *Coniferae* of western North America. Proc. 5th Pacific Sci. Congr. (Canada 1933) 4: 3289–3293.

06880 Kaempfer, E. (1712). Amoenitatum exoticarum politico-physico-medicarum fasciculi V, ... Lemgoviae (Lemgo, Lippe). [A list of Japanese plants, with 28 plates, on p. 765–912, among which are some conifers; for comments see R. A. Salisbury, 1817].

06900 Kanehira, R. (1925). The species of Coniferae and their distribution in eastern Asia. Trans. Nat. Hist. Soc. Taiwan 15: 109–117. (Japan.).

06920 Kanehira, R. (1933). On the ligneous flora of Formosa and its relationship to that of neighboring regions. Lingnan Sci. J. 12: 225–238. [Coniferales].

06940 Kapper, O. G. (1954). Khoinye porody. Lesovodstvennaîa kharakteristika. (Coniferous plants, forestry characteristics) 1–304. (Russ.). [Coniferales].

06961 Karlberg, S. (1960). Forests and Forestry in Turkey. Acta Horti Gothob. 23: 41–70. [Pinaceae, Cupressaceae, *Cedrus libani*, *Juniperus excelsa*, maps of Pinaceae].

07050 Keighery, G. J., L. I. Brighton & C. J. Robinson (1982). Notes on the biology and phytogeography of Western Australian Plants: part 27. Cycadaceae, Zamiaceae, Podocarpaceae and Cupressaceae. West Perth, Kings Park Board. [Coniferales, cycads].

07090 Keng, H. (1975). A new scheme of classification of the conifers. Taxon 24 (2–3): 289–292.

07130e Kerp, J. H. J. (1991). The study of fossil gymnosperms by means of cuticular analysis. Palaios 5: 548–569. [Coniferales, palaeobotany, ill.].

07131 Kerp, J. H. J., H. A. Poort, J. M. Swinkels & R. Verwer (1989). A Conifer-Dominated Flora from the Rotliegend of Oberhausen (Saar-Nahe Area). Cour. Forsch.-Inst. Senckenberg 109: 137–151. [ill., new concept of cone-morphology in fossil Walchiaceae].

07132 Kerp, J. H. J., H. A. Poort, J. M. Swinkels & R. Verwer (1989). Conifer dominated Rotliegend floras from the Saar-Nahe Basin (? Upper Carboniferous – Lower Permian; SW Germany) with special reference to the reproductive biology of early conifers. (Aspects of Permian palaeobotany and palynology IX). Rev. Palaeobot. Palynol. 62: 205–248. [compare C. R. Florin, 1951, 1954, J. A. Clement-Westerhof, 1984 and J. H. J. Kerp & al., 1989].

07137 Khan, A. G. & P. G. Valder (1972). Occurrence of root nodules in Ginkgoales, Taxales, and Coniferales. Proc. Linn. Soc. New South Wales 97: 35–41. [nodules found in *Sciadopitys*].

07330 Koch, K. H. E. (1861). J.G. Veitch's japanische Nadelhölzer oder Coniferen. Wochenschr. Vereines Beförd. Gartenbaues Königl. Preuss. Staaten 4: 86–88.

07400 Koidzumi, G. (1912). New classification of Coniferae. Bot. Mag. (Tokyo) 26: 116–119. (Japan.).

07450 Komarov, V. L. (1902). Coniferae of Manchuria. Trudy Imp. S.-Peterburgsk. Obsc. Estestvoisp., Vyp. 3, Otd. Bot. 32: 230–241. (Russ.).

07490 Komarov, V. L. (1934). A botanico-geographical sketch of the Gymnospermae of the U.S.S.R. Proc. 5th Pacific Sci. Congr. (Canada) 4: 3337–3339. [Coniferales of maritime provinces].

07520 Konar, R. N. & Y. P. Oberoi (1969). Recent work on reproductive structures of living conifers and taxads a review. Bot. Rev. (Lancaster) 35: 89–116.

07570 Kozloff, E. N. (1976). Plants and Animals of the Pacific Northwest. An Illustrated Guide to the Natural History of Western Oregon, Washington, and British Columbia. Univ. Washington Press, Seattle London.

07620 Ku, C. C. & Y. C. Cheo (1941). A preliminary survey of the forests in western China. Sinensia 12: 81–133. [Coniferales].

07630 Kuan, C. T. (Guan Zhong-tian) (1981). Fundamental features of the distribution of Coniferae in Sichuan. Acta Phytotax. Sin. 19 (4): 393–407. (Chin., Eng. summ., maps).

07691 Kurata, S. (1971). Illustrated important forest trees of Japan. Vols. 1–5. 2nd ed., Tokyo. [Cupressaceae, Pinaceae, Taxodiaceae in Vols. 1–2, bot. colour plates, overlay distr. maps; 1st ed. publ. 1968].

07741 Lam, H. J. (1952). De Coniferenkegel. Pharm. Weekblad 87: 619–626. (ill.).

07742 Lam, H. J. (1954). Again: the new morphology elucidated by the most likely phylogeny of the female coniferous cone. Svensk Bot. Tidskr. 48 (2): 347–360. (ill.).

07790 Lambert, A. B. (1803–24). A description of the genus *Pinus*; illustrated with figures, directions relative to the cultivation, and remarks on the uses of the several species ... Vol. 1–2, London. [Coniferales, *Pinus kaempferi* sp. nov. = Larix *kaempferi*, Pref. v, *P. banksiana* sp. nov., Vol. 1: 7, t. 3, *P. massoniana* sp. nov., Vol. 1: 17, t. 12, *P. taxifolia* sp. nov. (non Salisb., 1796) = *Pseudotsuga menziesii*, Vol. 1: 51, t. 33, 1803, *P. dammara* sp. nov.= *Agathis dammara*, Vol. 1: 61, t. 38, 1803, *Dombeya excelsa* sp. nov. = *Araucaria columnaris*, Vol. 1: 87, 1807; *Dammara australis*

D. Don, sp. nov., Vol. 2: 14, t. 6, *Juniperus squamata* Buch.-Ham. ex D. Don, sp. nov., Vol. 2: 17, *J. uvifera* D. Don, sp. nov. = *Pilgerodendron uviferum*, Vol. 2: 17, *Cupressus torulosa* D. Don, sp. nov., Vol. 2: 18, *C. nootkatensis* D. Don, sp. nov. = *Chamaecyparis nootkatensis*, Vol. 2: 18, *Thuja plicata* Donn ex D. Don, sp. nov., Vol. 2: [19], *Podocarpus oleifolius* D. Don, sp. nov., Vol. 2: [20], *P. salignus* D. Don, sp. nov., Vol. 2: [20], *P. glomeratus* D. Don, sp. nov., Vol. 2, p. [21], *P. neriifolius* D. Don, sp. nov., Vol. 2: [21], *Taxodium sempervirens* D. Don, sp. nov. = *Sequoia sempervirens*, Vol. 2, pp. 24-..., pl. 7, f. 1, *Dacrydium plumosum* D. Don, sp. nov. = *Libocedrus plumosa*, Vol. 2, App. 143, 1824].

07800   Lambert, A. B. (1828–37). A description of the genus *Pinus*; ... Vol. 1–3, Ed. 2, London. [*Podocarpus totara* G. Benn. ex D. Don, sp. nov., vol. 1: 184 (1828), *Thuja pensilis* D. Don, sp. nov. = *Glyptostrobus pensilis*, vol. 2: 115 (1828), *Dacrydium excelsum* sp. nov. = *Dacrycarpus dacrydioides*, App.: 143 (1828), *Araucaria cunninghamii* Aiton ex D. Don, sp. nov., vol. 3, t. 79 (1837), *Dacrydium taxifolium* Banks & Sol. ex D. Don = *Prumnopitys taxifolia*, vol. 3: 119 (1837)].

07810   Lambert, A. B. (1832). A description of the genus *Pinus*; ... Ed. 3, London. ("Ed. 8vo."). [Coniferales, *Pinus montezumae* Lamb., sp. nov., Vol. 1: 39, t. 22, *Thuya chilensis* D. Don, sp. nov. = *Austrocedrus chilensis*, Vol. 1: 128, *Thuja pendula* D. Don, sp. nov. = *Platycladus orientalis*, Vol. 2: 130, t. 67, *P. monticola* Dougl. ex D. Don, *P. sabiniana* Dougl. ex D. Don, *P. gerardiana* Wallich ex D. Don, *P. grandis* Dougl. ex D. Don = *Abies grandis*, *P. nobilis* Dougl. ex D. Don = *Abies procera*, spp. nov., Vol. 2, unnumb. pp. betw. 144–145, *Phyllocladus trichomanoides* D. Don, sp. nov., Vol. 2, p. 159, *Podocarpus ferrugineus* G. Benn ex D. Don, sp. nov. = *Prumnopitys ferruginea*, Vol. 2, p. [189]; for detailed ref. to the various eds. of Lambert's "Descr. of the Genus Pinus", see e.g. E. L. Little, Jr., 1949 and H. W. Renkema & J. Ardagh, 1930].

07862   Lanfer, K. (1934). Vergleichende Studien der weiblichen Coniferen-Blüten. 1. Anzeige (von R. Pilger), pp. 469–471 + pl.; 2. Beitrag zur Klärung und zum richtigen Verständnis der organogenetischen Untersuchungen der Coniferen-Zapfen von O. Hagerup. Bot. Jahrb. Syst. 66 (4): (469–)471–487. [see also O. Hagerup, 1933].

07960   Laubenfels, D. J. de (1953). The external morphology of coniferous leaves. Phytomorphology 3: 1–23. (ill.).

08000   Laubenfels, D. J. de (1978). The taxonomy of the Philippine Coniferae and Taxaceae. Kalikasan 7 (2): 127–152. [*Podocarpus rotundus* sp. nov., p. 136, *P. lophatus* sp. nov., p. 137, *P. macrocarpus* sp. nov., p. 140, *Taxus sumatrana* (Miq.) comb. nov., p. 151].

08024   Laubenfels, D. J. de (1992). Commentary: the organization of female fertile structures in conifers. Int. J. Plant Sci. 153 (4): vii–viii.

08041   Lawson, P. & E. J. Ravenscroft (1863–84). Pinetum Britannicum, containing a descriptive account of all hardy trees of the Pine tribe cultivated in Great Britain. Vol. 1–3. London. [Also known as "Lawson's Pinetum"; in the bound and page-numbered 3 vols. at Kew E. J. Ravenscroft is the only author/publisher mentioned on the title page. Contains 52 pl. + text & line drawings in very large format, orig. issued separately; authors of botanical descriptions are J. Lindley (pts. 1–3), A. Murray (pts. 4–37) & M. T. Masters (pts. 38–52); for dates of publ. of parts see B. B. Woodward in Gard. Chron., ser. 3, 36: 36–37. 1904. Coniferales, *Pinus porphyrocarpa* Mast. sp. nov., vol. 1: 83. 1884].

08130   Lemmon, J. G. (1893–98). Notes on west American Coniferae. Erythea 1 (1893): 48–52, 134–136, 224–231; 2 (1894): 102–104, 157(bis)–162, 173–177, 203; 5 (1897): 22–25; 6 (1898): 77–79. [Pinaceae, *Pinus contorta* var. *bolanderi* (Parl. in DC.) "comb. nov." (No. 2, p. 176), but see B. A. E. Koehne, 1893, Cupressaceae, Taxodiaceae].

08162b   Lewis, F. J. & E. S. Dowling (1924). The anatomy of the buds of Coniferae. Ann. Bot. 28: 217–228.

08163   Li, H. L. (1950). The Coniferales of Taiwan. Taiwania 1: 285–310. (Chin.).

08180   Li, H. L. (1953). Present distribution and habitats of the conifers and taxads. Evolution 7: 245–261.

08181   Li, H. L. (1956). An illustrated key of the genera of Conifers. Morris Arbor. Bull. 7 (4): 47–52.

08210   Li, H. L. (1982). The relict genera of conifers and taxads of eastern Asia and their geographical distribution. in: H. L. Li, Contributions to botany. pp. 216–235. Taiwan.

08220   Li, H. L. & H. Keng (1954). Icones gymnospermum Formosanarum. Taiwania 5: 25–83, pl. 1–29. (Japan., Eng.). [Coniferales, *Tsuga chinensis* var. *formosana* (Hayata) comb. et stat. nov., p. 64, *Juniperus squamata* var. *morrisonicola* (Hayata) comb. et stat. nov., p. 81, t. 28].

08229e   Li, R. W. & H. D. Zhang (2001). A floristic analysis on the gymnosperms of Sichuan. Guihaia 21 (3): 215–222. (Chin., Eng. summ.). [Coniferales].

08247   Lidholm, J., A. E. Szmidt, J.-E. Hallgren & P. Gustafsson (1988). The chloroplast genomes of conifers lack one of the rRNA-encoding inverted repeats. Mol. Gen. Genet. 212: 6–10.

08253a   Lin, J. X. & Y. X. Hu (eds.) (2000). Atlas of structure of gymnosperms. (Chin., Eng.). Science Press, Beijing. [gymnosperms, Coniferales, (wood) anatomy, LM + SEM ill.].

08330   Lindley, J. & G. Gordon (1850). A catalogue of the Coniferous Plants with their synonyms. J. Hort. Soc. London 5: 199–228. [*Abies lasiocarpa* Lindl. et Gord. (non Nutt.) = *A. grandis*, p. 210, *A. griffithiana* sp. nov. = *Larix griffithii*, p. 214].

08360   Link, J. H. F. (1841). Abietinae horti Regii botanici Berolinensis cultae, ... Linnaea 15: 481–545. [Coniferales, *Pinus, Picea, Abies, Larix, Cedrus, Cunninghamia, Araucaria* (*Eutacta* gen. nov.), *Picea excelsa* (Lam.) comb. nov. = *P. abies*, p. 517, *P. morinda* sp. nov. = *P. smithiana*, p. 522].

08570   Little, E. L., Jr. (1971). Atlas of United States trees; Vol. 1, Conifers and important hardwoods. U.S. Forest Serv. Misc. Publ. 1146. Washington, D.C.

08580   Little, E. L., Jr. (1975). Rare and local conifers in the United States. U.S. Forest Serv. Cons. Res. Rep. 19. Washington, D.C.

08702   Liu, T. S. (1966). Study on the Phytogeography of the Conifers and Taxads of Taiwan. Taiwan Forest. Res. Inst., Rep. 122. (maps).

08822   Lotsy, J. P. (1911). Vorträge über botanische Stammesgeschichte. Vol. 3. Jena. [Coniferales, classification].

08890   Madden, E. (1845). On Himalayan Conifers. J. Hort. Soc. London 5: 228–269. [Coniferales].

08961   Mahlert, A. (1885). Beiträge zur Kenntnis der Anatomie der Laubblätter der Coniferen mit besonderer Berücksichtigung des Spaltöffnungs-Apparates. Bot. Centralbl. 24: 54–59, 85–88, 118–122, 149–153, 180–185, 214–218, 243–249, 278–281, 310–312. (ill.).

09015   Mapes, G. (1987). Ovule inversion in the earliest conifers. Amer. J. Bot. 74: 1205–1210. [evolution, palaeobotany, palaeozoic conifers].

09017   Mapes, G. & G. W. Rothwell (1991). Structure and relationships of primitive conifers. Neues Jahrb. Geol. Palaeontol. 183: 267–287. [Coniferales, Emporiaceae fam. nov., *Emporia* gen. nov., *Utrechtia* (syn. *Lebachia*), *Walchia* (fossil "morpho-genus"), nomenclature, taxonomy, palaeobotany].

09018a   Mapes, G. & G. W. Rothwell (2003). Validation of the names Emporiaceae, *Emporia*, and *Emporia lockardii*. Taxon 52: 327–328. [Emporiaceae fam. nov., *Emporia* gen. nov., *E. lockardii* (Mapes & G. W. Rothwell) comb. nov., *Lebachia*, nom. illeg., palaeobotany].

09140   Martínez, M. (1945). Las pinaceas Méxicanas. Anales Inst. Biol. Univ. Nac. México 16: 1–345. [Cupressaceae, Pinaceae, *Taxodium mucronatum, Pinus pseudostrobus* var. *oaxacana* (nom. inval.: sine Lat. descr., p. 195) et var. *estevezi* (p. 196) var. nov. = *P. oaxacana, P. estevezi*; *P. lutea* Blanco (non Gord., nec Walter) sp. nov. (p. 233) = *P. cooperi*].

09191   Martínez, M. (1948). Las Coniferas Silvestres del Valle de México. Bol. Soc. Bot. Mex. 7: 1–20. [*Pinus, Abies, Taxodium mucronatum, Cupressus lindleyi* = *C. lusitanica, Juniperus deppeana, J. monticola* f. *compacta*, ill., key].

09240   Martínez, M. (1963). Las Pinaceas Méxicanas. Ed. 3, Univ. of Mexico, Mexico-city. (ed. 1 publ. 1953). [Cupressaceae, Pinaceae, *Taxodium mucronatum, Abies guatemalensis* var. *tacanensis* (Lundell) comb. et stat. nov., p. 129, *A. durangensis* var. *coahuilensis* (I. M. Johnst.) comb. et stat. nov., p. 139].

09242   Masamune, G. (1930). On the distribution of coniferous plants in Yakushima, prov. Osumi. J. Japan. Bot. 7: 39–43, 5 figs. (Japan.).

09270   Mason, H. L. (1927). Fossil records of some west American conifers. Publ. Carnegie Inst. Wash. 346: 139–158.

09340   Masters, M. T. (1881). On the conifers of Japan. J. Linn. Soc., Bot. 18: 473–524. [Pinaceae, Cupressaceae, Taxodiaceae, Podocarpaceae, other families, ill., esp. *Picea, Abies, Thuja obtusa* (Sieb. et Zucc.) comb. nov. (non Moench) = *Chamaecyparis obtusa*, p. 491, *Tsuga diversifolia* sp. nov., p. 514].

09440   Masters, M. T. (1887). Contributions to the history of certain species of conifers. J. Linn. Soc., Bot. 22: 169–212, pl. 2–10, f. 1–32. [*Abies amabilis, A. grandis, A. concolor, A. lasiocarpa, A. magnifica, A. procera, A. religiosa, A. fortunei* = *Keteleeria fortunei, Athrotaxis laxifolia, Picea omorika, Pinus peuce, Pseudolarix amabilis*].

09460   Masters, M. T. (1890). Review of some points in the comparative morphology, anatomy, and life-history of the Coniferae. J. Linn. Soc., Bot. 27: 226–332. [pp. 276–280: Cladodes or Phylloclades, *Sciadopitys*].

09465   Masters, M. T. (1892). List of conifers and taxads in cultivation in the open air in Great Britain and Ireland. J. Roy. Hort. Soc. London 14: 179–256. [*Agathis obtusa* (Lindl.) comb. nov. = *A. macrophylla*, p. 197, *A. moorei* (Lindl.) comb. nov., p. 197, *Tetraclinis articulata* (Vahl) comb. nov., p. 250.].

09470   Masters, M. T. (1893). Notes on the genera of Taxaceae and Coniferae. J. Linn. Soc., Bot. 30: 1–41. [Coniferales, *Tetraclinis* gen. nov., *T. articulata* (Vahl) "comb. nov.", p. 14, but earlier in J. Roy. Hort. Soc. 14: 250 (1892); keys and tables].

09530   Masters, M. T. (1901). Hybrid conifers. J. Roy. Hort. Soc. (London) 26: 97–110.

09540   Masters, M. T. (1902). Coniferae. in: F. B. Forbes & B. W. Hemsley (eds.). Enumeration of all the plants known from China... J. Linn. Soc., Bot. 26: 540–561. [Coniferales, Ginkgoales, Pinaceae: *Pinus henryi* sp. nov., p. 550, *Picea brachytyla* (Franch.) et *P. likiangensis* (Franch.) "comb. nov." (see G. A. Pritzel, 1901), *Keteleeria fabri* sp. nov. = *Abies fabri*, p. 555, *Keteleeria* "n. sp." (non nomen!), p. 555, *Tsuga brunoniana* var. *chinensis* (Franch.) comb. et stat. nov. = *T. chinensis*, p. 556, *T. chinensis* (Franch.) et *T. yunnanensis* (Franch.) "comb. nov." (see G. A. Pritzel, 1901), *Larix griffithii* Masters (non Hook. f. et Thomson) = *L. potaninii*, both on p. 558; short descr.].

09550   Masters, M. T. (1903). Chinese conifers collected by E. H. Wilson. J. Bot. 41: 267–270, 3 figs. [Cupressaceae, Pinaceae].

09620   Masters, M. T. (1908). On the distribution of the species of conifers in the several districts of China, and on the occurrence of the same species in neighbouring countries. J. Linn. Soc., Bot. 38: 198–205.

09635   Mathew, B. (ed.) (1999). Curtis's Botanical Magazine Vol. 16 (3) August 1999. [This volume is dedicated to conifers with papers by Farjon, Frankis, Gardner, Mill, Rushforth and others on a choice of species and their illustrations, in celebration of the 4th International Conifer Conference held 23–26 August 1999 in England].

09640   Matsumura, J. (1901). On Coniferae of Loochoo and Formosa. Bot. Mag. (Tokyo) 15: 137–141. [*Chamaecyparis formosensis* sp. nov. + other taxa].

09740   Mayer, H. & H. Aksoy (1986). Wälder der Turkei. Stuttgart – New York. [Coniferales, ecology].

09741   Mayr, H. (1889). Die Waldungen von Nordamerika, ihre Holzarten, deren Anbaufähigkeit und forstlichen Wert für Europa... München. (1890 on title p., but publ. Sept.–Nov. 1889). [Coniferales, *Pseudotsuga macrocarpa* (Vasey) comb. et stat. nov., p. 278, pl. 6, 8, 9; antedates similar comb. by J. G. Lemmon, 1890].

09860   Mehra, P. N. & T. N. Khoshoo (1956). Cytology of conifers I. J. Genet. 54: 165–180, pl. 5. [*Actinostrobus, Callitris, Cryptomeria, Cunninghamia, Cupressus, Juniperus, Taxodium, Tetraclinis, Thuja, Widdringtonia* (2n = 22); *Abies, Cedrus, Picea, Pinus* (2n = 24)].

09861   Mehra, P. N. & T. N. Khoshoo (1956). Cytology of conifers II. J. Genet. 54: 181–185. [supplement].

09865   Meijer Drees, E. (1940). The genus *Agathis* in Malaysia. Bull. Jard. Bot. Buitenzorg, ser. 3, 16: 455–474. [*Agathis latifolia*, p. 459, *A. hamii*, p. 462, *A. beckingii*, p. 463, *A. endertii*, p. 470, spp. nov., *A.* sect. *Macrobracteatae*, p. 457, *A.* sect. *Microbracteatae*, p. 461, sect. nov., 1 ill. of cross section of leaves].

09885   Melikyan, A. P. & A. V. F. C. Bobrov (2000). Morphology of female reproductive structures and an attempt of the reconstruction of [the] phylogenetic system of [the] orders Podocarpales, Cephalotaxales and Taxales. Bot. Žurn. 85: 50–67. (Russ., Eng. summ.). [classification raising genera to families and families to orders based on ovule orientation and integuments].

10061   Meyen, S. V. (1997). Permian conifers of Western Angaraland. Rev. Palaeobot. Palynol. 96: 351–447. [Coniferales, palaeo-botany, ill.].

10081   Michaux, A. (1810–13). Histoire des Arbres Forestiers de l'Amérique Septentrionale. vol. 1–3. Paris. [Coniferales, vol. 1: *Pinus* spp., (*Abies*) *Picea* spp., (*Abies*) *Tsuga canadensis, Abies balsamea*, vol. 3: (*Cupressus*) *Taxodium distichum*, (*Cupressus*) *Chamaecyparis thyoides, Thuja occidentalis, Larix laricina, Juniperus virginiana*, ill.].

10124a   Mill, R. R. (ed.). (2003). Proceedings of the Fourth International Conifer Conference. Conifers for the future? Wye College, England, 23ʳᵈ August – 26ᵗʰ August 1999. Acta Horticulturae 615. International Society for Horticultural Science (ISHS). [Coniferales, palaeobotany, phylogeny, taxonomy, ecology, DNA, horticulture, forestry, conservation, ill., maps, tables].

10190   Miller, C. N., Jr. (1977). Mesozoic Conifers. Bot. Rev. (Lancaster) 43 (2): 217–280. [Coniferales, palaeobotany, ill.].

10200   Miller, C. N., Jr. (1982). Current status of Paleozoic and Mesozoic conifers. Rev. Palaeobot. Palynol. 37: 99–114. [Coniferales, palaeobotany, ill.].

10211   Miller, C. N., Jr. (1988). The origin of modern conifer families. In: C. B. Beck, ed. Origin and Evolution of gymnosperms, pp. 448–486. [cladograms, see also C. B. Beck, ed. 1988].

10214   Miller, C. N., Jr. (1999). Implications of fossil conifers for the phylogenetic relationships of living families. Bot. Rev. (Lancaster) 65 (2): 239–277. [Coniferophytes, Coniferales, 'basal' conifers: Emporiaceae, Majonicaceae, Utrechtiaceae, 'higher' fossil conifers: Cheirolepidiaceae, Ulmanniaceae, Voltziaceae, extant conifers: Araucariaceae, Cephalotaxa-ceae, Cupressaceae, Pinaceae, Podocarpaceae, Sciadopityaceae, Taxaceae, Taxodiaceae, cladistic analysis].

10450   Miyabe, K. & Y. Kudo (1920). Icones of the essential forest trees of Hokkaido. I. Tokyo. [Cupressaceae, Pinaceae, *Abies mayriana* (Miyabe et Kudo) comb. nov., p. 9, t. 3, 4 = basion. : *A. sachalinensis* var. *mayriana* Miyabe et Kudo in Trans. Sapporo Nat. Hist. Soc. 7: 131, 1919, *A. wilsonii* (protologue: op. cit., 1919) = *A. sachalinensis* var. *nemorensis*, *Pinus himekomatsu* sp. nov. = *P. parviflora*, p. 29].

10463   Mohl, H. von (1845). Über die männlichen Blüten der Koniferen. Verm. Schriften Bot. Inhalts. [Coniferales, pollen cone morphology and anatomy, ill.].

10690   Muir, J. (1894, repr. 1961). The mountains of California. New York. [Coniferales: pp. 112–173].

10700e   Mundry, I. & M. Mundry (2001). Male cones in Taxaceae s.l. – an example of Wettstein's pseudanthium concept. Pl. Biol. 3: 405–416. [*Taxus baccata*, *Cephalotaxus harringtonii*, ontogeny, SEM ill.].

10700i   Muñoz Pizarro, C. (1973). Chile: plantas en extinción. Editorial Universitaria, Santiago de Chile. (in Bibliogr. Conif. ed. 1, 1990 under C. M. Pizarro, no. 11570). [Coniferales pp. 17–27, figs. 1–9, Podocarpaceae, *Araucaria araucana*, *Austrocedrus chilensis*, *Pilgerodendron uviferum*, *Fitzroya cupressoides*].

10730   Murray, A. (1862). Monographic sketch of the conifers of Japan. Proc. Roy. Hort. Soc. (London) 1862 (2): 265–292, 347–357, 409–432, 496–512, 633–652, 719–725, figs. [*Picea fortunei* sp. nov. (as "fortuni") = *Keteleeria fortunei*, p. 421].

10740   Murray, A. (1863). On the synonymy of various conifers. Proc. Roy. Hort. Soc. (London) 1863 (3): 140–150, 202–207, 308–322. [*Abies albertiana* sp. nov., p. 148 = *Tsuga heterophylla*, *A. magnifica* sp. nov., p. 318, f. 25–33, *A. bifolia* sp. nov., p. 320, f. 34–39 = *A. lasiocarpa*].

10792   Murray, M.D. (1989). Conifer forests in the San Gabriel Mountains. Fremontia 17 (3): 11–14. (ill.).

10880   Namboodiri, K. K. & C. B. Beck (1968). A comparative study of the primary vascular system of Conifers. 1. Genera with helical phyllotaxis. Amer. J. Bot. 55 (4): 447–457.

10890   Namboodiri, K. K. & C. B. Beck (1968). A comparative study of the primary vascular system of Conifers. 2. Genera with opposite and whorled phyllotaxis. Amer. J. Bot. 55 (4): 458–463.

10900   Namboodiri, K. K. & C. B. Beck (1968). A comparative study of the primary vascular system of Conifers. 3. Stelar evolution in gymnosperms. Amer. J. Bot. 55 (4): 464–472.

10970d   Nguyen Duc To Luu & P. I. Thomas (2004). Cây lá kim Viêt Nam/Conifers of Vietnam. An illustrative field guide for the most important forest trees. Hanoi.

11012   Noelle, W. (1910). Studien zur vergleichenden Anatomie und Morphologie der Koniferenwurzeln mit Rüchsicht auf die Systematik. Bot. Zeitung, 2. Abt. 68 (1): 169–266. (ill.).

11023a   Offler, C. E. (1984). Extant and fossil Coniferales of Australia and New Guinea. Part 1: A study of the external morphology of the vegetative shoots of the extant species. Palaeontographica Abt. B, Paläophytol. 193: 18–120. (ill.) [Araucariaceae, Cupressaceae, Podocarpaceae].

11023f   Ohsawa, T. (1997). Phylogenetic reconstruction of some conifer families: Role and significance of permineralized cone records. In: K. Iwatsuki & P. H. Raven (eds.). Evolution and diversification of land plants. (pp. 61–95, ill.). Springer Verlag, Tokyo, Berlin etc. [Coniferales, Cheirolepidiaceae, Cupressaceae, Sciadopityaceae, Taxodiaceae, palaeobotany].

11042   Oppenheimer, H. R. (1969). Properties explaining the survival of conifers in semi-arid climates. La-Ya'aran 19 (1): 1–16. [Cupressaceae, Pinaceae, ill., ecology].

11100   Orr, M. Y. (1937). On the value for diagnostic purposes of certain of the anatomical features of conifer leaves. Notes Roy. Bot. Gard. Edinburgh 19: 255–266, pl. 256–258.

11140c   Otto, A. & V. Wilde (2001). Sesqui-, di-, and triterpenoids as chemosystematic markers in extant conifers: a review. Bot. Rev. (Lancaster) 67 (2): 141–238.

11148   Outer, R. W. den (1967). Histological investigations of the secondary phloem of gymnosperms. Meded. Landbouwhogesch. Wageningen 67 (7): 1–119. [Coniferales, *Ginkgo*, wood anatomy, ill.].

11190   Page, C. N. (1979). The herbarium preservation of conifer specimens. Taxon 28 (4): 375–379.

11201   Page, C. N. (1990). Gymnosperms: Coniferophytina (Conifers and Ginkgoids). pp. 279–361 in: K. Kubitzki (ed.). The families and genera of vascular plants. Vol. 1: Pteridophytes and gymnosperms. (ed. K. U. Kramer & P. S. Green). Berlin – Heidelberg. [Coniferales, Ginkgoales, Araucariaceae, Cephalotaxaceae, Cupressaceae, Ginkgoaceae, Phyllocladaceae, Pinaceae, Podocarpaceae, Sciadopityaceae, Taxaceae, Taxodiaceae, all genera, ill.].

11202   Page, C. N. (2000). The ultramafic-rock conifers of New Caledonia. Int. Dendrol. Soc. Yearb. 1999: 48–55. (ill.).

11210 Page, C. N. & H. T. Clifford (1981). Ecological biogeography of Australian conifers and ferns. pp. 472–498 in: A. Keast (ed.). Ecological biogeography of Australia. (Monogr. Biol. vol. 41). The Hague.

11215 Page, C. N. & M. F. Gardner (1994). Conservation of rare temperate rainforest conifer tree species: a fast-growing role for arboreta in Britain and Ireland. In: A. R. Perry & G. Ellis (eds.). The common ground of wild and cultivated plants; chapter 17. National Museum of Wales, Cardiff.

11278 Pant, D. D. (1977). Early conifers and conifer allies. J. Indian Bot. Soc. 56: 23–37. [palaeobotany of conifers].

11290 Pardé, L. (1911–12). Conifères. Essais de tableaux dichotomiques pour la détermination des espèces. Bull. Soc. Dendrol. France 1911: 7–12; 1912: 23–27, 47–56. [keys].

11350 Parlatore, F. (1868). Coniferae. (Ordo CXCIX). in: A. P. de Candolle & Alph. de Candolle. Prodromus systematis naturalis regni vegetabilis. Vol. 16, part 2, pp. 361–521, 685. Paris. [Coniferales, Ginkgoales, *Pinus thunbergii* sp. nov., p. 388, *P. greggii* sp. nov., p. 396, *P. kasya* = *P. insularis*, p. 390, *P. quadrifolia* pro syn., p. 402, *P. cedrus* var. *atlantica* = *Cedrus atlantica*, p. 408, *Frenela subumbellata* sp. nov. = *Callitris sulcata*, p. 447, *Chamaecyparis pseudosquarrosa* sp. nov. = *C. thyoides*, p. 467, *Juniperus communis* var. *hemisphaerica* (J. et C. Presl) comb. et stat. nov., p. 479, *J. wallichiana* "Hook. f. herb. Kew", pro syn. (of *J. pseudosabina*) = *J. indica*, p. 482, *J. phoenicea* var. *turbinata* (Guss.) comb. et stat. nov., p. 487, *Dacrydium beccarii* sp. nov., p. 494, *D. kirkii* sp. nov. = *Halocarpus kirkii*, p. 495, *Dacrydium tetragonum* (Hook.) comb. nov. = *Microcachrys tetragona*, p. 496, *Phyllocladus trichomanoides* var. *alpinus* (Hook. f.) comb. nov., p. 498, *Podocarpus macrostachyus* sp. nov. = *P. oleifolius*, p. 510, *P. sprucei* sp. nov., p. 510, *P. aristulatus* sp. nov., p. 513, *P. parvifolius* sp. nov. = *P. alpinus*, p. 514, *P. cumingii* sp. nov. = *Dacrycarpus cumingii*, p. 521, *P. vieillardii* sp. nov. = *D. vieillardii*, p. 521, *Podocarpus falciformis* sp. nov. = *Falcatifolium falciforme*, p. 685].

11400 Patschke, W. (1913). Über die extra tropischen ostasiatischen Coniferen und ihre Bedeutung für die pflanzengeographische Gliederung Ostasiens. Bot. Jahrb. Syst. 48: 626–776, t. 8, f. 1–2. [*Picea pachyclada* sp. nov. = *P. brachytyla* s.l., p. 630, *P. ascendens* sp. nov. = *P. brachytyla* var. *rhombisquamea*, p. 632, *Larix* sect. *Multiserialis* (as "Multiseriales"), p. 651].

11412 Pax, F. (1922). Aufzählung der von Dr. Limpricht in Ostasien gesammelten Pflanzen. Repert. Spec. Nov. Regni Veg. Beih. 12: 299–515. [Pinaceae, Cupressaceae, pp. 304–305].

11413 Peirce, A. S. (1935). Types of pitting in Conifers. Trans. Illinois State Acad. Sci. 28 (2): 101–104. [wood-anatomy].

11440 Penhallow, D. P. (1896). The generic characters of the North American Taxaceae and Coniferae. Proc. & Trans. Roy. Soc. Canada, ser. 2 (sect. iv) 2: 33–57.

11508 Philippe, M., G. Zijlstra & M. Barbacka (1999). Greguss's morphogenera of homoxylous fossil woods: a taxonomic and nomenclatural review. Taxon 48 (4): 667–676. [Coniferales, nomenclature, palaeobotany, wood anatomy].

11510 Phillips, E. W. J. (1948). Identification of softwoods by their microscopic structure. Dept. Sci. Industr. Res. London Forest Prod. Res. Bull. 22, iii + 56 pp., 3 pls. (see also J. Linn. Soc., Bot. 52: 259–320, 1941). [Coniferales].

11520 Pilger, R. (1926). Phylogenie und Systematik der Coniferae. (general chapters pp. 121–199); Taxaceae. (pp. 199–211); Podocarpaceae (incl. *Phyllocladus*). (pp. 211–249); Araucariaceae. (pp. 249–266); Cephalotaxaceae. (pp. 267–271); Pinaceae. (pp. 271–342); Taxodiaceae. (pp. 342–360); Cupressaceae. (pp. 361–403); fossil conifers (pp. 403–407). In: A. Engler & K. A. E. Prantl (eds.). Die natürlichen Pflanzenfamilien XIII. 2. Aufl. Leipzig. (ill.). [*Acmopyle pancheri* (Brongn. & Gris) comb. nov., p. 240, Abietoideae subfam. nov., p. 311, *Larix gmelinii* var. *japonica* (Regel) et var. *principis-rupprechtii* (Mayr) comb. et stat. nov., p. 327, *Libocedrus uvifera* (D. Don) comb. nov., p. 389].

11541 Pillai, A. (1964). Root apical organization in gymnosperms. Some Conifers. Bull. Torrey Bot. Club 91 (1): 1–13. (ill.).

11710 Price, R. A. & J. M. Lowenstein (1989). An immunological comparison of the Sciadopityaceae, Taxodiaceae, and Cupressaceae. Syst. Bot. 14 (2): 141–149.

11731 Pritzel, G. A. (1900). Gymnospermae. in L. Diels. Die Flora von Central-China. Bot. Jahrb. Syst. 29 (2): 211–220. [Taxaceae, Pinaceae, Cupressaceae, *Picea brachytyla* (Franch.) comb. nov., p. 216, *P. likiangensis* (Franch) comb. nov., *Tsuga yunnanensis* (Franch) comb. nov. = *T. dumosa*, *T. chinensis* (Franch.) comb. nov. (basion. not fully cited, but not required at this date), p. 217, *Abies shensiensis* sp. nov. = *A. chensiensis*, p. 218].

11755a Quinn, C. J. & R. A. Price (2003). Phylogeny of the southern hemisphere conifers. In: R. R. Mill (ed.). Proceedings of the Fourth International Conifer Conference. Acta Horticulturae 615: 129–136. International Society for Horticultural Science (ISHS). [Araucariaceae, Cupressaceae, Podocarpaceae, Taxaceae, DNA, cladograms].

11755b Quinn, C. J., R. A. Price & P. A. Gadek (2002). Familial concepts and relationships in the conifers based on *rbc*L and *mat*K sequence comparisons. Kew Bull. 57 (3): 513–531. [Araucariaceae, Cephalotaxaceae, Cupressaceae, Pinaceae, Podocarpaceae, Sciadopytiaceae, Taxaceae, *Ginkgo* and other outgroups, DNA, cladistics, phylogeny, cladograms].

11770 Radais, M. L. (1894). Contribution à l'étude de l'anatomie comparée du fruit des conifères. Ann. Sci. Nat. Bot., sér. 7, 19: 165–368, pl. 1–15.

11798 Raubeson, L. A. & R. K. Jansen (1992). A rare chloroplast DNA structural mutation is shared by all conifers. Biochem. Syst. Ecol. 20: 17–24. [Conifers have a single copy of a large inverted repeat in the chloroplast genome; all other sampled seed plants, including Gnetales, have two copies. This is interpreted as a single loss shared by conifers].

11911 Rehder, A. (1929). A key to the Conifers based on leaf characters. J. Arnold Arbor. 10 (3): 196.

11970 Rehder, A. & E. H. Wilson (1914). Taxaceae; Pinaceae. in: C. S. Sargent, Plantae Wilsonianae II: 3–62. The Univ. Press, Cambridge, Mass. [Coniferales, Cephalotaxaceae, Taxaceae, Pinaceae, Taxodiaceae, Cupressaceae, *Cephalotaxus drupacea* var. *sinensis* (= *C.* sinensis) var. nov., p. 3, *Pinus*, *Larix mastersiana* sp. nov., p. 19, *Picea asperata* var. *notabilis* (= var. *heterolepis*) et var. *ponderosa*, var. nov., p. 23, *P. gemmata* sp. nov. = *P. retroflexa*, p. 24, *P. heterolepis* sp. nov. = *P. asperata* var. *heterolepis*, p. 24, *P. meyeri* sp. nov., p. 28, *P. balfouriana* sp. nov., *P. hirtella* "sp. nov." (non Loud.) = *P. likiangensis* var., pp. 30, 32, *P. sargentiana* sp. nov. = *P. brachytyla*, p. 35, *Tsuga*, *Keteleeria*, *Abies faxoniana* sp. nov. = *A. fargesii* var. *faxoniana*, p. 42, *A. beissneriana* sp. nov. = *A. recurvata* var. *ernestii*, p. 46, *A. sutchuenensis* (Franch.) comb. et stat. nov. = *A. fargesii* var. *sutchuenensis*, p. 48, *Cunninghamia*, *Cryptomeria*, *Thuja*, *Cupressus*, *Juniperus squamata* var. *fargesii*, var. nov., p. 59, *J. saltuaria* sp. nov., p. 61, *J. convallium* sp. nov., p. 62].

11980 Rehder, A. & E. H. Wilson (1928). Enumeration of the ligneous plants collected by J. F. Rock on the Arnold Arboretum expedition to north-western China and north-eastern Tibet. J. Arnold Arbor. 9: 4–27. [Coniferales, Cupressaceae, Pinaceae, *Juniperus chinensis* var. *arenaria* Wilson, var. nov., p. 20].

12030 Richard, L. C. M. (1826). Commentatio botanica de Conifereis et Cycadeis, ... (Mémoires sur les Conifères et les Cycadées). pp. i–xv, 1–212, pl. 1–29. Stuttgart. [*Podocarpus chilinus* sp. nov. = *P. salignus*, p. 11, pl. 1, f. 1, *P. coriaceus* Rich. & A. Rich., sp. nov., p. 14, t. 1, f. 3, *Callitris quadrivalvis* Rich. & A. Rich., sp. nov. = *Tetraclinis articulata*, p. 46, pl. 8, f. 1, *Callitris rhomboidea* R. Br. ex Rich. & A. Rich., sp. nov., p. 47, pl. 18, f. 1, *C. oblonga* Rich. & A. Rich., sp. nov., p. 49, pl. 18, f. 2, *Agathis dammara* (Lamb.) Rich. & A. Rich., comb. nov., p. 83; see also R. Brown, 1826, for a nom. cons.].

12191a Rothwell, G. W. (1982). New interpretations of the earliest conifers. Rev. Paleobot. Palynol. 37: 7–28.

12191f Rothwell, G. W. & G. Mapes (1988). Vegetation of a Paleozoic conifer community. In: G. Mapes & R. H. Mapes (eds.). Regional geology and palaeontology of [the] Upper Paleozoic Hamilton quarry area in southeastern Kansas. Guidebook 6, Kansas Geological Survey, Lawrence, Kansas, pp. 213–223. [walchian conifers, Cordaites and ferns/seed ferns formed higher-ground forests bordering lakes and streams].

12191g Rothwell, G. W. & G. Mapes (2003). Validation of the names Utrechtiaceae, *Utrechtia*, and *Utrechtia floriniformis*. Taxon 52: 329–330. [Utrechtiaceae Mapes & G. W. Rothwell, fam. nov., *Utrechtia* Mapes & G. W. Rothwell, gen. nov., *U. floriniformis* Mapes & G. W. Rothwell, sp. nov., *Lebachia*, nom. illeg., palaeobotany].

12191h Rothwell, G. W., G. Mapes & R. H. Mapes (1997). Late Paleozoic conifers of North America: structure, diversity and occurrences. Rev. Palaeobot. Palynol. 95: 95–113. [Permian fossils, e.g. the fossil "morpho-genus" *Walchia* and *Emporia* Mapes & Rothwell div. spp.].

12250 Rudloff, E. von (1975). Volatile leaf oil analysis in chemosystematic studies of North American conifers. Biochem. Syst. Ecol. 2: 131–167. [Pinaceae, Cupressaceae, Taxodiaceae].

12430a Sahni, B. (1931). Revisions of Indian fossil plants. Part II: Coniferales. Palaeontologica Indica, n.s., vol. 2. Calcutta. [Coniferales, palaeobotany, ill.].

12471 Salmon, J. T. (1980). The native trees of New Zealand. Rev. ed. 1986, 1996 (repr. 2001), Birkenhead, Auckland, N.Z. (short descr., numerous colour phot.). [Coniferales pp. 50–100, *Agathis australis*, *Libocedrus bidwillii*, *L. plumosa*, *Lagarostrobos*, *Lepidothamnus*, *Podocarpus*, *Phyllocladus*].

12480k Saporta, G. de (1865). La végétation du sud-est de la France à l'époque Tertiare. Part 2, No. 3. Ann. Sci. Nat. Bot., sér. 5, 4: 5–264, t. 1–13. [Cupressaceae, *Pinus*, Taxodiaceae ("Taxodineae") fam. nov., p. 44, palaeobotany, ill. of fossils].

12540 Sarlin, P. (1954). Bois et Forêts de la Nouvelle-Calédonie. Centre Techn. Forest. Tropical, Nogent-sur-Marne, France. [Coniferales, ill.].

12570   Sasaki, S. (1929). Conifers and taxads distributed in the Far East. Trans. Nat. Hist. Soc. Taiwan 19: 3–12. (Japan.). [Coniferales, Taxales, table].

12610   Sax, K. & H. J. Sax (1933). Chromosome numbers and morphology in the conifers. J. Arnold Arbor. 14: 356–374, pl. 75.

12660   Saxton, W. T. (1913). The classification of conifers. New Phytol. 12: 242–262. [Callitroideae subfam. nov., p. 253, Sciadopitoideae subfam. nov., Sequoioideae subfam. nov.; first class. uniting cupressaceous and taxodiaceous genera into one fam.Cupressaceae, based on anatomy of gametophyte and embryogeny].

12810   Schlechtendal, D. F. L. von (1864–65). Beiträge zur Kenntnis der Coniferen. Linnaea 33: 339–400, pl. 1–2, 693–705.

12894   Schmidt, P. A. (2002). Bäume und Sträucher Kaukasiens. Teil 1: Einführung und Gymnospermae (Nadelgehölze und sonstige Nacktsamer). Mitt. Deutsch. Dendrol. Ges. 87: 59–81. (ill.). [*Juniperus communis, J. excelsa, J. foetidissima, J. oxycedrus, J. sabina, Pinus nigra, P. sylvestris, P. brutia, Abies nordmanniana, Picea orientalis, Taxus baccata*].

12951   Schmucker, T. (1942). Die Baumarten der Nördlich-Gemässigten Zone und ihre Verbreitung. (Silvae Orbis 4). C. I. S., Berlin. [Coniferales, Pinaceae, maps].

12957   Schneckenburger, S. (1999). Einige bemerkenswerte Coniferen Neukaledoniens. Mitt. Deutsch. Dendrol. Ges. 84: 5–21. (coll. ill., maps). [*Agathis ovata, Araucaria humboldtensis, Dacrydium guillauminii, Neocallitropsis pancheri, Parasitaxus usta, Retrophyllum minus*].

13012   Schumann, K. (1902). Über die weiblichen Blüten der Coniferen. Verh. Bot. Ver. Prov. Brandenburg 44: 5–80. (ill.). [Coniferales, *Taxus*, Podocarpaceae, Pinaceae, Cupressaceae].

13033   Schweingruber, F. H. (1992). Baum und Holz in der Dendrochronologie. Morphologische, anatomische und jahrringanalytische Charakteristika häufig verwendeter Bäume. 2. Auflage. Eidgenössische Forschungsanstalt für Wald, Schnee und Landschaft, Birmersdorf, Schweiz/Switzerland. [Coniferales, Araucariaceae, Cupressaceae, Pinaceae, Taxodiaceae, ill., maps].

13040   Schweitzer, H.-J. (1974). Die "Tertiären" Koniferen Spitzbergens. Palaeontographica B, 149: 1–89, pl. 1–20. [Coniferales, *Cupressinocladus interruptus* (Newberry) Schweitzer, *Thuja ehrenswaerdii* (Heer) Schweitzer, palaeobotany].

13042   Schweitzer, H.-J. & M. Kirchner (1996). Die Rhäto-Jurassischen Floren des Iran und Afghanistans: 9. Coniferophyta. Palaeonto-

graphica B, 238 (4–6): 77–139, pl. 1–16. [Coniferales, Voltziaceae, *Compsostrobus brevirostratus* sp. nov., *Cycadocarpidium erdmannii* Nathorst, *Elatides* Heer, *E. thomasii* Harris, *Palissya* Endl., *P. oleschinskii* sp. nov., palaeobotany].

13053   Scott, A. (1974). The earliest conifer. Nature 252: 707–708. [Upper Carboniferous conifer leaves].

13053a   Scott, A. & W. G. Chaloner (1983). The earliest fossil conifer from the Westphalian B of Yorkshire. Proc. Roy. Soc. London B 220: 163–166.

13080   Sénéclauze, A. (1868). Les Conifères. Monographie descriptive et raisonnée classée,... Impr. gén. de Ch. Lahure, Paris. [*Pinus hondurensis* sp. nov. = *P. caribaea* var. *hondurensis*, p. 126].

13080a   Seoana, V. de (1998). Comparative study of extant and fossil conifer leaves from the Baquero Formation (Lower Cretaceous), Santa Cruz Province, Argentina. Rev. Palaeobot. Palynol. 99: 247–263. (ill.). [Cupressaceae s.l., Taxodiaceae, *Athrotaxis ungeri* (Halle) Archangelsky, palaeobotany].

13080d   Seward, A. C. (1919). Fossil plants. Vol. 4. Ginkgoales, Coniferales, Gnetales. (ill.). Cambridge University Press.

13162   Shimizu, T. (1997). Sciadopityaceae, Pinaceae. The World of Plants 128. (Japanese text, many colour photographs). [*Abies, Larix, Picea, Pinus, Sciadopitys verticillata*].

13240   Siebold, P. F. von (1830). Synopsis Plantarum Oeconomicarum Universi Regni Japonici. Ed. 1. Verh. Batav. Genootsch. Kunsten 12: I–IV, 1–74. [Coniferales, Pinaceae, Taxodiaceae, *Abies araragi* nom. nud. = *Tsuga sieboldii, A. momi* nom. nud., *A. torano* nom. nud., etc.].

13251   Siebold, P. F. von & J. G. Zuccarini (1846). Florae japonicae familiae naturales, adjectis generum et specierum exemplis selectis. Part 3. Abh. Math.-Phys. Cl. Königl. Bayer. Acad. Wiss. 4 (3): 123–240, pl. 3. [*Taxus cuspidata* sp. nov., p. 232, *Podocarpus macrophyllus* var. *maki* var. nov. = *P. chinensis* var. *chinensis*, p. 232, *Cephalotaxus drupacea* sp. nov. = *C. harringtonii*, p. 234, *C. pedunculata* sp. nov. = *C. harringtonii*, p. 234, *Torreya nucifera* (L.) comb. nov., p 234].

13290   Silba, J. (1984). An International Census of the Coniferae, I. Phytologia Memoirs VII. Plainfield, N.J. [This "revised checklist" contains numerous unexplained and therefore seemingly (?) wanton reductions, of which for nomenclatural reasons most new combinations are listed here: *Abies spectabilis* var. *densa* (Griff.) comb. et stat. nov., p. 10, *Actinostrobus pyramidalis* var. *arenarius* (C. A. Gardner) comb. et stat. nov., p. 11, *Callitris columellaris* var. *campestris* et var. *intratropica* (R. T. Baker et H. G. Smith), var. et comb. nov., p.

16, *C. preissii* var. *murrayensis* (J. Garden) et var. *verrucosa* (A. Cunn. ex Endl.), comb. et stat. nov., p. 17, *Dacrydium leptophyllum* (Wasscher) de Laub. ex Silba, p. 27, *J. ashei* var. *saltillensis* (M. T. Hall) comb. et stat. nov., p. 32, *J. convallium* var. *microsperma* (Cheng et Fu) comb. et stat. nov., p. 33, *J. excelsa* var. *polycarpos* (K. Koch) comb. et stat. nov., p. 34, *J. oxycedrus* var. *transtagana* (Franco) comb. et stat. nov., p. 35, *J. pinchotii* var. *erythrocarpa* (Cory) comb. et stat. nov., p. 35, *J. pingii* var. *wilsonii* (Rehder) comb. nov., p. 36, *J. pseudosabina* var. *turkestanica* (Komarov) comb. nov., p. 36, *Larix griffithii* var. *mastersiana* (Rehd. et Wils.) comb. et stat. nov., p. 39, *Pinus cembroides* var. *quadrifolia* (Parl. ex Sudw.) Laubenfels et Silba comb. et stat. nov. (publ. of basion. incorr. cit. (ICBN Art. 33.2), see for this E. L. Little, 1979, p. 197), p. 49, *P. densa* (Little et Dorman) Laubenfels et Silba comb. et stat. nov., p. 50, *P. insularis* var. *yunnanensis* (Franch.) comb. et stat. nov., p. 52, *P. occidentalis* var. *cubensis* (Griseb., non Sarg.!) comb. et stat. nov., p. 55, *P. wallichiana* var. *dalatensis* (Y. de Ferré) comb. et stat. nov., p. 59, *Podocarpus transiens* (Pilg.) de Laub. ex Silba, comb. nov., p. 68, *Taiwania cryptomerioides* var. *flousiana* (Gaussen) comb. et stat. nov., p. 72, *Taxus canadensis* var. *floridana* (Nutt. ex Chapm.) comb. et stat. nov., p. 72, *Torreya grandis* var. *fargesii* (Franch.) Silba, p. 74, *Tsuga argyrophylla* (Chun et Kuang) Laubenfels et Silba, comb. nov., p. 75].

13310   Silba, J. (1985). A supplement to the international census of Coniferae, I. Phytologia 58 (6): 365–370. [*Abies, Juniperus barbadensis* var. *urbaniana* (Pilger et Ekman) comb. et stat. nov., *J. flaccida* var. *martinezii* (Pérez de la Rosa) comb. et stat. nov., *Pinus brutia* var. *eldarica* (Medw.) comb. et stat. nov., *P. brutia* var. *pityusa* (Steven) comb. et stat. nov., etc.].

13341   Silba, J. (1990). A supplement to the International Census of the Coniferae, II. Phytologia 68 (1): 7–78. [Contains again numerous new combinations at infraspecific levels in many genera, some jointly with Farjon, but without his consent and in fact taken from the MS of Farjon, 1990; only a selected number is enumerated here: *Abies delavayi* var. *nukiangensis* (Cheng et Fu) Farjon et Silba, p. 13, *A. forrestii* var. *chengii* (Rushforth) Silba, p. 17, *A. forrestii* var. *ferreana* (Bord.-Rey et Gaussen) Farjon et Silba, p. 17, *A. hickelii* var. *oaxacana* (Martínez) Farjon et Silba, p. 20 (with errors in citation of place and date of the basionym, see Martínez, 1948), *A. nordmanniana* var. *bornmuelleriana* (Mattf.) Silba, p. 21, *Agathis macrophylla* var. *obtusa* (Lindl.) Masters, p. 23, *Amentotaxus yunnanensis* var. *formosana* (H. L. Li) comb. nov., p. 25, *Cupressus himalaica* var.

darjeelingensis Silba, var. nov. (= *C. cashmeriana*, p. 29, *Juniperus, Keteleeria, Larix griffithiana* var. *speciosa* (Cheng et Law) Silba, p. 37, *L. potaninii* var. *himalaica* (Cheng et Fu), Farjon et Silba, p. 37, *Picea likiangensis* var. *bhutanica* var. nov., p. 41, *P. likiangensis* var. *forrestii* var. nov., p. 41, *Pinus merkusii* var. *latteri* (F. Mason) Silba, p. 53, *P. radiata* var. *cedrosensis* (J. T. Howell) Silba, p. 60, *Podocarpus barretoi* de Laub. & Silba, p. 65, *Podocarpus costaricensis* de Laub. ex Silba, p. 67, *Pseudotsuga sinensis* var. *brevifolia* (Cheng et Fu) Farjon et Silba, p. 71, *P. sinensis* var. *gaussenii* (Flous) Silba, p. 71, *P. sinensis* var. *forrestii* (Craib) Silba, p. 72, *Torreya grandis* var. *yunnanensis* (W. C. Cheng & L. K. Fu) Silba, p. 72, *Tsuga chinensis* var. *forrestii* (Downie) Silba, p. 72, *Widdringtonia,* comb. et stat. nov., see also Farjon, 1990].

13385a   Siqueiros Delgado, M. E. (1999). Coniferas de Aguascalientes. (ill., maps). 2nd ed. Universidad Autonoma de Aguascalientes, Mexico.

13410   Skalicka, A. & V. Skalicky (1987). Taxonomic and nomenclatoric notes on the treatment of Coniferae for the Flora of Czech socialist republic. Novit. Bot. Delect. Seminum Horti Bot. Univ. Carol. Prag. 3: 53–62.

13420   Skottsberg, C. (1934). Phytogeography of conifers of western South America. Proc. 5th Pacific Sci. Congr. (Canada 1933) 4: 3265–3266.

13425   Slyper, E. J. (1933). Bestimmungstabelle für rezente und fossile Coniferenhölzer nach mikroskopischen Merkmalen. Recueil Trav. Bot. Néerl. 30: 482–513. [Coniferales, fossil and Recent, wood anatomy, determination key with line drawings].

13440   Smith, C. C. (1970). The co-evolution in pine-squirrels (Tamiasciurus) and conifers. Ecol. Monogr. 40: 349–371.

13481   Sneath, P. H. A. (1967). Conifer Distributions and Continental Drift. Nature 215: 467–470 (maps).

13560   Spach, E. (1841). Histoire naturelle des Végétaux. Phanérogames, Vol. 11. Librairie encyclopédique de Roret, Paris. [Coniferales, *Juniperus foetida* sp. nov., p. 314, *J. foetida* (var.) *squarrulosa* = *J. foetidissima,* p. 321, *Chamaecyparis* gen. nov., p. 329, *C. nootkatensis* (D. Don) comb. nov. (as "nutkatensis"), p. 333, *Platycladus* gen. nov., p. 333, *P. stricta* nom. nov. = *P. orientalis,* p. 335, *P. dolabrata* (Thunb. ex L. f.) comb. nov. = *Thujopsis dolabrata,* p. 337, *Pinus* sect. *Strobus* Sweet ex Spach sect. nov., p. 394, *P.* sect. *Cembra* sect. nov., p. 398, *Abies* sect. *Piceaster* sect. nov., p. 414, *A. nordmanniana* (Steven) comb. nov., p. 418, *A. spectabilis* (D. Don) comb. nov., p. 422].

13650   Stainton, J. D. A. (1972). Forests of Nepal. Hafner, London – New York. [Coniferales].

13791   Stefanov, B. (1941–42). Die geografische Verbreitung der Coniferen und die Formbildung in der Natur... Parts 1–3. God. Sofijsk. Univ. Agron. Fak., T. 20: 1–65. (Annuaire de l'Université de Sofia). (Bulgarian). (maps).

13792   Stefanovic, S., M. Jager, J. Deutsch, J. Broutin & M. Masselot (1998). Phylogenetic relationships of conifers inferred from partial 28S rRNA gene sequences.Amer. J. Bot. 85 (5): 688–697. [Coniferales, DNA-based phylogeny].

13797   Stein, N. (1978). Coniferen im westlichen Malayischen Archipel. Biogeographica 11. (ed. in chief J. Schmithüsen). The Hague, Boston, London. [Araucariaceae, Podocarpaceae, Cupressaceae].

13828   Stichting Nederlandse Plantentuinen (1996). The National Plant Collection: Coniferae. A catalogue of the Dutch Conifer Collection. Stichting Nederlandse Plantentuinen, Utrecht.

13832   Stockey, R. A. (1981). Some comments on the origin and evolution of conifers. Canad. J. Bot. 59: 1932–1940.

13833b   Stockey, R. A. (1990). Antarctic and Gondwana conifers. In: T. N. Taylor & E. L. Taylor (eds.). Antarctic paleobiology, its role in the reconstruction of Gondwana. Springer, New York. (pp. 179–191, ill.). [palaeobotany].

13879   Strasburger, E. (1869). Die Befruchtung bei den Coniferen. Jena. [*Tsuga canadensis, Picea abies, Pinus* spp., *Tetraclinis articulata, Juniperus virginiana, Platycladus orientalis*, ill.].

13880   Strasburger, E. (1872). Die Coniferen und die Gnetaceen. Eine morphologische Studie. Jena. (442 pp., ill. in sep. vol. 'Atlas').

13890   Studt, W. (1926). Die heutige und frühere Verbreitung der Koniferen und die Geschichte ihrer Arealgestaltung. Mitt. Inst. Allg. Bot. Hamburg 6 (2): 167–308.

13951   Su, Y. J. (1997). Comparative observation on leaf structure of Taxaceae, Cephalotaxaceae and Podocarpaceae. J. Wuhan Bot. Res. 15 (4): 307–316. (Chin., Eng. abstr., ill.).

14020   Sudworth, G. B. (1918). Miscellaneous conifers of the Rocky Mountain region. U.S.D.A. Bull. (1915–23) 680: 1–44.

14121   Szmidt, A. E. (1991). Phylogenetic and applied studies on chloroplast genome in forest conifers. Pp. 185–196 in: S. Fineschi *et al.* (eds.) Biochemical markers in the population genetics of forest trees. SPB Academic Publication, The Hague.

14260   Teng, S. C. (1947). The forest regions of Kansu and their ecological aspects. Bot. Bull. Acad. Sin. 1: 187–200. [Coniferales].

14341   Thomas, F. (1865). Zur vergleichenden Anatomie der Coniferen-Laubblätter. Jahrb. Wiss. Bot. 4: 23–63.

14344   Thompson, R. S., K. H. Anderson & P. J. Bartlein (1999). Atlas of relations between climatic parameters and distributions of important trees and shrubs in North America – Introduction and Conifers. U.S. Geol. Survey Prof. Pap. 1650-A, USGS, Denver, Colorado. (iv + 269 pp., ill, maps, tables). [Coniferales, Cupressaceae, Pinaceae, phytogeography & ecology].

14351   Thomson, R. B. (1940). The structure of the cone in the Coniferae. Bot. Rev. (Lancaster) 6: 73–84. (ill.) [Coniferales].

14460   Tournefort, J. P. de (1694). Institutiones rei herbariae. Paris. (Ed. 2, 1700; Ed. 3, (3 vols.) 1719). [Coniferales, with e.g. *Pinus, Abies, Larix, Thuja, Cupressus, Juniperus*].

14512   Tsumura, Y. & K. Ohba (1993). Molecular evolution of conifers. Gamma Field Symposia 32: 35–44. Inst. Radiation Breeding, NIAR, MAFF, Japan.

14513   Tsumura, Y., K. Yoshimura, N. Tomaru & K. Ohba (1995). Molecular phylogeny of conifers using RFLP analysis of PCR-amplified specific chloroplast genes. Theor. Appl. Genet. 91: 1222–1236.

14591   Ueno, J. (1959). Some palynological observations of Taxaceae, Cupressaceae, and Araucariaceae. J. Inst. Polytechn. Osaka City Univ., ser. D, Biol. 10: 75–87.

14592   Ueno, J. (1960). Studies on pollen grains of Gymnospermae: concluding remarks to the relationships between Coniferae. J. Inst. Polytechn. Osaka City Univ., ser. D, Biol. 11: 109–136, pl. 1.

14667   Valckenier Suringar, J. (1928). Personal ideas about the application of the international rules of nomenclature, or, as with the rules themselves, international deliberation? Some denominations of conifer-species. Meded. Rijks-Herb. 55: 1–75, 2 pl. [*Abies, Cedrus, Juniperus, Larix, Libocedrus, Picea, Pinus, Pseudolarix, Pseudotsuga, Tsuga*, nomenclature].

14673   Van Pelt, R. (2001). Forest giants of the Pacific coast. Global Forest Society/University of Washington Press, Vancouver/Seattle. [20 spp. of conifers, 117 individual trees, ill, maps, measurements].

14695   Vaucher, H. (1990). Baumrinden. Stuttgart. [530 colour photographs of outer patterns of bark on (mostly cultivated) trees, including conifers.].

14760   Vierhapper, F. (1910). Entwurf eines neuen Systems der Coniferen. Abh. K.K. Zool.-Bot. Ges. Wien 5 (4): 1–56.

14772   Visscher, H., J. H. J. Kerp & J. A. Clement-Westerhof (1986). Aspects of Permian Palaeobotany and Palynology. VI. Towards a flexible system of naming palaeozoic Conifers. Acta Bot. Neerl. 35 (2): 87–100.

14779   Vogellehner, D. (1967). Zur Anatomie und Phylogenie mesozoischer Gymnospermenhölzer. 5: Prodromus zu einer Monographie der Protopinaceae. I. Die protopinoiden Hölzer des Trias. Palaeontographica Abt. B, Paläophytol. 121: 30–51. [Coniferales].

14780   Vogellehner, D. (1967). Holzanatomie triassischer Gymnospermen und ihre Bedeutung für die Phylogenie. Ber. Deutsch. Bot. Ges. 80: 307–311. [Coniferales].

14790   Vogellehner, D. (1968). Zur Anatomie und Phylogenie mesozoischer Gymnospermenhölzer. 7: Prodromus zu einer Monographie der Protopinaceae. II. Die protopinoiden Hölzer des Jura. Palaeontographica Abt. B, Paläophytol. 124: 125–162. [Coniferales].

14800   Vorobjev, D. P. (1968). Wildgrowing trees and shrubs of the Far East. Nauka, Leningrad. (Russ.). [Coniferales, Pinaceae, Cupressaceae].

14810   Voss, A. (1907). Coniferen Nomenklatur Tabelle. Mitt. Deutsch. Dendrol. Ges. 1907 (16): 88–95. [*Thuja decurrens* (Torrey) comb. nov. = *Calocedrus decurrens*, p. 88, *T. macrolepis* (S. Kurz) comb. nov. = *Calocedrus macrolepis*, p. 88, *T. papuana* (F. Mueller) comb. nov. = *Papuacedrus papuana*, p. 88, *Picea glauca* (Moench) comb. nov., p. 93, *Pinus cembroides* var. *edulis* (Engelm.) comb. et stat. nov. (as "cembrodes") = *P. edulis*, p. 95, *P. cembroides* var. *monophylla* (Torrey et Frém.) comb. et stat. nov. = *P. monophylla*, p. 95, *P. cembroides* var. *parryana* var. nov. = *P. quadrifolia*, p. 95].

14910   Wang, C. (1968). The coniferous forests of Taiwan. Biol. Bull. Dept. Biol. Coll. Sci. Tunghai Univ. 34: 1–52.

14920   Wang, C. W. (1961). The forests of China; with a survey of grassland and desert vegetation. Publ. Maria Moors Cabot Found. Bot. Res. 5. (Harvard Univ.) Cambridge, Mass. [very comprehensive account of spp. composition and distribution of forest types, with many ecological data on conifers].

15070   Weck, J. (1958). Über Koniferen in den Tropen. Forstwiss. Zentralbl. 77: 193–256.

15110   Wells, P. V. (1983). Paleobiogeography of montane islands in the Great Basin since the last Glaciopluvial. Ecol. Monogr. 53: 341–382. [Coniferales, angiosperms].

15200   Wilson, E. H. (1912). Vegetation of Western China; A series of 500 photographs, with index and introduction by C. S. Sargent. Publ. Arnold Arbor. 2, London. (copy of 4 Vols. seen at P). [Coniferales, many habitus photographs].

15220   Wilson, E. H. (1916). The Conifers and Taxads of Japan. Publ. Arnold Arbor. 8, Univ. Press, Cambridge, Mass. [*Abies homolepis* var. *umbellata* (Mayr) comb. nov., p. 58, *Juniperus communis* var. *nipponica* var. nov. = *J. rigida* ssp. *nipponica*, p. 81, but esp. distribution, ecology, habitus photographs].

15250   Wilson, E. H. (1926). The Taxads and Conifers of Yunnan. J. Arnold Arbor. 7: 37–68. [*Picea yunnanensis* hort. ex Wilson, sp. nov. = *P. likiangensis*, p. 47].

15300   Wittaker, R. H. (1960). Vegetation of the Siskiyou Mountains, Oregon and California. Ecol. Monogr. 30: 279–338. [Coniferales, angiosperms, ecology].

15340   Wolfe, J. A. (1969). Neogene floristic and vegetational history of the Pacific Northwest. Madroño 20: 83–110. [Coniferales, angiosperms].

15353a   Woltz, P. (1987). Les Podocarpineae, étude des plantules et évolution: comparaison avec les familles de Conifères de l'hémisphère sud: Araucariaceae, Cupressaceae, Taxodiaceae, Taxaceae. Bull. Soc. Bot. France 134: 141–151. [Podocarpaceae s.l., Cupressaceae s.l., Taxaceae, seedling morphology, cotyledons].

15362   Worsdell, W. C. (1900). The structure of the female "flower" in Coniferae, a historical study. Ann. Bot. (London) 14: 39–82. (ill.). [Coniferales, review paper].

15426   Yañez Espinosa, L. & T. Terrazas (1998). Anatomía de la corteza de algunas Gimnospermas. Bol. Soc. Bot. México 62: 15–28. [Coniferales, anatomy of bark, ill.].

15440   Yao, B. J. & Y. S. Hu (1982). Comparative anatomy of conifer leaves. Acta Phytotax. Sin. 20 (3): 275–294. (Chin., ill., key).

15490   Ying, T. S. & L. Q. Li (1981). Ecological distribution of endemic genera of taxads and conifers in China and neighbouring area in relation to phytogeographical significance. (Chin., Eng. summ., maps). Acta Phytotax. Sin. 19 (4): 408–415.

15502   Young, D. J. & L. Watson (1969). Softwood structure and the classification of conifers. New Phytol. 68: 427–432.

15530   Zanoni, T. A. (1982). Pinophyta. in: S. P. Parker (ed.). Synopsis and Classification of Living Organisms. Part 1, pp. 349–356. New York. [Coniferales].

15692   Zuccarini, J. G. (1843). Beiträge zur Morphologie der Coniferen. Abh. Math.-Phys. Cl. Königl. Bayer. Akad. Wiss. 3: 751–805. (ill.). [*Taxus wallichiana* sp. nov., p. 803, t. 5].

ARAUCARIACEAE
(general titles)

01494   Burlingame, L. L. (1915). The origin and relationships of the Araucarians. Bot. Gaz. (Crawfordsville) 60 (2): 89–114. [Araucariaceae].

01497   Burrows, G. E. (1987). Leaf axil anatomy of the Araucariaceae. Austral. J. Bot. 35: 631–640. [*Agathis, Araucaria*].

02375   Cookson, I. C. & S. L. Duigan (1951). Tertiary Araucariaceae from southeastern Australia, with notes on living species. Austral. J. Sci. Res., ser. B, 4: 415–449. [*Agathis, Araucaria*, palaeobotany].

02860a   Dernbach, U., W. Jung & A. Selmeier (1992). *Araucaria.* Versteinerte Zapfen und versteinertes Holz vom Cerro Colorado, Argentiniën. Lorsch, Germany. [photography Konrad Götz, fossil and non-fossil Araucariaceae].

02865   Dettmann, M. E. & D. M. Jarzen (2000). Pollen of extant *Wollemia* (Wollemi Pine) and comparisons with other extant and fossil Araucariaceae. In: M. M. Harley, C. M. Morton & S. Blackmore (eds.). Pollen and spores: Morphology and Biology. Pp. 187–203. Royal Botanic Gardens, Kew, [*Agathis, Araucaria, Wollemia nobilis*].

04801   Gaussen, H. (1970). Les gymnospermes actuelles et fossiles. Fascicule 11: Araucariacées, Cephalotaxacées, Conclusion des Pinoidines, Additions et corrections. Faculté des Sciences, Toulouse (reprint from Trav. Lab. Forest. Toulouse, 1970). (ill.). [Araucariaceae, Cephalotaxaceae, Pinaceae, misc. groups and subjects].

04874   Gilmore, S. & K. D. Hill (1997). Relationships of the Wollemi Pine (*Wollemia nobilis*) and a molecular phylogeny of the Araucariaceae. Telopea 7 (3): 275–291. [*Agathis, Araucaria, Wollemia*, conifer phylogeny based on *rbc*L (DNA) data].

05876   Hill, R. S. & A. J. Bigwood (1987). Tertiary gymnosperms from Tasmania: Araucariaceae. Alcheringa 11: 325–335. [*Araucarioides*, palaeobotany, ill.].

07135   Kershaw, P. & B. Wagstaff (2001). The southern conifer family Araucariaceae: history, status and value for palaeo-environmental reconstruction. Ann. Rev. Ecol. Syst. 32: 397–414.

11507   Philippe, M. (1993). Nomenclature générique des trachéidoxyles fossiles mésozoïques à champs araucarioïdes. Taxon 42 (1): 74–80. [Araucariaceae].

12200a   Rouane, M. L. & P. Woltz (1979). Apport de l'étude des plantules pour la taxonomie et l'évolution des Araucariées. Bull. Soc. Bot. France 126, Actual. Bot. 3: 67–76. [Araucariaceae, *Agathis, Araucaria*].

12200b   Rouane, M. L. & P. Woltz (1980). Intéret des plantules pour la systématique et l'évolution des Araucariées. Bull. Soc. Hist. Nat. Toulouse 116: 120–136. [Araucariaceae, *Agathis, Arauraria*].

13833b   Stockey, R. A. (1982). The Araucariaceae: an evolutionary perspective. Rev. Palaeobot. Palynol. 37: 133–154.

13833c   Stockey, R. A. (1994). Mesozoic Araucariaceae: morphology and systematic relationships. J. Plant Res. 107: 493–502. [Araucariaceae, *Agathis, Araucaria*, palaeobotany, SEM of cuticles and stomata].

CUPRESSACEAE
(general titles)

00153a   Alvin, K. L., D. H. Dalby & F. A. Oladele (1981). Numerical analysis of cuticular characters in Cupressaceae. In: D. F. Cutler, K. L. Alvin & C. E. Price (eds.). The plant cuticle. Linnean Society Symposium Series No. 10. London.

00530   Baker, R. T. & H. G. Smith (1910). A Research on the Pines of Australia. N. S. W., Technical Education Series, No. 16. Sydney. [Cupressaceae, *Callitris, C. tuberculata* R. Br. ex R. T. Baker & H. G. Smith, p. 99, *C. propinqua* R. Br. ex R. T. Baker & H. G. Smith, p. 112, *C. glauca* R. Br. ex R. T. Baker et H. G. Smith, p. 118 (but see B. & S., 1908, p. 145), *C. intratropica*, sp. nov. = *C. columellaris*, p. 172, *Actinostrobus*].

01580   Campo-Duplan, M. van (1951). Recherches sur la phylogénie des Cupressacées d'après leurs grains de pollen. Trav. Lab. Forest. Toulouse, T. 2, sect. 7, vol. 4, art. 3: 1–20.

02651   Daguillon, A. (1899). Observations morphologiques sur les feuilles des Cupressinées. Rev. Gén. Bot. 11: 168–204. [Araucariaceae, *Araucaria araucana*, Cupressaceae, *Chamaecyparis lawsoniana, Cupressus, Thuja occidentalis, Cryptomeria japonica, Platycladus orientalis, Sequoia sempervirens, Sequoiadendron giganteum, Taxodium distichum, Juniperus communis*, anatomy, seedling characters, ill.].

02921   Dogra, P. D. (1984). The embryology, breeding systems, and seed sterility in Cupressaceae – a monograph. in: P. K. K. Nair (ed.). Glimpses in plant research, 6, pp. 1–113, figs. 1–264. New Delhi.

03270   Eckenwalder, J. E. (1976). Re-evaluation of Cupressaceae and Taxodiaceae: a proposed merger. Madroño 23 (5): 237–256.

03621   Erdtman, H. & T. Norin (1966). The chemistry of the order Cupressales. Fortschr. Chem. Organ. Naturstoffe 24: 206–287. [Cupressaceae, Taxodiaceae, *Chamaecyparis nootkatensis* distinct from other spp., important review].

03696m   Farjon, A. (2005). A monograph of Cupressaceae and *Sciadopitys*. Royal Botanic Gardens, Kew, Richmond. [Comprehensive treatment of all species incl. former *Taxodiaceae*, introductory chapters on morphology, phylogeny and classification, palaeobotany (by invited authors R. Stockey,

G. Rothwell, J. & Z. Kvačeck and R. Hill), line drawings by the author, SEM illustrations, exsiccata, maps etc. *Sciadopitys verticillata* included as sister taxon to the family].

03843   Ferré, Y. de (1942). Cotylédons et évolution chez les Cupressinées. Bull. Soc. Hist. Nat. Toulouse 77: 145–160.

03951   Ferré, Y. de & H. Gaussen (1968). Les Cupressacées australes. Compt. Rend. Hebd. Séances Acad. Sci. (Paris) 267: 483–487. [Cupressaceae in the southern hemisphere].

04660a Gadek, P. A., D. L. Alpers, M. M. Heslewood & C. J. Quinn (2000). Relationships within Cupressaceae sensu lato: a combined morphological and molecular approach. Amer. J. Bot. 87 (7): 1044–1057. [Cupressaceae, Taxodiaceae, DNA, morphology, phylogeny].

04660b Gadek, P. A., D. Alpers & C. J. Quinn (1997). Combined morphological and molecular analysis of Cupressaceae s.l. Amer. J. Bot. 84 (Suppl.-abstr.): 195–196.

04661   Gadek, P. A. & C. J. Quinn (1983). Biflavones of the subfamily Callitroideae, Cupressaceae. Phytochemistry 22: 969–972. [chemotaxonomy].

04662   Gadek, P. A. & C. J. Quinn (1985). Biflavones of the subfamily Cupressoideae, Cupressaceae. Phytochemistry 24: 267–272. [chemotaxonomy].

04664   Gadek, P. A. & C. J. Quinn (1989). Affinities of southern Cupressaceae. Austral. Syst. Bot. Soc. Newsl. 60: 20–21. [chemotaxonomy].

04666   Gadek, P. A. & C. J. Quinn (1993). An analysis of relationships within the Cupressaceae sensu stricto based on rbcL sequences. Ann. Missouri Bot. Gard. 80: 581–586.

04800   Gaussen, H. (1968). Les Gymnospermes actuelles et fossiles. Fascicule 10: Les Cupressacées. Faculté des Sciences, Toulouse (reprint from Trav. Lab. Forest. Toulouse, 1968). (ill.). [taxonomic treatment of all genera and species of Cupressaceae s. str.].

05200   Hair, J. B. (1968). The chromosomes of Cupressaceae 1. *Tetraclineae* and *Actinostrobeae* (Callitroideae). New Zealand J. Bot. 6 (3): 277–284.

05839   Hickel, R. (1905). Les variations du type chez les Cupressinées. La feuille des jeunes naturalistes, ser. 4, 35: 1–7.

05877   Hill, R. S. & R. Carpenter (1989). Tertiary gymnosperms from Tasmania: Cupressaceae. Alcheringa 13: 89–102. [*Austrocedrus, A. tasmanica* sp. nov. (p. 94), *Libocedrus, L. jacksonii* sp. nov. (p. 95), *L. mesibovii* sp. nov. (p. 94), *L. morrisonii* sp. nov. (p. 97), *Papuacedrus, P. australis* sp. nov. (p. 97), palaeobotany (Oligocene–Miocene), ill.].

06124   Horaninov, P. F. (1847). Characteres essentiales familiarum ac tribuum regni vegetabilis... Petropoli (St. Petersburg). [Cupressaceae, fam. descr., p. 26; see also F. G. Bartling & H. L. Wendland, 1830 and F. W. Neger, 1907].

06644   Jagel, A. (2001). Morphologische und morphogenetische Untersuchungen zur Systematik und Evolution der Cupressaceae s.l. (Zypressengewächse). Dissertation, Fakultät Biologie, Ruhr-Universität Bochum. [Cupressaceae, Taxodiaceae, Cupressoideae Rich. ex Sweet, Callitroideae Saxton, taxodioid Cupressaceae, ovulate cone ontogeny, SEM and other ill., diagrams].

06814   Jiang, Z. P. & H. R. Wang (1997). Taxonomy of the Cupressaceae: subfamilies, tribes and genera. Acta Phytotax. Sin. 35 (3): 236–248. (Chin., Eng. summ.). [recognition of 2 subfam.: Cupressoideae and Callitroideae, with resp. 4 and 3 tribes, in Cupressaceae s. str.].

07290   Klemm, P. M. (1886). Ueber den Bau der beblätterten Zweige der Cupressineen. Inaugural-Dissertation Univ. Leipzig. Berlin. (46 pp., 4 pl.).

07697   Kurmann, M. H. (1994). Pollen morphology and ultrastructure in the Cupressaceae. Acta Bot. Gallica 141 (2): 141–147. (ill.).

07710a Kusumi, J., Y. Tsumura, H. Yoshimaru & H. Tachida (2000). Phylogenetic relationships in Taxodiaceae and Cupressaceae *sensu stricto* based on *mat*K gene, *chl*L gene, *trn*L-*trn*-F IGS region, and *trn*L intron sequences. Amer. J. Bot. 87 (10): 1480–1488. [all spp. of Taxodiaceae, 5 Japanese spp. of Cupressaceae analysed, DNA].

08142   Lemoine-Sebastian, C. (1969). La vascularisation du complexe bractée-écaille dans le cône femelle des Cupressacées. Bot. Rhedonica, sér. A. 7: 1–27.

08162k Li, C. X. & Q. Yang (2002). Divergence time estimates for major lineages of Cupressaceae (s. l.). Acta Phytotax. Sin. 40 (4): 323–333. (Chin., Eng. summ.). [DNA 'molecular clock ' phylogeny].

08170   Li, H. L. (1953). A Reclassification of *Libocedrus* and Cupressaceae. J. Arnold Arbor. 34: 17–34, pl. I–II. [*Heyderia formosana* (Florin) comb. nov., *H. macrolepis* (S. Kurz) comb. nov. = *Calocedrus* spp., p. 23, *Papuacedrus*, gen. nov., p. 25, *P. arfakensis* (Gibbs), *P. papuana* (F. Mueller), *P. torricellensis* (Schlechter ex Lauterb.), comb. nov., p. 25].

08227e Li, L. C. (1998). A cytotaxonomic study on the family Cupressaceae. Acta Bot. Yunnan. 20 (2): 197–203. (Chin., Eng. summ.). [Cupressaceae s. str., karyology, all spp. 2n = 22].

11026   Oladele, F. A. (1983). Inner surface sculpture patterns of cuticles in Cupressaceae. Canad. J. Bot. 61: 1222–1231.

11027   Oladele, F. A. (1983). Scanning electron microscope study of stomatal-complex configuration in Cupressaceae. Canad. J. Bot. 61: 1232–1240.

11430   Peirce, A. S. (1937). Systematic anatomy of the woods of the Cupressaceae. Trop. Woods 49: 5–21, f. 1–16.

11540   Pillai, A. (1963). Structure of the shoot apex in some Cupressaceae. Phyton (Horn) 10 (3–4): 261–271. (ill.).

11740   Propach-Gieseler, C. (1936). Zur Morphologie und Entwicklungsgeschichte der weiblichen Blütenzapfen der Cupressaceen. [Die Blüten der Coniferen II (ed. M. Hirmer)] Bibl. Bot. 114 (2): 1–56. (ill.). [Cupressaceae].

12200   Rouane, P. (1973). Étude comparée de la répartition des ramifications au cours de l'ontogenèse de quelques Cupressacées. Trav. Lab. Forest. Toulouse, T. 1, vol. 9, art. 3: 1–277.

14159   Tarbaeva, V. M. (1992). (Anatomo-morphological structure of seeds of Taxodiaceae and Cupressaceae species). Series of pre-prints "Scientific Reports" issue 279 (18 p.). Komi Science Centre of the Ural Division of the Russian Acad. Sci. (Russ.). [several spp., mainly Asiaic, ill.].

14159a   Targioni-Tozzetti, O. (1810). Observationum botanicarum decas tertia, quarta, & quinta. Tabulae IX, X, XI. [earliest observations of early conifer cone development in Cupressaceae]. Repr. in: Ann. Mus. Imp. Fis. Stor. Nat. Firenze 2 (1), t. 9–11.

14693   Vaudois, N. & C. Privé (1971). Révision des bois fossiles de Cupressaceae. Palaeontographica, B 134: 61–86. (ill.).

PHYLLOCLADACEAE
(general titles)

00881   Bessey, C. E. (1907). A synopsis of plant phyla... Nebraska Univ. Stud. 7: 275–373. [Phyllocladaceae fam. nov., p. 325].

00957   Bobrov, A. V. F. C., A. P. Melikyan & E. Y. Yembaturova (1999). Seed morphology, anatomy and ultrastructure of *Phyllocladus* L. C. and A. Rich. ex Mirb. (Phyllocladaceae (Pilg.) Bessey) in connection with the generic system and phylogeny. Ann. Bot. (UK) 83 (6): 601–618. (ill.). [authors propose family rank in Taxales].

07080   Keng, H. (1963–79). A monograph of the genus *Phyllocladus* (coniferae). Taipeh. [proposes acceptance of family rank].

11754a   Quinn, C. J. (1987). The Phyllocladaceae Keng – a critique. Taxon 36 (3): 559–565. [includes genus in Podocarpaceae].

14433c   Tomlinson, P. B., J. E. Braggins & J. A. Rattenbury (1997). Contrasted pollen capture mechanisms in Phyllocladaceae and certain Podocarpaceae (Coniferales). Amer. J. Bot. 84 (2): 214–223. [*Phyllocladus*, Podocarpaceae, different mechanisms].

14433e   Tomlinson, P. B. & T. Takaso (1989). Developmental shoot morphology in *Phyllocladus* (Podocarpaceae). Bot. J. Linn. Soc. 99: 223–248. [no inferences on systematics].

14433f   Tomlinson, P. B., T. Takaso & J. A. Rattenbury (1989). Cone and ovule ontogeny in *Phyllocladus* (Podocarpaceae). Bot. J. Linn. Soc. 99: 209–221. [no inferences on systematics].

PINACEAE
(general titles)

00156   Andersen, Ø. M. (1992). Anthocyanins from reproductive structures in Pinaceae. Biochem. Syst. Ecol. 20: 145–148. [Pinaceae, *Abies, Larix, Picea, Pinus*].

00450   Bailey, I. W. (1909). The structure of wood in the Pinaceae. Bot. Gaz. (Crawfordsville) 48: 47–55, pl. 5.

00560   Bannan, M. W. (1936). Vertical resin ducts in the secondary wood of the Abietineae. New Phytol. 35: 11–46, pl. 1. [Pinaceae].

01010   Bobrov, E. G. (1983). On intergeneric hybridization in the family Pinaceae. Bot. Žurn. (Moscow & Leningrad) 68: 857–865. (Russ.).

01170   Boureau, E. (1939). Recherches anatomiques et expérimentales sur l'ontogénie des plantules des Pinacées et ses rapports avec la phylogénie. Ann. Sci. Nat. Bot., sér. 11, 1: 1–219.

01350   Buchholz, J. T. (1920). Polyembryony among Abietineae. Bot. Gaz. (Crawfordsville) 69: 153–167. [Pinaceae].

01460   Budkevich, E. V. (1961). Wood of Pinaceae: Anatomical structure and keys for identification of genera and species. Izd. Akad. Nauk SSSR, Leningrad. (Russ.).

01550   Campo-Duplan, M. van (1948). Considérations biometriques sur les grains de pollen des Abietinées. Trav. Lab. Forest. Toulouse, T. 1, vol. 4, art. 18. [Pinaceae, *Abies, Pinus, Picea, Cedrus, Larix, Pseudotsuga*].

01560   Campo-Duplan, M. van (1950). Recherches sur la phylogénie des Abiétinées d'après leurs grains de pollen. Trav. Lab. Forest. Toulouse, T. 2, sect. 1, vol. 4, art. 1: 1–182. [Pinaceae, ill.].

01831   Caspary, R. (1860). Abietinearum floris feminei structura morphologica. Ann. Sci. Nat. Bot., sér. 4, 14: 200–209. [Pinaceae, ill.].

01842 Čelakovsky, L. J. (1882). Zur Kritik der Ansichten von der Fruchtschuppe der Abietineen. Abh. Königl. Böhm. Ges. Wiss. (Prag) VI, Band 2. [Pinaceae; see also A. W. Eichler, 1881 and 1882].

02471 Creber, G. T. (1967). Notes on some petrified cones of the Pinaceae from the Cretaceous. Proc. Linn. Soc. London 178 (2): 147–152.

03140 Dubois, G. (1938). Pollen et phylogénie chez les Abiétinées. Bull. Soc. Hist. Nat. Toulouse 72: 125–145, f. 1–39. [Pinaceae].

03380 Eichler, A. W. (1882). Entgegnung auf Herrn L. Čelakowsky's Kritik meiner Ansicht über die Fruchtschuppe der Abietineen. Sitzungsber. Ges. Naturf. Freunde Berlin 20 Juni 1882, pp. 77–92, figs. [Pinaceae; see also L. J. Čelakovsky, 1882].

03540 Engelmann, G. (1880). Abietinae. in: S. Watson. Botany of California 2: 117–128. Univ. Press, Cambridge, Mass. [Pinaceae, *Pseudotsuga douglasii* var. *macrocarpa* (Vasey) comb. nov. = *P. macrocarpa*, p. 120, *Tsuga* [Sect.] *Hesperopeuce*, sect. nov., p. 121, *Tsuga pattoniana* (Andr. Murray) comb. nov. (the latter also in Gard. Chron., ser. 3, 2: 756, 1879), *Pinus contorta* var. *murrayana* (Andr. Murray) comb. nov., p. 126, *P. ponderosa* var. *scopulorum* var. nov., p. 126, *P insignis* var. *binata* var. nov. = *P. radiata* var. *binata* (Engelm.) Lemmon, p. 127–128].

03690 Farjon, A. (1990). Pinaceae: drawings and descriptions of the genera *Abies, Cedrus, Pseudolarix, Keteleeria, Nothotsuga, Tsuga, Cathaya, Pseudotsuga, Larix* and *Picea*. (Regnum Veg. Vol. 121), Königstein, Fed. Rep. Germany. (ill., maps, keys, tables, dendrogram). [*Abies forrestii* var. *georgei* (Orr), *A. sachalinensis* var. *gracilis* (Komarov), *A. sibirica* ssp. *semenovii* (B.A. Fedtsch.), *A. vejarii* ssp. *mexicana* (Martínez), *Picea koraiensis* var. *pungsanensis* (Uyeki ex Nakai) Schmidt-Vogt ex Farjon, comb. et stat. nov.].

03695 Farjon, A. (1993). Names in current use in the Pinaceae (Gymnospermae) in the ranks of genus to variety. In: W. Greuter (ed.). NCU-2. Names in current use in the families Trichocomaceae, Cladoniaceae, Pinaceae, and Lemnaceae. (Regnum Veg. Vol. 128), Königstein, Fed. Rep. Germany. [448 names, 266 typified (60%), 38 lectotypifications in this list].

03832 Ferré, Y. de (1939). Cotylédons et évolution chez les Abiétinées. Bull. Soc. Hist. Nat. Toulouse 73: 299–314. [Pinaceae, ill.].

03840 Ferré, Y. de (1941). La place des caneaux résinifères dans les feuilles des Abiétinées. Trav. Lab. Forest. Toulouse, T. 1, vol. 3, art. 12. (also in: Bull. Soc. Hist. Nat. Toulouse 76: 199–204, 1941). [Pinaceae, ill.].

03844 Ferré, Y. de (1943). L'évolution parallèle des Taxodinées et des Abietinées. Bull. Soc. Hist. Nat. Toulouse 78: 71–83. [Pinaceae, Taxodiaceae, ill.].

03855 Ferré, Y. de (1946). Quelques anomalies de structure chez les Abiétinées. Trav. Lab. Forest. Toulouse 1, 4 (9): 1–12. (repr. from Bull. Soc. Hist. Nat. Toulouse 81, 1946). [Pinaceae, *Abies, Picea, Pinus*].

03860 Ferré, Y. de (1947). Anomalies de structure dans la famille des Abiétinées. Bull. Soc. Hist. Nat. Toulouse 81: 87–96, f. 1–16. [Pinaceae, e.g. *Abies koreana*].

03890 Ferré, Y. de (1952). Les formes de jeunesses des Abiétinées, ontogenie-phylogenie. Trav. Lab. Forest. Toulouse, T. 2, vol. 3, art. 1: 1–284. [Pinaceae, ill.].

04230 Flous, F. (1936). Classification et évolution d'une groupe d'Abiétinées. Trav. Lab. Forest. Toulouse, T. 2, vol. 2, art. 17: 1–286. [Pinaceae].

04240 Flous, F. (1937). Caractères évolutifs du cône des Abiétinées. Compt. Rend. Hebd. Séances Acad. Sci. (Paris) 204: 511–513. [Pinaceae].

04510 Frankis, M. P. (1989). Generic inter-relationships in Pinaceae. Notes Roy. Bot. Gard. Edinburgh 45 (3): 527–548. (1988 on t. p.).

04540 Franklin, J. F., T. Maeda, Y. Oshumi, M. Matsui & H. Yagi (1979). Subalpine Coniferous Forests of Central Honshu, Japan. Ecol. Monogr. 49 (3): 311–334. [Pinaceae: *Abies mariesii, A. veitchii, Picea jezoensis* var. *hondoensis, Tsuga diversifolia, Larix leptolepis = L. kaempferi*].

04771 Gaussen, H. (1965). Le polyphylétisme des Abiétacées. Compt. Rend. Hebd. Séances Acad. Sci. (Paris) 261: 5585–5590. [Pinaceae].

05166 Gugerli, F., C. Sperisen, U. Büchler, I. Brunner, S. Brodbeck, J. D. Palmer & Y. L. Qiu (2001). The evolutionary split of Pinaceae from other conifers: evidence from an intron loss and a multigene phylogeny. Mol. Phylogen. Evol. 21 (2): 167–175.

06123c Horaninov, P. F. (1834). Primae Linae Systematis Naturae, ... Class III Strobiliferae, Ordo 1. Pinaceae, Ordo 2. Zamiaceae. (pp. 45–45). Petropoli (St. Petersburg).

06271 Hu, Y. S., K. Napp-Zinn & D. Winne (1989). Comparative anatomy of seed scales of female cones of Pinaceae. Bot. Jahrb. Syst. 111 (1): 63–85. (ill., tables, key).

06661 Jain, K. K. (1976). Evolution of wood structure in Pinaceae, Israel. J. Bot. 25: 28–33.

06680 Jain, K. K. (1976). Morphology of the female cone in Pinaceae. Phytomorphology 26 (2): 189–200.

06770   Jeffrey, E. C. (1905). The comparative anatomy and phylogeny of the Coniferales. Part 2. The Abietineae. Mem. Boston Soc. Nat. Hist. 6 (1): 1–37. [Pinaceae].

07140   Khoshoo, T. N. (1959). Polyploidy in gymnosperms. Evolution 13: 24–39. [Pinaceae].

07181   Kim, Y. S., S. C. Ko & B. H. Choi (1981). Distribution atlas of plants of Korea: 4. Atlas of Pinaceae in Korea. J. Korean Plant Taxon. 11 (1–2): 53–75. (Korean, Eng. summ., maps).

07670   Kung, H. W. (1934). Pinaceae collected from Hsiaopeshan (Kirin). Contr. Inst. Bot. Natl. Acad. Beijing 2: 107–114, pl. 4–6. (Eng., Chin. summ.).

08144   Lemoine-Sebastian, C. (1975). La vascularisation de l'écaille séminale chez les Abiétacées. Bull. Soc. Bot. France 122: 225–242. [Pinaceae, ill.].

08145c   LePage, B. A. (1991). The use of the bract as a diagnostic feature in the taxonomy of the Pinaceae. Amer. J. Bot. 78: 117–118.

08145g   LePage, B. A. (2003). The evolution, biogeography and palaeoecology of the Pinaceae based on fossil and extant representatives. In: R. R. Mill (ed.). Proceedings of the Fourth International Conifer Conference. Acta Horticulturae 615: 29–52. International Society for Horticultural Science (ISHS). [*Larix, Picea, Pseudolarix, Tsuga*, ill., maps].

08227d   Li, L. C. (1995). Studies on the karyotype and phylogeny of the Pinaceae. Acta Phytotax. Sin. 33 (5): 417–432. (Chin., Eng. abstr., tables, ill.). [*Abies, Cathaya, Cedrus, Keteleeria, Larix, Picea, Pinus, Pseudolarix, Pseudotsuga, Tsuga*].

08229   Li, N. (1995). Studies on the geographic distribution, origin and dispersal of the family Pinaceae Lindl. Acta Phytotax. Sin. 33 (2): 105–130. (Chin., Eng. abstr., ill., maps). [*Abies, Cathaya, Cedrus, Keteleeria, Larix, Picea, Pinus, Pseudolarix, Pseudotsuga, Tsuga*].

08229d   Li, N., L. K. Fu & Z. D. Zhu (1996). Studies on [the] systematic[s] of the family Pinaceae Lindl. (I). Bull. Bot. Res. 16 (1): 32–45. (Chin., Eng. summ.). [classification, dendrogram, cladogram of 10 genera].

08251   Lin, J. X., Y. H. Hu & X. P. Wang (1996). The conservation status of conifers endemic to China. In: D. Hunt (ed.). Temperate trees under threat. Proceedings of an IDS symposium on the conservation status of temperate trees, University of Bonn, 30 September – 1 October 1994, pp. 73–79. International Dendrology Society. [ill. of *Cupressus gigantea*].

08421   Little, E. L., Jr. (1948). David Douglas' new species of conifers. Phytologia 2 (11): 485–490. [Pinaceae, nomenclature; see also J. R. Griffin, 1964].

08510   Little, E. L., Jr. (1956). Pinaceae of Nevada. Contributions toward a Flora of Nevada 40: 1–40. U.S.D.A. Agric. Res. Serv. Beltville, Md.

08820   Lotova, L. I. (1975). On the correlation of the anatomical features of the wood and phloem in the Pinaceae. Vestn. Moskovsk. Univ., Ser. 6, Biol. 6 (1): 41–51. (Russ.).

09320   Masters, M. T. (1879–80). Japanese conifers. Gard. Chron., n. s., 12 (1): 198–199, (2): 556, (3): 588–589, (4): 788, (5): 823 (= 1879); 13 (6): 115, (7): 212–213, (8): 233–234, (9): 275, (10): 300, (11): 363, (12[10]): 589 (= 1880). (ill). [Pinaceae: *Abies mariesii* sp. nov., 4: 788; *A. sachalinensis* (Fr. Schmidt) comb. nov., 3: 588; *A. veitchii* var. *nephrolepis* (Maxim.) comb. nov., 3: 589; *Picea ajanensis* "Fischer" (with veg. parts) = *P. brachytyla*, *P. ajanensis* var. *microsperma* (Lindl.) comb. et stat. nov., 6: 115; *P. alcockiana*,ill., 7: 212–213; *P. glehnii* (Fr. Schmidt) comb. nov., 10: 300; *P. maximowiczii* sp. nov., 11: 363].

09580   Masters, M. T. (1903–06). Chinese conifers. Gard. Chron., ser. 3, 33: 34, 66, 84, 116–117, 133, 194, 227–228 (= 1903); 39: 146–147, 212, 236, 299 (= 1906). (ill.). [Pinaceae: *Picea wilsonii* sp. nov., 33: 133; *Abies squamata* sp. nov., 39: 299; *Keteleeria evelyniana* sp. nov., 33: 194; *Picea complanata* sp. nov., 39: 146; *P. neoveitchii* sp. nov., 33: 116; *P. montigena* sp. nov., 39: 146; *Tsuga yunnanensis* "comb. nov.", 39: 236, with basion. *Abies yunnanensis* Franchet in J. Bot. (Morot) 189: 258, but see G. A. Pritzel, 1901].

09600   Masters, M. T. (1906). On the conifers of China. J. Linn. Soc., Bot. 37: 410–424. [Pinaceae: *Pinus densata* sp. nov. = *P. tabuliformis* var. *densata*, p. 416, *P. prominens* sp. nov., p. 417, *Picea purpurea* sp. nov., p. 418, *P. watsoniana* sp. nov. = *P. wilsonii*, p. 419, *P. asperata* sp. nov., p. 419, *P. aurantiaca* sp. nov., p. 420, *P. retroflexa* sp. nov., p. 420, *Abies recurvata* sp. nov., p. 423].

09611   Masters, M. T. (1907). Coniferae chinenses novae. Repert. Spec. Nov. Regni Veg. 4: 108–111. [Latin diagn. of earlier descr. new spp. of Pinaceae].

09750   Mayr, H. (1890). Monographie der Abietineen des Japanischen eiches, (Tannen, Fichten, Tsugen, Lärchen und Kiefern). München. [Pinaceae, *Abies, Picea, Tsuga, Larix, Pinus, A. umbellata* sp. nov. = *A. homolepis* var. *umbellata*, p. 34, *A. veitchii* var. *nikkoensis* var. nov. = *A. veitchii* var. *veitchii*, p. 39, *A. sachalinensis* var. *nemorensis* var. nov., et f. *typica*, p. 42, *Picea* sect. *Casicta* sect. nov., p. 44, *Picea bicolor* (Maxim.) comb. nov. = *P. alcockiana*, p. 49, *P. hondoensis* sp. nov. = *P. jezoensis* ssp. *hondoensis*, p. 51, *Larix kurilensis* sp. nov. = *L. gmelinii* var. *japonica*, p. 66, *Pinus pentaphylla* sp. nov. = *P. parviflora* var. *pentaphylla*, p. 78; under non-endemics: *Pseudolarix fortunei* nom. nov., p. 99 = *P. amabilis*].

10110   Miki, S. (1957). Pinaceae of Japan, with special reference to its remains. J. Inst. Polytechn. Osaka City Univ., ser. D., Biol. 8: 221–272, pl. 1–10, f. 1–12. [map, mainly palaeobotany].

10180   Miller, C. N., Jr. (1976). Early Evolution in the Pinaceae. Rev. Palaeobot. Palynol. 21: 101–117.

10620   Moss, E. H. (1953). Forest communities in northwestern Alberta. Canad. J. Bot. 31: 212–252. [Pinaceae, *Picea*, *Pinus*].

10750   Murray, A. (1863). The Pines and Firs of Japan. Illustrated by upwards of 200 woodcuts. pp. 1–123, f. 1–224. London. (revisions of papers in Proc. Roy. Hort. Soc. (London) 1862, 2). [Pinaceae, *Larix japonica* sp. nov. = *L. kaempferi*, p. 105].

10911   Napp-Zinn, K. & Y. S. Hu (1989). Anatomical studies on the bracts in Pinaceous female cones. III. Comparative study of (mostly Chinese) representatives of all genera. Bot. Jahrb. Syst. 110: 461–478. [Pinaceae, ill., tables].

10950   Nelson, E. J. (pseudon. Senilis) (1866). Pinaceae; being a handbook of the firs and pines. 223 pp. London. [for comment, see e.g. H. St. John & R. W. Krauss, 1954].

10961   Neustadt, M. I. (1957). Istorija Lesov i paleogeografija SSSR v golotsene. (History of Forests and palaeogeography of the USSR during the Holocene). Akad. Nauk SSSR, Inst. Geogr. Moscow, pp. 242–246, figs. 167–170. (Russ.). [recent and fossil distr. of Pinaceae].

10962   Neustadt, M. I. (1966). Present achievements in a palynological Study of the Quarternary Period in the U.S.S.R. The Paleobotanist 15 (1–2): 213–227. [Pleistocene distr. of Pinaceae, maps].

11000   Niemann, G. J. & H. H. van Genderen (1980). Chemical relationships between Pinaceae. Biochem. Syst. Ecol. 8: 237–240.

11200   Page, C. N. (1989). New and maintained genera in the Conifer families Podocarpaceae and Pinaceae. Notes Roy. Bot. Gard. Edinburgh 45 (2): 377–395 (1988 on t. p.). [Podocarpaceae: *Sundacarpus* (J. Buchholz & N. E. Gray) comb. nov., p. 378, *S. amarus* (Blume) comb. nov., p. 378, *Retrophyllum* gen. nov., p. 379, *R. comptonii* (J. Buchholz) comb. nov., p. 380, *R. minus* (Carrière) comb. nov., p. 380 ("*minor*"), *R. piresii* (Silba) comb. nov., p. 380, *R. rospigliosii* (Pilg.) comb. nov., p. 380, *R. vitiense* (Seem.) comb. nov., p. 380, *Afrocarpus* (J. Buchholz & N. E. Gray) comb. nov., p. 383, *A. dawei* (Stapf) comb. nov., p. 384, *A. falcatus* (Thunb.) comb. nov., p. 383, *A. gaussenii* (Woltz) comb. nov., p. 384, *A. gracilior* (Pilg.) comb. nov., p. 383, *A. mannii* (Hook. f.) comb. nov., p. 384, *A. usambarensis* (Pilg.) comb. nov., p. 384; Pinaceae: *Hesperotsuga*, nothogen. nov., × *H. jeffreyi* (Henry) comb. nov. = *Tsuga*

mertensiana ssp. *mertensiana* var. *jeffreyi*, p. 389, *Nothotsuga*, gen. nov., *N. longibracteata* (Cheng) comb. nov., p. 390].

11670   Prager, E. M., D. P. Fowler & A. C. Wilson (1976). Rates of evolution in conifers (Pinaceae). Evolution 30: 637–649.

11700   Price, R. A. (1989). The genera of Pinaceae in the southeastern United States. J. Arnold Arbor. 70: 247–305. [with extensive bibliography, ill.].

11720   Price, R. A., J. Olsen-Stojkovich & J. M. Lowenstein (1987). Relationships among the Genera of Pinaceae: An Immunological Comparison. Syst. Bot. 12 (1): 91–97.

12080   Ritchie, I. C. (1984). Past and present vegetation of the far northwest of Canada. Toronto. [Pinaceae].

12450   Saida, K. (1895). On Japanese Pinaceae. Bot. Mag. (Tokyo) 9: 17–19. (Japan., key).

12510   Sargent, C. S. (1902–13). Trees and shrubs, illustrations of new or little known ligneous plants... Vol. 1 + 2. Boston – New York. [*Picea morindoides* Rehder, sp. nov. = *P. spinulosa*, Vol. 1, part 2: 95, t. 48, 1903; *Pinus terthrocarpa* (Griseb.) Shaw, comb. nov. = *P. tropicalis* (basion.: *P. cubensis* var. *terthrocarpa* Griseb., Cat. Pl. Cub. Wright. 217, 1866), Vol. 1, part 3: 149, t. 75, 1903; *P. altamiranii* Shaw, sp. nov. = *P. lawsonii*, Vol. 1, part 4: 209, t. 99, 1905; *P. pringlei* Shaw, sp. nov., Vol. 1, part 4: 211, t. 100, 1905].

12590   Satake, Y. (1934). On the systematic importance of the vascular course in the cone scales of the Japanese Pinaceae, comprising the genera *Abies*, *Tsuga*, *Larix*, *Pinus* and *Picea*. I. J. Japan. Bot. 10: 703–716, f. 1–74; II. ibid. : 794–808, f. 75–131. (Japan., Eng. summ.).

13400   Sivak, J. (1975). Les caractères de diagnose des graines de pollen à ballonets. Pollen Spores 17: 349–421. (ill.) [Pinaceae, Podocarpaceae].

13600   Srivastava, L. M. (1963). Secondary phloem in the Pinaceae. (Univ. Calif. Publ. Bot. 36), Univ. of California Press, Berkeley – Los Angeles – London. (142 pp., ill.).

14156   Tarbaeva, V. M. (1990). (Anotomo-morphological seed structure of Pinaceae species). Series of pre-prints "Scientific Reports" issue 252 (24 pp.), Komi Science Centre of the Ural Division of the USSR Acad. Sci. (Russ.). [Pinaceae, 10 spp., ill., tab.].

14270   Teng, S. C. (1948). Forest geography of the east-Tibetan plateau. Bot. Bull. Acad. Sin. 2: 62–66. [Pinaceae].

14590   Ueno, J. (1958). Some palynological observations of Pinaceae. J. Inst. Polytechn. Osaka City Univ., ser. D., Biol. 9: 163–177, pl. 1–3.

14730 Viereck, L. A. (1973). Wildfire in the taiga of Alaska. J. Quartern. Res. 3 (3): 465–495. [Pinaceae, *Larix, Picea, Pinus*].

15404a Xi, Y. Z. & J. C. Ning (1995). Pollen morphology of Pinaceae and its evolutionary implications. Cathaya 7: 75–97. [10 genera studied, ill.].

15450 Yao, Q. & P. Huang (1980). A research of the seeds of genera in Pinaceae. J. Nanjing Techn. Coll. Forest. Prod. 1: 28–39. (Chin., ill., key).

PODOCARPACEAE
(general titles)

00956 Bobrov, A. V. F. C. & A. P. Melikyan (1998). Specific structures of seed coat in Podocarpaceae Endlicher, 1847 and a possibility of using them in family systematics. Bjull. Moskovsk. Obšč. Isp. Prir., Otd. Biol. (Bull. Moscow Soc. Naturalists) 103 (1): 56–62. [*Bracteocarpus* gen. nov. = *Dacrycarpus*, p. 58, *Margbensonia* gen. nov. = *Podocarpus* spp., p. 59].

01430 Buchholz, J. T. & N. E. Gray, (1948). A taxonomic revision of *Podocarpus*. I. The sections of the genus and their subdivisions, with special reference to leaf anatomy. J. Arnold Arbor. 29: 49–63. [*Podocarpus* s.l., *Podocarpus* sect. *Afrocarpus* sect. nov., p. 57; see for II–VI: J. Arnold Arbor. 29: 64–76, 117–151, J. Arnold Arbor. 32: 82–97].

01431 Buchholz, J. T. & N. E. Gray, (1948). A taxonomic revision of *Podocarpus*. II. The American species of *Podocarpus*: Section *Stachycarpus*. J. Arnold Arbor. 29: 64–76. [*Podocarpus* s.l., *Podocarpus standleyi* sp. nov. = *Prumnopitys standleyi*, p. 72].

02345 Conran, J. G., G. M. Wood, P. G. Martin, J. M. Dowd, C. J. Quinn, P. A. Gadek & R. A. Price (2000). Generic relationships within and between the gymnosperm families Podocarpaceae and Phyllocladaceae based on an analysis of the chloroplast gene *rbc*L. Austral. J. Bot. 48: 715–724. [*Phyllocladus, Podocarpus* and all genera except *Parasitaxus*, DNA, cladistics].

03047 Doweld, A. B. & J. L. Reveal (1998). Validation of new suprageneric names in Pinophyta. Phytologia 84 (5): 363–367. [Microcachrydaceae fam. nov., Saxegothaeaceae fam. nov. (validated here) = Podocarpaceae, new class and order names].

03048 Doweld, A. B. & J. L. Reveal (2001). Validation of some suprageneric names in Podocarpopsida. Novon 11 (4): 395–397. [suprageneric names for the genera *Microcachrys, Microstrobos* and *Pherosphaera* (the latter is syn. of both)].

04802 Gaussen, H. (1973). Les gymnospermes actuelles et fossiles. Fascicule 12: Les Podocarpines, Etude générale. Faculté des Sciences, Toulouse (reprint from Trav. Lab. Forest. Toulouse, 1973). (ill.). [Podocarpaceae].

04803 Gaussen, H. (1974). Les gymnospermes actuelles et fossiles. Fascicule 13: Les Podocarpines sauf les *Podocarpus*. Faculté des Sciences, Toulouse (reprint from Trav. Lab. Forest. Toulouse, 1974). (ill.). [Podocarpaceae].

04804 Gaussen, H. (1976). Les gymnospermes actuelles et fossiles. Fascicule 14: Genre *Podocarpus*, Conclusion des Podocarpines. Faculté des Sciences, Toulouse (reprint from Trav. Lab. Forest. Toulouse, 1976). (ill.). [Podocarpaceae].

04990 Gray, N. E. (1953–62). A taxonomic revision of *Podocarpus* VII–XIII. J. Arnold Arbor. 34: 67–76, 163–175; 36: 199–206; 37: 160–172; 39: 424–477; 41: 36–39; 43: 67–79. [*Podocarpus* s.l., detailed below; see also J. T. Buchholz & N. E. Gray, 1948 (I) and N. E. Gray & J. T. Buchholz, 1948–1951 (III–VI)].

04990a Gray, N. E. (1953). A taxonomic revision of *Podocarpus* VII. The African species of *Podocarpus*: section *Afrocarpus*. J. Arnold Arbor. 34: 67–76. [= *Afrocarpus* spp.].

04990b Gray, N. E. (1953). A taxonomic revision of *Podocarpus* VIII. The African species of section *Eupodocarpus*, subsections A and E. J. Arnold Arbor. 34: 163–175. [= section *Podocarpus*].

04990c Gray, N. E. (1955). A taxonomic revision of *Podocarpus* IX. The South Pacific species of section *Eupodocarpus*, subsection F. J. Arnold Arbor. 36: 199–206. [= section *Podocarpus, P. decumbens* sp. nov., p. 202, *P. decipiens* sp. nov. = *P. neriifolius*, p. 204].

04990d Gray, N. E. (1956). A taxonomic revision of *Podocarpus* X. The South Pacific species of section *Eupodocarpus*, subsection D. J. Arnold Arbor. 37: 160–172. [= section *Podocarpus*].

04990e Gray, N. E. (1958). A taxonomic revision of *Podocarpus* XI. The South Pacific species of section *Podocarpus*, subsection B. J. Arnold Arbor. 39: 424–477. [*P. gibbsiae* sp. nov., p. 429 ("*gibbsii*"), *P. novae-caledoniae* var. *colliculatus* var. nov. = *P. sylvestris*, p. 432, *P. ridleyi* (Wasscher) comb. nov., p. 435, *P. idenburgensis* sp. nov., p. 447, *P. annamiensis* sp. nov., p. 451, *P. archboldii* sp. nov., p. 452, *P. archboldii* var. *crassiramosus* var. nov., p. 453, *P. neriifolius* var. *degeneri* var. nov., p. 467, *P. macrophyllus* var. *chingii* var. nov., p. 474, ill., key].

04990f Gray, N. E. (1960). A taxonomic revision of *Podocarpus* XII. Section Microcarpus. J. Arnold Arbor. 41: 36–39. [*Podocarpus ustus* = *Parasitaxus usta*, ill.].

04990g Gray, N. E. (1962). A taxonomic revision of *Podocarpus* XIII. Section *Polypodiopsis* in the South Pacific. J. Arnold Arbor. 43: 67–79. [*Retrophyllum* spp., *Podocarpus filicifolius* sp. nov. = *Retrophyllum vitiense*, p. 74].

04990h Gray, N. E. & J. T. Buchholz (1948). A taxonomic revision of *Podocarpus*. III. The American species of *Podocarpus*: section *Polypodiopsis*. J. Arnold Arbor. 29: 117–122. (ill.). [*Podocarpus rospigliosii*].

04990i Gray, N. E. & J. T. Buchholz (1948). A taxonomic revision of *Podocarpus* IV. The American species of Section *Eupodocarpus*, subsections C and D. J. Arnold Arbor. 29: 123–151. (ill.). [*Podocarpus* sect. *Podocarpus* (sect. *Eupodocarpus* = nom. illeg.), *Podocarpus pittieri* sp. nov. = *P. salicifolius*, p. 130, *P. reichei* sp.nov. = *P. matudae*, p. 131, *P. magnifolius* sp. nov., p. 133, *P. steyermarkii* sp. nov., p. 133, *P. rusbyi* sp. nov., p. 134, *P. tepuiensis* sp. nov., p. 134, *P. trinitensis* sp. nov., p. 135, *P. pendulifolius* sp. nov., p. 138, *P. cardenasii* sp. nov. = *P. glomeratus*, p. 142].

04990j Gray, N. E. & J. T. Buchholz (1948). A taxonomic revision of *Podocarpus* V. The South Pacific species of *Podocarpus*: section *Stachycarpus*. J. Arnold Arbor. 32: 82–92. (ill.). [*Podocarpus distichus* J. Buchholz, sp. nov. = *Prumnopitys ferruginoides*, p. 89].

07055 Kelch, D. G. (1997). The phylogeny of the Podocarpaceae based on morphological evidence. Syst. Bot. 22 (1): 113–131. [Podocarpaceae s.l., *Dacrydium*, *Phyllocladus*, *Podocarpus*, *Saxegothaea*, etc.].

07056 Kelch, D. G. (1998). Phylogeny of Podocarpaceae: comparison of evidence from morphology and 18S rDNA. Amer. J. Bot. 85 (7): 986–996. [Podocarpaceae s.l., *Dacrydium*, *Phyllocladus*, *Podocarpus*, *Saxegothaea* etc.].

07965 Laubenfels, D. J. de (1962). The primitiveness of polycotyledony considered with special reference to the cotyledonary condition in Podocarpaceae. Phytomorphology 12 (3): 296–300.

07980 Laubenfels, D. J. de (1969). A revision of the Malesian and Pacific Rainforest Conifers, 1: Podocarpaceae. J. Arnold Arbor. 50 (2–3): 274–369. [*Dacrydium nausoriense* sp. nov. ("*nausoriensis*"), p. 287, *D. pectinatum* sp. nov., p. 289, *D. nidulum* sp. nov., p. 292, *D. magnum* sp. nov., p. 299, *D. spathoides* sp. nov., p. 299, *Falcatifolium* gen. nov., p. 308, *F. falciforme* (Parl.) comb. nov., p. 309, *F. taxoides* (Brongn. & Gris) comb. nov., p. 310, *F. angustum* sp. nov., p. 312, *F. papuanum* sp. nov., p. 312, *Dacrycarpus* gen. nov., p. 315, *D. imbricatus* (Blume) comb. nov., p. 317, *D. imbricatus* var. *patulus* var. nov., p. 320, *D. imbricatus* var. *robustus* var. nov., p. 323, *D. imbricatus* var. *curvulus* (Miq.) comb. nov., p. 326, *D. vieillardii* (Parl.) comb. nov., p. 326, *D. cumingii* (Parl.) comb. nov., p. 329, *D.*

kinabaluensis (Wasscher) comb. nov., p. 330, *D. cinctus* (Pilg.) comb. nov., p. 332, *D. expansus* sp. nov., p. 334, *D. compactus* (Wasscher) comb. nov., p. 336, *D. dacrydioides* (A. Rich.) comb. nov., p. 337, *Decussocarpus* gen. nov. = *Retrophyllum* (p.p.), p. 340, *D. vitiensis* (Seem.) comb. nov. = *Retrophyllum vitiense*, p. 342, *D. comptonii* (J. Buchholz) comb. nov. = *R. comptonii*, p. 344, *D. minus* (Carrière) comb. nov. = *R. minus*, p. 346, *D. rospigliosii* (Pilg.) comb. nov. = *R. rospigliosii*, p. 347, *D. wallichianus* (C. Presl) comb. nov. = *Nageia wallichiana*, p. 349, *D. motleyi* (Parl.) comb. nov. = *N. motleyi*, p. 352, *D. maximus* sp. nov. = *N. formosensis*, p. 353, *D. fleuryi* (Hickel) comb. nov. = *N. fleuryi*, p. 355, *D. nagi* (Thunb.) comb. nov. = *N. nagi*, p. 357].

11512b Pilger, R. (1916). Die Taxaceen Papuasiens. Bot. Jahrb. Syst. 54: 207–211. [Podocarpaceae, *Dacrydium*, *Dacrycarpus*, *Podocarpus*, *P. ledermannii* sp. nov., p. 210, *Phyllocladus*, *P. major* sp. nov. = *P. hypophyllus*, p. 211].

11753b Quinn, C. J. (1969). Generic boundaries in the Podocarpaceae. Proc. Linn. Soc. New South Wales 94: 166–172.

13385 Sinnott, E. W. (1913). The morphology of the reproductive structures in the Podocarpineae. Ann. Bot. (London) 27: 39–82, t. 6–7. [Podocarpaceae; seed scale in Pinaceae, 'ligule' in Araucariaceae, and epimatium in Podocarpaceae are all homologous and vestiges of an axillary shoot].

14138a Takaso, T. & P. B. Tomlinson (1992). Aspects of cone morphology and development in Podocarpaceae (Coniferales). Int. J. Plant Sci. 153: 572–588. [anatomy and ontogeny of ovulate cones, ill., LM, SEM].

14433a Tomlinson, P. B. (1994). Functional morphology of saccate pollen in conifers with special reference to Podocarpaceae. Int. J. Plant Sci. 155 (6): 699–715. [Pinaceae, Podocarpaceae, ill.].

14433b Tomlinson, P. B., J. E. Braggins & J. A. Rattenbury (1991). Pollination drop in relation to cone morphology in Podocarpaceae: a novel reproductive mechanism. Amer. J. Bot. 78: 1289–1303. (ill.).

14433d Tomlinson, P. B. & T. Takaso (2003). Seed cone structure in conifers in relation to development and pollination: a biological approach. Canad. J. Bot. 80: 1250–1273. (ill.). [Coniferales, Cupressaceae, Taxodiaceae, Taxaceae, Podocarpaceae].

15105 Wells, P. M. & R. S. Hill (1989). Leaf morphology of the imbricate-leaved Podocarpaceae. Austral. Syst. Bot. 2: 369–386. (ill.).

15172a Wilde, M. H. (1944). A new interpretation of coniferous cones I. Podocarpaceae (*Podocarpus*). Ann. Bot. (London), n.s. 8: 1–41. (ill.).

15501   Young, M. S. (1910). The morphology of the Podocarpineae. Bot. Gaz. (Crawfordsville) 50: 81–103. [Podocarpaceae s.l.].

TAXACEAE
(general titles)

01983   Chaw, S. M., H. Long, B. S. Wang, A. Zharkikh & W. H. Li (1993). The phylogenetic position of Taxaceae based on 18S rRNA sequences. J. Mol. Evol. 37: 624–630. [Coniferales, Ginkgoales, Pinaceae, Podocarpaceae, Taxaceae; cladistic analysis of limited sample of four species + outgroup (*Zamia pumila*)].

02373   Cope, E. A. (1998). Taxaceae: the genera and cultivated species. Bot. Rev. (Lancaster) 64 (4): 291–322. [*Amentotaxus*, *Austrotaxus*, *Pseudotaxus*, *Taxus*, *Torreya*].

03800   Ferguson, D. K. (1978). Some current research on fossil and recent Taxads. Rev. Palaeobot. Palynol. 26: 213–226. [Taxaceae s.l.].

04092   Florin, C. R. (1948). On the morphology and relationships of the Taxaceae. Bot. Gaz. (Crawfordsville) 110 (1): 31–39. (ill.).

04829   Gaussen, H. (1979). Les Gymnospermes actuelles et fossiles. Fasc. 15. Les Taxines. Faculté des Sciences, Toulouse (reprinted from Trav. Lab. Forest. Toulouse, 1979). [Taxaceae, *Amentotaxus*, *Austrotaxus*, *Pseudotacus*, *Taxus*].

04991e   Gray, S. M. (1821). A natural arrangement of British plants, ... (2 vols.). London. [Taxaceae fam.nov., vol. 2: 222, 226 (nom. cons.)].

11511e   Pilger, R. (1903). Taxaceae. In: A. Engler (ed.). Das Pflanzenreich IV.5 [18]. Berlin. [Taxaceae incl. Podocarpaceae s.l., *Dacrydium falciforme* (Parl.) comb. nov. = *Falcatifolium falciforme*, p. 45, *D. biforme* (Hook.) comb. nov. = *Halocarpus biformis*, p. 45, *Podocarpus nagi* (Thunb.) comb. nov. = *Nageia nagi*, p. 60, *Podocarpus harmsianus* sp. nov. = *Prumnopitys harmsiana*, p. 68, *P. usambarensis* sp.nov. = *Afrocarpus usambarensis*, p. 70, *P. gracilior* sp. nov. = *A. gracilior*, p. 71, *P. longifoliolatus* sp. nov., p. 79, *P. parlatorei* sp. nov., p. 86, *P. lambertii* var. *transiens* var. nov. = *P. transiens*, p. 86, *P. sellowii* var. *angustifolius* var. nov., p. 88, *P. urbanii* sp. nov., p. 89, *Torreya nucifera* var. *grandis* (Fortune ex Lindl.) comb. et stat. nov. = *T. grandis*, p. 107, *Taxus baccata* ssp. *cuspidata* (Siebold & Zucc.) comb. et stat. nov. = *T. cuspidata*, p. 112, *T. baccata* ssp. *cuspidata* var. *chinensis* var. nov. = *T. chinensis*, p. 112, *T. baccata* ssp. *cuspidata* var. *latifolia* Pilg. = *T. cuspidata*, p. 112, *T. baccata* ssp. *wallichiana* (Zucc.) comb. et stat. nov. = *T. wallichiana*, p. 112, *T. baccata* ssp. *canadensis* (Marshall) comb. et stat. nov. = *T. canadensis*, p. 113, *T. baccata* ssp. *floridana* (Nutt. ex Chapm.) comb. et stat. nov. = *T. floridana*, p. 113, *T. baccata* ssp. *brevifolia* (Nutt.) comb. et stat. nov. = *T. brevifolia*, p. 113, *Acmopyle* gen. nov., p. 117].

11512a   Pilger, R. (1916). Kritische Übersicht über die neuere Literatur betreffend die Familie der Taxaceae. Bot. Jahrb. Syst. 54: 1–43. [Taxaceae incl. Phyllocladaceae, Podocarpaceae, *Amentotaxus* gen. nov., *A. argotaenia* (Hance) comb. nov., p. 41].

11705   Price, R. A. (1990). The genera of Taxaceae in the southeastern United States. J. Arnold Arbor. 71: 69–91. [*Taxus, Torreya*].

13903   Stützel, T. & I. Röwekamp (1999). Female reproductive structures in Taxales. Flora 194: 145–157. [*Taxus baccata, Torreya californica, T. nucifera*, Cephalotaxaceae not studied].

14965   Wang, F. H., Z. K. Chen & Y. S. Hu (1979). On the systematic position of Taxaceae from the embryological and anatomical studies. Acta Phytotax. Sin. 17 (3): 1–7. (Chin., Eng. summ.).

15172b   Wilde, M. H. (1975). A new interpretation of microsporangiate cones in Cephalotaxaceae and Taxaceae. Phytomorphology 25: 434–450. (ill.).

TAXODIACEAE
(general titles)

01570   Campo-Duplan, M. van (1951). Recherches sur la phylogénie des Taxodiacées d'après leurs grains de pollen. Trav. Lab. Forest. Toulouse, T. 2, sect. 7, vol. 4, art. 2: 1–14. [Taxodiaceae, ill.].

01973b   Charturvedi, S. (1998). Micromorphology and vegetative anatomy of Taxodiaceae L. [sic!] Geophytology 26 (2): 43–56.

02916   Dogra, P. D. (1966). Embryogeny of the Taxodiaceae. Phytomorphology 16: 125–141.

03870   Ferré, Y. de (1947). Cotyledons et évolution chez les Taxodinées. Bull. Soc. Hist. Nat. Toulouse 82: 214–224. [Taxodiaceae, 7 genera, ill.].

04665   Gadek, P. A. & C. J. Quinn (1989). Biflavones of Taxodiaceae. Biochem. Syst. Ecol. 17 (5): 365–372. [chemotaxonomy].

04790   Gaussen, H. (1967). Les Gymnospermes actuelles et fossiles. Fasc. 9: Additions et corrections aux Abiétacées; Les Taxodiacées. Faculté des Sciences, Toulouse (reprint from Trav. Lab. Forest. Toulouse, 1967). [Taxodiaceae, ill.].

05461   Harris, T. M. (1953). Conifers of the Taxodiaceae from the Wealden Formation of Belgium. Mém. Inst. Roy. Sci. Nat. Belgique 126: 1–44. [palaeobotany, ill.].

05463   Harris, T. M. (1979). The Yorkshire Jurassic Flora V: Coniferales. British Museum (Natural History), London. (Ill.). [including a system of 8 form genera for foliage remains of Mesozoic conifers, mostly Taxodiaceae; see also Stewart & Rothwell, ed. 2: 422–423 (1993)].

05860 Hida, M. (1957). The comparative study of Taxodiaceae from the standpoint of the development of the cone scale. Bot. Mag. (Tokyo) 70: 44–51. (Japan., Eng. summ.) [classification].

08133 Lemoine-Sebastian, C. (1968). La vascularisation du complexe bractée-écaille chez les Taxodiacées. Trav. Lab. Forest. Toulouse, T. 1, vol. 7, art. 1: 1–22. [Taxodiaceae, ill.].

08141 Lemoine-Sebastian, C. (1968). Sexualité, strobiles prolifères et hermaphrodites chez les Taxodiacées. Bot. Rhedonica, sér. A. 5: 2–19. [Taxodiaceae, ill.].

08226a Li, L. C. (1989). Studies on the cytotaxonomy and systematic evolution of Taxodiaceae Warming. Acta Bot. Yunnan. 11 (2): 113–131. (Chin., Eng. summ.). [chromosome numbers].

08227 Li, L. C. (1990). Two evolutionary lines of Taxodiaceae. Acta Phytotax. Sin. 28 (1): 1–9. (Chin., Eng. summ.). [cytology, karyotype evolution].

08930 Maekawa, F. (1954). Phylogenetic considerations on conifer taxonomy. (1). J. Japan. Bot. 29: 307–313, f. 1–3. (Japan.). [Taxaceae, Taxodiaceae, *Sciadopitys*, palynology, evolution (not phylogeny)].

11420 Peirce, A. S. (1936). Anatomical inter-relationships of the Taxodiaceae. Trop. Woods 46: 1–15, f. 1–12. [wood anatomy].

12580 Satake, Y. (1934). On the systematic importance of the course of vascular bundles in the cone scales of the Japanese Taxodiaceae (preliminary report). Bot. Mag. (Tokyo) 48: 186–205, f. 1–4. (Japan., Eng. summ.). [*Sciadopitys, Taiwania, Cryptomeria, Cunninghamia*].

12755 Schlarbaum, S. E. & T. Tsuchiya (1984). Cytotaxonomy and phylogeny in certain species of Taxodiaceae. Plant Syst. Evol. 147: 29–54. [*Cryptomeria, Cunninghamia, Metasequoia, Sciadopitys, Sequoia, Sequoiadendron, Taiwania, Taxodium*].

13593 Srinivasan, V. & E. M. Friis (1989). Taxodiaceous conifers from the Upper Cretaceous of Sweden. Biol. Skr. 35. (57 pp., tables, plates). [Cupressaceae s.l., Taxodiaceae, *Elatidopsis* gen. nov., fossil and extant genera].

15503 Yu, Y. F. (1994). Taxonomic studies on the family Taxodiaceae. Bull. Bot. Res. (Harbin) 14 (4): 369–382. [*Athrotaxis, Cryptomeria, Cunninghamia lanceolata, Glyptostrobus, Metasequoia, Sequoia, Sequoiadendron, Taiwania cryptomerioides, Taxodium*].

15504 Yu, Y. F. (1995). Origin, evolution and distribution of the Taxodiaceae. Acta Phytotax. Sin. 33 (4): 362–389. (Chin., Eng. summ., maps). [*Athrotaxis, Cryptomeria, Cunninghamia, Glyptostrobus, Metasequoia, Sequoia, Sequoiadendron, Taiwania, Taxodium*].

15505 Yu, Y. F. & L. K. Fu (1996). Phylogenetic analysis of the family Taxodiaceae. Acta Phytotax. Sin. 34 (2): 124–141. (Chin., Eng. summ.). [*Athrotaxis, Cryptomeria, Cunninghamia, Glyptostrobus, Metasequoia, Sequoia, Sequoiadendron, Taiwania, Taxodium, Sciadopitys* = outgroup, subdivision into five tribes].

TAXA BELOW FAMILY RANK

00010a Abdoun, F. & M. Beddiaf (2002). *Cupressus dupreziana* A. Camus: répartition, dépérissement et régénération au Tassili n'Ajjer, Sahara central. C. R. Biologies 325: 617–627. (ill., maps).

00011 Abrams, L. R. (1919). [*Cupressus nevadensis* sp. nov.) in: Torreya 19: 92. [= *C. arizonica* var. *nevadensis*; not seen].

00020 Adamovich, E. I. (1948). The Siberian pine (*Pinus sibirica*) an ornamental, fruit-bearing, and resin-producing tree. Trudy Molotov Sel'skohoz. Inst. 12: 203–247, f. 1–31. (Russ.). [monographic paper].

00030 Adams, R. (1985). Distribution of *Callitris* in Victoria and some relic populations close to Melbourne. Victoria Naturalist 102 (2): 48–51.

00033 Adams, R. & D. Simmons (1987). A chemosystematic study of *Callitris* (Cupressaceae) in south-eastern Australia using volatile oils. Austral. Forest Res. 17: 113–125. [*C. columellaris, C. endlicheri, C. preissii, C. rhomboidea, C. verrucosa*].

00040 Adams, R. P. (1972). Chemosystematic and numerical studies of natural populations of *Juniperus pinchotii* Sudw. Taxon 21: 407–427.

00050 Adams, R. P. (1973). Re-evaluation of the biological status of *Juniperus deppeana* var. *sperryi* Correll. Brittonia 25 (3): 284–289. (ill.). [red. to forma].

00060 Adams, R. P. (1975). Numerical-chemosystematic studies of infraspecific variation in *Juniperus pinchotii*. Biochem. Syst. Ecol. 3: 71–74.

00061 Adams, R. P. (1983). Infraspecific terpenoid variation in *Juniperus scopulorum*: Evidence for Pleistocene refugia and recolonization in western North America. Taxon 32 (1): 30–46. [distr. maps, *Juniperus* spp.].

00062 Adams, R. P. (1983). The junipers (*Juniperus*; Cupressaceae) of Hispaniola: comparisons with other Caribbean species and among collections from Hispaniola. Moscosoa 2 (1): 77–89. (map, tables, figs.). [*Juniperus bermudiana, J. ekmanii, J. gracilior, J. lucayana, J. urbaniana*, terpene chemistry, taxonomy].

00070 Adams, R. P. (1986). Geographic variation in *Juniperus silicicola* and *J. virginiana* of the southeastern United States: multivariate analyses of morphology and terpenoids. Taxon 35 (1): 61–75. [*J. virginiana* var. *silicicola* (Small) Silba accepted].

00071 Adams, R. P. (1989). Biogeography and evolution of the junipers of the West Indies. in: Charles A. Woods (ed.). Biogeography of the West Indies, pp. 167–190. Gainsville, Fla. (map, tables, figs.). [*Juniperus bermudiana, J. ekmanii, J. gracilior, J. lucayana, J. saxicola, J. virginiana* var. *silicicola, J. urbaniana*, terpene chemistry, taxonomy].

00072 Adams, R. P. (1990). *Juniperus procera* of East Africa: Volatile leaf oil composition and putative relationship to *J. excelsa*. Biochem. Syst. Ecol. 18 (4): 207–210.

00072a Adams, R. P. (1993). Nomenclatural note: *Juniperus coahuilensis* (Martínez) Gaussen ex R. P. Adams. Phytologia 74 (5): 413.

00073 Adams, R. P. (1994). Geographic variation and systematics of monospermous *Juniperus* (Cupressaceae) from the Chihuahua Desert based on RAPD's and terpenes. Biochem. Syst. Ecol. 22 (7): 699–710. [*J. angosturana* nom. nov., p. 704, *J. coahuilensis, J. coahuilensis* var. *arizonica* var. nov., p. 708, *J. erythrocarpa, J. monosperma* var. *gracilis, J. pinchotii*].

00074 Adams, R. P. (1995). Revisionary study of Caribbean species of *Juniperus* (Cupressaceae). Phytologia 78 (2): 134–150. [*J. barbadensis, J. barbadensis* var. *lucayana* (Britton) comb. nov., p. 145, *J. gracilior, J. gracilior* var. *ekmanii* (Florin) comb. nov., p. 144, *J. gracilior* var. *urbaniana* (Pilger et Ekman) comb. nov., p. 144, *J. saxicola, J. bermudiana*, terpenes].

00075 Adams, R. P. (1998). The leaf essential oils and chemotaxonomy of *Juniperus* sect. *Juniperus*. Biochem. Syst. Ecol. 26: 637–645. [*Juniperus brevifolia, J. cedrus, J. communis, J. conferta, J. formosana, J. navicularis = J. oxycedrus* ssp. *transtagana, J. oblonga = J. communis, J. oxycedrus, J. rigida, J. sibirica = J. communis* var. *saxatilis*].

00075a Adams, R. P. (1999). Systematics of multi-seeded eastern hemisphere *Juniperus* based on leaf essential oils and RAPD DNA fingerprinting. Biochem. Syst. Ecol. 27: 709–725. [*Juniperus* ssp., chemotaxonomy, DNA-PCR analysis, *Juniperus erectopatens* (W. C. Cheng & L. K. Fu) R. P. Adams, comb. nov.].

00075b Adams, R. P. (2000). Systematics of smooth leaf margin *Juniperus* of the western hemisphere based on leaf essential oils and RAPD DNA finger printing. Biochem. Syst. Ecol. 28: 149–162. [*Juniperus* spp., chemotaxonomy, DNA-PCR analysis, *J. mucronata* sp. nov., p. 158].

00075c Adams, R. P. (2000). Systematics of *Juniperus* section *Juniperus* based on leaf essential oils and random amplified polymorphic DNAs (RAPDs). Biochem. Syst. Ecol. 28: 515–528. [Recognises as full species *J. oblonga, J. badia, J. macrocarpa, J. navicularis, J. sibirica, J. lutchuensis* on the basis of these data].

00075d Adams, R. P. (2000). Systematics of the one seeded *Juniperus* of the eastern hemisphere based on leaf essential oils and random amplified polymorphic DNAs (RAPDs). Biochem. Syst. Ecol. 28: 529–543. [Recognises as full species *J. carinata* comb. et stat. nov., *J. coxii, J. microsperma* comb. et stat. nov., *J. morrisonicola*, and *J. indica* and *J. wallichiana* as distinct species on the basis of these data].

00075e Adams, R. P. (2000). The serrate leaf margined *Juniperus* (section *Sabina*) of the western hemisphere: systematics and evolution based on leaf essential oils and Random Amplified Polymorphic DNAs (RAPDs). Biochem. Syst. Ecol. 28: 975–989. [17 spp. analysed].

00075f Adams, R. P. (2001). Geographic variation in leaf essential oils and RAPDs of *Juniperus polycarpos* K. Koch in central Asia. Biochem. Syst. Ecol. 29: 609–619. [*J. excelsa* ssp. *polycarpos, J. seravschanica, J. turcomanica, J. procera*, in central Asia *J. polycarpos* recognised].

00075j Adams, R. P., J. Altarejos, C. Fernandez & A. Camacho (1999). The leaf essential oils and taxonomy of *Juniperus oxycedrus* L. subsp. *oxycedrus*, subsp. *badia* (H. Gay) Debeaux, and subsp. *macrocarpa* (Sibth. et Sm.) Ball. J. Essent. Oil Res. 11 (2): 167–172. [chemotaxonomy].

00075m Adams, R. P., A. D. Dembitsky & S. Shatar (1998). The leaf essential oils and taxonomy of *Juniperus centrasiatica* Kom., *J. jarkendensis* Kom., *J. pseudosabina* Fisch. & Mey. and Ave Lall., *J. sabina* L. and *J. turkestanica* Kom. from Central Asia. J. Essent. Oil. Res. 10 (5): 489–496. (maps).

00076 Adams, R. P. & T. Demeke (1993). Systematic relationships in *Juniperus* based on random amplified polymorphic DNAs (RAPDs). Taxon 42 (3): 553–571. [*Juniperus* sect. *Juniperus*, sect. *Caryocedrus*, sect. *Sabina*, analysis of 44 taxa from N America and Eurasia; analysis of some Central Asian taxa based on misidentified plants in cultivation, see Adams & Turuspekov, 1998].

00076a Adams, R. P. & A. Hagerman (1976). A comparison of the volatile oils of mature versus long leaves of *Juniperus scopulorum*: chemosystematic significance. Biochem. Syst. Ecol. 4 (2): 75–79.

00077 Adams, R. P. & L. Hogge (1983). Chemosystematic studies of the Caribbean junipers based on their volatile oils. Biochem. Syst. Ecol. 11 (2): 85–89. [*Juniperus barbadensis, J. bermudiana, J. ekmanii, J. gracilior, J. lucayana, J. silicicola, J. virginiana*].

00077a Adams, R. P., C. F. Hsieh, J. Murata & R. N. Pandey (2002). Systematics of *Juniperus* from eastern Asia based on Random Amplified Polymorphic DNAs (RAPDs). Biochem. Syst. Ecol. 30: 231–241. [*Juniperus chinensis* var. *taiwanensis*, var. nov., *J. formosana* var. *mairei* comb. nov.].

00078 Adams, R. P., C. E. Jarvis, V. Slane & T. A. Zanoni (1987). Typification of *Juniperus barbadensis* L. and *J. bermudiana* L. and rediscovery of *J. barbadensis* from St. Lucia, BWI (Cupressaceae). Taxon 36 (2): 441–445. [ill. of neotype and circumstantial material p. 442].

00079 Adams, R. P. & J. R. Kistler (1991). Hybridization between *Juniperus erythrocarpa* Cory and *Juniperus pinchotii* Sudworth in the Chisos Mountains, Texas. Southw. Natlst. 36 (3): 295–301.

00079a Adams, R. P. & R. N. Pandey (2003). Analysis of *Juniperus communis* and its varieties based on DNA fingerprinting. Biochem. Syst. Ecol. 31 (11): 1271–1278.

00079b Adams, R. P., [R.] N. Pandey, S. Rezzi & J. Casanova (2002). Geographic variation in Random Amplified Polymorphic DNAs (RAPDs) of *Juniperus phoenicea, J. p.* var. *canariensis, J. p.* subsp. *eumediterranea*, and *J. p.* var. *turbinata*. Biochem. Syst. Ecol. 30: 223–229. [*J. phoenicea* subsp. *eumediterranea* (nom. illeg.) = *J. phoenicea* subsp. *phoenicea*].

00079c Adams, R. P., R. N. Pandey, J. W. Leverenz, N. Dignard, K. Hoegh & T. Thorfinnsson (2003). Pan-Arctic variation in *Juniperus communis*: historical biogeography based on DNA fingerprinting. Biochem. Syst. Ecol. 31 (2): 181–192. (maps).

00080 Adams, R. P. & B. L. Turner (1970). Chemosystematic and numerical studies of natural populations of *Juniperus ashei* Buch. Taxon 19: 728–751.

00085 Adams, R. P. & Y. Turuspekov (1998). Taxonomic reassessment of some Central Asian and Himalayan scale-leaved taxa of *Juniperus* (Cupressaceae) supported by random amplification of polymorphic DNA. Taxon 47 (1): 75–83. [*Juniperus centrasiatica, J. turkestanica = J. pseudosabina, J. indica*, RAPD analysis using PCO analysis].

00090 Adams, R. P. & T. A. Zanoni (1979). The distribution, synonymy and taxonomy of the junipers of southwestern United States and northern Mexico. Southw. Naturalist 24 (2): 323–329. [*Juniperus* spp.)

00090a Adams, R. P., T. A. Zanoni, A. Lara, F. Barrero & L. G. Cool (1997). Comparisons among *Cupressus arizonica* Greene, *C. benthamii* Endl., *C. lindleyi* Klotzsch ex Endl. and *C. lusitanica* Mill. Using leaf essential oils and DNA fingerprinting. J. Essent. Oil Res. 9: 303–309.

00091 Adams, R. P., T. A. Zanoni, E. von Rudloff & L. Hogge (1981). The South-western U.S.A. and Northern Mexico One-seed Junipers: their volatile oils and evolution. Biochem. Syst. Ecol. 9 (2–3): 93–96. [*Juniperus erythrocarpa, J. monosperma, J. pinchotii*].

00092 Adams, R. P. & T. A. Zanoni (1993). *Juniperus monticola* (Cupressaceae) revisited. Taxon 42 (1): 85–86. [*J. monticola* f. *compacta*, J. monticola f. *orizabensis*].

00100 Adolphi, K. (1980). Zur Unterscheidung von *Tsuga canadensis* (L.) Carr. und *Tsuga heterophylla* (Raf.) Sarg. Göttinger Florist. Rundbriefe 13 (4): 90–91.

00106 Afzal-Raii, Z. & R. S. Dodd (1994). Biometrical variability of foliage and cone characters in *Cupressus bakeri* (Cupressaceae). Plant Syst. Evol. 192: 151–164.

00110 Aguilar, G. J. I. (1961). Pinos de Guatemala. 3rd. Ed. Ministerio de Agric., Direcc. Gen. Forestal. Guatemala City. [*Pinus* spp.; ed. 1 (typescript) issued 1953, on p. 17: *Pinus quichensis* Aguilar, cited as publ. in Catalógo Mus. Nac. Ciencias Nat. La Aurora (1942): 64; no seen, but without Latin descr.?, not in IK; descr. in 1953 issue in Spanish = *P. montezumae*].

00115 Ahuja, M. R. & D. B. Neale (2002). Origins of polyploidy in Coast Redwood (*Sequoia sempervirens* (D. Don) Endl.) and relationship of Coast Redwood to other genera of Taxodiaceae. Silvae Genet. 51 (2–3): 93–100. [Cupressaceae s.l., Taxodiaceae, cytology/karyology, palaeobotany, phylogeny].

00130 Aldén, B. (1987). Taxonomy and geography of the genus *Picea* Int. Dendrol. Soc. Yearb. 1986: 85–96.

00140 Alexander, R. R. (1958). Silvical characteristics of Engelmann spruce. U.S. Forest Serv., Rocky Mt. Forest and Range Exp. Stat. Paper 31. [*Picea engelmannii*].

00141 Aleksandrovsky, E. S. (1972). Biology of blooming and fruiting of *Juniperus turcomanica* Fedtsch. Lesovedenie 1972 (3): 76–84. (Russ., Eng. summ.). [*J. excelsa* ssp. *polycarpos*].

00146 Ali, I. F., D. B. Neale & K. A. Marshall (1991). Chloroplast DNA restriction fragment length polymorphism in *Sequoia sempervirens* (D. Don) Endl., *Pseudotsuga menziesii* (Mirb.) Franco, *Calocedrus decurrens* Torr., and *Pinus taeda* L. Theor. Appl. Genet. 81: 83–89.

00150   Allen, G. S. & J. N. Owens (1972). The life history of Douglas-fir. Information Canada, Ottawa. (139 p., ill.). [*Pseudotsuga menziesii*].

00150c   Allnutt, T. R., P. Thomas, A.C. Newton & M. F. Gardner (1998). Genetic variation in *Fitzroya cupressoides* cultivated in the British Isles, assessed using RAPDs. Edinburgh J. Bot. 55 (3): 329–341.

00150d   Allnutt, T. R., A. C. Newton, A. Lara, J. J. Armesto, A. Premoli, R. Vergara & M. Gardner (1999). Genetic variation in *Fitzroya cupressoides* (alerce), a threatened South American conifer. Mol. Ecol. 8: 975–987.

00151   Alvin, K. L. (1960). Further conifers of the Pinaceae from the Wealden formation of Belgium. Institut Royal des Sciences Naturelles de Belgique; Mémoires 146. Bruxelles. [Pinaceae, *Pinus*, palaeobotany, ill.].

00151a   Alvin, K. L. (1977). The conifers *Frenelopsis* and *Manica* in the Cretaceous of Portugal. Palaeontology 20: 387–404. [Cheirolepidiaceae, *Frenelopsis* Schenk, *Manica* J. Watson = *Pseudofrenelopsis* Nathorst, palaeobotany, ill.].

00152   Alvin, K. L. (1983). Reconstruction of a Lower Cretaceous conifer. Bot. J. Linn. Soc. 86: 169–176. [cupressoid/taxodioid fossil conifer, Cheirolepidiaceae, *Pseudofrenelopsis parceramosa*, ill.].

00153   Alvin, K. L. & M. C. Boulter (1974). A controlled method of comparative study for taxodiaceous leaf cuticles. Bot. J. Linn. Soc. 69: 277–286, pl. 1–5. [SEM comparisons of *Athrotaxis*, *Cryptomeria*, *Cunninghamia*, *Glyptostrobus*, *Sequoia*, *Sequoiadendron* and *Taxodium*].

00154   Alvin, K. L. & J. J. C. Pais (1978). A *Frenelopsis* with opposite decussate leaves from the Lower Cretaceous of Portugal. Palaeontology 21: 873–879. [cupressoid fossil conifer, Cheirolepidiaceae, *Frenelopsis ramosissima*].

00156a   Anderson, H. M. (1978). *Podozamites* and associated cones and scales from the Upper Triassic Molteno Formation, Karoo Basin, South Africa. Palaeontol. Africana 21: 57–77. (ill.) [Araucariaceae, palaeobotany].

00156b   Anderson, K. B. & B. A. LePage (1995). Analysis of fossil resins from Axel Heiberg Island, Canadian Arctic. In: K. B. Anderson & J. C. Crelling (eds.). Amber, resinite, and fossil resins. American Chemical Society Symposium Series 617: 170–192. [*Metasequoia*, *Pseudolarix*, *Pinus* sp., palaeobotany, chemistry].

00157   Anderson, R. H. (1968). The trees of New South Wales. 4th ed., Sidney. [*Callitris columellaris*, *C. endlicheri*, p. 16, *C. muelleri*, p. 121–122, key, p. 366–367, ill. of *C. muelleri*].

00158   Andersson, E. (1965). Cone and seed studies in Norway spruce. Stud. Forest. Suec. 23: 1–214. [*Picea abies*].

00160   André, E. (1900). Les *Keteleeria*. Rev. Hort. 1900: 201–205, f. 97–100. [*K. fortunei*].

00170   Andresen, J. W. (1964). The taxonomic status of *Pinus chiapensis*. Phytologia 10: 417–421. [*Pinus chiapensis* (Martínez) comb. nov., p. 417].

00180   Andresen, J. W. (1966). A multivariate analysis of the *Pinus chiapensis – monticola – strobus* phylad. Rhodora 68: 1–24.

00190   Andresen, J. W. & J. H. Beaman (1961). A new species of *Pinus* from Mexico. J. Arnold Arbor. 42: 437–441. [*P. culminicola* sp. nov., p. 437].

00200   Andresen, J. W. & R. J. Steinhoff (1971). The taxonomy of *Pinus flexilis* and *P. strobiformis*. Phytologia 22 (2): 57–70.

00200a   Anguinagalde, I., F. Llorente & G. Benito (1997). Relationships among five populations of European Black pine (*Pinus nigra* Arn.) Using morphometric and isozyme markers. Silvae Genet. 46 (1): 1–5. [*Pinus nigra* ssp. *laricio*, *P. nigra* ssp. *nigra*, *P. nigra* ssp. *salzmannii*, genetic variation].

00200b   Anonymous (1914). Decades kewenses LXXXI–LXXXII. Bull. Misc. Inf. R.B.G. Kew 1914: 323–332. [*Agathis flavescens* Ridl., p. 332].

00201   Anonymous (1950). Native trees of Canada. Bulletin 61. 4th Ed. Part 1. Coniferous trees. (pp. 1–86). Dept. of Resources and Development, Forestry Branch. Ottawa. [Pinaceae, Cupressaceae, photographs, maps].

00202   Anonymous (1957). Forest Trees of Australia. Forestry and Timber Bureau, Dept. of the Interior, Canberra. [*Araucaria bidwillii* p. 204, *Athrotaxis selaginoides* p. 216, *Callitris* spp. pp. 210–214, maps].

00203   Anonymous (1962). Seminar and study tour of Latin-American conifers. (Engl. ed. transl. from Spanish; written by various participants). Instituto Nacional de Investigaciones Forestales (I.N.I.F.)/Food and Agriculture Organization (F.H.O.). [*Abies religiosa*, *Cupressus lindleyi*, *C. benthamii* (= *C. lusitanica*), *Pinus* spp.].

00220   Anonymous (1978). The genus *Juniperus* in British Columbia. Davidsonia 9 (4): 98–110.

00221   Anonymous (2001). Proceedings of the International Symposium: Problems of Juniper forests: looking for solutions, methods, techniques. Osh, Kyrgyzstan, 6th–11th August 2000. (Eng., Russ.). LESIC. Forest & Walnut Research Institute, National Academy of Science of the Kyrgyz Republic & Kyrgyz-Swiss Forestry Support Programme. [*Juniperus* spp., ecology, forestry, management, taxonomy].

00231 Antoine, F. (1857). Die Cupressineen-Gattungen: *Arceuthos, Juniperus* und *Sabina*. Wien. (78 pp., 92 figs.). [*Juniperus brevifolia* (Seub.) comb. nov., p. 16, pl. 20–22 (basion. : *J. oxycedrus* var. *brevifolia* Seub. in Flora azorica, 1844), *J. tenella* sp. nov. = *J. oxycedrus*, p. 20, pl. 27–29, *J. microphylla* sp. nov., p. s. n., pl. 31–32, *J. hochstetteri* sp. nov. = *J. procera*, p. s. n., pl. 33, *Sabina vulgaris* nom. nov. (typus) = *J. sabina*, p. 58, t. 80, *Sabina* spp. = *Juniperus* spp., nomenclaturally significant e.g. *Sabina osteosperma* (Torrey) comb. nov., p. 51 antedating *Juniperus utahensis* (Engelm.) Lemm. at species rank, see Little, 1948].

00240 Antoine, F. (1864). *Pinus leucodermis* Ant. Österreich. Bot. Zeitschr. 14: 366–368. [*P. leucodermis* sp. nov., p. 366].

00241 Antoine, F. & T. Kotschy (1853). Eine neue Tanne vom Taurus- Gebirge. *Pinus* (*Abies*) *Cilicica* Antoine et Kotschy. Österreich. Bot. Wochenbl. 3 (52): 409–410. [*Pinus* (*Abies*) *cilicica* sp. nov. = *Abies cilicica*].

00242 Antoine, F. & T. Kotschy (1854). *Arceuthos* Antoine et Kotschy (Genus e tribu Cupressinearum). Österreich. Bot. Wochenbl. 4 (31): 249–250. [*Arceuthos drupacea* (Labill.) gen. et comb. nov. = *Juniperus drupacea*].

00243 Antoine, F. & T. Kotschy (1855). Coniferen des cilicischen Taurus. Wien. (7 p., 1 pl.). [*Abies cilicica, Cedrus, Pinus*; see also D. F. L. von Schlechtendal in Bot. Zeit. 17 (8): 75, 1859].

00245 Archer, W. (1850). Note on *Microcachrys*, Hook. fil., and on a new allied genus of *Coniferae* of Van Diemen's Land. Hooker's J. Bot. Kew Gard. Misc. 2: 51–52. [*Pherosphaera hookeriana* gen et sp. nov. (syn. *Microstrobos niphophilus* J. Garden & L. A. S. Johnson, nom. inval., Art. 36)].

00246 Arévalo, A. G. (1993). Morphological variations in *Pinus praetermissa* (Pinaceae) from Durango, Mexico. Phytologia 75 (3): 243–246.

00247 Arévalo, A. G. & M.-F. Passini (1993). Distribucion y ecologia de *Pinus johannis* M.-F. Robert. Phytologia 74 (2): 125–127.

00247e Arista, M., P. L. Ortiz & S. Talavera (1997). Reproductive isolation of two sympatric subspecies of *Juniperus phoenicea* (Cupressaceae) in southern Spain. Plant Syst. Evol. 208: 225–237. [*J. phoenicea* ssp. *phoenicea, J. phoenicea* ssp. *turbinata*, phenology, ill. of cone development and embryology].

00248 Arista, M. & S. Talavera (1994). Phenology and anatomy of the reproductive phase of *Abies pinsapo* Boiss. (Pinaceae). Bot. J. Linn. Soc. 116: 223–234.

00250 Armstrong, W. P. (1978). Southern California's vanishing cypresses. Fremontia 6 (2): 24–29. [*Cupressus* spp.].

00260 Armstrong, W. P. (1983). Patriarchs of the Southwest. Environment Southwest 503: 20–23. [ill. of *Juniperus occidentalis* var. *australis*].

00261 Arnborg, T. & E. Edlund (1962). Lärkskogar i Siberien. Norrlands Skogsvårdsför. Tidskr. 1: 10–29?, maps. [*Larix sibirica, L. gmelinii, L. decidua* var. *polonica*].

00270 Arno, S. F. & J. R. Habeck (1972). Ecology of Alpine larch (*Larix lyallii* Parl.) in the Pacific Northwest. Ecol. Monogr. 42 (4): 417–450.

00271 Arno, S. F. & R. J. Hoff (1989). Silvics of Whitebark Pine (*Pinus albicaulis*).U.S.D.A. Forest Service Gen. Tech. Rep. INT-253. Intermountain Research Station, Ogden, Utah.

00280 Arnold, J. F. X. (1785). Reise nach Mariazell in Steyermark. Wien. [*Pinus nigra* sp. nov., p. 8, cum tab.].

00283 Arnold, C. A. & J. S. Lowther (1955). A new Cretaceous conifer from northern Alaska. Amer. J. Bot. 42: 522–528. [Taxodiaceae, *Parataxodium* (fossil) gen. nov., *Metasequoia, Taxodium*].

00284 Arnoldi, W. (1899). Beiträge zur Morphologie einiger Gymnospermen. I. Die Entwicklung des Endosperms bei *Sequoia sempervirens*. Bull. Soc. Imp. Nat. Moscou, n.s. 13: 329–341. (ill.). [embryology].

00285 Arnoldi, W. (1899). Beiträge zur Morphologie einiger Gymnospermen. II. Über die Corpuscula und Pollenschläuche bei *Sequoia sempervirens*. Bull. Soc. Imp. Nat. Moscou, n.s. 13: 405–422. (ill.). [embryology].

00286 Arnoldi, W. (1900). Beiträge zur Morphologie einiger Gymnospermen. III. Embryogenie von *Cephalotaxus fortunei*. Flora 87: 46–63. (ill.). [embryology].

00287 Arnoldi, W. (1900). Beiträge zur Morphologie einiger Gymnospermen. IV. Was sind die "Keimbläschen" oder "Hofmeisters-Körperchen" in der Eizelle der Abietineen? Flora 87: 194–204. (ill.) [Pinaceae, embryology].

00288 Arnoldi, W. (1901). Beiträge zur Morphologie einiger Gymnospermen. V. Weitere Untersuchungen der Embryologie in der Familie der Sequoiaceen. Bull. Soc. Imp. Nat. Moscou, n.s. 14: 449–476. (ill.). [Taxodiaceae, *Cryptomeria, Cunninghamia, Sequoia, Taxodium, Sequoiadendron*, Sciadopityaceae, *Sciadopitys*, embryology].

00289 Arnott, G. A. W. (1838). On the genus *Torreya*. Ann. Mag. Nat. Hist., ser. 1, 1: 126–132. [*Torreya* gen. nov. (nom. cons.), *T. taxifolia* sp. nov., p. 130].

00290 Ascherson, P. & P. Graebner (1897). Synopsis der mitteleuropäischen Flora 1. Leipzig. [Cupressaceae, Pinaceae pp. 185–255, *Pinus nigra* var. *pallasiana* (Lamb.) comb. nov., p. 214].

00292   Ash, J. (1985). The rings in tropical *Callitris macleayana* F. Muell. Austral. J. Bot. 31: 277–281. [wood anatomy].

00295   Ashworth, V. E. T. M., B. C. O'Brien & E. A. Friar (2001). Survey of *Juniperus communis* (Cupressaceae) L. varieties from the western United States using RAPD fingerprints. Madroño 48 (3): 172–176. [*J. communis* var. *jackii*, var. *montana*, var. *saxatilis*, var. *sibirica*, var. *depressa* not distinct based on RAPD (DNA) distance measures.].

00313   Aulenback, K. & B. A. LePage (1998). *Taxodium wallisii* sp. nov.: first occurrence of *Taxodium* from the Upper Cretaceous. Int. J. Plant Sci. 159 (2): 367–390. [*Taxodium wallisii* Aulenback et LePage, sp. nov., p. 371, *Cryptomeria*, *Glyptostrobus* and related taxodiaceous macrofossils, ill.].

00314   Aune, P. S. (tech. coord.) (1994). Proceedings of the symposium on Giant Sequoias: their place in the ecosystem and society, June 23–25, 1992, Visalia, California. U.S.D.A. Forest Service, Pacific Southwest Research Station, Gen. Tech. Rep. PSW-GTR-151. [*Sequoiadendron giganteum*, taxonomy, ecology, forestry, conservation].

00314a   Averyanov, L. V., Nguyen Tien Hiep, D. K. Harder & Phan Ke Loc (2002). The history of discovery and natural habitats of *Xanthocyparis vietnamensis* (Cupressaceae). Turczaninowia 5 (4): 31–39.

00315   Avila, J. A., E. Garcia & J. A. Reyes (1992). Registro de *Pinus discolor* Bailey et Hawksworth en la Sierra de Monte Grande, San Luis Potosí, México. Acta Bot. Mexicana 20: 9–12.

00321   Axelrod, D. I. (1959). Late Cenozoic evolution of the Sierran Bigtree Forest. Evolution 13: 9–23. [Coniferales, *Sequoiadendron*].

00323   Axelrod, D. I. (1962). A Pliocene *Sequoiadendron* forest from western Nevada. Univ. California Publ. Geol. Sci. 39: 195–268. (ill.). [palaeobotany].

00325   Axelrod, D. I. (1967). Evolution of the Californian closed-cone pine forest. In: R. N. Philbrick (ed.). Proceedings of the Symposium on the biology of the Californian Islands, pp. 93–149. Santa Barbara Bot. Gard., Santa Barbara, California. [*Pinus* spp. (recent and fossil), ill., map].

00330   Axelrod, D. I. (1976). Evolution of the Santa Lucia fir (*Abies bracteata*) ecosystem. Ann. Missouri Bot. Gard. 63 (1): 24–41.

00350   Axelrod, D. I. (1980). History of the maritime closed-cone pines, Alta and Baja California. Geol. Sci. 120: 1–143. [*Pinus*, fossils and living, of subsect. *Oocarpae*].

00355   Axelrod, D. I. (1983). New Pleistocene conifer records, coastal California. Geol. Sci. 127 (Univ. California Publ.). [*Pinus muricata*, *P. muricata* var. *borealis* var. nov., p. 76, *P. muricata* var. *stantonii* var. nov., p. 77. [non-fossil taxa, no types indicated, ill. on pp. 85–89].

00358   Axelrod, D. I. (1986). The Sierra Redwood (*Sequoiadendron*) forest: end of a dynasty. Geophytology 16: 25–36. [*Sequoiadendron giganteum*].

00360   Axelrod, D. I. (1986). Cenozoic history of some western American pines. Ann. Missouri Bot. Gard. 73: 565–641. (ill., maps). [*Pinus*].

00360a   Axelrod, D. I. & J. Cota (1993). A further contribution to closed-cone pine (*Oocarpae*) history. Amer. J. Bot. 80 (7): 743–751. [*Pinus* subsect. *Oocarpae*, *P. radiata*, *P. hazenii* (fossil), *P. masonii* (fossil), *P. pieperi* (fossil), *P. verdiana* Axelrod, fossil sp. nov., p. 750, ill.].

00360b   Axelrod, D. I. & F. Govean (1996). An early Pleistocene closed-cone pine forest at Costa Mesa, southern California. Int. J. Plant Sci. 157 (3): 323–329. (ill.). [*Pinus muricata*, *P. radiata*, *P. remorata*, *Cupressus macrocarpa*, palaeobotany].

00361   Axelsson, R. (1966). En frökaraktär av värde vid särskiljande av frö från *Larix decidua* och *Larix leptolepis*. Svensk Bot. Tidskr. 60 (2): 310–314, maps 2–3. [*Larix decidua*, *L. kaempferi*, ill.].

00361a   Axsmith, B. J. & T. N. Taylor (1997). The Triassic seed cone *Glyptolepis*. Rev. Palaeobot. Palynol. 96: 71–79. [Coniferales, *Glyptolepis richteri* sp. nov., palaeobotany, ill.].

00361c   Axsmith, B. J., T. N. Taylor & E. L. Taylor (1998). A new fossil conifer from the Triassic of North America: implications for models of ovulate cone scale evolution. Int. J. Plant Sci. 159 (2): 358–366. [*Conewagia longiloba* gen. et sp. nov., morphology and evolution of ovulate cones].

00361d   Axsmith, B. J., T. N. Taylor & E. L. Taylor (1998). Anatomically preserved leaves of the conifer *Nothophytum krauselii* (Podocarpaceae) from the Triassic of Antarctica. Amer. J. Bot. 85 (5): 704–713. [Coniferales, *Nothophytum*, *Heidiphyllum* and *Telemachus* (cone) are possibly conspecific, suggesting a podocarpaceous conifer with multi-veined leaves and multi-scaled, long bractaceous seed cones was widespread in the Triassic].

00362   Aytug, B. (1960). Contribution à l'étude anatomique de quatre espèces de Sapins (*Abies* Tourn.). Bull. Mus. Hist. Nat. (Paris), sér. 2, 32: 436–444. [*Abies alba*, *A. cilicica*, *A. cephalonica*, *A. nordmanniana*].

00371   Bader, F. J. W. (1965). Some boreal and subantarctic elements in the flora of the high mountains of tropical Africa and their relation to other intertropical continents. Webbia 19 (2): 531–544. [*Widdringtonia*, *Podocarpus*, *Juniperus*, world map].

00371a   Bagci, E. & M. T. Babac (2003). A morphometric and chemosystematic study on the *Abies* Miller (fir) species in Turkey. Acta Bot. Gallica 150 (3): 355–367. [*Abies cilicica, A. nordmanniana*, subspecies *isaurica, equi-trojani, bornmuelleriana*, morphology, chemistry, hybridization, ill.].

00372   Baikovskaja, T. N. (1956). (Composition of the upper Cretaceous Floras of Northern Asia, with respect to the systematic composition, ecological type and geological age). Trudy Bot. Inst. Komarova Akad. Nauk SSSR, Ser. 8, Paleobotanika 2: 105–139., Moscow Leningrad. (Russ.). [maps with fossil distr. of *Agathis, Araucaria, Phyllocladus* and *Widdringtonia* in U.S.S.R., p. 117].

00380   Bailey, D. K. (1970). Phytogeography and taxonomy of *Pinus* subsection *Balfourianae*. Ann. Missouri Bot. Gard. 57: 210–249. [*P. longaeva* sp. nov., *P. aristata, P. balfouriana*].

00385   Bailey, D. K. (1975). *Pinus albicaulis*. Kew Mag. (Curtis's Bot. Mag.) n.s., t. 691: 141–147.

00390   Bailey, D. K. (1983). A new allopatric segregate from and a new combination in *Pinus cembroides* Zucc. at its southern limits. Phytologia 54 (2): 89–99. [*Pinus cembroides* ssp. *orizabensis* ssp. nov., p. 89, *P. cembroides* ssp. *lagunae* (Rob.-Pass.) D. K. Bailey stat. nov., p. 98].

00400   Bailey, D. K. (1987). A study of *Pinus* subsection *Cembroides* I: The single-needle Pinyons of the Californias and the Great Basin. Notes Roy. Bot. Gard. Edinburgh 44 (2): 275–310. [*Pinus californiarum* sp. nov., p. 278, *P. californiarum* ssp. *fallax* (Little) comb. nov., p. 279].

00405   Bailey, D. K. (1988). The single-needle Pinyons – one taxon or three? In: C. D. Hall & V. A. Jones (eds.). Plant biology of eastern California. pp. 69–91. Proc. Mary DeDecker Symp., White Mountain Res. Stat., Univ. California, Bishop. [*Pinus* subsect. *Cembroides*, *P. monophylla, P. edulis* var. *fallax, P. californiarum* ssp. *fallax*].

00410   Bailey, D. K. (1990). Phytogeography and taxonomy of the pinyon pines *Pinus* subsection *Cembroides*. Simposio Nac. sobre pinos piñoneros. [in press].

00420   Bailey, D. K. & F. G. Hawksworth (1979). Pinyons of the Chihuahuan Desert Region. Phytologia 44: 129–133. [*Pinus discolor* nom. nov., *P. remota* (Little) comb. nov.].

00430   Bailey, D. K. & F. G. Hawksworth (1983). Pinaceae of the Chihuahuan Desert Region. Phytologia 53: 226–234. [*Pinus* spp., *Abies durangensis* var. *coahuilensis, Pseudotsuga menziesii* var. *glauca*].

00431   Bailey, D. K. & F. G. Hawksworth (1988). Phytogeography and Taxonomy of the Pinyon Pines, *Pinus* subsection *Cembroides*. In: M.-F. Passini, D. Cibrian Tovar & T. Eguiluz Piedra (comp.). II simposio nacional sobre pinos piñoneros, 6-7-8 de agosto de 1987, pp. 41–64. Chapingo & México (D.F.).

00434   Bailey, D. K. & F. G. Hawksworth (1992). Change in status of *Pinus cembroides* ssp. *orizabensis* (Pinaceae) from Central Mexico. Novon 2 (4): 306–307. [*P. orizabensis* (D. K. Bailey) comb. nov.].

00437   Bailey, D. K., K. Snajberk & E. Zavarin (1982). On the question of natural hybridization between *Pinus discolor* and *Pinus cembroides*. Biochem. Syst. Ecol. 10 (2): 111–119. [*Pinus cembroides* var. *bicolor*].

00440   Bailey, D. K. & T. Wendt (1979). New pinyon records for northern Mexico. Southw. Naturalist 24: 389–390. [*Pinus* subsect. *Cembroides*].

00441   Bailey, F. M. (1883). A synopsis of the Queensland Flora;... Brisbane. [*Callitris endlicheri* (Parl.) comb. nov., p. 497, *Agathis robusta* (C. Moore ex F. Muell.) comb. nov., p. 498].

00441a   Bailey, F. M. (1891). Contributions to the Flora of Queensland. Dept. Agric. Brisbane Bull., Bot. 9 (offprint). [*Agathis palmerstonii* (F. Muell.) comb. nov. = *A. robusta*, p. 17].

00442   Bailey, F. M. (1902). The Queensland Flora. Part V. Order 123. Coniferae; pp. 1494–1500. Brisbane. [*Callitris robusta* (A. Cunn. ex Parl.) comb. nov. = *C. preissii*, p. 1496].

00443   Bailey, F. M. (1905). Contributions to the Flora of Queensland. Queensland Agric. J. 15 (8): 897–899. (ill.). [*Podocarpus ladei* sp. nov. = *Prumnopitys ladei*].

00460   Bailey, I. W. (1910). Anatomical characters in the evolution of *Pinus*. Amer. Naturalist 44: 284–293. (ill.).

00470   Bailey, I. W. & A. F. Faull (1934). The cambium and its derivative tissues. IX. Structural variability in the redwood, *Sequoia sempervirens*, and its significance in the identification of fossil woods. J. Arnold Arbor. 15: 233–254.

00475   Bailey, J. F. & C. T. White (1916). Contributions to the flora of Queensland. Contr. Queensland Fl. Bot. Bull. 18. [*Agathis microstachya* sp. nov., p. 13].

00480   Bailey, L. H. (1933). The cultivated conifers of North America. New York. (404 pp., 158 pls.). [*Juniperus silicicola* (Small) comb. nov., p. 18, with basion. *Sabina silicicola, Picea glauca* var. *densata* var. nov. = *P. glauca*, p. 108].

00490   Baillaud, L. & Y. Courtot (1960). Observations sur le rythme de la ramification du *Chamaecyparis nootkatensis*. (VII Conferenza Int. Soc. Studio Ritmi Biol. Basimetria; Siena 5–7 Sept. 1960). Inst. Bot. Fac. Sci. Besançon (France). (7 pp.).

00500 Baillaud, L. & Y. Courtot (1961). Nouvelles remarques sur le rythme de la répartition des rameaux du *Chamaecyparis nootkatensis*. Ann. Sci. Univ. Besançon, sér. 2, 17: 63–68.

00520 Bailly, E. (1896). Le *Thuiopsis standishii*. Rev. Hort. 1896: 160–163, 1 pl. [*Thuja standishii*].

00521 Baird, A. M. (1953). The life-history of *Callitris*. Phytomorpology 3 (3): 258–284. [anatomy, morphology, pollination, embryology, ill.].

00526 Baker, J. G. (1885). Further contributions to the Flora of Madagascar – second and final part. J. Linn. Soc., Bot. 21: 407–455. [*Podocarpus madagascariensis* sp. nov., p. 447].

00527 Baker, R. T. (1903). On a new species of *Callitris* from eastern Australia. Proc. Linn. Soc. New South Wales 28 (4): 839–841, pl. 45. [*Callitris gracilis* sp. nov. = *C. preissii* Miq., p. 839].

00528 Baker, R. T. (1907). Contribution to a knowledge of the Flora of Australia. Proc. Linn. Soc. New South Wales 31 (4): 711–721, pl. 65–67. [*Callitris morrisonii* sp. nov., p. 717, pl. 67].

00529 Baker, R. T. & H. G. Smith (1908). On the pines of Australia, No. 1. – *Callitris glauca* R. Br., "White or Cypress Pine". Proc. Roy. Soc. New South Wales 42: 145–183, pl. 15–29. [*C. glauca* R. Br., nom. nud. = *C. columellaris* F. Muell., photogr. of leaf and wood anatomy].

00532 Bakhuizen van den Brink, R. C. (1955). Nomenclatural note on *Dammara* Lmk. and *Agathis* Salisb. Taxon 4 (8): 195–196. [*Dammara* Gaertner, 1790, *Agathis* Salisb., 1806 nom. cons.].

00540 Bakuzis, E. V. & H. L. Hansen (1965). Balsam Fir. *Abies balsamea* (Linnaeus) Miller. Univ. of Minnesota Press, Minneapolis.

00541 Ball, J. (1878). Spicilegium florae maroccanae. J. Linn. Soc., Bot. 16: 569–772. [*Juniperus oxycedrus* ssp. *macrocarpa* (Sibth. et Smith) comb. et stat. nov., p. 670].

00542 Ball, J. (1886). Notes on the botany of western South America. J. Linn. Soc., Bot. 22: 137–168. [*Dacrydium fonkii* (Phil.) Ball = *Lepidothamnus fonkii*, p. 168]

00545 Bamber, R. K. (1959). Anatomy of the barks of five species of *Callitris* Vent. Proc. Linn. Soc. New South Wales 84: 375–381.

00561 Bannan, M. W. (1941). Wood structure in *Thuja occidentalis*. Bot. Gaz. (Crawfordsville) 103: 295–309.

00562 Bannan, M. W. (1952). The microscopic wood structure of North American species of Chamaecyparis. Can. J. Bot. 30: 170–187. [*C. lawsoniana, C. nootkatensis, C. thyoides*].

00568 Bannister, M. H. (1958). Evidence of hybridization between *Pinus attenuata* and *P. radiata* in New Zealand. Trans. Roy. Soc. New Zealand 85 (2): 1–9. [repr. as Techn. Pap. 13 of the Forest Res. Inst., New Zealand Forest Service; good drawings of cones].

00569 Bannister, M. H. (1958). Specimens of two pine trees from Guadalupe Island, Mexico. New Zealand J. Forest. 7 (5): 81–87. [*Pinus radiata*].

00570 Bannister, M. H. & I. R. C. McDonald (1983). Turpentine composition of the pines of Guadalupe and Cedros Islands, Baja California. New Zealand J. Bot. 21: 373–377. [*Pinus radiata, P. radiata* var. *cedrosensis*].

00570e Barbero, M., P. Lebreton & P. Quezel (1994). Sur les affinités biosystématiques et phytoécologiques de *Juniperus thurifera* L. et de *Juniperus excelsa* Bieb. Ecol. Mediterranea 20 (3/4): 21–37.

00571 Barbey, A. (1934). Une relique de la sapinière méditerranéenne, Le Mont Babor; monographie de l'*Abies numidica* Lann. [*A. numidica, Cedrus atlantica*, insects, 80 pp., 33 tables). Paris.

00580 Barnes, R. D. & B. T. Styles (1983). The closed cone pines of Mexico and Central America. Commonw. Forest. Rev. 62: 81–84.

00588 Barrett, W. H. G. (1972). Variacion de Characteres Morfologicos en poblaciones naturales de *Pinus patula* en México. IDIA. Suplemento Forestal 7: 9–35.

00590 Barrett, W. H. G. & L. Golfari (1962). Descripción de dos nuevas variedades del Pino del Caribe. Caribbean Forest. 23: 59–71. [*Pinus caribaea* var. *bahamensis* (Griseb.) comb. nov., *P. caribaea* var. *hondurensis* (Sénécl.) comb. nov.].

00590a Barrón, E. & A. Buades (2002). Aportaciones al estudio de las epidermis foliares en las especies vivientes de la familia Taxodiaceae (Coniferales, Coniferophyta). Bol. R. Soc. Esp. Hist. Nat. (Sec. Biol.) 97 (1–4): 5–18. (LM ill.). [*Athrotaxis, Cryptomeria, Cunninghamia, Glyptostrobus, Metasequoia, Sequoia, Sequoiadendron, Taiwania, Taxodium*].

00591 Barry, J.-P., B. Belin, J.-C. Celles, D. Dubost, L. Faurel & P. Hethener (1973). Essai de monographie du *Cupressus dupreziana* A. Camus, Cyprès endémique du Tassili des Ajjer (Sahara central). Trav. Lab. Forest. Toulouse, T. 1, vol. 9, art. 2: 95–178. (orig. publ. in: Bull. Soc. Hist. Nat. Afrique N. 61 (1–2): 95–178; ill., map, tables).

00596 Bartel, J. A. (1991). Nomenclatural changes in *Dudleya* (Crassulaceae) and Cupressus (Cupressaceae). Phytologia 70 (4): 229–230. [*C. goveniana* ssp. *pygmaea* (Lemmon) comb. superfl., see A. Camus, 1914].

00597b Bartel, J. A., R. P. Adams, S. A. James, L. E. Mumba & R. N. Pandey (2003). Variation among *Cupressus* species from the western hemisphere based on random amplified polymorphic DNAs. Biochem. Syst. Ecol. 31: 693–702. [*Cupressus* spp., *Chamaecyparis nootkatensis*, RAPDs, taxonomy].

00600   Bartholomew, B., D. E. Boufford & S. A. Spongberg (1983). *Metasequoia glyptostroboides*: its present status in Central China. J. Arnold Arbor. 64 (1): 105–128.

00600a   Bartlett, A. W. (1913). Note on the occurrence of an abnormal bisporangiate strobilus of *Larix europea*, DC. Ann. Bot. (London) 27: 575–576. [*Larix decidua*].

00600b   Bartlett, H. H. (1935). A method of procedure for field work in tropical American phytogeography based upon a botanical reconnaisance in parts of British Honduras and the Peten Forest of Guatemala. Publ. Carnegie Inst. Washington 461: 1–25. (ill.). [*Podocarpus pinetorum* sp. nov. = *P. guatemalensis*, p. 21 in adnot.].

00601   Bartling, F. G. (Th.) (1830). Ordines naturales plantarum eorumque characteres et affinitates adjecta generum enumeratione. Göttingen. [Cupressaceae, fam. nov., as "Ordo 65. Cupressinae", p. 90].

00605   Basinger, J. F. (1981). The vegetative body of *Metasequoia milleri* from the Middle Eocene of British Columbia. Canad. J. Bot. 59: 2379–2410. [wood anatomy, morphological comparison with *Taxodium distichum*, *Glyptostrobus pensilis*, *Sequoia sempervirens*, *Sequoiadendron giganteum*, *Metasequoia glyptostroboides*, palaeobotany, ill.].

00606   Basinger, J. F. (1984). Seed cones of *Metasequoia milleri* from the Middle Eocene of southern British Columbia. Canad. J. Bot. 62: 281–289. [*M. milleri* Rothwell & Basinger 1979 emend., ill. of anatomy sections].

00610   Batalin, A. T. (1894). Notae de plantis asiaticis XLIX–LXXI. Trudy Imp. S.-Petersburgsk. Bot. Sada 13: 385. [*Larix potaninii* sp. nov.; see also L. Beissner, 1899, p. 115].

00630   Bauhin, C. (K.) (1623). Pinax Theatri botanici. Ludovicus Rex, Basiliae (Basel). [*Pinus*].

00635   Bayer, A. (1908). Zur Deutung der weiblichen Blüten der Cupressineen nebst Bemerkungen über *Cryptomeria*. Bot. Centralbl. Beih. 23 (1): 27–44. [*Juniperus communis*, *J. sabina*, *Chamaecyparis*, *Thuja*, *Cryptomeria japonica*, ill.].

00640   Beals, E. W. (1965). The remnant Cedar forests of Lebanon. J. Ecol. 53 (3): 679–694. [*Cedrus libani*].

00645   Beaman, J. H. & J. W. Andresen (1966). The vegetation, floristics and phytogeography of the summit of Cerro Potosi, Mexico. Amer. Midl. Naturalist 75 (1): 1–33. [*Pinus culminicola*, *P. hartwegii*, *Juniperus monticola*].

00650   Bean, W. J. (1920). The Formosan redwood; *Taiwania cryptomerioides*, Hayata. Gard. Chron., ser. 3, 68: 213, f. 99.

00680   Beaven, G. F. & H. J. Oosting (1939). Pocomoke swamp: A study of a cypress swamp on the Eastern Shore of Maryland. Bull. Torrey Bot. Club 66: 367–389. [*Taxodium distichum*].

00690   Becherer, A. (1934). Der wissenschaftliche Name der Fichte und der Weißtanne. Schweiz. Z. Forstwesen 85: 5–8. [*Picea abies*, *Abies alba*, nomenclature].

00717   Behling, H. (1997). Late Quarternary vegetation, climate and fire history of the *Araucaria* forest and campos region from Serra Campos Gerais, Paraná State (South Brazil). Rev. Palaeobot. Palynol. 97: 109–121. [*Araucaria angustifolia*, palynology].

00720   Beissner, L. (1887). Handbuch der Coniferen-Benennung. Leipzig. [*Abies cephalonica* var. *apollinis* (Link) comb. et stat. nov. = *A. cephalonica*; this publ. of Oct. 1887 was preceded by the "Systematische Eintheilung der Coniferen" of April 1887, with essentially the same content, distr. for the Conifer Conference to be held at Dresden; the date of publ. of new names therefore is April 1887].

00740   Beissner, L. (1896). Neues und Interessantes auf dem Gebiete der Nadelholzkunde. Mitt. Deutsch. Dendrol. Ges. 1896 (5): 52–69; 2nd. ed. (1909) (1–7): 199–216. [*Pseudotsuga japonica* (Shirasawa) comb. nov., p. 209, *Larix chinensis* sp. nov., p. 215].

00750   Beissner, L. (1897). Conifères de Chine. Nuovo Giorn. Bot. Ital. ser. 2, 4: 183–187, pl. 5. [*Larix chinensis* "sp. nov."; coll. J. Giraldi].

00760   Beissner, L. (1898). Conifères de Chine. Bull. Soc. Bot. Ital. 7: 166–170. [contin. of L. Beissner, 1897].

00770   Beissner, L. (1899). Interessantes über Coniferen. Mitt. Deutsch. Dendrol. Ges. 8: 106–127. [misc. notes, repr. of several Lat. diagn., e.g. *Juniperus foetidissima* Willd., *Pinus pindica* Formánek, *Larix potaninii* Batalin].

00780   Beissner, L. (1898–1902). Conifères de Chine. Bull. Soc. Bot. Ital. 8: 2–38 (1898); 8: 309–311 (1899); 10: 357–361 (1901); 11: 90–92 (1902). [coll. J. Geraldi].

00790   Beissner, L. (1906). Mitteilungen über Koniferen. Mitt. Deutsch. Dendrol. Ges. 15: 82–100, 144. [Coniferales, *Abies concolor* var. *brevifolia* var. nov., p. 144].

00805   Bekessy, S. A., T. R. Alnutt, A. C. Premoli, A. Lara, R. A. Ennos, M. A. Burgman, M. Cortes & A. C. Newton (2002). Genetic variation in the vulnerable and endemic Monkey Puzzle tree, detected using RAPDs. Heredity 88: 243–249. [*Araucaria araucana*, genetics, conservation].

00810   Belder, J. & D. O. Wijnands (1979). *Metasequoia glyptostroboides*. Dendroflora 15–16: 24–35.

0810c Bennett, J. J., R. Brown & T. Horsfield (1838). Plantae javanicae rariores, descriptae iconibus illustratae, quas in insula Java, annis 1802–1818, legit et investigavit Thomas Horsfield,... Part. 1. London. [*Podocarpus spicatus* R. Br. = *Prumnopitys taxifolia*, p. 40].

00820 Bentham, G. (1839–57). Plantae Hartwegianae. London. [Coniferales pp. 57–58, 92, 337, *Pinus tenuifolia* "sp. nov." p. 92 (1842) = *P. maximinoi* H.E. Moore,; see also H.E. Moore, Jr., 1966].

00840 Berezin, E. L. (1970). Die Artenzusammensetzung der Fichtenpopulationen in den Wäldern des Tien-Shan und des Dsungarischen Ala-Tau. Bot. Žurn. (Moscow & Leningrad) 50: 491–498. (Russ., Germ. summ.). [*Picea morinda* ssp. *tianschanica* (Rupr.) comb. et stat. nov. = *P. schrenkiana* ssp. *tianschanica*, p. 493].

00850 Berezin, E. L. (1970). Zur Systematik und Formenmannigfaltigkeit der Schrenks-Fichte. in: Lesnaja genetika, selekcija i semenovodstvo. Izd. "Karelija", Petrozavodsk, pp. 199–202. (Russ., Germ. summ.). [*Picea schrenkiana*].

00851 Berg, O. (1860). Beschreibung einiger neuen Droguen vom Cap. Bonplandia 8 (12): 190–192. [*Widdringtonia caffra* sp. nov. = *W. nodiflora*, p. 190].

00860 Berger, W. (1951). *Pinus stellwagi* Kink. aus dem Unterpliozän (Pannon) von Wien. Österreich. Bot. Zeitschr. 98 (1–2): 138–141.

00860a Bergmann, F. & E. M. Gillet (1997). Phylogenetic relationships among *Pinus* species (Pinaceae) inferred from different numbers of 6PGDH loci. Plant. Syst. Evol. 208 (1–2): 25–34. [*Pinus* spp., *P.* subsect. *Cembrae*, limited phylogeny derived from this isozyme analysis due to low diversity in subgen. *Pinus* and variation within the monophyletic subsect. *Strobi*].

00862 Bernath, E. L. (1937). Coniferous forest trees of Chile. Trop. Woods 52: 19–26. [Araucariaceae, *Austrocedrus chilensis*].

00863 Bernhard, T. (1931). Die Kiefern Kleinasiens. Mitt. Deutsch. Dendrol. Ges. 43: 29–50. [*Pinus brutia, P. nigra, P. pinea, P. sylvestris, P. halepensis* culta, ill., maps].

00863c Berry, E. W. (1912). Notes on the genus *Widdringtonites*. Bull. Torrey Bot. Club 39: 341–348. [*Widdringtonites* Endl., palaeobotany].

00864 Bertoloni, A. (1862). Miscellanea botanica 23. Bologna. (20 pp., 6 pl.). [*Juniperus indica* sp. nov., p. 16, *Taxus orientalis* sp. nov. = *T. wallichiana*, p. 17].

00870 Bertrand, M. C. E. (1871). Note sur le genre *Abies*. Bull. Soc. Bot. France 18: 376–382. (also publ. in: Bull. Soc. Philom. Paris, séance du 9-12-1871: 245–254). [*A. firma* var. *brachyphylla* (Maxim.) comb. et stat nov. = *A. homolepis*, p. 380].

00886 Bigwood, A. J. & R. S. Hill (1985). Tertiary araucarian macrofossils from Tasmania. Austral. J. Bot. 33: 645–656. [SEM photogr. pp. 649–650, 654, *Araucarioides* gen. nov., p. 647, *A. linearis* sp. nov., p. 647, *A. sinuosa* sp. nov., p. 648, *A. annulata* sp. nov., p. 651, early Eocene].

00890 Billain, B. von (1853). *Cryptomeria Lobbiana.* Allg. Gartenzeitung 21: 233–234. [*Cryptomeria fortunei* (nom. rej.), *C. lobbiana* (sp. nov.)= *C. japonica*; a short note on this paper in: Gard. Chron. 1853: 582–583, 1853].

00896 Biondi, E. & E. Brugiapaglia (1991). *Taxodioxylon gypsaceum* in the fossil forest of Dunarobba. Flora Mediterranea 1: 111–120. [*Taxodioxylon gypsaceum* (Göppert) Kräusel, *Sequoia sempervirens*, palaeobotany, Pliocene, wood anatomy, ill.].

00900 Birks, J. S. & R. D. Barnes (1985). Multivariate analysis of data from international provenance trials of *Pinus oocarpa/ Pinus patula* subspecies *tecunumanii*. Commonw. Forest. Rev. 64: 367–374.

00901 Birks, J. S. & P. J. Kanowski (1988). Interpretation of the composition of coniferous resin. Silvae Genet. 37 (1): 29–39. [Coniferales, *Pinus*, chemistry].

00910 Bisse, J. (1975). Nuevos arboles de la flora de Cuba (*Pinus maestrensis* sp. nov.). Botanica 10 (2): 1–3.

00930 Blake, S. T. (1959). New or noteworthy plants, chiefly from Queensland, 1. Proc. Roy. Soc. Queensland 70 (6): 33–46. [Cupressaceae pp. 34–39. *Callitris columellaris, C. glauca, C. hugelii* (incertae sedis), *C. intratropica, C. canescens* (Parl.) comb. nov., p. 39; of first 4 only *C. columellaris* recognized].

00940 Blanco, C. E. (1938). Los pinos de México. Bol. Dept. Forest. México 3 (11): 237–255. [*Pinus, Abies, Pseudotsuga*].

00950 Blanco, C. E. (1950). *Pinus cooperi* Blanco, sp. nova. Anales Inst. Biol. Univ. Nac. México 20: 183–187. [*P. cooperi*, nom. nov. + *P. cooperi* var. *ornelasi* (Martínez) comb. nov., p. 185; substituted for *P. lutea* Blanco, a later homonym of *P. lutea* Walter].

00955a Blume, C. L. von (1836–49). Rumphia, sive commentationes botanicae imprimis de plantis Indiae orientalis, ... (4 volumes). Leiden, etc. [*Podocarpus discolor* sp. nov., *P. neglectus* sp. nov. = *P. neriifolius*, vol. 3: 213, *P. thevetiifolius* Zipp. ex Blume = *P. polystachyus*, vol. 3: 213, *P. leptostachyus* sp. nov. = *P. neriifolius*, vol. 3: 214, *P. rumphii* sp. nov., vol. 3: 214 (1847)].

00960 Bobrov, E. G. (1970). Generis *Picea* Historia et Systematica. Novosti Sist. Vyssih Rast. (Novit. Syst. Pl. Vasc.) 7: 7–39. Inst. Bot. V. L. Komarov, Leningrad. (Russ., Lat.).

00970 Bobrov, E. G. (1972). Die introgressive Hybridization in der Gattung *Picea* A. Dietr. Symp. Biol. Hungarica 12: 141–148.

00980 Bobrov, E. G. (1972). Generis *Larix* Mill. Historia et Systematica. Komarovskie Cteniya (Moscow-Leningrad) 25. (Russ., 96 pp., ill., maps).

00990 Bobrov, E. G. (1972). Synopsis Speciarum Generis *Larix*. Novosti Sist. Vysših Rast. (Novit. Syst. Pl. Vasc.) 7: 4–15. Inst. Bot. V. L. Komarov, Leningrad. (Russ., Lat.).

01000 Bobrov, E. G. (1978). Forest forming Coniferae of the USSR. Nauka, Leningrad. (Russ., Eng. summ.). [Coniferales, Pinaceae].

01023 Bock, W. (1954). *Primaraucaria*, a new araucarian genus from the Virginia Triassic. J. Paleontol. 28: 32–42. (ill.).

01024 Boddi, S., L. M. Bonzi & R. Calamassi (2002). Structure and ultrastructure of *Pinus halepensis* primary needles. Flora 197 (1): 10–23. (SEM ill.).

01025 Boer, P. de (1866). Specimen botanicum inaugurale de Coniferis archipelagi Indici, ... Ph.D. Thesis, Utrecht Univ. pp. 1–54, pl. 1–3. Utrecht. [*Pinus merkusii*, *Dammara alba* = *Agathis dammara*, *Podocarpus* spp., *Dacrydium elatum*, pl. of *Podocarpus* spp.].

01030 Boerner, F. (1972). Bemerkungen zur "Tigerschwanzfichte". Kurzmitt. Deutsch. Dendrol. Ges. 9: 19–21. [*Picea torano*].

01040 Boissier, P. E. (1842). Voyage botanique dans le Midi de l'Espagne pendant l'année 1837. Vol. 2, fasc. 19: 577–608. Paris. [*Abies pinsapo* sp. nov., p. 584, pl. 160?; name first publ. in P. E. Boissier, Notice sur l'*Abies pinsapo*: 8 (1838); Biblioth. Universelle Genève 13: 406 (1838); Ann Sci. Nat. (Paris) 9: 167–172, 1838].

01060 Boivin, B. (1954). *Pseudotsuga menziesii* (Mirbel) Franco versus *Pseudotsuga taxifolia* (Poiret) Britton. Bol. Soc. Brot., Sér. 2, 28: 63–64.

01070 Boivin, B. (1959). *Abies balsamea* (Linné) Miller et ses variations. Naturaliste Canad. 86: 219–223. [*Abies balsamea* ssp. *lasiocarpa* (Hook.) comb. et stat. nov.].

01080 Bokhari, M. H. (1975). A new record of *Cupressus* from SW Iran. Notes Roy. Bot. Gard. Edinburgh 33 (3): 445–447.

01080c Bond, R. (2002). *Pinus krempfii*: a rare pine from Vietnam. Conifer Quart. 19 (1): 10–13.

01081 Bongard, A. G. H. (1832). Observations sur la végétation de l'île de Sitcha. Mém. Acad. Imp. Sci. St.-Petersbourg, Sér. 6, Sci. Math. 2: 119–178. [*Pinus mertensiana* sp. nov. = *Tsuga mertensiana*, p. 163, *Pinus sitchensis* sp. nov. = *Picea sitchensis*, p. 164, *Thuja excelsa* sp. nov. = *Chamaecyparis nootkatensis*, p. 164; date of publ. Aug. 1832].

01081d Bonisegna, J. A. & R. L. Holmes (1985). *Fitzroya cupressoides* yields 1534-year long South American chronology. Tree-ring Bull. 45: 37–42.

01090 Boratyn'ski, A. & K. Browicz (1983). *Juniperus drupacea* in Greece. Arbor. Kórnickie 27: 3–16.

01095 Borchert, M. (1985). Serotiny and cone-habit variation in populations of *Pinus coulteri* (Pinaceae) in the southern Coast Ranges of California. Madroño 32 (1): 29–48.

01100 Bordères, O. & H. Gaussen (1939). Une nouvelle espèce de Thuya chinois, *T. chengii*. Bull. Soc. Hist. Nat. Toulouse 73: 280–290, f. 1–3. [repr. in: Trav. Lab. Forest. Toulouse, T. 1, vol. 3, art. 6: 1–14, f. 1–3 (1939?); see also H. Gaussen, 1968 (p. 236), where this taxon has been placed (tentatively?) under *Biota*, without full ref. to the basionym].

01110 Bordères-Rey, O. & H. Gaussen (1944). A Propos des *Abies* des confins du Chen-si, du Se-Tchouan et du Hou-Pé. Trav. Lab. Forest. Toulouse, T. 1, vol. 4, art. 5: 1–14. [*A. chensiensis*, *A. fargesii* et var., *A. kansouensis* sp. nov., p. 6].

01120 Bordères-Rey, O. & H. Gaussen (1947). Trois espèces nouvelles des Sapins chinois: *Abies ferreana*, *A. minensis*, *A. salouenensis*. Trav. Lab. Forest. Toulouse, T. 1, vol. 4, art. 15: 110.

01130 Bordères-Rey, O. & H. Gaussen (1948). Deux espèces nouvelles des Sapins chinois: *Abies rolii* et *A. yuana*. Bull. Soc. Hist. Nat. Toulouse 83: 210–216. [= *A. forrestii* var. *ferreana*].

01140 Borel, A. & J.-L. Polidori (1983). Le génévrier thurifère (*Juniperus thurifera* L.) dans le Parc National du Mercantour (Alpes Maritimes). Bull. Soc. Bot. France (Lettres bot.) 130 (3): 227–242.

01150 Bormann, F. H. & R. B. Platt (1958). A disjunct stand of hemlock in the Georgia Piedmont. Ecology 39: 16–23. (ill., map). [*Tsuga canadensis*].

01157 Bortenschlager, S. (1990). Aspects of pollen morphology in the Cupressaceae. Grana 29: 129–137. (SEM ill.). [*Callitris*, *Chamaecyparis nootkatensis* =*Xanthocyparis nootkatensis*, *Cupressus*, *Juniperus*, *Thuja*, *Thujopsis*].

01160 Bory de Saint-Vincent, J. B. G. M. *et al.* (1823). Dictionnaire classique d'histoire naturelle. Vol. 3. [see A. Richard, 1823].

01166 Boscherini, G., M. Morgante, P. Rossi & G. G. Vendramin (1994). Allozyme and chloroplast DNA variation in Italian and Greek populations of *Pinus leucodermis*. Heredity 73: 284–290. [*P. leucodermis* = *P. heldreichii*].

01167 Bose, M. N. (1975). *Araucaria haastii* Ettingshausen from Shag Point, New Zealand. Palaeobotanist 22: 76–80. [Araucariaceae, palaeobotany].

01168   Bose, M. N. & S. B. Manum (1990). Mesozoic conifer leaves with "*Sciadopitys*-like" stomatal distribution. Norsk Polarinst. Skrifter 192: 1–81. [re-evaluation based on fossils from Spitsbergen, Greenland and Baffin Island, ill.].

01169   Bose, M. N. & S. B. Manum (1991). Addition to the family Miroviaceae (Coniferae) from the Lower Cretaceous of West Greenland and Germany: *Mirovia groenlandica* n. sp., *Tritaenia crassa* (Seward) comb. nov., and *Tritaenia linkii* Mägdefrau et Rudolf emend. Polar Research 9 (1): 9–20. ["*Sciadopitys*-like" fossils classified in the fossil family Miroviaceae Reymanówna].

01169a   Boulter, M. C. (1969). *Cryptomeria* – a significant component of the European Tertiary. Paläobotanik, B, 3 (3–4): 279–287.

01169b   Boulter, M. C. (1970). Lignified guard cell thickenings in the leaves of some modern and fossil species of Taxodiaceae (Gymnospermae). Biol. J. Linn. Soc. 2: 41–46.

01171   Boutelje, J. B. (1955). The wood anatomy of *Libocedrus* Endl. s.l. and *Fitzroya* J. D. Hooker. Acta Horti Berg. 17 (6): 177–216. [*Austrocedrus, Calocedrus* (*Heyderia*), *Libocedrus, Papuacedrus, Pilgerodendron*].

01171a   Bowen, M. R. & T. C. Whitmore (1980). A second look at *Agathis*. C.F.I. Occ. Pap. 13. Dept. Forestry, Commonwealth Forestry Inst., Univ. Oxford. (19 pp.).

01174   Boydak, M. & I. Erdogrul (1999). A new variety of Cilician fir (*Abies cilicica* Carr.) from Anatolia. Istanbul Univ. Orman Fak. Dergisi, A. 49 (2): 17–26. (Eng., Turk.). (ill.). [*Abies cilicica* subsp. *isaurica* var. *pyramidalis* var. nov.].

01180   Bradshaw, K. E. (1941). Field characters distinguishing *Pinus ponderosa* and *P. jeffreyi*. Madroño 6: 15–18.

01190   Brand, D. G. (1987). Estimating the surface area of spruce and pine foliage from displaced volume and length. Canad. J. Forest Res. 17 (10): 1305–1308. [*Picea, Pinus*].

01200   Braun, H. J., S. Carlquist, P. Ozenda & I. Roth, eds. (1961–78). Handbuch der Pflanzenanatomie. Berlin Stuttgart. [see H. von Guttenberg, 1961; H. Singh, 1978].

01202   Braun, U. (1998). Typusmaterial des Herbariums der Martin-Luther-Universität Halle-Wittenberg (HAL). Teil 1. Pteridophyta und Gymnospermae. Schlechtendalia 1: 15–18. [*Juniperus, Pinus, Taxus, Prumnopitys*, Schlechtendal's herbarium].

01210   Braun-Blanquet, J. (1927). *Callitris articulata* (Vahl) Murb. Pflanzenareale, 1 (3): 44, map 30. [= *Tetraclinis articulata* (Vahl) Masters].

01215   Brayshaw, T. C. (1997). Washoe and Ponderosa Pines on Promontory Hill near Merritt, B.C., Canada. Ann. Naturhist. Mus. Wien 99B: 673–680. [*Pinus washoensis, P. ponderosa* var. *ponderosa*].

01220   Bredemeier, H. (1886). Einige interessante Coniferen-Zapfen. Deutsche Gart.-Zeitung (Wittmack) 1: 487–489, f. 105, 495–496, f. 106–107. [*Picea jezoensis, Sciadopitys verticillata*].

01221   Brem, M. (1934). Anatomical method for determining the wood of the Spruce and the Larch. Bull. Acad. Polon. Sci., Cl. Sci. Math., Sér. B 1, Bot. 1: 103–111. [*Picea, Larix*, ill.].

01230   Brennan, M. & J. C. Doyle (1956). The gametophytes and embryogeny of *Athrotaxis*. Sci. Proc. Roy. Dublin Soc. 27 (7): 193–252, figs., pl. 4–8.

01231   Britton, N. L. (1908). North American Trees; being descriptions and illustrations of the Trees growing independently of cultivation in North America, north of Mexico and the West Indies... With the assistance of J. A. Shafer. (894 pp., ill.). London. [*Juniperus lucayana* sp. nov. = *J. barbadensis* var. *lucayana* (Britton) R. P. Adams (*J. barbadensis* auct., non L.), p. 121; lectotypification of genus *Pinus* L. with *P. sylvestris* L.].

01232   Britton, N. L. (1923). Studies of West Indian plants... XI. 64. Undescribed species from Cuba. Bull. Torrey Bot. Club 50: 35–51. [*Juniperus saxicola* Britton et Wilson, sp. nov., p. 35].

01240   Britton, N. L. (1926). The swamp cypresses. J. New York Bot. Gard. 27: 205–207. [*Taxodium distichum, T. mucronatum*].

01250   Britton, N. L., E. E. Sterns & J. F. Poggenburg (1888). Preliminary catalogue of Anthophyta and Pteridophyta reported as growing spontaneously within one hundred miles of New York City. (Torrey Bot. Club) New York. [*Chamaecyparis thyoides* (L.) comb. nov., *Picea mariana* (Mill.) comb. nov., p. 71].

01260   Brongniart, A. (1833). Note sur quelques Conifères de la tribu des Cupressinées. Ann. Sci. Nat. (Paris), sér. 1, 30: 176–191. [*Taxodium ascendens* sp. nov. = *T. distichum* var. *imbricatum*, p. 182, *T. microphyllum* sp. nov. = *T. distichum*, p. 182, *T. japonicum* (Thunb. ex L.f.) Brongn. = *Cryptomeria japonica*, p. 183, *T. japonicum* var. *heterophyllum* var. nov. "(*Taxodium heterophyllum*?)" = *Glyptostrobus pensilis*, p. 184, *Pachylepis* Brongn. non Lessing = *Widdringtonia*, p. 190].

01261   Brongniart, A. (pres.) (1854). (Séance du 14 Juin 1854) "M. Decaisne présente des échantillons de plusieurs Conifères gigantesques..." Bull. Soc. Bot. France 1: 70–71. [*Taxodium montezumae* Decne., nom. nud. = *T. mucronatum*, p. 71].

01263   Brongniart, A. & A. Gris (1866). Sur quelques Conifères de la Nouvelle Calédonie. Bull. Soc. Bot. France 13: 422–427. [*Podocarpus ustus* (Vieill.) comb. nov. = *Parasitaxus usta*, p. 426].

01263a   Brongniart, A. & A.Gris (1866). Observations sur diverses plantes nouvelles ou peu connus de la Nouvelle-Calédonie. Ann. Sci. Nat. Bot., sér. 5, 6: 238–266. [*Dacrydium araucarioides* sp. nov., p. 244].

01264   Brongniart, A. & A. Gris (1869). Nouvelle note sur les Conifères néocalédoniennes. Bull. Soc. Bot. France 16: 325–331. [*Frenela balansae* sp. nov. = *Callitris sulcata*, p. 327, *Dacrydium balansae* sp. nov., p. 328, *D. lycopodioides* sp. nov., p. 329, *D. pancheri* sp. nov. = *Acmopyle pancheri*, p. 330].

01269   Brongniart, A. & A. Gris (1871). Observations sur diverses plantes nouvelles ou peu connus de la Nouvelle Calédonie. Ann. Sci. Nat. Bot., sér. 5, 13: 340–404. [*Araucaria balansae* sp. nov. = *A. subulata*, p. 351, *A. cookii* R. Br. ex D. Don var. *luxurians* var. nov.= *A. luxurians*, p. 354, *A. montana* sp. nov., p. 358, *A. muelleri* (Carrière) comb. nov., p. 362].

01270   Brongniart, A. & A. Gris (1871). Supplément aux Conifères de la Nouvelle Calédonie. Bull. Soc. Bot. France 18: 130–141. [*Libocedrus austrocaledonica* sp. nov. (as "austrocaledonicus"), p. 140].

01270a   Brophy, J. J., R. J. Goldsack, M. Z. Wu, C. J. R. Fookes & P. I. Forster (2000). The steam volatile oil of *Wollemia nobilis* and its comparison with other members of the Araucariaceae (*Agathis* and *Araucaria*). Biochem. Syst. Ecol. 28: 563–578.

01270d   Browicz, K. (1996). *Juniperus macrocarpa* Sibth. & Sm. in the area of the "Flora of Turkey and East Aegean Islands". Karaca Arbor. Mag. 3 (3): 117–121. (ill., map). [= *Juniperus oxycedrus* ssp. *macrocarpa*].

01270e   Browicz, K. & J. Zielinski (1974). Nowe formy jalowca halnego z Pilska. Arbor. Kórnickie 19: 39–43. [*Juniperus communis* (ssp. *nana*) var. *recta*, var. nov., p. 41 et f. *crispa*, f. nov., p. 42].

01274   Brown, C. A. & G. N. Montz (1986). Baldcypress, the tree unique, the wood eternal. Baton Rouge, Louisiana. (9 colour, 179 black & white phot.). [*Taxodium distichum*].

01280   Brown, N. E. (1909). *Pinus Bungeana*. Bot. Mag. 135: pl. 8240.

01289   Brown, R. (1826). Character and description of Kingia, a new genus of plants found on the south-west coast of New Holland: with observations on the structure of its unimpregnated ovulum; and on the female flower of Cycadeae and Coniferae. In: Appendix to Captain Phillip P. King: Narrative of a survey of the intertropical and western coasts of Australia. John Murray, London. [first description of gymnospermy]

01290   Brown, R. (1826). [*Cunninghamia* gen. nov., p. 149, pl. 18; *C. sinensis* sp. nov., p. 80, pl. 18, f. 3) in: L.C.M. Richard. Commentatio botanica de Conifereis et Cycadeis,... Stuttgart. (nom. cons., see W. Greuter *et al.*, 1988; *C. sinensis* = *C. lanceolata*].

01300   Brown, R. (1868). A monograph of the coniferous genus *Thuja* Linn., and of the North American species of the genus *Libocedrus*, Endl. Trans. Bot. Soc. Edinburgh 9: 358–378. [*Thuja occidentalis* L., *T. gigantea* Nutt. = *T. plicata* D. Don, *Libocedrus decurrens* Torrey = *Calocedrus decurrens* (Torrey) Florin].

01310   Brown, R. W. (1936). The genus *Glyptostrobus* in America. J. Washington Acad. Sci. 26: 353–357, f. 1–7. [*Glyptostrobus dakotensis* sp. nov., *G. oregonensis* sp. nov., palaeobotany].

01311   Brown, R. W. (1940). [*Abies concoloroides*) (*Pseudolarix americana*) new species and names... J. Washington Acad. Sci. 30: 347. [palaeobotany].

01312   Brown, S. (1907). [Shorter notes] *Picea albertiana* sp. nov. Torreya 7 (6): 125–126. [= *P. glauca* var. *albertiana*, p. 126].

01317   Bruederle, L. P., D. F. Tomback, K. K. Kelly & R. C. Hardwick (1998). Population genetic structure in a bird-dispersed pine, *Pinus albicaulis* (Pinaceae). Canad. J. Bot. 76: 83–90.

01320   Brügger, Chr. G. (1886). Mitteilungen über neue und kritische Pflanzenformen. Jahresber. Naturf. Ges. Graubünden, ser. 2, 29: 46–177. [*Abies alpestris* sp. nov. = *Picea abies* var. *alpestris* (Bruegger) P. Schmidt = *P. abies* var. *abies*, p. 167, see also Stein, 1887].

01323   Brummitt, R. K., R. R. Mill & A. Farjon (2004). The significance of 'it' in the nomenclature of three Tasmanian conifers: *Microcachrys tetragona* and *Microstrobos niphophilus* (Podocarpaceae), and *Diselma archeri* (Cupressaceae). Taxon 53 (2): 529–539. [*Microcachrys tetragona*, *Microstrobos niphophilus* = *Pherosphaera hookeriana*, *M. fitzgeraldii* = *P. fitzgeraldii* (Podocarpaceae), *Diselma archeri* (Cupressaceae), nomenclature, typification, ill.].

01330   Buch, C. L. von (1828). Physicalische Beschreibung der Canarischen Inseln. Berlin. (1825 on title page, but see Stafleu & Cowan, TL-2, Vol. 1, 1976). [*Juniperus grandifolius* Link in Buch, sp. nov., *Pinus canariensis* C. Smith in Buch, sp. nov., p. 159].

01340   Buchholz, J. T. (1918). Suspensor and early embryo of *Pinus*. Bot. Gaz. (Crawfordsville) 66: 185–228, pl. 6–10.

01352   Buchholz, J. T. (1930). The Ozark white cedar. Bot. Gaz. (Crawfordsville) 90 (3): 326–332, 2 figs. [*Juniperus ashei* sp. nov., p. 329].

01360   Buchholz, J. T. (1931). The pine embryo and the embryos of related genera. Trans. Illinois State Acad. Sci. 23: 117–125. [*Pinus*].

01360a Buchholz, J. T. (1932). The suspensor of *Cryptomeria japonica*. Bot. Gaz. (Crawfordsville) 93: 221–226. [embryology].

01362 Buchholz, J. T. (1938). Cone formation in *Sequoia gigantea*. Amer. J. Bot. 25: 296–305. [*Sequoiadendron giganteum*].

01370 Buchholz, J. T. (1939). The morphology and embryogeny of *Sequoia gigantea*. Amer. J. Bot. 26: 93–101. [*Sequoiadendron giganteum*].

01375 Buchholz, J. T. (1939). The embryogeny of *Sequoia sempervirens* with a comparison of the sequoias. Amer. J. Bot. 26: 248–256. [*Sequoia, Sequoiadendron*].

01380 Buchholz, J. T. (1939). The generic segregation of the Sequoias. Amer. J. Bot. 26: 535–538. [*Sequoia* Endl., *Sequoiadendron* Buchholz, gen. nov., *S. giganteum* (Lindl.) comb. nov., p. 536].

01390 Buchholz, J. T. (1942). A comparison of the embryogeny of *Picea* and *Abies*. Madroño 6: 156–167.

01410 Buchholz, J. T. (1949). Additions to the coniferous flora of New Caledonia. Bull. Mus. Hist. Nat. (Paris) sér. 2, 21: 279–286. [*Araucaria humboldtensis* sp. nov., *A. biramulata* sp. nov., p. 279, *A. bernieri* sp. nov., p. 280, *Dacrydium guillauminii* sp. nov., p. 282, *Libocedrus chevalieri* sp. nov., p. 283, *Podocarpus comptonii* sp. nov. = *Retrophyllum comptonii*, p. 284, *P. palustris* sp.nov. = *R. minus*, p. 284, *P. sylvestris* sp. nov., p. 285].

01420 Buchholz, J. T. (1951). A flat-leaved pine from Annam, Indo-China. Amer. J. Bot. 38: 245–252. [*Pinus krempfii*].

01429 Buchholz, J. T. & N. E. Gray (1947). A Fijian *Acmopyle*. J. Arnold Arbor. 28: 141–143, pl. 1. [*Acmopyle pancheri*, *A. sahniana* sp. nov., p. 142].

01440 Budkevich, E. V. (1934). Anatomie de quelques espèces du genre *Juniperus*. Sovetsk. Bot. 1934 (6): 116–124. [*Juniperus davurica*, key to spp.)

01450 Budkevich, E. V. (1958). Anatomical structure of wood of *Ducampopinus krempfii* (Lecomte) A. Chevalier. Bot. Žurn. (Moscow & Leningrad) 43: 1156–1160. (Russ.). [= *Pinus krempfii*].

01461 Bui, N. S. (1962). Matériaux pour la "Flore du Cambodge, du Laos et du Vietnam". Les Abiétacées. Adansonia, sér. 2, 2 (2): 329–342. [Pinaceae, *Keteleeria evelyniana*, *Pinus dalatensis*, *P. khasya* = *P. kesiya*, *P. krempfii*, *P. merkusii* = *P. latteri*].

01465 Bukovac, M. J. & F. B. Widmoyer (1980). Observations on leaf characteristics of Afghanistan pine. J. Amer. Soc. Hort. Sci. 105 (3): 293–297. [*Pinus brutia*, SEM ill.].

01470 Bulard, C. (1947). Additions et corrections à l'étude du genre *Picea*. I. *P. sikangensis* Cheng. Bull. Soc. Hist. Nat. Toulouse 81: 131–136, 1 f.

01480 Bulard, C. (1947). Quatre épicéas de la Chine centrale: *P. ascendens*, *P. brachytyla*, *P. complanata*, *P. sargentiana*. Bull. Soc. Hist. Nat. Toulouse 82: 177–186. (ill.).

01481 Bulard, C. (1948). Les canaux résinifères de la feuille de *Picea omorica*. Bull. Soc. Hist. Nat. Toulouse 83 (3–4): 169–172. [*Picea omorika*].

01485 Bullock, A. A. (1957). The typification of the generic name *Callitris* Vent. Taxon 6 (8): 227–228. [*C. rhomboidea* as lectotype, *C. cupressiformis*, *C. quadrivalvis*, *Frenela*].

01486 Bullock, A. A. & D. R. Hunt (1966). The generic name of the golden larch. Taxon 15: 240–241. [*Pseudolarix amabilis*].

01490 Burgh, J. van der (1973). Hölzer der niederrheinischen Braunkohlenformation, 2. Hölzer der Braunkohlengruben "Maria Theresia" zu Herzogenrath, "Zukunft West" zu Eschweiler und "Victor" (Zülpich mitte) zu Zülpich. Nebst einer systematisch-anatomischen Bearbeitung der Gattung *Pinus* L. Rev. Palaeobot. Palynol. 15 (2–3): 73–275. [*Pinus* subsect. *Aristatae*, p. 90, subsect. *Nelsoniae*, p. 92, sect. *Leiophylla*, p. 92, sect. *Lumholtzii*, p. 92, subsect. *Halepenses*, p. 92, subsect. *Pseudostrobi*, p. 93, subsect. *Attenuatae*, p. 93, subsect. *Torreyanae*, p. 94, sect. et subsect. nov.].

01490a Burgh, J. van der & J. J. F. Meijer (1996). *Taxodioxylon gypsaceum* and its botanical affinities. Current Science 70 (5): 373–378. [palaeobotany, wood anatomy].

01491 Burgsdorf, F. A. L. von (1787). Anleitung zur sicheren Erziehung und zweckmässigen Anpflanzung der einheimischen und fremden Holzarten... Vol. 1–2. Berlin. [*Juniperus sibirica* sp. nov. = *J. communis* var. *saxatilis*, Vol. 1, p. 124, n. 272; contemp. German floras recognize this taxon as a distinct sp. in the subalpine comm. of central Europe].

01494 Burlingame, L. L. (1913). The morphology of *Araucaria brasiliensis* I. The staminate cone and male gametophyte. Bot. Gaz. (Crawfordsville) 55 (2): 97–113. [*Araucaria angustifolia*].

01495 Burlingame, L. L. (1914). The morphology of *Araucaria brasiliensis* II. The ovulate cone and female gametophyte. Bot. Gaz. (Crawfordsville) 57 (6): 490–508. [*Araucaria angustifolia*].

01496 Burlingame, L. L. (1915). The morphology of *Araucaria brasiliensis* III. Fertilization, the embryo, and the seed. Bot. Gaz. (Crawfordsville) 59 (1): 1–38. [*Araucaria angustifolia*].

01498 Burrows, G. E. (1999). Wollemi pine (*Wollemia nobilis*, Araucariaceae) posesses the same unusual leaf axil anatomy as the other investigated members of the family. Austral. J. Bot. 47 (1): 61–68.

01499   Burrows, G. E. & S. Bullock (1999). Leaf anatomy of Wollemi pine (*Wollemia nobilis*, Araucariaceae). Austral. J. Bot. 47: 795–806. [*Agathis, Araucaria, Wollemia*, ill.].

01500   Burschel, P. (1965). Die Omorikafichte *Picea omorika* (Pančić) Purkyne. Eine Literaturübersicht. Forstarchiv 36: 113–131.

01502a   Bush, E. W. & M. F. Doyle (1997). Taxonomic description of *Acmopyle sahniana* (Podocarpaceae): additions, revisions, discussion. Harvard Pap. Bot. 2: 229–233.

01503   Busing, R. T., C. B. Halpern & T. A. Spies (1995). Ecology of Pacific yew (*Taxus brevifolia*) in western Oregon and Washington. Conserv. Biol. 9 (5): 199–1207.

01505   Businský, R. (1989). Beitrag zur Taxonomie und Nomenklatur von *Pinus heldreichii* Christ und P. *leucodermis* Antoine sowie des Kultivars 'Smidtii'. Mitt. Deutsch. Dendrol. Ges. 79: 91–106.

01506   Businský, R. (1999). Study of *Pinus dalatensis* Ferré and of the enigmatic "Pin du Moyen Annam". Candollea 54: 125–143. [*Pinus dalatensis* var. *bidoupensis* var. nov., p. 127, *P. dalatensis* subsp. *procera* subsp. nov., p. 133].

01507   Businský, R. (1999). Taxonomic revision of Eurasian pines (genus *Pinus*) – survey of species and infraspecific taxa according to latest knowledge. Acta Pruhoniciana 68: 7–86. [*Pinus* subgenus *Pinus, P.* subgenus *Strobus*, 43 spp. recognized in Eurasia, 10 new combinations at infraspecific rank, ill. of cones].

01508   Businský, R. (1999). Taxonomická studie agregátu *Pinus mugo* ajeho hybridních populací. (Taxonomic essay of the *Pinus mugo* complex and its hybrid populations). Acta Pruhoniciana 68: 123–143. (Czech, Eng. & German summ.).

01509   Businský, R. (2003). Taxonomy and biogeography of Chinese hard pine, *Pinus hwangshanensis* W. Y. Hsia. Bot. Jahrb. Syst. 125 (1): 1–17. [*P. hwangshanensis* ssp. *transfluminea* ssp. nov., p. 7, *P. massoniana, P. taiwanensis, P. × cerambycifera* nothosp. nov., p. 13, ill.].

01509a   Businský, R. (2003). A new hard pine (*Pinus*, Pinaceae) from Taiwan. Novon 13 (3): 281–288. [*Pinus fragilissima* sp. nov., p. 282, similar to *P. taiwanensis* and *P. luchuensis*, ill.].

01515   Bůžek, Č., F. Holý & Z. Kvaček (1968). Die Gattung *Doliostrobus* Marion und ihr Vorkommen im nordböhmischen Tertiär. Palaeontographica, B 123 (1–6): 153–172. [Araucariaceae, Cupressaceae s.l., Taxodiaceae, *Doliostrobus gurnardii* (Florin) Bůžek, Holý & Kvaček, *D. selseyensis* (Chandler) Bůžek, Holý & Kvaček, *D. certus* Bůžek, Holý & Kvaček, *D. hungaricus* (Rasky) Bůžek, Holý & Kvaček, *D. taxiformis* (Sternberg) Kvaček, palaeobotany, ill.].

01520   Bykov, B.A. (1950). Fichtenwälder des Tien-Schan, ihre Geschichte, Besonderheiten und Typologien. Izv. Akad. Nauk. Kazakhsk. SSR, Ser. Bot. 5: 1–128. (Russ.). [*Picea schrenkiana* ssp. *tianschanica* (Rupr.) comb. et stat. nov., p. 22].

01525   Caballero D., M. (1967). Estudio comparativo de dos especies de pinos mexicanos (*Pinus pseudostrobus* Lindl. y *Pinus montezumae* Lamb.) con base en caracteristicas de plantula y semilla. Boletin Tecnico 20. Secr. Agric. & Ganaderia, Subsecr. Forest. Fauna, Inst. Nac. Invest. Forest. (INIF), Mexico.

01540   Campo, M. van (1955). Quelques pollens d'hybrides d'Abiétacées. Silvae Genet. 4: 123–126. [*Tsuga longibracteata = Nothotsuga longibracteata, Tsuga mertensiana, Abies* art. hybr.].

01590   Campo-Duplan, M. van & H. Gaussen (1948). Sur quatre hybrides des genres chez les Abietinées. Trav. Lab. Forest. Toulouse, T. 1, vol. 4, art. 24: 24–28. [Pinaceae, × *Tsugo-Keteleeria longibracteata*, × *Tsugo-Picea hookeriana*, × *Tsugo-Piceo-Picea crassifolia*, × *Tsugo-Piceo-Tsuga jeffreyi*, nothogen. et comb. nov.: nomina illeg. ICBN Art. H 6.2; for comment see also J. W. Duffield, 1950].

01595   Campos Dias, J. L. (1993). Claves para la determinación de los pinos mexicanos. Apoyos Académicos No. 22. Univ. Autonoma Chapingo. [*Pinus* spp., keys on morphology].

01600   Campos Ortega, L. R. (1927). Boceto histórico sobre el ahuehuete de El Tule, Oaxaca. (127 pp.). Bibliotheca SMGE. [*Taxodium mucronatum*].

01610   Camus, A. (1914). Les Cyprès. Encyclopédie économique de Sylviculture 2. Paul Lechevalier, Paris. [*Cupressus, C. goveniana* ssp. *pygmaea* (Lemmon) comb. nov., p. 50, *C. duclouxiana* Hickel, sp. nov., p. 91; treates *Chamaecyparis* spp. under *Cupressus*, ill.].

01611   Camus, A. (1926). Le *Cupressus dupreziana* A. Camus, Cyprès nouveau du Tassili. Bull. Soc. Dendrol. France 58: 39–44. [sp. nov. first publ. in: Bull. Mus. Hist. Nat. (Paris) 32 (1): 101, 1926].

01620   Candolle, A.P. de & Alph. de Candolle (1867). Prodromus systematis naturalis regni vegetabilis. Vol. 16, part 2. Paris. [see F. Parlatore, 1867].

01630   Candolle, A. P. de & J. B. A. P. M. de Lamarck (1805). Flore française, ou descriptions succinctes de toutes les plantes qui croissent naturellement en France,... Vol. 3. Ed. 3, Paris. [*Abies pectinata* (Lam.) Lam. et DC. = *A. alba*, p. 276, *Larix europaea* Lam. et DC. = *L. decidua*, p. 277; *Pinus uncinata* Ramond ex DC, p. 726; see also L. F. E. de C. Ramond, 1805].

01640 Canright, J. E. (1972). Evidence of the existence of *Metasequoia* in the Miocene of Taiwan. Taiwania 17 (2): 222–228.

01644 Cantrill, D. J. (1991). Broad leafed coniferous foliage from the Lower Cretaceous Otway Group, southeastern Australia. Alcheringa 15: 177–190. [Araucariaceae, *Podozamites taenioides*, sp. nov.; palaeobotany (Albian)].

01645 Cantrill, D. J. (1992). Araucarian foliage from the lower Cretaceous of southern Victoria, Australia. Int. J. Plant Sci. 153 (4): 622–645. (ill.). [*Araucaria lanceolatus* sp. nov., p. 624, *A. acutifolius* sp. nov., p. 626, *A. carinatus* sp. nov., p. 631, *A. falcatus* sp. nov., p. 631, *A. otwayensis* sp. nov., p. 635, *A. victoriensis* sp. nov., p. 635; palaeobotany (Albian)].

01650 Carlisle, A. (1958). A guide to the named variants of Scots Pine (*Pinus sylvestris* Linnaeus). Forestry (Oxford) 31: 203–224.

01654 Carlson, C. E. & L. J. Theroux (1993). Cone and seed morphology of western larch (*Larix occidentalis*), alpine larch (*Larix lyallii*), and their hybrids. Canad. J. Forest. Res. 23: 1264–1269.

01660 Carrière, E. A. (1854). *Thuia gigantea* et autres conifères de la Californie et du Mexique septentrional. Rev. Hort. 4 (3): 223–229. (ill.). [*Thuja gigantea* = *T. plicata, Pinus chihuahuana, P. engelmannii* sp. nov., p. 227, *P. edulis, P. strobiformis, Taxus boursieri* sp. nov. = *T. brevifolia*, p. 228].

01661 Carrière, E. A. (1854). Description d'une nouvelle espèce de Genévrier, *Juniperus californica* Carr. Rev. Hort. 4 (3): 352–353, f. 21.

01670 Carrière, E. A. (1855). *Biota orientalis, Biota pyramidalis* et *Taxus adpressa*. Rev. Hort. 4 (4): 93–96, f. 6–8. [*Biota* = *Thuja, Platycladus orientalis*].

01671 Carrière, E. A. (1855). Les *Cupressus macrocarpa* et *lambertiana*. Rev. Hort. 1855: 232–235. [*C. macrocarpa, C. lambertiana, C. hartwegii* sp. nov. = *C. macrocarpa*, p. 233].

01681 Carrière, E. A. (1856). [*Abies selinusia* sp. nov. = *A. cilicica*, p. 69, *Larix kaempferi* (Lamb.) comb. nov., p. 97) in: Fl. Serres Jard. Eur. (Ghent) Vol. 11.

01690 Carrière, E. A. (1861). *Biota orientalis*. Rev. Hort. 1861: 229–231, f. 45. [= *Platycladus orientalis*].

01700 Carrière, E. A. (1866). *Abies numidica* De Lannoy ex Carrière, sp. nov. Rev. Hort. 1866 (37): 106–107.

01703 Carrière, E. A. (1866). Fructification du *Libocedrus doniana*. Rev. Hort. 1866 (37): 230. (ill.). [=*Libocedrus plumosa*].

01704 Carrière, E. A. (1866). *Eutacta rulei* [var.] *polymorpha*. Rev. Hort. 1866 (37): 350, pl. [=*Araucaria rulei*].

01705 Carrière, E. A. (1866). Quelques *Eutacta* de la Nouvelle-Calédonie. Rev. Hort. 1866 (37): 392, pl. [*Araucaria rulei, A. muelleri*, new comb. not validly publ., Art. 33.1, but validated in Index to Traité Gén. Conif., ed. 2, 1867].

01710 Carrière, E. A. (1866). *Keteleeria fortunei*. (A. Murray bis) Carrière, gen. et comb. nov. Rev. Hort. 1866 (37): 449–451. [basion.: *Picea fortunei* A. Murray bis, 1862].

01730 Carrière, E. A. (1868). *Keteleeria fortunei*. Rev. Hort. 1868 (40): 132–133, 1 pl.

01731 Carrière, E. A. (1868). Anomalies presentées par des feuilles de *Sciadopitys verticillata*. Rev. Hort. 1868 (40): 150–151.

01732 Carrière, E. A. (1868). *Pseudotsuga lindleyana* (Roezl) Carrière, comb. nov. Rev. Hort. 1868 (40): 152, pl. [basion.: *Tsuga lindleyana* Roezl, 1857 = *Pseudotsuga menziesii* var. *glauca*].

01740 Carrière, E. A. (1868). *Pseudolarix kaempferi*. Rev. Hort. 1868: 332–333, 1 pl. [*Pseudolarix amabilis*].

01750 Carrière, E. A. (1869). *Chamaecyparis obtusa*. Rev. Hort. 1869: 97, f. 25.

01760 Carrière, E. A. (1870). *Cryptomeria nigricans*. Rev. Hort. 1870: 119–120. [*C. nigricans* sp. nov. = *C. japonica*].

01770 Carrière, E. A. (1871). *Pseudolarix kaempferi*. Rev. Hort. 1870–71: 608–609, f. 80–81. [*Pseudolarix amabilis*].

01780 Carrière, E. A. (1873). *Pseudotsuga davidiana*. Rev. Hort. 1873: 37–38, f. 3–5. [= *Keteleeria davidiana*, sic!].

01790 Carrière, E. A. (1874). Un grand arbre nanisé. Rev. Hort. 1874: 272–273, f. 34. [*Pinus densiflora* var. *albiflora* = cultivar].

01800 Carrière, E. A. (1880). Revue du genre *Retinispora*. Rev. Hort. 1880: 35–39, f. 4–8, 93–97, f. 17–22, 177–179. [*Retinispora* = *Chamaecyparis, Thuja, Widdringtonia*, mostly describing and illustrating cultivars under this genus as circumscribed by the author].

01810 Carrière, E. A. (1884). Fructification du *Sciadopitys verticillata*. Rev. Hort. 1884: 16–18, f. 1–4.

01820 Carrière, E. A. (1887). *Keteleeria fortunei*. Rev. Hort. 1887: 207–212, f. 42, 45; pp. 246–248

01821 Carrière, E. A. (1890). *Cryptomeria araucarioides*. Rev. Hort. 1890 (Vol. 62): 518. [*C. araucarioides* sp. nov. = *C. japonica* cultivar].

01830 Carvajal, S. (1986). Notas sobre la flora fanerogamica de Nueva Galicia, III. Phytologia 59 (2): 127–147. [*Pinus novo-galiciana* nom. nov. (syn. : *P. ayacahuite* var. *brachyptera* Shaw), p. 131, *P.* subsect. *Rzedowskianae* subsect. nov., p. 134, *P. macvaughii* sp. nov., p. 139, et var. nov.; ill., map, key].

01832d Castor, C., J. G. Cuevas, M. T. Kalin Arroyo, Z. Rafii, R. Dodd & A. Peñaloza (1996). *Austrocedrus chilensis* (D. Don) Pic.-Ser. et Bizz. (Cupressaceae) from Chile and Argentina: monoecious or dioecious? Rev. Chilena Hist. Nat. 69: 89–95. [only dioecious plants found in 19 wild populations].

01832e Cavaleiro, C., S. Rezzi, L. Salgueiro, A. Bighelli, J. Casanova & A. Proenca da Cunha (2001). Infraspecific chemical variability of the leaf essential oil of *Juniperus phoenicea* var. *turbinata* from Portugal. Biochem. Syst. Ecol. 29 (11): 1175–1183.

01833 Ceballos, L. F. C. & M. Bolaños (1928). Flora y mapa forestal. El Pinsapo y el Abeto de Marruecos. Bol. Inst. Nac. Invest. Exp. Agron. Forest. 1 (2): 1–57. [*Abies pinsapo* var. *marocana* (Trabut) comb. et stat. nov., p. 18].

01834 Ceballos, L. & F. Ortuño (1947). Notas sobre Flora Canariense. Año xviii, Núm. 33, Madrid. [*Juniperus phoenicea*, pp. 7–9, pl. 1–2, ill. of transv. sect. of ultim. branchlets].

01840 Čelakovsky, L. J. (1867). Prodromus der Flora von Böhmen. Prag. [*Pinus montana* ssp. *uncinata*, (Ramond ex DC.) comb. nov. = *P. mugo* ssp. *uncinata* (Ramond ex DC.) Domin].

01870 Cesalpino, A. (1583). De Plantis. Libri XVI. (Coniferales in: Liber III, Cap. 52, pp. 129–132: "Pinus, Picea, Abies, Cupressus, et altera..."). Apud Georgium Marescottum, Florence.

01877 Chadwick, L. C. & R. A. Keen (1976). A study of the genus *Taxus*. Ohio Agric. and Res. Bull. 1086. Columbus, Ohio.

01880 Challinor, D. & D. B. Wingate (1971). The struggle for survival of the Bermuda cedar. Biol. Conservation (Barking) 3 (3): 220–222. [*Juniperus bermudiana* L.].

01885 Chaloner, W. G. & J. Lorch (1960). An opposite-leaved conifer from the Jurassic of Israel. Palaeontology 2 (2): 236–242.

01888 Chamberlain, C. J. (1908). Development of *Juniperus*. In: Current literature – notes for students. Bot. Gaz. (Crawfordsville) 46: 237. [embryology, abstract].

01891 Chambers, T. C., A. N. Drinnan & S. McLoughlin (1998). Some morphological features of Wollemi pine (*Wollemia nobilis*: Araucariaceae) and their comparison to Cretaceous plant fossils. Int. J. Plant Sci. 159 (1): 160–171. [*Agathis, Araucaria, Wollemia*, araucarian fossils, ill.].

01895 Chan, L. L. (1985). The anatomy of the bark of *Libocedrus* in New Zealand. I.A.W.A. Bull., n.s. 6 (1): 23–34. [*L. bidwillii, L. plumosa*].

01898 Chandler, M. E. J. (1922). *Sequoia couttsiae* Heer, at Hordle, Hants: a study of the characters which serve to distinguish *Sequoia* from *Athrotaxis*. Ann. Bot. (London) 36 (143): 385–390. (ill.). [morphology, palaeobotany].

01900 Chaney, R. W. (1948). The bearing of the living *Metasequoia* on problems of Tertiary paleobotany. Proc. Nation. Acad. U.S.A. 34: 503–515.

01905 Chaney, R. W. (1949). Redwoods – occidentale and orientale. Science 110: 551–552. [*Metasequoia, Sequoia, Sequoiadendron*].

01910 Chaney, R. W. (1951). A revision of fossil *Sequoia* and *Taxodium* in Western North America based on the recent discovery of *Metasequoia*. Trans. Amer. Philos. Soc., n.s. 40: 169–263.

01920 Chaney, R. W. (1954). A new pine from the Cretaceous of Minnesota and its paleo-ecological significance. Ecology 35: 145–151. [*Pinus clementsii*, fossil sp. nov.].

01939 Chang, C.S., J.I. Jeon & H. Kim (2000). A systematic reconsideration of *Abies koreana* Wilson using flavonoids and allozyme analysis. Korean J. Pl. Taxon. 30 (3): 215–234. [*Abies koreana, A. nephrolepis*, hybridization].

01940 Chang, T. C. & C. G. Chen (1981). A new variety of *Sabina vulgaris* Ant. Acta Phytotax. Sin. 19 (2): 263. [*S. vulgaris* var. *yulinensis*). [*Juniperus sabina* L.].

01960 Chapman, H. H. (1952). The place of fire in the ecology of pines. Bartonia 26: 39–44. [*Pinus*].

01970 Chapman, J. D. (1961). Some notes on the taxonomy, distribution, ecology and economic importance of *Widdringtonia*, with particular reference to *W. whytei*. Kirkia 1: 138–154. [= *W. nodiflora*].

01971 Chapman, J. D. (1994). Notes on Mulanje cedar – Malawi's national tree. Commonw. Forest. Rev. 73 (4): 235–242. [*Widdringtonia nodiflora*].

01971a Chapman, J. D. (1995). The Mulanje cedar, Malawi's national tree. The Society of Malawi, Blantyre. [*Widdringtonia whytei, W. nodiflora*, ill.].

01972 Chapman, J. D. & F. White (1970). The evergreen forests of Malawi. Commonw. Forest. Inst. Univ. Oxford. [*Juniperus procera*, phot. 16, *Widdringtonia whytei*, phot. 51, 53–56].

01972c Charlet, D. A. (1996). Atlas of Nevada conifers, a phytogeographic reference. University of Nevada Press, Reno. (ill.). [*Juniperus, Pinus, Abies concolor*].

01973c Chaturvedi, S. V. (1999). Leaf morpho-anatomy of six species of *Pinus* L. (Abietaceae). Philippine J. Sci. 127 (1): 49–64. [*Pinus densiflora, P. echinata, P. wallichiana, P. koraiensis, P. sylvestris, P. tabuliformis*, ill.].

01974 Chaudhri, J. J. (1963). Distribution of gymnosperms in West-Pakistan. Vegetatio 11 (5–6): 372–382. [e.g. *Juniperus*].

01977 Chen, K. Y. (1983). Chromosome number of *Fokienia*. Acta Bot. Sin. 25: 120–122. (Chin., Eng. summ., ill.). [*Fokienia hodginsii*, 2n = 24, count needs confirmation].

01977a Chen, K. Y. (1995). Karyotype and sex chromosomes in *Cephalotaxus chinensis*. Acta Bot. Sin. 37 (2): 159–161. (Chin., Eng. summ.).

01977b Chen, W. L. (1991). The pine forests in western China. pp. 5–16 in: N. Nakagoshi & F. B. Golley (eds.). Coniferous forest ecology from an international perspective. Den Haag. [*Pinus*].

01978 Chen, Z. K. & F. H. Wang (1981). The early embryogeny of the genus *Fokienia*, with a note on its systematic position. Acta Phytotax. Sin. 19: 23–28, pl. 1–2. [*Fokienia hodginsii*].

01978b Chen, Z. K., J. H. Zhang & F. Zhou (1995). The ovule structure and development of [the] female gametophyte in *Cathaya* (Pinaceae). Cathaya 7: 165–176. [*Cathaya argyrophylla*].

01979 Chen, Z. X. & Z. Q. Li (1989). A new variety of *Podocarpus macrophyllus* (Thunb.) D. Don. Bull. Bot. Res. North-East Forest. Inst. 9 (3): 69. [*P. macrophyllus* var. *piliramulus* var. nov.].

01980 Cheng, W. C. (1930). A study of Chinese pines. Contr. Biol. Lab. Chin. Assoc. Advancem. Sci., Sect. Bot. 6: 5–21. [*Pinus*, 9 spp. descr., key].

01990 Cheng, W. C. (1931). A new spruce from western China. Contr. Biol. Lab. Chin. Assoc. Advancem. Sci., Sect. Bot. 6: 33–34, pl. 1–2. [*Picea sikangensis* sp. nov. = *P. likiangensis* var. *balfouriana*, see for comment: C. Bulard, 1947].

02000 Cheng, W. C. (1932). A new *Tsuga* from southwestern China. Contr. Biol. Lab. Chin. Assoc. Advancem. Sci., Sect. Bot. 7: (1): 1–3. [*T. longibracteata* sp. nov. = *Nothotsuga longibracteata*].

02015 Cheng, W. C. (1934). An enumeration of vascular plants from Chekiang, III. Gymnospermae. Contr. Biol. Lab. Sci. Soc. China, Sect. Bot. 9 (3): 240–242. (ill.). [*Taxus chienii* sp. nov. = *Pseudotaxus chienii*].

02030 Cheng, W. C. (1940). Une nouvelle espèce de *Juniperus* chinois, *J. gaussenii*. Trav. Lab. Forest. Toulouse, T. 1, vol. 3, art. 8: 1–4.

02040 Cheng, W. C. (1947). New Chinese trees and shrubs. Notes Forest. Inst. Natl. Centr. Univ. Nanking, Dendrol. Ser. 1: 1–4. (Eng., Lat., Chin. summ.). [*Pseudotaxus* gen. nov., *P. chienii* (W. C. Cheng) comb. nov., *Juniperus pingii* sp. nov., *Picea balfouriana* var. *hirtella* (Cheng ex Chen) comb. nov.; see also Trav. Lab. Forest. Toulouse, T. 5, sect. 1, vol. 1, art. 2: 93 (1940), there without Lat. descr.].

02060 Cheng, W. C., L. K. Fu & C. Y. Cheng (1975). Species Novae in Gymnospermae Sinicae. Acta Phytotax. Sin. 13 (4): 56–90 (Chin., Lat., ill.). [*Keteleeria oblonga*, p. 82, *K. pubescens*, p. 82, *K. calcarea*, p. 82, *Abies chayuensis*, p. 83, *A. nukiangensis*, p. 83, *A. delavayi* var. *motuoensis*, p. 83, *Pseudotsuga brevifolia*, p. 83, *Tsuga chinensis* var. *robusta*, p. 83, *T. chinensis* var. *oblongisquamata*, p. 83, *Picea likiangensis* var. *linzhiensis*, p. 83, *Larix speciosa*, p. 84, *L. himalaica*, p. 84, *L. potaninii* var. *macrocarpa* Law, p. 84, *Pinus dabeshanensis* Cheng et Law, p. 85, *Pinus* div. var. nov., p. 85, *Cupressus gigantea*, p. 85, *Sabina* (= *Juniperus*) div. var. nov., p. 86, *Cephalotaxus lanceolata* K. M. Feng, sp. nov., p. 86, *C. sinensis* var. *latifolia*, p. 86, *Taxus yunnanensis* sp. nov., p. 86, *Torreya yunnanensis* sp. nov., p. 87; all new taxa if not otherwise cited are by W. C. Cheng & L. K. Fu].

02070 Chevalier, A. (1944). Notes sur les conifères de l'Indochine. Rev. Int. Bot. Appl. Agric. Trop. 24: 7–34. [Coniferales, *Pinus langbianensis* sp. nov., p. 25, *Ducampopinus* gen. nov., *D. krempfii* (Lecomte) comb. nov. = *Pinus krempfii* Lecomte, p. 30].

02100 Chii, K. L. & W. S. Cooper (1950). An ecological reconnaissance in the native home of *Metasequoia glyptostroboides*. Ecology 31: 260–278, f. 1–6.

02100a Ching, K. K. (1959). Hybridization between Douglas-fir and Big-cone Douglas-fir. Forest Sci. 5: 246–254. [*Pseudotsuga menziesii*, *P. macrocarpa*].

02101 Chou, Y. L., ed. (1986). The Ligneous Flora of Heilongjiang. [*Larix olgensis* var. *heilingensis* (Yang et Chou) comb. nov. = *L. gmelinii* var. *gmelinii*, p. 40, *Picea koraiensis* var. *intercedens* (Nakai) comb. nov. = *P. koraiensis* var. *koraiensis*, p. 49].

02110 Chowdhury, C. R. (1961). The morphology and embryology of *Cedrus deodara* (Roxb.) Loud. Phytomorphology 11: 283–304.

02120a Chowdhury, K. A. (1974). *Abies* and *Picea*, morphological studies. Bot. Monogr. 8. CSIR, New Delhi. (ill.). [*A. delavayi*, *A. densa*, *A. pindrow*, *A. spectabilis*, *P. smithiana*, *P. spinulosa*].

02121 Christ, H. (1863). Uebersicht der Europäischen Abietineen (*Pinus*, Linn.). Verh. Naturf. Ges. Basel (n. s.) 3: 540–557. [*Pinus heldreichii* sp. nov., p. 549].

02122 Christ, H. (1863–67). Beiträge zur Kenntnis europaischer *Pinus* Arten. Flora 46: 368–380 (1863); 47: 147–158 (1864); 48: 257–260 (1865); 50: 81–84 (1867). [*Pinus* × *rhaetica* Brügger, 47: 150].

02130 Christensen, E. M. & M. J. Hunt (1965). A bibliography of Engelmann spruce. U.S. Forest Serv. Res. Paper INT-19. Intermountain Forest and Range Exp. Stat., Ogden, Utah. (37 pp.). [*Picea engelmannii*].

02140 Christensen, K. I. (1985). *Juniperus communis* ssp. *alpina* (Smith) Čelakovsky (Cupressaceae): a nomenclatural comment. Taxon 34 (4): 686–688.

02141 Christensen, K. I. (1987). A morphometric study of the *Pinus mugo* Turra complex and its natural hybridization with *P. sylvestris* L. (Pinaceae). Fedde's Repert. 98 (11–12): 623–635. [figs. + tables with numerical data].

02142 Christensen, K. I. (1987). Atypical cone and leaf character states in *Pinus mugo* Turra, *P. sylvestris* L. and *P.* × *rhaetica* Bruegger (Pinaceae). Gleditschia 15 (1): 1–5. (ill., map)].

02150 Christensen, K. I. (1987). Taxonomic revision of the *Pinus mugo* complex and *P.* × *rhaetica* (*P. mugo* × *sylvestris*) (Pinaceae). Nordic J. Bot. 7 (4): 383–408. [*P. mugo* ssp. *mugo, P. mugo* ssp. *uncinata*; very extensive synon. and bibl.].

02151 Christensen, K. I. (1993). Comments on the earliest validly published varietal name for the Corsican pine. Taxon 42 (3): 649–653. [*Pinus nigra* var. *corsica*, non var. *maritima*].

02153 Christensen, K. I. & G. H. Dar (1997). A morphometric analysis of spontaneous and artificial hybrids of *Pinus mugo* × *sylvestris* (Pinaceae). Nord. J. Bot. 17 (1): 77–86. [*Pinus mugo, P. sylvestris, P.* × *rhaetica*, hybridization].

02160 Christophel, D. C. (1973). *Sciadopitophyllum canadense* gen. et spec. nov.: a new conifer from western Alberta. Amer. J. Bot. 60 (1): 61–66. [fossil from Upper-Cretaceous to Palaeogene].

02170 Chu, C. (1981). A Brief Introduction to the Chinese Species of the Genus *Pseudotsuga*. Davidsonia 12 (1): 15–17.

02180 Chu, C. C. & C. S. Sun (1981). Chromosome numbers and morphology in *Cathaya*. Acta Phytotax. Sin. 19: 444–446. (Chin.).

02189 Chun, W. Y. (1925). Two new trees from Chekiang. J. Arnold Arbor. 6: 144–145. [*Torreya jackii* sp. nov.].

02190 Chun, W. Y. (1963). Materials for the Flora of Hainan (I) Pinaceae 1. *Keteleeria* Carrière: *Keteleeria hainanensis* Chun et Tsiang sp. nov. Acta Phytotax. Sin. 8 (3): 7–9.

02200 Chun, W. Y. & K. Z. Kuang (1958). Genus novum Pinacearum ex sina australi et occidentali. Bot. Žurn. (Moscow & Leningrad) 43 (4): 461–476. (Lat. + Russ.). [*Cathaya argyrophylla* gen. et sp. nov., *C. nanchuanensis* sp. nov. (type of genus not indicated), *C. loehrii* (Engelhardt et Kinkelin) comb. nov. ("*loehri*", fossil, with err. cit. of *Keteleeria loehrii* as basion.) pp. 464–465].

02201 Chun, W. Y. & K. Z. Kuang (1962). De genere *Cathaya* Chun & Kuang. Acta Bot. Sin. 10 (3): 245–247, pl. 1–3. (Chin., ill.). [validation of earlier published names by designation of the type of the genus on p. 245: *C. argyrophylla, C. nanchuanensis* cited in synonymy with type sp. on p. 246].

02202 Chung, L. W. (1951). The taxonomic revision and phytogeographical study of chinese Pines. Acta Phytotax. Sin. 5: 131–164. [*Pinus* spp.].

02230 Chylarecki, H. & M. Giertych (1969). Variability of *Picea abies* (L.) Karst. cones in Poland. Arbor. Kórnickie 14: 39–71.

02235 Ciesla, W. M. (2002). Juniper forests – a special challenge for sustainable forestry. Forests, Trees and Livelihoods 12: 195–207. [*Juniperus* spp., ecology, forestry, conservation].

02250 Clausen, R. T. (1939). Contributions to the Flora of New Jersey. Torreya 39: 125–133. [*Pinus rigida* ssp. *serotina* stat. nov.].

02251 Clement-Westerhof, J. A. (1984). The conifer *Ortiseia* Florin from the Val Gardena Formation of the Dolomites and the Vicentinian Alps (Italy) with special reference to a revised concept of the Walchiaceae (Goeppert) Schimper. (Aspects of Permian palaeobotany and palynology IV). Rev. Palaeobot. Palynol. 41: 51–166. [compare e.g C.R. Florin, 1951, 1954 and H. Kerp *et al.*, 1989].

02251a Clement-Westerhof, J. A. (1987). The Majonicaceae, a new family of Late Permian conifers. (Aspects of Permian palaeobotany and palynology VII). Rev. Palaeobot. Palynol. 52: 375–402.

02251b Clement-Westerhof, J. A. & J. H. A. van Konijnenburg-van Cittert (1991). *Hirmeriella muensteri*: New data on the fertile organs leading to a revised concept of the Cheirolepidiaceae. Rev. Palaeobot. Palynol. 68: 147–179. [Cheirolepidiaceae, *Hirmeriella muensteri* (Schenk) Jung, palaeobotany, ill.].

02270 Clinton-Baker, H. (1916). *Fokienia hodginsii* (A. Henry and H.H. Thomas). Gard. Chron., ser. 3, 59: 72, f. 30–31.

02282 Coates Palgrave, K. (1990). Trees of Southern Africa. Ed 2. Cape Town. [*Podocarpus* spp., pp. 55–58, *Widdringtonia* spp., pp. 59–61, *Juniperus procera*, p. 61, ill.].

02290 Codd, L. E., B. de Winter & H. B. Rycroft, eds. (1966). Flora of Southern Africa Vol. 1. [see J. A. Marsh, 1966].

02291 Coincy, M. de (1898). Remarques sur le *Juniperus thurifera* L. et les espèces voisines du Bassin de la Méditerranée. Bull. Soc. Bot. France 45 (5): 429–433. [*J. thurifera* var. *gallica*, var. nov., p. 430; see also in Bull. Soc. Bot. France 44 (4): 232, 1897, without descr.].

02298 Colenso, W. (1884). An account of visits to, and crossings over, the Ruahine Mountain Range, Hawke's Bay, New Zealand, and of the natural history of that region, 1845–1847. Napier. [*Podocarpus cunninghamii* sp. nov., p. 58].

02300 Colleau, C. (1968). Anatomie comparée des feuilles de *Picea*. Cellule 67: 185–253, pl. 1–3.

02303 Collingwood, G. H., W. D. Brush & D. Butcher (rev. ed.) (1984). Knowing your trees. Ed. 6. (32nd printing). The American Forestry Association, Washington, D.C. (ill., maps). [*Pinus* spp., *Larix* spp., *Picea* spp., *Tsuga* spp., *Pseudotsuga menziesii*, *Abies* spp., *Sequoia*, *Sequoiadendron*, *Taxodium distichum*, *Calocedrus decurrens*, *Thuja* spp., *Cupressus* spp., *Chamaecyparis* spp., *Juniperus* spp. *Taxus*].

02305 Collins, D., R. R. Mill & M. Möller (2003). Species separation of *Taxus baccata*, *T. canadensis*, and *T. cuspidata* (Taxaceae) and origins of their reputed hybrids inferred from RAPD and cpDNA data. Amer. J. Bot. 90 (2): 175–182.

02310 Coltman-Rogers, C. (1919). *Abies forrestii*. Gard. Chron., ser. 3, 65: 150–151 (sp. nov., protologue!); 80: 427, f. 191 (1926).

02320 Comber, J. (1930). *Thuya dolabrata*. Gard. Chron., ser. 3, 87: 324. [= *Thujopsis dolabrata*].

02330 Compton, R. H. (1922). Gymnosperms. in: R. H. Compton *et al.* A Systematic Account of the Plants collected in New Caledonia and the Isle of Pines by Mr. R. H. Compton, M.A., in 1914 Part II, gymnosperms and Cryptogams. J. Linn. Soc., Bot. 45: 421–434, pl. 26–27. [*Podocarpus ferruginoides* sp. nov. = *Prumnopitys ferruginoides*, p. 424, *Austrotaxus spicata* gen.et sp. Nov., p. 427, *Callitropsis* gen. nov., *C. araucarioides* sp. nov. = *Neocallitropsis pancheri*, p. 432, pl. 27].

02336 Conkle, M. T. & W. B. Critchfield (1988). Genetic variation and hybridization of Ponderosa pine. In: D. M. Baumgartner & J. E. Lotan (eds.). Ponderosa pine – the species and its management. Symp. proc. pp. 27–43. Washington State Univ., Pullman, Washington. [*Pinus ponderosa*, *P. ponderosa* var. *scopulorum*, *P. arizonica*, *P. engelmannii*, *P. washoensis*, *P. jeffreyi*].

02340 Conkle, M. T., G. Schiller & C. Grunwald (1988). Electrophoretic analysis of diversity and phylogeny of *Pinus brutia* and closely related taxa. Syst. Bot. 13: 411–424.

02348 Contreras-Medina, R. & I. Luna Vega (2002). On the distribution of gymnosperm genera, their areas of endemism and cladistic biogeography. Austral. Syst. Bot. 15: 193–203. (area cladogram).

02350 Conzatti, C. (1921). Monografía del árbol de Santa María del Tule, México. (65 pp., 21 ill.). [*Taxodium mucronatum*].

02365 Cool, L. G., Z. L. Hu & E. Zavarin (1998). Foliage terpenoids of Chinese *Cupressus* species. Biochem. Syst. Ecol. 26 (8): 899–913. [*Cupressus* spp., chemistry].

02370 Cooling, E. N. G. & H. Gaussen (1970). In Indochina: *Pinus merkusiana* sp. nov. et non *P. merkusii* Jungh. et De Vriese. Trav. Lab. Forest. Toulouse, T. 1, vol. 8, art. 7: 1–5.

02375a Coppen, J. J. W., C. Gay, D. J. James, J. M. Robinson & L. J. Mullin (1993). Xylem resin composition and chemotaxonomy of three varieties of *Pinus caribaea*. Phytochemistry 33: 1103–1111.

02378 Corbasson, M. (1968). Une nouvelle espèce d'*Araucaria* de Nouvelle Calédonie. Adansonia, sér. 2, 8: 467–468. [*Araucaria laubenfelsii* Corbasson].

02380 Cordes, J. W. H. (1866). Het geslacht *Pinus* in 't Zuidelijk halfrond. Natuurk. Tijdschr. Ned.-Indië 29: 130–135. [*P. merkusii*].

02380d Corner, E. J. H. (1939). Notes on the systematy and distribution of Malayan phanerogams III. (ill.). Gard. Bull. Straits Settlem. 10 (2): 239–329. [*Dacrydium comosum* sp. nov., p. 244, t. 10].

02390 Cory, V. L. (1936). Three junipers of western Texas. Rhodora 38: 182–187. [*Juniperus mexicana* Spreng. var. *monosperma* (Engelm.) comb. nov. = *J. monosperma*, p. 183, *J. gymnocarpa* (Lemmon) comb. et stat. nov., p. 184, *J. erythrocarpa*, sp. nov., p. 186].

02410 Coulter, J. M. & J. H. Rose (1886). Synopsis of North American pines based upon leaf-anatomy. I, II. Bot. Gaz. (Crawfordsville) 11: 256–262, 302–309.

02420 Courtot, Y. & L. Baillaud (1955). Sur la répartition des sexes chez un *Chamaecyparis*. Ann. Sci. Univ. Besançon, sér. 2, 6: 75–81. (ill.). [*C. nootkatensis*].

02430 Courtot, Y. & L. Baillaud (1961). Sur la ramification d'un *Cupressus*. Ann. Sci. Univ. Besançon, sér. 2, 17: 69–72. [*C. macrocarpa* var. *guadelupensis* = *C. guadelupensis*].

02440 Cozar, S. S. (1946). El *Abies* del Tazaot (*A. tazaotana*). Revista Real Acad. Ci. Madrid 15 (3): 449–469, pl. [= *A. pinsapo* var. *tazaotana*; new sp. descr. without Lat. diagn.].

02450 Crabtree, D. R. (1983). *Picea wolfei*, a new species of petrified cone from the Miocene of northwestern Nevada. Amer. J. Bot. 70 (9): 1356–1364. [*Picea wolfei*, fossil sp. nov.].

02460 Craib. W. G. (1919). A New Chinese *Pseudotsuga*. Notes Roy. Bot. Gard. Edinburgh 11 (55): 189–190, pl. [*P. forrestii* sp. nov.].

02470 Craib, W. G. (1919). *Abies delavayi* in Cultivation. Notes Roy. Bot. Gard. Edinburgh 11 (55): 277–280, 3 pl. [*Abies fabri* (Masters) comb. nov. (as "*faberi*"), p. 278].

02470f Crane, P. R., S. R. Manchester & D. L. Dilcher (1990). A preliminary survey of fossil leaves and well-preserved reproductive structures from the Sentinel Butte Formation (Paleocene) near Almont, North Dakota. Fieldiana, Geology, new ser. 20: 1–63. (ill.). [angiosperms, Ginkgo, *Parataxodium*, *Metasequoia*].

02476 Crespo, J. H. (1963). Anatomia de la madera de 12 especies de coniferas Mexicanas. Bol. Tecn. 8; Inst. Nac. Invest. Forest. (INIF), Mexico. [*Abies religiosa*, *Cupressus lusitanica*, *Juniperus monosperma* var. *gracilis*, *Pinus* spp., wood anatomy, ill.].

02480 Critchfield, W. B. (1957). Geographic variation in *Pinus contorta*. Publ. Maria Moors Cabot Found. Bot. Res. 3. (Harvard. Univ., 118 pp., ill.). [*Pinus contorta* ssp. *bolanderi* (Parl. in DC.) comb. nov., p. 106, *P. contorta* ssp. *murrayana* (Andr. Murray) comb. nov., p. 106, *P. contorta* ssp. *latifolia* (Engelm.) comb. nov., p. 107].

02490 Critchfield, W. B. (1963). Hybridization of the southern pines of California. Southern Forest. Tree Improvement Commiss. Publ. 22: 40–48. [*Pinus* subsect. *Australes*].

02500 Critchfield, W. B. (1966). Crossability and relationships of the California big-cone pines. U.S. Forest Serv. Res. Paper NC-6: 36–44. [*Pinus coulteri*, *P. sabiniana*, *P. jeffreyi*].

02501 Critchfield, W. B. (1966). Phenological notes on Latin American *Pinus* and *Abies*. J. Arnold Arbor. 47 (4): 313–318.

02510 Critchfield, W. B. (1967). Crossability and relationships of the closed cone pines. Silvae Genet. 16 (3): 89–97. [*Pinus* subsect. *Oocarpae*, *P. attenuata*, *P. muricata*, *P. radiata*].

02520 Critchfield, W. B. (1975). Interspecific hybridization in *Pinus*: a summary review. In: D. P. Fowler & C. W. Yeatman, eds. Proc. 14th Meeting Canad. Tree Improv. Assoc. 2: 99–105. Canadian Forest Service, Ottawa.

02530 Critchfield, W. B. (1977). Hybridization of foxtail and bristlecone pines. Madroño 24: 193–212. [*Pinus balfouriana*, *P. aristata*, *P. longaeva*].

02535 Critchfield, W. B. (1980). Genetics of lodgepole pine. U.S.D.A. Forest Service Res. Paper WO-37. Washington, D.C. [*Pinus contorta*, *P. contorta* ssp. *latifolia*, *P. contorta* ssp. *murrayana*, biosystematics].

02540 Critchfield, W. B. (1984). Crossability and relationships of Washoe pine. Madroño 31: 144–170. [*Pinus washoensis*, *P. ponderosa*].

02550 Critchfield, W. B. (1984). Impact of the Pleistocene on the genetic structure of North American conifers. in: R. M. Lanner, ed. Proc. of the Eighth North American Forest Biology Workshop. Logan, Utah. (pp. 70–118). [*Abies*, *Picea*, *Pinus*, *Pseudotsuga*].

02560 Critchfield, W. B. (1985). The late Quarternary history of lodgepole and jack pines. Canad. J. Forest Res. 15: 749–772. [*Pinus contorta*, *P. banksiana*].

02570 Critchfield, W. B. (1986). Hybridization and classification of the white pines (*Pinus* section *Strobus*). Taxon 35: 647–656.

02580 Critchfield, W. B. (1988). Hybridization of the California firs. Forest Sci. 34: 139–151. [*Abies*].

02590 Critchfield, W. B. & B. B. Kinlock (1986). Sugar pine and its hybrids. Silvae Genet. 35: 138–145. [*Pinus lambertiana*].

02611 Croom, H. B. (1837). A Catalogue of the Plants of New Bern, North Carolina, ... Ed. 2. New York. [*Taxodium distichum* var. *imbricatum* (as "*imbricarium*") (Nutt.) comb. nov., No. 3048].

02611a Cross, G. L. (1939). A note on the morphology of the deciduous shoot of *Taxodium distichum*. Bull. Torrey Bot. Club 66: 167–172.

02611d Csató, J. de (1886). *Juniperus kanitzii* (*Juniperus sabina* × *communis*). Magyar Növényt. Lapok 10: 145–146. [isotype: J. Csató 1841 (L!) = *J. sabina* L. var. *sabina*].

02612 Cucchi, C. (1955). Note sulla morfologica e fitogeografia delle Thujopsidineae. Delpinoa [8] 25: 195–238. [Cupressaceae, *Biota* Endl. = *Platycladus* Spach, *Libocedrus* Endl., *Thuja* L., *Thujopsis* Endl.].

02620 Cucchi, C. (1958). Indagine geobotanica sui ginepri europei. Delpinoa [11] 28: 171–222. [*Juniperus*].

02624 Cullen, P. J. (1987). Regeneration patterns in populations of *Athrotaxis selaginoides* D. Don (Taxodiaceae) from Tasmania. J. Biogeogr. 14: 39–51.

02625 Cullen, P. J. & J. B. Kirkpatrick (1988). The ecology of *Athrotaxis* D. Don (Taxodiaceae). I. Stand structure and regeneration of *A. cupressoides*. Austral. J. Bot. 36: 547–560.

02626 Cullen, P. J. & J. B. Kirkpatrick (1988). The ecology of *Athrotaxis* D. Don (Taxodiaceae). II. The distribution and ecological differentiation of *A. cupressoides* and *A. selaginoides*. Austral. J. Bot. 36: 561–573.

02628 Cunningham, A. (1838). Florae insularum Novae Zelandiae precursor; or a specimen of the botany of the islands of New Zealand. Ann. Nat. Hist. (London) 1: 210–216. [Coniferales, *Dammara australis* = *Agathis australis*, *Phyllocladus trichomanoides*, *Podocarpus ferrugineus* = *Prumnopitys ferruginea*, *Podocarpus totara*, *Dacrydium excelsum* = *Dacrycarpus dacrydioides*, *Dacrydium cupressinum*, *Dacrydium plumosum* = *Libocedrus plumosa*, *Dacrydium mai* = *Prumnopitys taxifolia*].

02630 Curtis, J. D. & D. W. Lynch (1957). Sylvics of ponderosa pine. U.S. Forest Serv. Misc. Publ. 12. Intermountain Forest and Range Exp. Stat., Ogden, Utah. [*Pinus ponderosa*].

02648 Czaja, A. (2000). *Pseudotsuga jechorekiae* sp. nova, der erste fossile Nachweis der Gattung *Pseudotsuga* Carrière nach Zapfen aus dem Miozän der Oberlausitz, Deutschland. Feddes Repert. 111 (3–4): 129–134. (ill.).

02650 Czeczott, H. (1954). The past and present distribution of *Pinus halepensis* Mill. and *P. brutia* Ten. 8th Int. Congr. Bot., Paris and Nice, Papers 8 (sect. 2, 4–6): 196–197.

02660 Dale, I. R. & P. J. Greenway (1961). Kenya trees and shrubs. Nairobi – London. [*Juniperus procera*, *Podocarpus* spp., pp. 3–8, hab. phot. nos. 5–12].

02670 Dallimore, W. (1916). Coniferous timbers III. The Asiatic pines. Bull. Misc. Inform. 1916: 254–259. [*Pinus* spp.].

02680 Dallimore, W. (1921). The yellow pines of North America. Bull. Misc. Inform. 1921: 330–335. [*Pinus* spp.].

02690 Dallimore, W. (1931). The cypresses. Empire Forest. J. 10: 37–47. [*Cupressus* spp.].

02700 Dallimore, W. (1934). The lace-bark pine of China (*Pinus bungeana* Zuccarini). J. Roy. Hort. Soc. (London) 59: 249–250, pl. 94.

02710 Dallimore, W. (1934). *Larix potaninii*. Bot. Mag. 157: pl. 9338.

02711 Dallimore, W. (1948). *Metasequoia glyptostroboides*. Quart. J. Forest. 42 (3): 150–151.

02731 Danert, S., S. Geier & P. Hanelt (1961). Vegetationskundliche Studien in Nordostchina (Mandschurei) und der Inneren Mongolei. Fedde's Repert. Beih. 139 (4) (Beiträge zur Vegetationskunde): 5–144. [*Abies*, *Larix*, *Picea*, *Pinus*, *Juniperus*, ill., maps].

02740 Daniels, J. D. (1969). Variation and intergradation in the grand fir-white fir complex. Ph.D. diss., 235 pp. Univ. Idaho at Moscow, Idaho. [*Abies grandis*, *A. concolor*].

02741 Dansereau, P. (1957). The geographical distribution of the genus *Libocedrus*. Proc. 8th Pacific Sci. Congr. (Manila 1953) 4: 431–... [map with *Libocedrus* s.l.].

02744 Dar, G. H. & K. I. Christensen (2003). Gymnosperms of the western Himalaya. 1. The genus *Juniperus* (Cupressaceae). Pakistan J. Bot. 35 (3): 283–311. [*Juniperus recurva*, *J. squamata*, *J. semiglobosa*, *J. polycarpos* = *J. excelsa* ssp. *polycarpos*, *J. wallichiana* = *J. indica*, *J. pseudosabina*].

02745 Dark, S. O. S. (1932). Chromosomes of *Taxus*, *Sequoia*, *Cryptomeria*, and *Thuya*. Ann. Bot. 46: 965–977. [*Thuya* = *Thuja*].

02746 Darrow, W. K. & T. Zanoni (1991). Hispaniolan pine (*Pinus occidentalis* Swartz) a little known sub-tropical pine of economic potential. Commonw. Forest. Rev. 69 (2): 133–146.

02750 Daubenmire, R. (1968). Some geographic variations in *Picea sitchensis* and their ecologic interpretation. Canad. J. Bot. 46: 787–798. (ill., map).

02751 Daubenmire, R. (1968). Taxonomic and ecological relationships between *Picea glauca* and *Picea sitchensis* and their ecological interpretation. Canad. J. Bot. 46: 787–798.

02760 Daubenmire, R. (1972). On the relation between *Picea pungens* and *Picea engelmannii* in the Rocky Mountains. Canad. J. Bot. 50 (4): 733–742.

02770 Daubenmire, R. (1974). Taxonomic and ecologic relationships between *Picea glauca* and *Picea engelmannii*. Canad. J. Bot. 52: 1545–1560. (ill.).

02780 Daubenmire, R. F. (1943). Vegetational zonation in the Rocky Mountains. Bot. Rev. (Lancaster) 9: 325–393. [Coniferales, Pinaceae].

02781 Davis, P. H. (1949). [*Cedrus libani* ssp. *stenocoma* (O. Schwarz) comb. nov.) in: A journey in south-west Anatolia. J. Roy. Hort. Soc. (London) 74: 113.

02800 Davis, P. H., J. Cullen, M. J. E. Coode & I. C. Hedge (1965). Materials for a Flora of Turkey: X. Notes Roy. Bot. Gard. Edinburgh 26 (2): 165–167. [Pinaceae, *Abies cilicica* ssp. *isaurica* ssp. nov., *A. nordmanniana* ssp. *bornmuelleriana* (Mattf.) comb. et stat. nov., *A. nordmanniana* ssp. *equi-trojani* (Aschers. et Sint. ex Boiss.) comb. et stat. nov., p. 167].

02810 Dayton, W. A. (1943). The names of the giant sequoia. Leafl. W. Bot. 3: 209–219. [*Sequoiadendron giganteum*].

02820 Dayton, W. A. (1952). Some notes on United States tree names. Rhodora 54 (639): 67–79. [*Pinus* spp., *P. bracteata* D. Don (= *Abies bracteata*), date of publ. shown to be before July 9, 1836].

02827 Debreczy, Z. & I. Rácz (1995). New species and varieties of conifers from Mexico. Phytologia 78 (3): 1–24. (ill.). [*Abies hidalgensis* sp. nov., p. 4, *A. neodurangensis* sp. nov., p. 7, *A. zapotekensis* sp. nov., p. 9, *A. guatemalensis* var. *longibracteata* var. nov., p. 11, *A. lowiana* var. *viridula* var. nov., p. 13, *Pinus yecorensis* sp. nov., p. 15, *P. yecorensis* var. *sinaloensis* var. nov., p. 18, *P. oaxacana* var. *diversiformis* var. nov., p. 19, *P. lawsonii* var. *gracilis* var. nov., p. 19, *Pseudotsuga menziesii* var. *oaxacana* var. nov., p. 21].

02828 Debreczy, Z. & I. Rácz (1995). El Arbol del Tule: The ancient giant of Oaxaca. Arnoldia 58 (1): 1–11. [*Taxodium mucronatum*].

02829 Debreczy, Z. & I. Rácz (1999). The prostrate form of the Phoenician Juniper: *Juniperus phoenicea* L. f. *prostrata*, f. nov. Studia Bot. Hungarica 29: 87–94 (1998).

02830 Decaisne, J. (1854). Without title, but beginning: "L'un de ces échantillons..." Bull. Soc. Bot. France 1: 70–71, session of June 28, 1854. [*Sequoia gigantea* comb. nov. = *Sequoiadendron giganteum*].

02830c Delevoryas, T. & R. C. Hope (1973). Fertile coniferophyte remains from the Late Triassic Deep River Basin, North Carolina. Amer. J. Bot. 60: 810–818. [Coniferales, Voltziaceae, palaeobotany, ill.].

02830d Delevoryas, T. & R. C. Hope (1975). *Voltzia andrewsii* n. sp., an Upper Triassic seed cone from North Carolina, U.S.A. Rev. Palaeobot. Palynol. 20: 67–74. [Coniferales, Voltziaceae, palaeobotany, ill.].

02830e Delevoryas, T. & R. C. Hope (1987). Further observations on the Late Triassic conifers *Compsostrobus neotericus* and *Voltzia andrewsii*. Rev. Palaeobot. Palynol. 51: 59–64. [Coniferales, Voltziaceae, palaeobotany, ill.].

02831 Delevoy, G. (1948). A propos de *Pinus heldreichii* Christ. Trav. Stat. Rech. Groenendaal, ser. B., No. 2: 1–9.

02840 Delevoy, G. (1949). A propos de la systématique de *Pinus nigra* Arnold. Trav. Stat. Rech. Groenendaal, ser. B., No. 12: 1–37.

02860 Denevan, W. M. (1961). The upland pine forests of Nicaragua. Univ. Calif. Publ. Geogr. 12: 251–320. [*Pinus caribaea*].

02861 Desfontaines, R. L. (1829). Catalogus plantarum horti regii Parisiensis. Ed. 3. Paris. [Coniferales, *Abies taxifolia* Desf. (first publ. in Tabl. École Bot. Mus. Nat.: 206, 1804 (not seen); not based on *Pinus taxifolia* Lamb., nor on *P. taxifolia* Salisb.) = *Abies alba*, p. 356].

02862 Desole, L. (1960). *Pinus pinaster* Sol. in Sardegna. Nuovo Giorn. Bot. Ital. (n. s.) 67 (1–2): 24–62. (Ital., Eng. summ., ill.). [leaf anatomy, ecology].

02863 Déterville, (publ.) (1802–04). Nouveau dictionnaire d'histoire naturelle,... Vols. 1–24 (in 8 books), Ed. 1. Paris. [see M. Du Tour, 1803].

02870 DeVore, J. E. (1972). Fraser fir in the Unicoi Mountains. Castanea 37: 148–149. [*Abies fraseri*].

02870c Dickson, A. (1860). Observations on some bisexual cones occurring in the spruce fir (*Abies excelsa*). Trans. Edinburgh Bot. Soc. 6: 418–422. [*Picea abies*, teratological formation in ovuliferous cones].

02870d Dickson, A. (1866). On the phylloid shoots of *Sciadopitys verticillata* Sieb. et Zucc. J. Bot., British & Foreign 4: 224–225.

02871 Diels, J. (1901). Die Flora von Central-China, nach der vorhandenen Literatur und neu mitgeteilten Original-Materiale. Bot. Jahrb. Syst. 29 (2): 169–659. [see G. A. Pritzel, 1901].

02880 Dietrich, A. (1824). Flora der Gegend um Berlin, oder Aufzählung und Beschreibung der in der Mittelmark wildwachsenden und angebauten Pflanzen. Pars II. Berlin. [*Picea* gen. nov., p. 794].

02890 Doak, C. C. (1935). Evolution of foliar types, dwarf shoots, and cone scales of *Pinus*, with remarks concerning similar structures in related forms. Illinois Biol. Monogr. 13: 1–106. (ill.).

02891 Doak, C. C. (1937). Morphology of *Cupressus arizonica*: gametophytes and embryogeny. Bot. Gaz. (Crawfordsville) 98: 808–815.

02900 Dode, M. (1912). Deux genres nouveaux pour la Chine: *Pseudotsuga sinensis* espèce nouvelle et *Carya sinensis* espèce nouvelle. Bull. Soc. Dendrol. France 23/24: 58–59.

02901 Dode, M. (1914). Observations sur *Abies nebrodensis*. Bull. Soc. Dendrol. France 9: 58–59.

02910 Doerksen, A. H. & K. K. Ching (1972). Karyotypes in the genus *Pseudotsuga*. Forest Sci. 18 (1): 66–69. [2n = 24 in Asian spp.].

02915a Dogra, P. D. (1964). Gymnosperms of India – II. Chilgoza pine (*Pinus gerardiana* Wall.). Bull. Nat. Bot. Gard. 109 (Lucknow, India). [*Pinus gerardiana*, monograph, ecology, forestry, ill.].

02917 Dogra, P. D. (1966). Observations on *Abies pindrow* with a discussion on the question of occurrence of apomixis in gymnosperms. Silvae Genet. 15: 11–20.

02922 Dogra, P. D. & S. Tandon (1984). Observations on the embryology of *Juniperus procera* Hochst. in: P. K. K. Nair (ed.). Glimpses in plant research 6, pp. 114–124, figs. 1–53. New Delhi.

02930 Doi, T. (1915). A new species of *Thuja*. Bot. Mag. (Tokyo) 29: 421–422. [*T. odorata* Doi, sp. nov., non *Thuja odorata* Marshall].

02940 Doi, T. & K. I. Morikawa (1929). An anatomical study of the leaves of the genus *Pinus*. J. Kyushu Imp. Univ., Dept. Agric. 2: 149–198. (ill.).

02945 Doludenko, M. P. (1963). New species of *Sciadopitytes* from the Jurassic of West Ukraine. Palaeontol. Žurn. 1963 (1): 123–126. (Russ.). [Sciadopityaceae, palaeobotany].

02946 Doludenko, M. P. & E. I. Kostina (1987). On the coniferous genus *Elatides*. Palaeontol. Žurn. 1987 (1): 110–114. (Russ., Eng. summ.). [Cupressaceae s.l., Taxodiaceae, palaeobotany].

02947   Doludenko, M. P., E. I. Kostina & I. A. Shilkina (1988). A new genus *Kanevia* (Taxodiaceae), Late Albian conifer from the Ukraine. Bot. Žurn. 73 (4): 465–476. (Russ., Eng. summ.). [Cupressaceae s.l., Taxodiaceae, *Kanevia* gen. nov., palaeobotany].

02950   Domin, K. (1931). Schedae ad floram Cechoslovenicam exsiccatam. Centuria III.; Acta Bot. Bohem. 10. [Pinaceae: pp. 3–7, *Larix carpatica* (Domin) comb. et stat. nov., *L. decidua* f. *tatrensis* f. nov.].

02960   Domin, K. (1935). Plantarum Cechoslovakiae Enumeratio. Preslia 13–15: 1–305. [*Pinus mugo* ssp. *uncinata* (Ramond ex DC) Domin, comb. nov., p. 13].

02970   Domin, K. (1940). Nekolik poznámek o modrínu v Alpách, vychodních Sudetech a Karpatech. Preslia 18–19: 53–85. (Czech.). [*Larix decidua* s.l., *L. decidua* ssp. *europaea*, ssp. *sudetica*, ssp. *polonica* (var. *eupolonica*, var. *carpatica*), ssp. *sibirica*, ssp., comb., var. + formae nov.; var. *carpatica* earlier in Sborn. Výzk. Úst. Zem. RČS 65: 149 (1930), ssp. *polonica* earlier in Acta Bot. Bohem. 10: 6 (1931), nomenclature].

02980   Don, D. (1825). Prodromus Florae Nepalensis,... II. Dicotyledones; Coniferae. London. [Coniferales, descr. of new taxa pp. 54–55: *Juniperus recurva* Buch.-Ham. ex D. Don, sp. nov., p. 55, *Cupressus torulosa* "sp. nov.", but see D. Don in Lambert, 1824, p. 55, *Pinus dumosa* sp. nov. = Tsuga dumosa, p. 55, *P. spectabilis* sp. nov. = *Abies spectabilis*, p. 55].

02990   Don, D. (1836). Descriptions of five new species of the genus *Pinus* discovered by Dr. Coulter in California. Trans. Linn. Soc. London 17: 439–444. [*Pinus coulteri* sp. nov., p. 440, *P. muricata* sp. nov., p. 441, *P. radiata* sp. nov., p. 442, *Pinus tuberculata* sp. nov., = *P. attenuata*, p. 442, *P. bracteata* sp. nov. = *Abies bracteata*, p. 443].

02991   Don, D. (1836). Descriptions of two species of the genus *Pinus* from the Himalaya Alps. London & Edinburgh Philos. Mag. & J. Sci. 8: 255. [*Pinus pindrow* sp. nov. = *Abies pindrow, P. smithiana* sp. nov. = *Picea smithiana*].

02999   Don, D. (1838). [Linnean Society of London, session 17th April 1838]. ...describing two new genera of the natural family of plants called Coniferae. Ann. Nat. Hist. 1: 233–235. [*Cryptomeria* gen. nov., *Athrotaxis* gen. nov.].

03000   Don, D. (1839). Descriptions of the two new genera of the natural family of plants called Coniferae. (read 17th April, 1838). Trans. Linn. Soc. London 18: 163–179, pl. 13–14. [*Cryptomeria* p. 166, *C. japonica* (basion.: *Cupressus japonica* L. f.) p. 167, *Athrotaxis* p. 171, *A. selaginoides* p. 172, *A. cupressoides* p. 173. N.B.: an earlier publication of the same new taxa appeared in Ann. Nat. Hist. 1; respectively on pp: 233, 234, 234, 235, 235 in 1838, except for the combination *Cryptomeria japonica* which first appeared here].

03001   Don, D. (1844). *Cryptomeria japonica* "sp. nov." in: W. J. Hooker. Icones plantarum;... Ser. 2, Vol. 7, t. 668. [see D. Don, 1839].

03012   Donahue, J. K. & J. Lopez Upton (1999). A new variety of *Pinus greggii* (Pinaceae) in Mexico. Sida 18 (4): 1083–1093. [*Pinus greggii* var. *australis* var. nov., ill.].

03013   Donahue, J. K. & C. Mar Lopez (1995). Observations on *Pinus maximartinezii* Rzed. Madroño 42 (1): 19–25. (map).

03015   Donoso, C., R. Grez & V. Sandoval (1990). Caracterización del tipo forestal alerce. Bosque 11 (1): 21–34. [*Fitzroya cupressoides*].

03016   Donoso, C., V. Sandoval, R. Grez & J. Rodriguez (1993). Dynamics of *Fitzroya cupressoides* forests in southern Chile. J. Veg. Sci. 4: 303–312.

03020   Doorenbos, J. (1949?). De soorten van het geslacht *Picea*. Jaarb. Ned. Dendrol. Ver. 17: 25–53. (Eng. summ., key).

03022   Dorado, O., G. Avila, D. M. Arias, R. Ramirez, D. Salinas & G. Valladares (1996). The Arbol del Tule (*Taxodium mucronatum* Ten.) Is a single genetic individual. Madroño 45 (4): 445–452. [analysis of RAPD's indicate the famous tree is a single individual].

03030   Douglas, D. (1827). An Account of a new Species of Pinus, native of California: in a letter to Joseph Sabine,... Trans. Linn. Soc. London 15: 497–500. (read Nov. 6, 1827, publ. 1827). [*Pinus lambertiana* sp. nov., p. 500].

03031   Douglas, D. (1833). Description of a new Species of the Genus *Pinus*. (read April 3, 1832). Trans. Linn. Soc. London 16: 747–749. [*Pinus sabiniana* "sp. nov.", but earlier publ. by D. Don in A. B. Lambert, 1832].

03032   Douglas, D. (1914). Journal kept by D. Douglas during his travels in North America 1823–1827. London. [Cupressaceae, Pinaceae, collecting, discovery].

03040   Douglass, M. M. (1958). Intraspecific variation in *Pinus flexilis*. J. Colorado-Wyoming Acad. Sci. 4: 30–31.

03046   Doweld, A. B. (2001). De genere *Libocedrus* Endl. (Cupressaceae). Novosti Sist.Vyssh. Rast. 33: 41–44. (Russ., Latin diagnosis). [*Stegocedrus* gen. nov., *S. austrocaledonica* (Brongn. & Gris) Doweld, *S. chevalieri* (Buchholz) Doweld, *S. yateensis* (Guillaumin) Doweld = *Libocedrus* spp. from New Caledonia].

03050   Downie, D. G. (1923). Chinese Species of *Tsuga*. Notes Roy. Bot. Gard. Edinburgh 14 (67): 13–19. [*T. calcarea, T. dura, T. wardii* spp. nov. = *T. dumosa, T. patens* sp. nov. = *T. chinensis, T. forrestii* sp. nov., p. 18].

03050a Dowsett-Lemaire, F. & F. White (1990). New and noteworthy plants from the evergreen forests of Malawi. 2. The genus *Podocarpus* in Malawi. Bull. Jard. Bot. Natl. Belge 60 (1/2): 73–110 [86–87]. [*Podocarpus falcatus* (*Afrocarpus falcatus*), *P. henkelii, P. latifolius*].

03051 Doyle, J. C. (1926). The ovule of *Larix* and *Pseudotsuga*. Sci. Proc. Roy. Dublin Soc. 18: 170–180.

03060 Doyle, J. C. (1934). The columella in the cone of *Diselma*. Ann. Bot. (London) 48 (184): 307–308, figs. [*Diselma archeri*].

03071 Doyle, J. C. (1945). Naming of the redwoods. Nature 155: 254–265. [*Sequoia sempervirens, Sequoiadendron giganteum*, discussion of taxonomy and nomenclature, criticising the segregation of Buchholz, 1939 (ref. 01380)].

03080 Doyle, J. C. (1963). Proembryogeny in *Pinus* in relation to that in other conifers, a survey. Proc. Roy. Irish Acad. 62, sect. B: 181–216.

03110 Doyle, J. C. & M. O'Leary (1934). Abnormal cones of *Fitzroya* and their bearing on the nature of the conifer strobilus. Sci. Proc. Roy. Dublin Soc. 21 (3): 23–35, figs., pl. 1. [*F. cupressoides*].

03120 Doyle, J. C. & M. O'Leary (1935). Pollination in *Pinus*. Sci. Proc. Roy. Dublin Soc. 2 (21): 181–190, pl. 3.

03130 Doyle, J. C. & M. O'Leary (1935). Pollination in *Tsuga, Cedrus, Pseudotsuga* and *Larix*. Sci. Proc. Roy. Dublin Soc. 2 (21): 191–204, pl. 4–5.

03131 Doyle, J. C. & W. T. Saxton (1933). Contributions to the life-history of *Fitzroya*. Proc. Roy. Irish Acad. 41, sect. B: 191–217. [*F. cupressoides*].

03132 Drawson, M. C. (1972). A record Western Juniper in Oregon. Int. Dendrol. Soc. Yearb. 1972: 45–47. [*Juniperus occidentalis*].

03133 Druce, G. C. (1917). *Glyptostrobus lineatus* comb. nov. in: Bot. Soc. Exch. Club Brit. Isles 1916: 624. [basion. : *Thuja lineata* Poir., 1817 = *Taxodium distichum* cultivar?].

03141 Dubois-Ladurantie, G. (1941). Révision de quelques espèces d'Abiétinées fossiles. Trav. Lab. Forest. Toulouse, T. 2, vol. 6, art. 1: 1–41. (also publ. in: Bull. Soc. Hist. Nat. Toulouse 76: 363–402, 1941). [*Abies*, fossil and recent, ill.].

03150 Dudley, T. R. (1988). Chinese Firs: Particularly *Abies beshanzuensis*. Amer. Conifer Soc. Bull. 5 (4): 84–93.

03160 Duffield, J. W. (1950). Review of "Sur quatre hybrides des genres chez les Abiétinées" [Van Campo-Duplan & Gaussen, 1948]. J. Forest. (Washington) 48: 440.

03170 Duffield, J. W. (1951). Interrelationships of the California closed-cone pines with special reference to *Pinus muricata* D. Don. Ph.D. diss., Univ. California, Berkeley.

03180 Duffield, J. W. (1952). Relationships and species hybridization in the genus *Pinus*. Z. Forstgenet. Forstpflanzenzücht (Silvae Genet.) 1: 93–97.

03190 Duffield, J. W. & W. C. Cumming (1949). Does *Pinus ponderosa* occur in Baja California? Madroño 10: 22–24. [= *P. jeffreyi*].

03200 Duhamel du Monceau, H. L. (1755). Traité des Arbres et arbustes qui se cultivent en France en pleine terre. Tome premier: *Cedrus* Duhamel = *Cupressus* (p. 139, t. 52); Tome second: *Pinus* Tourn. & Linn. (pp. 121–169). Paris. [*P.* sect. *Quinquefoliis* sect. nov., p. 124].

03209 Dümmer, R. A. (1912). New or noteworthy plants. Gard. Chron., ser. 3, 52: 295. (ill.). [*Podocarpus formosensis* sp. nov. = *Nageia formosensis*].

03211 Dümmer, R. A. (1914). Three conifers. J. Bot. 52: 236–241. [*Callitris neo-caledonica* sp. nov., p. 239, *Thuja* (*Biota*) *orientalis* var. *mexicana* = cultivar (?), *Podocarpus motleyi* (Parl.) comb. nov. = *Nagaeia motleyi*, p. 241].

03215 Dunal, F. (1851). Description du *Pinus salzmanni* de la forêt de Saint-Guilhem-le-Désert. Mem. Acad. Sci. Lettres Montpellier, sect. Sci. 2: 81–95, t. 3–4. [*Pinus salzmannii* sp. nov., p. 83].

03220 Dungern, F. (1910). Dendrologisches aus dem Tian-Schan. Mitt. Deutsch. Dendrol. Ges. 19: 226–229, 2 figs. [*Picea schrenkiana*].

03221 Dunn, S. T. (1908). A Botanical Expedition to Central Fokien – Enumeration and descriptions of new species. J. Linn. Soc., Bot. 38: (350–)353–373. [*Cupressus* (§ *Chamaecyparis*) *hodginsii* sp. nov. = *Fokienia hodginsii*, p. 367].

03230 Durland, W. D. (1931). Mexico's pine forests. Amer. Forests 37: 348–349, 364. [*Pinus*, ecology, phytogeography].

03231 Du Roi, J. P. (1800). Die Harbkesche wilde Baumzucht... Vol. 1–2. 2. Aufl. (ed. by J. F. Pott). Braunschweig. [*Pinus intermedia* Pott, sp. nov. = *Larix laricina*, Vol. 2, p. 114].

03232 Durrieu, A. (1961). Étude chromosomique de *Pseudolarix kaempferi* (Lindl.) Gord. Compt. Rend. Hebd. Séances Acad. Sci. (Paris) 252: 773–775. [= *Pseudolarix amabilis*, 2n = 44].

03233 Durrieu, A. (1971). Étude chromosomique des espèces du genre *Pseudotsuga*. Trav. Lab. Forest. Toulouse, T. 1, vol. 8, art. 18: 1–21 + pl. [*Pseudotsuga menziesii* s.l., 2n = 26].

03234 Du Tour, M. (1803). *Pinus sibirica* sp. nov. in: Déterville, (publ.). Nouveau dictionnaire d'histoire naturelle,... Vol. 18: 18. Paris.

03235 Dvorak, W. S. & J. K. Donahue (1988). *Pinus maximinoi* seed collections in Mexico and Central America. CAMCORE Bull. Trop. Forest. 4. North Carolina State University, Raleigh, N.C. (report).

03235a Dvorak, W. S., A. P. Jordon, G. P. Hodge & J. L. Romero (2000). Assessing evolutionary relationships of pines in the *Oocarpae* and *Australes* subsections using RAPD markers. New Forests 20: 163–192. [*Pinus* subsect. *Australes*, subsect. *Oocarpae*, *P. caribaea* var. *hondurensis*, *P. echinata*, *P. oocarpa*, *P. palustris*, *P. taeda*, *P. teocote* (18 taxa analysed, DNA)].

03235b Dvorak, W. S., J. A. Pérez de la Rosa, M. Mápula & V. J. Reyes (1998). The ecology and conservation of *Pinus jaliscana*. Forest Genet. Resources 26: 13–19.

03236 Dvorak, W. S. & R. H. Raymond (1991). The taxonomic status of closely related closed-cone pines in Mexico and Central America. New Forests 4: 291–307 (*Pinus* spp.].

03239 Dylis, N. V. (1945). Novye dannye po sistematike I istorii sibirskoy listvennitsy. (New data on the systematics and history of the Siberian larch). Khvoinye Porody 50: 489–492. (Russ.). [*Larix sukaczewii* nom. nov. = *L. archangelica* Lawson (1836) = *L. sibirica*].

03240 Dylis, N. V. (1947). Contribution to the systematics, geography, and history of the Siberian larch. Ref. Naucno-Issl. Rabot, 1945 Otd. Biol. Nauk, Akad. Nauk SSSR, pp. 101–104. (Russ.). [*Larix sibirica*, *L. sukaczewii* = *L. sibirica*].

03250 Dylis, N. V. (1947). Sibirskaîa listvennitsa. Materialy k sistematike, geografii i istorii. (Siberian larch. Materials on the systematics, geography and history). Mater. Pozn. Fauny Fl. SSSR, Otd. Bot., n.s. 2: 1–138, f. 1–140. (Russ., Eng. summ.). [*Larix sibirica*, *L. sukaczewii* = *L. sibirica*, monographic].

03255 Eames, A. J. (1913). The morphology of *Agathis australis*. Ann. Bot. 27: 1–37.

03276 Ecroyd, C. E. (1982). Biological Flora of New Zealand 8. *Agathis australis* (D. Don) Lindl. (Araucariaceae) kauri. New Zeal. J. Bot. 20: 17–36. (map).

03293 Edwards, S. W. (1984). New light on Alaskan and Port Orford *Chamaecyparis* Cedars. The Four Seasons 7 (2): 4–15. (ill.). [*C. nootkatensis*, *C. lawsoniana*, foliar morphology].

03294 Edwards, S. W. (1992). Foliar morphology of *Chamaecyparis* and *Thuja*, with notes on seeds. The Four Seasons 9 (2): 4–29. [= J. Regional Parks Bot. Gard., East Bay Regional Park District, California].

03296 Edwards-Burke, M. A., J. L. Hamrick & R. A. Price (1997). Frequency and direction of hybridization in sympatric populations of *Pinus taeda* and *P. echinata* (Pinaceae). Amer. J. Bot. 84 (7): 879–886. [allozymes, cpDNA].

03300 Eggeling, W. J. (1952). The indigenous trees of the Uganda Protectorate. Entebbe – London. [*Juniperus procera*, pp. 101–103, hab. phot. 14].

03308 Eguiluz Piedra, T. (1982). Clima y distribución del género *Pinus* en México. Ci. Forest. [7] 38: 30–44. (map, tables).

03309 Eguiluz Piedra, T. (1982). Natural variation and taxonomy of *Pinus tecunumanii* from Guatemala. Ph.D. thesis, N.C.S.U. – Forestry, Raleigh, N.C. (unpubl.).

03310 Eguiluz Piedra, T. (1984). Geographic variation in needles, cones and seeds of *Pinus tecunumanii* in Guatemala. Silvae Genet. 33: 72–79.

03315 Eguiluz Piedra, T. (1986). Taxonomic relationships of *Pinus tecunumanii* from Guatemala. Commonw. Forest. Rev. 65: 303–313. [*P. tecunumanii*, *P. patula*, *P. oocarpa*, *P. maximinoi*].

03320 Eguiluz Piedra, T. (1988). Distribución natural de los pinos en México. Notas Técn. Centro Genét. Forest. Chapingo 1: 1–6. [52 species of *Pinus*].

03330 Eguiluz Piedra, T. & J. P. Perry, Jr. (1983). *Pinus tecunumanii*: una especie nueva de Guatemala. Revista Ci. Forest. 8: 3–22. [colour photogr.; see also F. Schwerdtfeger, 1953 and B. T. Styles, 1984)

03350 Ehrendorfer, F., ed. (1988). Woody Plants Evolution and Distribution Since the Tertiary. Spec. Ed. of Plant Syst. Evol., Vol. 162. Wien – New York. [see W. Klaus, 1988].

03370 Eichler, A. W. (1882). Ueber Bildungsabweichungen bei Fichtenzapfen. Sitzungsber. Königl. Preuss. Akad. Wiss. Berlin 1882: 1–20, 1 pl. [*Picea abies*].

03391 Eliçin, G. (1974). Etude anatomique chez *Arceuthos drupacea* Ant. et Kotschy et son aire naturelle en Turquie. Istanbul Univ. Orman Facült. Dergisi Ser. A. 24 (2): 194–208. [= *Juniperus drupacea*, wood anatomy, phytogeography, ill., map].

03400 El-Kassaby, Y. A., A. M. Colangelli & O. Sziklai (1983). A numerical analysis of karyotypes in the genus *Pseudotsuga*. Canad. J. Bot. 61: 536–544.

03410a Elliott, B. (1994). The identity crisis of the Douglas fir. The New Plantsman 1 (1): 20–28. [nomenclature of *Pseudotsuga menziesii* (Mirbel) Franco].

03410b Elliott, C. G. (1951). Some notes on *Athrotaxis*. Proc. Linn. Soc. New South Wales 76: 36–40.

03450d Emig, W. H. (1931). The megagametophyte of *Pinus*. Science 74: 337–338.

03455 Endlicher, S. L. (1842). Catalogus horti academici Vindobonensis Vol. 1. Wien. (publ. date Jul–Aug 1842). [*Widdringtonia* "gen. nov.", *W. cupressoides* (L.) "comb. nov." = *W. nodiflora*, p. 209; see also S. L. Endlicher in Gen. Pl. Suppl. II, 1842, with descr.].

03461 Endtmann, K. J., D. H. Mai, E. Lange, G. Hofmann & E. Schwartz (1191). Die Kiefer (*Pinus sylvestris*). Berichte aus Forschung und Entwicklung 24. Forschungsanstalt für Forst- und Holzwirtschaft Eberswalde, Deutschland. [taxonomy, phytogeography, palaeobotany, ecology, *Pinus* spp., ill.].

03470 Engelmann, G. (1848). Sketch of the botany of Dr. Wislizenus' expedition, Appendix to Wislizenus, (1848); pp. 87–115. [*Pinus edulis* sp. nov., p. 88, *P. strobiformis* sp. nov., p. 102, *P. chihuahuana* sp. nov., p. 103].

03480 Engelmann, G. (1863). On *Pinus aristata*, a new species of pine, discovered by Dr. C. C. Parry in the alpine regions of Colorado Territory, and on some other pines of the Rocky Mountains. Trans. St. Louis Acad. Sci. 2: 205–210 (publ. 5 May 1863). [*Pinus albicaulis* sp. nov.; *P. aristata* sp. nov. first publ. in: Amer. J. Sci. Arts ser. 2 (34): 331, 1862].

03490 Engelmann, G. (1863). *Picea engelmannii* Parry ex Engelm., sp. nov. in: Trans. St. Louis Acad. Sci. 2: 212 (publ. 5 May 1863, see 03480; based on *Abies engelmanni* Parry in: Trans. St. Louis Acad. Sci. 2: 122 = nom. nud.; descr. and diagn. by Engelmann (as *Abies* (*Picea*) *engelmannii*) also in: Gard. Chron., n.s., 5: 1035, 1863; *Picea menziesii* = *P. pungens* on p. 214].

03500 Engelmann, G. (1878). Coniferae (of Wheeler's expedition.) in: G. M. Wheeler. Report upon U.S. geographical surveys west of the 100th meridian in charge of G.M. Wheeler. Vol. VI. Botany. Washington, D.C. (pp. 255–264). [*Pinus flexilis* var. *serrulata*, var. *macrocarpa*, var. *reflexa* var. nov., p. 258, *P. arizonica* sp. nov., p. 260].

03510 Engelmann, G. (1878). The American junipers of the section *Sabina*. Trans. St. Louis Acad. Sci. 3: 583–592. [*Juniperus californica* var. *utahensis*, var. nov. = *J. osteosperma*, p. 588, *J. occidentalis* ß (var.) *monosperma*, var. nov. = *J. monosperma*, p. 590, *J. tetragona* Schltdl. var. *oligosperma* var. nov. = *J. ashei*, p. 591].

03520 Engelmann, G. (1878). A Synopsis of the American Firs (*Abies* Link). Trans. St. Louis Acad. Sci. 3: 593–602. [*Abies* sect. *Balsamea*, sect. *Grandis*, sect. *Bracteata*, sect. *Nobilis*, sect. nov.; *A. subalpina* = *A. lasiocarpa* (first descr. by Engelm. in Amer. Naturalist 10: 555, 1876, not seen), p. 597].

03530 Engelmann, G. (1879). The American spruces. Gard. Chron., n. s., 11: 334. (contrib. by J. D. Hooker). [*Picea pungens* sp. nov.].

03550 Engelmann, G. (1880). Revision of the genus *Pinus*, and a description of *Pinus elliottii*. Trans. St. Louis Acad. Sci. 4: 161–190. [*P. elliottii* sp. nov., p. 186, pls. 1–3].

03551 Engelmann, G. (1881). *Tsuga caroliniana*, n. sp. in: Bot. Gaz. (Crawfordsville) 6: 223–224. [*T. caroliniana* sp. nov., p. 223].

03560 Engelmann, G. (1882). *Pinus latisquama* n. sp. Gard. Chron., n. s., 18: 712–713. [= *P. pinceana* Gord., p.p.].

03561 Engelmann, G. (1882). Notes on western conifers. Bot. Gaz. (Crawfordsville) 7: 4. [*Pinus reflexa* sp. nov.].

03566 Engels, F. M. & M. Gianordoli (1983). The basic autonomy of *Metasequoia* female gametophytes. Acta Bot. Neerl. 32 (4): 295–305. [*M. glyptostroboides*, light & electron micrographs].

03593 Enright, N. J. (1982). The *Araucaria* forests of New Guinea. In: J. L. Gressitt (ed.). Biogeography and Ecology of New Guinea. Monogr. Biol. 42: 381–400. The Hague. (maps).

03594a Enright, N. J., J. Ogden & L. S. Rigg (1999). Dynamics of forests with Araucariaceae in the western Pacific. J. Veg. Sci. 10: 793–804. [*Agathis australis*, *Araucaria hunsteinii*, *A. laubenfelsii*, ecology, temporal stand replacement model of Ogden].

03600 Epling, C. & W. Robinson (1940). *Pinus muricata* and *Cupressus forbesii* in Baja California. Madroño 5: 248–250. [*C. forbesii* = *C. guadelupensis* var. *forbesii* (Jepson) Little].

03600a Epperson, B. K., M. G. Chung & F. W. Telewski (2003). Spatial pattern of allozyme variation in a contact zone of *Pinus ponderosa* and *P. arizonica* (Pinaceae). Amer. J. Bot. 90 (1): 25–31.

03601 Erdtman, G. (1933). Granens utbredning i Europa. Svensk Bot. Tidskr. 27: 115–116. [*Picea abies*, map].

03619 Erdtman, H., B. Kimland & T. Norin (1966). Pine phenolics and pine classification. Bot. Mag. (Tokyo) 79: 499–505. [*Pinus*].

03620 Erdtman, H., B. Kimland & T. Norin (1966). Wood constituents of *Ducampopinus Krempfii* (Lecomte) Chevalier (*Pinus Krempfii* Lecomte). Phytochemistry 5: 927–931.

03631 Ern, H. (1986). *Pinus flexilis* James und *Pinus strobiformis* Engelm. im Botanischen Garten Berlin-Dahlem. Mitt. Deutsch. Dendrol. Ges. 76: 67–70, f. 1–3.

03632 Ern, H. (1989). Aus dem Botanischen Garten Berlin-Dahlem – Die Heimat von *Cupressus cashmeriana* Carrière ist Bhutan! Mitt. Deutsch. Dendrol. Ges. 79: 117–118. [*C. cashmeriana*, *C. corneyana*].

03633 Ernst, S. G., J. W. Hanover & D. E. Keathley (1990). Assessment of natural interspecific hybridization of Blue and Engelmann spruce in southwestern Colorado. Canad. J. Bot. 68: 1489–1496. [*Picea pungens*, *P. engelmannii*].

03645   Everett, R. L. (comp.) (1987). Proceedings — Pinyon-Juniper Conference, Reno, Nevada, 13–16 January 1986. USDC Report No. GTR-INT-215, Intermountain Research Station, Ogden, Utah. [*Juniperus* spp., *Pinus* subsect. *Cembroides* spp., ecology, management, conservation, phytogeography, ill., tables etc.].

03648   Ewel, K. C. & H. T. Odum, eds. (1984). Cypress swamps. Gainesville, Florida. [ecology of *Taxodium distichum* swamps; paper by C. A. Brown (pp. 16–24): Morphology and biology of cypress trees].

03650   Ewers, F. W. & R. Schmid (1981). Longevity of needle fascicles of *Pinus longaeva* (bristlecone pine) and other North American pines. Oecologia 51: 107–115.

03654   Fady, B., M. Arbez & A. Marpeau (1992). Geographic variability of terpene composition in *Abies cephalonica* Loudon and *Abies* species around the Aegean: hypotheses for their possible phylogeny from the Miocene. Trees Struct. Funct. 6: 162–171.

03656   Falder, A. B., G. W. Rothwell, G. Mapes, R. H. Mapes & L. A. Doguzhaeva (1998). *Pityostrobus milleri* sp. nov., a pinaceous cone from the Lower Cretaceous (Aptian) of southwestern Russia. Rev. Palaeobot. Palynol. 103 (3–4): 253–261. [Pinaceae, *Pinus*, fossils].

03656a   Falder, A. B., R. A. Stockey & G. W. Rothwell (1999). *In situ* fossil seedlings of a *Metasequoia*-like taxodiaceous conifer from Palaeocene river floodplain deposits of central Alberta, Canada. Amer. J. Bot. 86 (6): 900–902. [palaeobotany].

03657   Fan, G. S. & C. J. Hsueh (1993). Distributional characteristics of *Pinus* L. in southeast Yunnan. Guihaia 13 (4): 349–354. (Chin., Eng. summ.).

03660   Farjon, A. (1984). Pines: drawings and descriptions of the genus *Pinus*. Leiden. [*Pinus* spp., ill., maps, tables, cladograms].

03670   Farjon, A. (1988). Taxonomic notes on Pinaceae I. Proc. Kon. Ned. Akad. Wetensch. Bot., ser. C, 91 (1): 3–42. [*Abies, Pseudotsuga, Tsuga mertensiana* ssp. *grandicona*, ssp. nov., p. 39, ill.].

03680   Farjon, A. (1989). A second revision of the genus *Keteleeria* Carrière. (Taxonomic notes on Pinaceae II). Notes Roy. Bot. Gard. Edinburgh 46 (1): 81–99. (ill., map). [*K. fortunei, K. davidiana, K. evelyniana* retained].

03691   Farjon, A. (1992). *Cathaya loehrii*, a misnomer for a Pliocene conifer cone. Taxon 41 (4): 721–723.

03693   Farjon, A. (1993). Nomenclature of the Mexican cypress or "cedar of Goa", *Cupressus lusitanica* Mill. (Cupressaceae). Taxon 42 (1): 81–84. [*C. lusitanica, C. lusitanica* var. *benthamii*].

03694   Farjon, A. (1993). The taxonomy of multiseed junipers (*Juniperus* sect. *Sabina*) in Southwest Asia and East Africa. (Taxonomic notes on Cupressaceae I). Edinburgh J. Bot. 49 (3): 251–283. [*J. excelsa, J. excelsa* ssp. *polycarpos, J. foetidissima, J. procera, J. semiglobosa*].

03694a   Farjon, A. (1994). *Cupressus cashmeriana* (Cupressaceae). Kew Mag. 11 (4): 156–166. [neotypification, nomenclature, taxonomy].

03696   Farjon, A. (1994). The rare pines of Mexico. Oxford Plant Syst. (OPS-newsletter) 2: 12–13. [*Pinus* spp., e.g. *P. cembroides* subsp. *lagunae, P. cembroides* subsp. *orizabensis, P. maximartinezii, P. nelsonii, P. pinceana, P. rzedowskii*, ill.].

03696a   Farjon, A. (1994). Bigcone pinyon pine (*Pinus maximartinezii*). Pp. 1072–1073 in: M. Emanoil (ed.). Encyclopedia of Endangered Species. Gale Research Inc., Detroit, in assoc. with I.U.C.N. – The World Conservation Union.

03696b   Farjon, A. (1995). Threats to conifers in New Caledonia. Species 23: 25–26. [*Dacrydium guillominii, Neocallitropsis pancheri, Retrophyllum minus*, conservation].

03696c   Farjon, A. (1995). Typification of *Pinus apulcensis* Lindley (Pinaceae), a misinterpreted name for a Latin American pine. Novon 5: 252–256. [*Pinus apulcensis, P. pseudostrobus* var. *apulcensis, P. oaxacana*].

03696e   Farjon, A. (1996). Biodiversity of *Pinus* (Pinaceae) in Mexico: speciation and palaeo-endemism. Bot. J. Linn. Soc. 121 (4): 365–384. [*Pinus* spp., maps, ill.].

03696h   Farjon, A. (1999). Introduction to the conifers. Curtis's Bot. Mag. 16 (3): 158–172.

03696i   Farjon, A. (1999). *Cryptomeria japonica* (Cupressaceae). Curtis's Bot. Mag. 16 (3): 212–228, pl. 371.

03696j   Farjon, A. (1999). Research into conifers at Kew. Curtis's Bot. Mag. 16 (3): 240–244.

03696k   Farjon, A. (2000). Cupressaceae. In: Güner & al. (eds.). Flora of Turkey. Vol. 11 (Supplement 2): 8–10. [*Juniperus*].

03700   Farjon, A. & P. Bogaers (1985). Vegetation zonation and primary succession along the Porcupine River in interior Alaska. Phytocoenologia 13 (4): 465–504. [*Picea glauca*, ecology].

03701   Farjon, A. & M. P. Frankis (1998). Proposal to conserve the name *Pinus maximinoi* (Pinaceae) against three competing binomials. Taxon 47 (3): 733–734.

03701a   Farjon, A. & M. P. Frankis (2002). *Pinus pungens* (Pinaceae). Curtis's Bot. Mag. 19 (2): 97–103, pl. 442.

03701f Farjon, A., Nguyen Tien Hiep, D. K. Harder, Phan Ke Loc & L. Averyanov (2002). A new genus and species in Cupressaceae (Coniferales) from northern Vietnam, *Xanthocyparis vietnamensis*. Novon 12 (2): 179–189. [*Xanthocyparis* Farjon & Hiep, gen. nov., p. 179, *X. vietnamensis* Farjon & Hiep, sp. nov., p. 180, *X. nootkatensis* (D. Don) Farjon & D. K. Harder, comb. nov., p. 188, ×*Cuprocyparis* Farjon, nom. nov., p. 188, × *C. leylandii* (A. B. Jackson & Dallimore) Farjon, comb. nov., p. 188, × *C. notabilis* (Mitchell) Farjon, comb. nov., p. 188, × *C. ovensii* (Mitchell) Farjon, comb. nov., p. 188].

03702 Farjon, A. & D. R. Hunt (1994). Proposal to conserve *Thujopsis* Endl. against *Dolophyllum* Salisb. (Cupressaceae). Taxon 43 (2): 291–292. [*Thujopsis dolabrata*].

03703 Farjon, A., G. Miehe & S. Miehe (2001). The taxonomy, distribution and ecology of Junipers in High Asia. (Eng., Russ.). In: Anonymus. Proceedings of the International Symposium: Problems of Juniper Forests, pp. 70–90. Osh, Kyrgyzstan, 6th–11th August 2000. LESIC. National Acad. Sci. Kyrgyz Republic & Kyrgyz-Swiss Forestry Support Programme. [*Juniperus* spp., maps].

03704 Farjon, A. & R. R. Mill (1999). Proposals to conserve the current spelling of two generic names of conifers. Taxon 48 (1): 151–154. [*Fitzroya, Saxegothaea*].

03706 Farjon, A., J. A. Pérez de la Rosa & B. T. Styles (1997). A field guide to the pines of Mexico and Central America. Royal Botanic Gardens, Kew. (ill., maps). [*Pinus*, 47 spp., based on Flora Neotropica Monograph 75 (Farjon & Styles, 1997)].

03707 Farjon, A., J. A. Pérez de la Rosa & B. T. Styles (1997). Guía de campo de los pinos de México y América Central. Royal Botanic Gardens, Kew. (ill., maps). [*Pinus*, 47 spp., based on Flora Neotropica Monograph 75 (Farjon & Styles, 1997)].

03710 Farjon, A. & K. D. Rushforth (1989). A classification of *Abies* Miller (Pinaceae). Notes Roy. Bot. Gard. Edinburgh 46 (1): 59–79. [*Abies* subsect. *Delavayianae*, subsect. *Hickelianae*, subsect. *Holophyllae*, subsect. nov., pp. 70–71, bibliography].

03710a Farjon, A. & S. Ortiz Garcia (2002). Towards the minimal conifer cone: ontogeny and trends in *Cupressus, Juniperus* and *Microbiota* (Cupressaceae s. str.). Bot. Jahrb. Syst. 124 (2): 129–147. [*Cupressus goveniana, Juniperus phoenicea, J. virginiana, J. indica, Microbiota decussata*, evolution and phylogeny, SEM ill.].

03710b Farjon, A. & S. Ortiz Garcia (2003). Cone and ovule development in *Cunninghamia* and *Taiwania* (Cupressaceae sensu lato) and its significance for conifer evolution. Amer. J. Bot.

90 (1): 8–16. [Cupressaceae, Taxodiaceae, *Cunninghamia, Taiwania, Sciadopitys*, Majonicaceae, ontogeny, evolution, SEM ill.].

03711 Farjon, A. & B. T. Styles (1997). Flora Neotropica Monograph 75. *Pinus* (Pinaceae). (with accounts of wood anatomy by I. D. Gourlay, pollen morphology by M. H. Kurmann and monoterpenes by J. S. Birks). The New York Botanical Garden, New York. (ill., maps). [*Pinus* in Mexico, Central America and the Caribbean, revision, 47 spp. recognized, full synonymy, types, exsiccatae, bibliography, etc.; *Pinus arizonica* var. *cooperi* (C. E. Blanco) Farjon, p. 107].

03712 Farjon, A., P. Thomas & Nguyen Duc To Luu (2004). Conifer conservation in Vietnam; three potential flagship species. Oryx 38 (3): 1–9. [*Pinus krempfii, Taiwania cryptomerioides, Xanthocyparis vietnamensis*, ill., tables].

03713 Farjon, A., Q. P. Xiang & X. Q. Zhang (2004). The comparative study on the cuticle micromorphology of *Pilgerodendron uviferum* (Cupressaceae) and its relatives. Acta Phytotax. Sin. 42 (5): 427–435. [*Austrocedrus chilensis, Libocedrus plumosa, Papuacedrus papuana, Pilgerodendron uviferum*, SEM ill.].

03720a Fassett, N. C. (1943). The validity of *Juniperus virginiana* var. *crebra*. Amer. J. Bot. 30: 469–477, f. 1–31. [descr. and range given by Fernald & Griscom (1935) amended].

03721 Fassett, N. C. (1944). *Juniperus virginiana, J. horizontalis* and *J. scopulorum* I. The specific characters. Bull. Torrey Bot. Club 71 (4): 410–418. (tables).

03722 Fassett, N. C. (1944). *Juniperus virginiana, J. horizontalis* and *J. scopulorum* II. Hybrid swarms of *J. virginiana* and *J. scopulorum*. Bull. Torrey Bot. Club 71 (5): 475–483. (ill.).

03723 Fassett, N. C. (1945). *Juniperus virginiana, J. horizontalis* and *J. scopulorum* III. Possible hybridization of *J. horizontalis* and *J. scopulorum*. Bull. Torrey Bot. Club 72 (1): 42–46. (ill.) [*J. scopulorum* var. *patens* var. nov., p. 46].

03724 Fassett, N. C. (1945). *Juniperus virginiana, J. horizontalis* and *J. scopulorum* IV. Hybrid swarms of *J. virginiana* and *J. horizontalis*. Bull. Torrey Bot. Club 72 (4): 379–384. (ill.). [*J. virginiana* var. *ambigens* var. nov., p. 380].

03730 Fassett, N. C. (1945). *Juniperus virginiana, J. horizontalis* and *J. scopulorum* V. Taxonomic treatment. Bull. Torrey Bot. Club 72 (5): 480–482. (map). [*J. scopulorum* var. *columnaris* var. nov., p. 482].

03735 Favela L., S. (1991). Taxonomia de *Pinus pseudostrobus* Lindl., *Pinus montezumae* Lamb., y *Pinus hartwegii* Endl. Rep. Cientifico 26, 30 pp. Linares, Mexico.

03740 Fechner, G. H. & R. W. Clark (1969). Preliminary observations on hybridization of Rocky Mountain spruces. Proc. Commiss. Forest Tree breeding in Canada 11: 237–247. [*Picea*].

03750 Fedtschenko, B. A. (1898). *Abies semenovii* mihi, eine neue Tanne aus Zentralasien. Bot. Zentralbl. 73 (7): 210–211. [*A. semenovii* sp. nov. = *A. sibirica* ssp. *semenovii*].

03765 Felger, R. S., M. B. Johnson & M. F. Wilson (2001). The trees of Sonora, Mexico. OUP, Oxford. [Angiosperms, Conifers (pp. 35–58), Cupressaceae, *Cupressus*, *Juniperus*, *Taxodium*, Pinaceae, *Abies*, *Pinus*, *Pseudotsuga*, floristics, ecology, taxonomy, ill.].

03770 Ferguson, C. W. (1968). Bristlecone pine: science and esthetics. Science 159: 839–846. [*Pinus aristata*, *P. longaeva*].

03780 Ferguson, C. W. & R. A. Wright (1963). Tree rings in the western Great Basin. Nevada State Mus. Anthrop. Papers 9: 10–16. (ill.). [*Pinus longaeva*].

03810 Ferguson, D. K. (1985). A new species of *Amentotaxus* (Taxaceae) from northeastern India. Kew Bull. 40 (1): 115–119.

03815 Ferguson, D. K. (1989). On Vietnamese *Amentotaxus* (Taxaceae). Adansonia, sér. 4, 11 (3): 315–318. (ill.). [*A. poilanei* (Ferré & Rouane) comb. et stat. nov., p. 316].

03820 Ferguson, D. K., H. Jähnichen & K. L. Alvin (1978). *Amentotaxus* Pilger from the European Tertiary. Fedde's Repert. 89 (7–8): 379–410.

03830 Fernald, M. L. (1909). A new variety of *Abies balsamea*. Rhodora 11: 201–203. [*Abies balsamea* var. *phanerolepis*, var. nov.].

03831 Fernald, M. L. (1919). Lithological factors limiting the ranges of *Pinus banksiana* and *Thuja occidentalis*. Rhodora 21 (243): 41–67. (map).

03831a Fernald, M. L. & L. Griscom (1935). Three days of botanizing in southeastern Virginia. Rhodora 37 (436): 129–157. [*Juniperus virginiana* var. *crebra* var. nov., p. 133, pl. 332–333].

03831b Fernald, M. L. & C. A. Weatherby (1932). *Picea glauca*, forma *parva*. Rhodora 34: 187–189. [lectotypification of *Pinus canadensis* L. = *Tsuga canadensis* with J. Clayton 547, BM].

03850 Ferré, Y. de (1944). Une nouvelle espèce de *Pseudolarix*: *P. Pourteti*. Trav. Lab. Forest. Toulouse, T. 1, vol. 4, art. 4: 1–9, f. 1–13. (also publ. in: Bull. Soc. Hist. Nat. Toulouse 79, 1944). [= *P. amabilis*].

03880 Ferré, Y. de (1948). Quelques particularités anatomiques d'un pin indochinois: *P. Krempfii*. Trav. Lab. Forest. Toulouse, T. 1, vol. 4, art. 25: 1–6, f. 1–4. [*Pinus krempfii*].

03900 Ferré, Y. de (1952). Additions et corrections 1. *Keteleeria roulletii*. Bull. Soc. Hist. Nat. Toulouse 87: 340–342. (ill.).

03910 Ferré, Y. de (1953). Division du genre *Pinus* en quatre sous-genres. Compt. Rend. Hebd. Séances Acad. Sci. (Paris) 236: 226–228.

03920 Ferré, Y. de (1960). Une nouvelle espèce de pin au Viet-Nam. *Pinus dalatensis*. Bull. Soc. Hist. Nat. Toulouse 95: 171–180. (also publ. in: Trav. Lab. Forest. Toulouse, T. 1, vol. 6, art. 4: 1–10).

03930 Ferré, Y. de (1965). Structure des plantules et systématique du genre *Pinus*. Bull. Soc. Hist. Nat. Toulouse 100: 230–280. (ill.; also publ. in: Trav. Lab. Forest. Toulouse, T. 2, vol. 3, art. 2: 1–50, 1965).

03931 Ferré, Y. de (1966). Validité de l'espèce *Pinus pumila* et affinités systématiques. Trav. Lab. Forest. Toulouse, T. 1, vol. 6, art. 26: 1–6. [leaf anatomy of *P. pumila*, *P. parviflora*, *P. sibirica*, ill., map].

03940 Ferré, Y. de & M.-Th. Augère (1943). De la présence de *Larix* dans la province du Kiang-si. Bull. Soc. Hist. Nat. Toulouse 78 (Art. 24): 137–140. [*Larix leptolepis* var. *louchanensis* var. nov. = *L. kaempferi* culta].

03950 Ferré, Y. de & H. Gaussen (1945). Le rameau phyletique: *Pinus*, *Pseudolarix*, *Keteleeria*. Trav. Lab. Forest. Toulouse, T. 1, vol. 4, art. 8: 1–11. (repr. from Bull. Soc. Hist. Nat. Toulouse 80, 1945). [Pinaceae, character matrix, evolution, phylogeny].

03951a Ferré, Y. de & M. L. Rouane (1978). A propos du genre *Amentotaxus* Pilger. Trav. Lab. Forest. Toulouse T. 1 (9, 1): 1–6. [*A. yunnanensis* var. poilanei Ferré & Rouane, var. nov., p. 3].

03951b Ferré, Y. de, M. L. Rouane & P. Woltz (1978). Systematique et anatomie comparée des feuilles de Taxaceae, Podocarpaceae, Cupressaceae de Nouvelle-Calédonie. Cahiers Pacifique 20: 241–266. [*Libocedrus*, *Neocallitropsis*].

03952 Ferré, Y. de & J. Soual (1940). Feuille primordiale et feuille fasciculée du Pin des Canaries. Bull. Soc. Hist. Nat. Toulouse 75: 139–151. [*Pinus canariensis*, ill.].

03960 Fieschi, V. & H. Gaussen (1932). La classification des pins maritimes. Trav. Lab. Forest. Toulouse, T. 1, art. 19: 1–10. [*Pinus mesogeensis* sp. nov. (also publ. in: Bull. Soc. Hist. Nat. Toulouse 64: 440, 1932). [*P. maritima* = *P. pinaster*].

03960a Finckh, M. & A. Paulsch (1995). Die ökologische Strategie einer Reliktkonifere (The ecological strategy of *Araucaria araucana*). Flora 190: 365–382.

03961   Fischer, F. E. L. & C. A. Meyer (1842). Index seminum, quae Hortus botanicus imperialis petropolitanus pro mutua commutatione offert. Index octavus (8): Animadversiones botanicae. St. Petersburg. [*Juniperus pseudosabina* sp. nov., p. (15) 65].

03962   Fischer, F. E. L. & C. A. Meyer (1842). (Animadversiones botanicae: *Picea schrenkiana* sp. nov.) in: Bull. Sci. Acad. Imp. Sci. Saint.-Pétersbourg 10: 253.

03970   Fitschen, J. (1926). Beitrag zur Kenntnis der in Deutschland anbauwürdigen Fichten. Mitt. Deutsch. Dendrol. Ges. 37: 35–56. [*Picea*, morphology, key].

03980   Fitschen, J. (1929). Die Gattung *Tsuga*. Mitt. Deutsch. Dendrol. Ges. 1929: 1–12. (ill.).

04010   Flake, R. H., E. von Rudloff & B. L. Turner (1969). Quantitative study of clinal variation in *Juniperus virginiana* using terpenoid data. Proc. Natl. Acad. U.S.A. 64: 487–494. [*J. virginiana* compared with *J. ashei*].

04011   Florin, C. R. (1922). On the geological history of the Sciadopitineae. Svensk Bot. Tidskr. 16 (2): 260–270. [*Sciadopitys*, fossil and recent].

04020   Florin, C. R. (1927). Gymnospermae. in: H. Smith. Plantae sinenses. Acta Horti Gothob. 3: 1–10, pl. 1–4. [Pinaceae, Cupressaceae, *Juniperus komarovii* sp. nov., p. 3, *J. glaucescens* sp. nov., p. 5, *J. ramulosa* sp. nov., p. 5, *J. distans* sp. nov., p. 6, Ephedraceae, pl. of new spp. in *Juniperus*].

04030   Florin, C. R. (1930). Die Koniferengattung *Libocedrus* Endl. in Ostasien. Svensk Bot. Tidskr. 24 (1): 117–131, pl. I–II. [*Libocedrus macrolepis* = *Calocedrus macrolepis*, *L. formosana* sp. nov. = *C. formosana*, p. 126, figs., 2 plates].

04040   Florin, C. R. (1930). *Pilgerodendron*, eine neue Koniferengattung aus Süd-Chile. Svensk Bot. Tidskr. 24 (1): 132–135. [*Pilgerodendron uviferum*, gen. et comb. nov., pp. 132–133].

04091   Florin, C. R. (1948). On *Nothotaxus*, a new genus of the Taxaceae, from Eastern China. Acta Horti Berg. 14 (8): 385–395. (ill.). [*Nothotaxus chienii* (1948) = *Pseudotaxus chienii* (1947)].

04110   Florin, C. R. (1952). On *Metasequoia*, living and fossil. Bot. Not. 1 (105): 1–29.

04130   Florin, C. R. (1956). Nomenclatural notes on genera of living gymnosperms. Taxon 5 (8): 188–192. [*Calocedrus decurrens* (Torrey), *C. formosana* (Florin), comb. nov. (basion.: *Libocedrus* spp.), p. 192].

04137   Florin, C. R. (1960).Die frühere Verbreitung der Koniferengattung *Athrotaxis* D. Don. Senck. Leth. 41: 199–207. [phytogeography, palaeobotany; many fossils attributed to this genus are in fact not this taxon, see

(Coniferales) No. 05876a Hill & Brodribb, 1999, also (Coniferales) No. 07684 Kunzmann, 1999].

04150   Florin, C. R. & J. B. Boutelje (1954). External morphology and epidermal structure of leaves in the genus *Libocedrus*, s. lat. Acta Horti Berg. 17 (2): 7–37, pl. I–X. [*Austrocedrus chilensis* gen. et comb. nov., p. 28–29, with cit. of *L. chilensis* (D. Don) Endl., not the basion. : see R. E. G. Pichi Sermolli & M. P. Bizzarri, 1978].

04151   Florschütz, F. (1925). On *Pseudolarix Kaempferi* Gord. from the clay of Reuver. Recueil Trav. Bot. Néerl. 22 (3–4): 269–273. [fossil *Pseudolarix* cf. *amabilis*, ill.].

04160   Flous, F. (1934). Diagnoses d'espèces et variétés nouvelles de *Pseudotsuga* americains. Bull. Soc. Hist. Nat. Toulouse 66: 211, 219, 330–342, 366, 388., pls. [*P. guinierii*, p. 211, *P. macrolepis*, p. 219, *P. californica*, p. 330, *P. flahaultii*, p. 332, *P. globulosa*, p. 334, *P. vancouverensis*, p. 340, *P. guinieri* var. *mediostrobus*, p. 342, *P. guinieri* var. *parvistrobus*, p. 342, *P. merrillii*, p. 366, *P. rehderi*, p. 388, spp. et var. nov.; also publ. in Trav. Lab. Forest. Toulouse, T. 1, vol. 2, art. 2: 1–14, 1934; see for comment: E. L. Little, Jr., 1952].

04170   Flous, F. (1936). Espèces nouvelles de *Keteleeria*. Bull. Soc. Hist. Nat. Toulouse 69: 399–408, figs. [*K. chien-peii*, p. 400–402, f. 1–11, *K. cyclolepis*, p. 402–403, f. 1–11, *K. dopiana*, p. 404–406, f. 1–11, *K. roulletii*, p. 406–408, f. 1–13, spp. nov.; also publ. in Trav. Lab. Forest. Toulouse, T. 1, vol. 2, art. 14: 1–8, 1936].

04180   Flous, F. (1936). Révision du genre *Keteleeria*. Trav. Lab. Forest. Toulouse, T. 2, vol. 4, art. 1: 1–76. [see also A. Farjon, 1989].

04190   Flous, F. (1936). Espèces nouvelles du genre *Tsuga*. Bull. Soc. Hist. Nat. Toulouse 69: 409–415, pls. [*T. blaringhemii*, p. 410, *T. crassifolia*, p. 412, *T. tchekiangensis*, p. 414, spp. nov.; also publ. in Trav. Lab. Forest. Toulouse, T. 1, vol. 2, art. 15: 1–6, 1936)

04200   Flous, F. (1936). Révision du genre *Tsuga*. Trav. Lab. Forest. Toulouse, T. 2, vol. 4, art. 3: 1–136.

04210   Flous, F. (1936). Espèces nouvelles de *Pseudotsuga* asiatiques. Bull. Soc. Hist. Nat. Toulouse 69: 416–420. [*P. gaussenii*, p. 417, *P.salvadorii*, p. 419, spp. nov.].

04220   Flous, F. (1936). Révision du genre *Pseudotsuga*. Trav. Lab. Forest. Toulouse, T. 2, vol. 4, art. 2: 1–132.

04250   Flous, F. & H. Gaussen (1932). Une nouvelle espèce de Sapin du Mexique: *Abies hickeli*. Trav. Lab. Forest. Toulouse, T. 1, vol. 1, art. 17: 1–7. [*Abies hickelii* sp. nov.; also publ. in Bull. Soc. Hist. Nat. Toulouse 44: 24, fig., 1932].

04261 Fomin, A. V. (1914). Zur Systematik der Arten und Rassen der Gattung *Pinus* aus dem Kaukasus und der Krim. Vestn. Tiflissk. Bot. Sada (Monit. Jard. Bot. Tiflis) 34: 15–24. (Russ.). [*Pinus sylvestris* ssp. *hamata* (Steven) stat. nov., p. 16, *P. sylvestris* div. ssp. & var. comb./stat. nov., *P. stanckewiczii* (Sukatchev) comb. nov., p. 18].

04270 Foote, M. J. (1983). Classification, Description, and Dynamics of Plant Communities After Fire in the Taiga of Interior Alaska. U.S. Forest Serv. Res. Paper PNW-307. Pacific Northwest Forest and Range Exp. Stat., Portland, Oregon.

04280 Forbes, J. (1839). Pinetum woburnense: or, a catalogue of coniferous plants, in the collection of the Duke of Bedford, at Woburn Abbey; systematically arranged. London. (ill.). [*Abies pichta* sp. nov. = *A. sibirica*, p. 113, pl. 39, *A. amabilis* sp. nov., p. 125, pl. 44, *Larix microcarpa* sp. nov. = *L. laricina*, p. 139, t. 47, *Taxodium sinense* sp. nov. = *Glyptostrobus pensilis*, p. 179, *Cupressus articulata* (Vahl) comb. nov. = *Tetraclinis articulata*, p. 191, *Juniperus hudsonica* sp. nov. = *J. horizontalis*, p. 208, *Podocarpus chinensis* (Roxb.) Wall. ex Forbes, p. 212].

04290 Forde, M. B. (1963). Variation in the natural populations of Monterey pine (*Pinus radiata* Don) in California. Unpubl. Ph. D. diss., Univ. California, Davis, Ca.

04291 Forde, M. B. (1964). Variation in natural populations of *Pinus radiata* in California. Part 3: Cone characters, 4: Discussion. New Zealand J. Bot. 2 (4): 459–501.

04292 Formanek, E. (1896). Zweiter Beitrag zur Flora von Serbien, Macedonien und Thessalien. Verh. Naturf. Vereines Brünn 34: 255–365. [*Juniperus foetidissima* var. *pindicola* var. nov., p. 272].

04297 Forster, J. G. A. (1786). De plantis esculentis insularum oceani australis commentatio botanica. Berlin. [*Dacrydium cupressinum* Sol. ex G. Forst., p. 80].

04300 Fortune, R. (1854). Le Cyprès pleureur de la Chine ou *Cupressus funebris*. Hort. Belge 4: 343–344, pl. 55. [syn. *Chamaecyparis funebris*].

04310 Fosberg, F. R. (1959). *Pinus contorta* and its variations. Baileya 7: 7–10.

04315 Fountain, D. W., J. M. Holdsworth & H. A. Outred (1989). The dispersal unit of *Dacrycarpus dacryoides* (A. Rich.) de Laubenfels (Podocarpaceae) and the significance of the fleshy receptacle. Bot. J. Linn. Soc. 99: 197–207. (ill.).

04330 Fowler, D. P. (1983). The hybrid black × Sitka spruce, implications to phylogeny of the genus *Picea*. Canad. J. Forest Res. 13: 108–115. [*Picea mariana, P. sitchensis*].

04340 Foxworthy, F. W. (1911). Phillipine gymnosperms. Phillipine J. Sci., C. Botany 6 (3): 149–176. [gymnosperms, *Pinus merkusii, P. insularis, Podocarpus brevifolius* (Stapf) comb. nov., p. 160, *P. philippinensis* sp. nov., p. 163].

04370 Francini, E. (1958). Ecologia comparata di *Pinus halepensis* Mill., *P. pinaster* Sol. e *P. pinea* L. sulla base del comportamento del gametofito femminale. Acad. Ital. Sci. Forest., Firenze 7: 107–172.

04380 Franco, J. do A. (1941). O género *Chamaecyparis* Spach. Agros 24: 91–99. [*Chamaecyparis funebris* (Endl.) comb. nov., p. 93 (basion.: *Cupressus funebris* Endl. in Syn. Conif., 1847)].

04390 Franco, J. do A. (1942). Sub-géneros e secçaes do género *Abies* Mill. Bol. Soc. Portug. Ci. Nat. 13 (Suppl. 2): 163–170, f. 1–5.

04395 Franco, J. do A. (1945). A *Cupressus lusitanica* Miller, notas acêrca da sua história e sistemática. Agros (Lisbon) 28: 3–27. (ill.).

04400 Franco, J. do A. (1949). *Pinus thunbergiana* et *Pinus clusiana* var. *corsicana*. Anais Inst. Super. Agron. (Lisboa) 16: 129–132. [nomenclature: nom. et comb. nov.].

04410 Franco, J. do A. (1949). Notas nomenclaturais. Bol. Soc. Brot., Sér. 2, 23: 159–176. [*Abies amabilis, A. excelsior* nom. nov. = *A. grandis*, p. 162, *Pinus canariensis*].

04420 Franco, J. do A. (1949). Notas sobre a nomenclatura de algumas coníferas. Portugaliae Acta Biol., Sér. B, Sist. Vol. "Júlio Henriques" pp. 19–35. [*Abies spectabilis, Agathis, Araucaria, Callitris rhomboidea, Cedrus atlantica, Dacrydium, Platycladus orientalis* (L.) comb. nov., p. 33, *Tetraclinis articulata*].

04430 Franco, J. do A. (1950). De coniferarum duarum nominibus. Lisboa. (7 pp.). [*Cedrus libanensis* Mirb. = *C. libani* A. Richard, *Pseudotsuga menziesii* (Mirb.) Franco comb. nov.; also in: Bol. Soc. Brot., Sér. 2, 24: 73–77, new comb. on p. 74].

04440 Franco, J. do A. (1950). *Cedrus libanensis* et *Pseudotsuga menziesii*. Bol. Soc. Brot., Sér. 2, 24: 73–77. [*P. menziesii* var. *glauca* (Beissn.) comb. nov., *P. menziesii* var. *caesia* (Schwerin) comb. nov. (= var. *glauca*), p. 77].

04450 Franco, J. do A. (1950). Abetos. Lisboa. [*Abies*, monogr., *A.* sect. *Momi* sect. nov., *A.* sect. *Oiamel* sect. nov.].

04470 Franco, J. do A. (1954). On the legitimacy of the combination *Pseudotsuga menziesii* (Mirb.) Franco. Bol. Soc. Brot., Sér. 2, 28: 115–116.

04480 Franco, J. do A. (1962). Taxonomy of the common juniper. Bol. Soc. Brot., Sér. 2, 36: 101–120. [*Juniperus communis*, div. ssp. descr., *J. communis* ssp. *depressa* (Pursh) comb. et stat. nov., p. 117, *J. rigida* Siebold & Zucc. ssp. *nipponica* (Maxim.) comb. nov., p. 119, ill.].

04481 Franco, J. do A. (1963). Taxonomic Notes on *Juniperus oxycedrus* L. and *J. macrocarpa* Sm. Fedde's Repert. Spec. Nov. Regni Veg. 68 (3): 163–167. [*Juniperus brevifolia*, *J. cedrus*, *J. oxycedrus* ssp. *macrocarpa*, *J. oxycedrus* ssp. *transtagana* ssp. nov., map].

04490 Franco, J. do A. (1968). On Himalayan-Chinese Cypresses. Portugaliae Acta Biol., Sér. B, Sist. 9: 183–195. [*Cupressus fallax* sp. nov., p. 190, *C. corneyana* auct. non Carrière].

04491 Franco, J. do A. (1986). Pinaceae-Ephedraceae. (ed. G. López Gonzáles) in: M. Laínz, S. Castroviejo, G. López Gonzáles *et al.* (eds.). Flora Iberica. Madrid. (pp. 163–195, ill.). [*Abies alba*, *A. pinsapo*, *Pinus nigra* ssp. *salzmannii*, *Pinus* spp., *Cupressus* spp., *Tetraclinis articulata* (f. 60), *Juniperus communis*, *J. oxycedrus*, *J. navicularis* (syn.: *J. oxycedrus* ssp. *transtagana*), *J. phoenicea*, *J. thurifera*, *J. sabina*].

04495 Frank, A. B. (1873). Über den Einfluss des Lichtes auf den bilateralen Bau der symmetrischen Zweige der *Thuja occidentalis.* Jahrb. Wiss. Bot. 9: 147–190, t. 16. [stomata will shift position in new leaves when plagiotropic shoots with leaves are turned upside down].

04500 Frankis, M. P. (1989). Some interesting, unusual and recently described pines for Australian gardens part 1: "Soft" Pines (subgenus *Strobus*). The Conifer Soc. of Australia, Newsletter No. 5: 12–15.

04511 Frankis, M. P. (1990). More pines for Australian gardens. The Conifer Soc. of Australia, Newsletter No. 7: 7–9. [*Pinus devoniana*, *P. gordoniana*, *P. martinezii*, *P. herrerae*].

04511a Frankis, M. P. (1991). Fire-climax pines: there's more to it than you thought! The Conifer Soc. of Australia, Newsletter No. 9: 8–9. [*Pinus* spp.].

04512 Frankis, M. P. (1993). Morphology and affinities of *Pinus brutia*. In: O. Taskin (ed.). Papers International Symp. on *Pinus brutia* Ten. Part A. Botanical characteristics and natural distribution; pp. 11–18. Ministry of Forestry, Ankara.

04513 Frankis, M. P. (1993). Nootka Cypress: *Chamaecyparis* or *Cupressus*? The Conifer Soc. of Australia, Newsletter No. 12: 9–10. [*Chamaecyparis nootkatensis* = *Xanthocyparis nootkatensis*].

04514 Frankis, M. P. (1999). *Pinus brutia* (Pinaceae). Curtis's Bot. Mag. 16 (3): 173–184, pl. 367.

04515 Frankis, M. P. (2000). Pinaceae. In: Güner & al. (eds.). Flora of Turkey. Vol. 11 (Supplement 2): 5–7. [*Pinus*].

04517 Frankis, M. P. & F. Lauria (1994). The maturation and dispersal of cedar cones and seeds. Int. Dendrol. Soc. Yearb. 1993: 43–46. [*Cedrus deodara*, *C. libani* s.l.].

04519 Franklin, D. A. (1968). Biological Flora of New Zealand 3. *Dacrydium cupressinum* Lamb. (Podocarpaceae) rimu. New Zeal. J. Bot. 6: 493–513. (maps).

04520 Franklin, J. F. (1974). *Abies* Mill. in: Seeds of woody plants in the United States. (pp. 168–183, ill.). U.S.D.A. Agric. Handb. 450. Washington, D.C.

04550 Freiherr, H. (1935). Zur Systematik der Gattung *Larix.* Mitt. Deutsch. Dendrol. Ges. 47: 1–7.

04560 Frémont, J. C. (1845). Report of the exploring expedition to the Rocky Mountains in the year 1842,... Washington. [see J. Torrey & J. C. Frémont, 1845].

04571 Fries, E. M. (1846–49). Summa vegetabilium Scandinaviae. Holmiae (Stockholm). (viii + 572 pp.). [*Pinus sylvestris* var. *lapponica* var. nov., p. 58, 1848, nom. nud.; see C. J. Hartman, 1849–50].

04572 Friis, I. (1992). Forests and forest trees of northeast tropical Africa. their natural habitats and distribution patterns in Ethiopia, Djibouti and Somalia. Kew Bulletin Additional Series XV. Royal Botanic Gardens, Kew; Her Majesty's Stationery Office, London. [*Juniperus procera* on pp. 91–93].

04590 Frothingham, E. H. (1909). Douglas fir: a study of the Pacific Coast and Rocky Mountain forms. Dept. Agric. Forest. Div. Circ. 150. (38 p., ill.). [*Pseudotsuga menziesii*].

04600 Fry, W. & J. R. White (1938). Big trees. Stanford Univ. Press, Stanford, Calif. [*Sequoiadendron giganteum*, 126 pp., many ill.].

04605 Fu, D. Z. (1992). Nageiaceae – a new gymnosperm family. Acta Phytotax. Sin. 30: 515–528. [*Nageia nagi* (Podocarpaceae), *Nageia* s.l.].

04610 Fu, L. K., Y. J. Lu & S. L. Mo (1980). The genus *Abies* discovered for the first time in Guangxi and Hunan. Acta Phytotax. Sin. 18 (2): 205–210. (Chin. + Lat.). [*A. yuanbaoshanensis*, sp. nov., p. 206, *A. ziyuanensis*, sp. nov., p. 208].

04611 Fu, L. K., R. R. Mill & N. J. Turland (1999). Validation of the name *Cephalotaxus latifolia* (Cephalotaxaceae), a species from southeast China. Novon 9 (2): 185–186. [*Cephalotaxus latifolia* sp. nov.].

04615 Fueyo, G. M. del (1999). Cone and ovule development in the *Podocarpus* species from Argentina. Phytomorphology 49 (1): 49–60. [*Podocarpus lambertii*, *P. nubigenus*, *P. parlatorei*, LM & SEM ill.].

04620 Fukarek, P. (1951). Danasnje rasprostranjenje Panciceve omorike (*Picea omorika* Pančić) i neki podaci o njenim sastojinama. Yearb. Biol. Inst. Sarajewo 1951: 141–198. (Serb.-Kroat.). [Present distribution of *P. omorika* and some remarks on its locations].

04621   Fukarek, P. (1958). Die Standortsrassen der Schwarzföhre (*Pinus nigra* sens. lat.). Centralbl. Gesammte Forstwesen 75: ? (maps 1–3 on pp. 204–206).

04622   Fukarek, P. (1967). Neue Standorte der Panzerkiefer (*Pinus heldreichii* Christ emend. Markgraf). Bot. Jahrb. Syst. 86 (1–4): 449–462. (ill., maps).

04630   Fukarek, P. (1973). Die Tannen der Balkanhalbinsel. in: Istanbul Universitesi: Orman Fakültesi. Kazdagi göknari ve Turkiye florasi uluslarasi simpozyomu bildirileri Istanbul 1973: 83–94. [symposium proceedings with other papers on *Abies*].

04640   Fulling, E. H. (1934). Identification, by leaf-structure, of the species of *Abies* cultivated in the United States. Bull. Torrey Bot. Club 61: 497–524. (ill.).

04650   Fulling, E. H. (1936). *Abies intermedia*, the Blue Ridge fir, a new species. Castanea 1: 91–94. [= *A.* × *phanerolepis* (Fern.) Liu].

04660   Fulling, E. H. (1976). *Metasequoia* Fossil and Living. (An initial thirty-year (1941–1970) annotated and indexed bibliography with an historical introduction). Bot. Rev. (Lancaster) 42 (3): 215–315.

04662a   Gabila, E. M. & H. L. Mogensen (1973). Foliar initiation and the fate of the dwarf-shoot apex in *Pinus monophylla*. Amer. J. Bot. 60 (7): 671–677.

04663   Gadek, P. A. & C. J. Quinn (1987). Biflavones and the affinities of *Cupressus funebris*. Phytochemistry 26: 2551–2552. [chemo-taxonomy: biflavonyl profile does not support transfer to *Chamaecyparis*].

04667   Gaertner, J. (1788). De fructibus et seminibus plantarum. Vol. 1. Stuttgart. [*Nageia* gen. nov., p. 191].

04668   Gajardo, R., P. Woltz, M. Gondran & J. Marguerier (1996). Xylologie des Conifères endemiques des Andes méridionales au MEB I. Saxegothaeaceae. Rev. Cytol. Biol. Végét.-Bot. 19: 31–45. [Podocarpaceae, *Saxegothaea conspicua*, SEM ill. of wood anatomy].

04670   Gambles, R. L. & N. G. Dengler (1974). The leaf anatomy of hemlock, *Tsuga canadensis*. Canad. J. Bot. 52: 1049–1056. (ill.).

04670a   Gamisans, J., D. Jeanmonod, P. Regato & M. Gruber (1994). Notes et contributions à la flore de Corse: 10B – Contribution 30. Le genévrier thurifère (*Juniperus thurifera* L.) en Corse. Candollea 49 (2): 600–607. (maps, tables).

04671   Gandoger, M. (1910). Notes sur la flore hispano-portugaise. Quatrième voyage en Portugal; IX. Bull. Soc. Bot. France 57: 54–62. [*Juniperus navicularis* sp. nov., p. 55].

04672   Gandolfo, M. A., K. C. Nixon & W. L. Crepet (2001). Turonian Pinaceae of the Raritan Formation, New Jersey. Plant Syst. Evol. 226: 187–203. [*Pinus*, *Prepinus*, Upper Cretaceous fossils, ill.].

04677   Garcia, A. & S. Gonzalez (1991). Flora y vegetacion de la cima del Cerro Potosi, Nuevo Leon, México. Acta Bot. Mexicana 13: 53–74. [*Pinus culminicola*, *P. hartwegii*, *Juniperus ashei*].

04680   Garden, J. (1957). A Revision of the Genus *Callitris* Vent. Contr. New South Wales Natl. Herb. 2 (5): 363–392. [*C. preissii* ssp. *murrayensis* ssp. nov., p. 373, *C. preissii* ssp. *verrucosa* (Endl.) comb. et stat. nov., p. 375 = *C. preissii* Miq. et *C. verrucosa* (Endl.) F. Mueller, resp. (see J. Venning, 1986), *C. monticola* sp. nov., p. 385].

04681   Garden, J. & L. A. S. Johnson (1951). *Microstrobos*, a new name for a podo-carpaceous genus. Contr. New South Wales Natl. Herb. 1: 315–316. [= *Pherosphaera* W. Archer bis, *M. fitzgeraldii* (F. Muell.) comb. nov. = *P. fitzgeraldii* (F. Muell.) Hook. f., *M. niphophilus* sp. nov. (nom. inval., Art. 36) = *P. hookeriana* W. Archer bis].

04690   Gardner, C. A. (1964). Contributiones Florae Australiae Occidentalis XIII. Cupressaceae. J. Roy. Soc. W. Austral. 47: 54. [*Actinostrobus arenarius* sp. nov.].

04691   Gardner, H. H. (1926). East African Pencil Cedar. Emp. Forest. J. 5: 39–53. [*Juniperus procera*].

04692   Gardner, M. F., P. Thomas, A. Lara & B. Escobar (1999). *Fitzroya cupressoides* (Cupressaceae). Curtis's Bot. Mag. 16 (3): 229–240, pl. 372.

04700   Garfitt, J. E. (1966). The Cyprus Cedar. Quart. J. Forest. 60 (3): 185–189. [*Cedrus brevifolia*].

04710   Garman, E. H. (1957). The occurrence of spruce in the interior of British Columbia. Canad. Dept. Lands Forests, British Columbia Forest Serv. Techn. Publ. T. 49. (31 p., ill.). [*Picea glauca*, *P. engelmannii*, *P. mariana*].

04720   Gates, F. C. (1938). Layering in black spruce. Amer. Midl. Naturalist 19: 589–594. [*Picea mariana*].

04724   Gauquelin, T., J.-F. Asmodé & G. Largier (eds.). (2000). Le Genévrier thurifère (*Juniperus thurifera* L.) dans la bassin occidental de la Méditerranée: systematique, écologie, dynamique et gestion. (proceedings of an international symposium, 26–27 September 1997, Marignac, Haute-Garonne, France). Les Dossiers Forestiers No. 6.

04725   Gauquelin, T., V. Bertaudière, N. Montes, W. Badri & J.-F. Asmodé (1999). Endangered stands of thuriferous juniper in the western Mediterranean basin: ecological status, conservation and management. Biodiv. Cons. 8 (11): 1479–1498. [*Juniperus thurifera*, maps, ill.].

04726   Gauquelin, T., M. Idrissi Hassani & P. Lebreton (1988). Le Genévrier thurifère, *Juniperus thurifera* L. (Cupressacées): analyse biométrique et biochimique: propositions systématiques. Ecol. Mediterranea 14 (3–4): 31–42. [infraspecific taxa invalidly published].

04727   Gauquelin, T. & P. Lebreton (1998). Systématique de *Juniperus thurifera* L.: le cas de la population pyrénéenne de la Montagne de Rié. (Haute-Garonne, France). J. Bot. Soc. France 5: 105–109. (map).

04730   Gaussen, H. (1928). Une nouvelle espèce de Sapin. *Abies pardei*. Trav. Lab. Forest. Toulouse, T. 1, vol. 1, art. 2: 1–10. [*A. pardei* sp. nov., p. 5 (= sp. dub., see also A. Farjon, 1988), figs.].

04740   Gaussen, H. (1939). Une nouvelle espèce de *Taiwania*, *T. flousiana*. Trav. Lab. Forest. Toulouse, T. 1, vol. 3, art. 2: 1–9. [*T. flousiana* sp. nov., p. 6 = *T. cryptomerioides*].

04742   Gaussen, H. (1950). Espèces nouvelles de Cyprès: *Cupressus atlantica* au Maroc, *Cupressus lereddei* aux Ajjers. Le Monde des Plantes, Toulouse 45: 55–56. [*C. atlantica* sp. nov. = *C. dupreziana* var. *atlantica*, *C. lereddei* sp. nov. = *C. dupreziana*].

04750   Gaussen, H. (1955). Classification des pins diplostélés. Compt. Rend. Hebd. Séances Acad. Sci. (Paris) 251: 1366–1369. [*Pinus* subgen. *Pinus*].

04760   Gaussen, H. (1960). Les Gymnospermes actuelles et fossiles. Fascicule 6: Chapitre XI Généralités, Genre *Pinus*. Faculté des Sciences, Toulouse (reprint from Trav. Lab. Forest. Toulouse, 1960). (ill.).

04770   Gaussen, H. (1964). Les Gymnospermes actuelles et fossiles. Fascicule 7: Genres *Pinus* (suite), *Cedrus* et *Abies*. Faculté des Sciences, Toulouse (reprint from Trav. Lab. Forest. Toulouse, 1960). (ill.).

04780   Gaussen, H. (1966). Les Gymnospermes actuelles et fossiles. Fascicule 8: *Pseudolarix, Keteleeria, Larix, Pseudotsuga, Pityites, Picea, Cathaya, Tsuga*. Faculté des Sciences, Toulouse (reprint from Trav. Lab. Forest. Toulouse, 1966). (ill.).

04781   Gaussen, H. (1967). La classification des Genévriers (*Juniperus*). Compt. Rend. Hebd. Séances Acad. Sci. (Paris) 265: 954–957.

04810   Gaussen, H. (1971). Les *Cathaya* ne sont pas des *Pseudotsuga*. Compt. Rend. Hebd. Séances Acad. Sci. (Paris) 273: 1098–1099.

04815   Gaussen, H. (1974). Un nouveau *Podocarpus* de Madagascar: *P. woltzii*. Bull. Soc. Hist. Nat. Toulouse 110 (1–2): 122–124. [*P. woltzii* sp. nov., p. 123].

04840   Gaussen, H. & M. Lacassagne (1930). Les épicéas du Kamchatka. Bull. Soc. Hist. Nat. Toulouse 59: 190–202. [*Picea kamtchatkensis* = *P. jezoensis*, map].

04840a   Gaussen, H. & P. Woltz (1975). Anatomie foliaire de quelques *Podocarpus* malgaches de haute montagne. Bull. Soc. Hist. Nat. Toulouse 111 (3–4): 319–321. [*P. perrieri* sp. nov., p. 319].

04840b   Ge, S., H. Q. Wang, C. M. Zhang & D. Y. Hong (1997). Genetic diversity and population differentiation of *Cathaya argyrophylla* in Bamian Mountain. Acta Bot. Sin. 39 (3): 266–271. (Chin., Eng. summ.). [analysis of allozyme data suggests high interpopulation diversity].

04840c   Geada López, G., K. Kamiya & K. Harada (2002). Phylogenetic relationships of diploxylon pines (subgenus *Pinus*) based on plastid sequence data. Int. J. Plant Sci. 163 (5): 737–747. [*Pinus* subgenus *Pinus*, chloroplast DNA, cladograms].

04840d   Ge, S., D. Y. Hong, H. Q. Wang, Z. Y. Liu & C. M. Zhang (1998). Population genetic structure and conservation of an endangered conifer, *Cathaya argyrophylla* (Pinaceae). Int. J. Plant Sci. 159 (2): 351–357. [allozyme variation in and among 8 populations].

04840e   Gedney, D. R., D. L. Azuma, C. L. Bolsinger & N. McKay (1999). Western Juniper in eastern Oregon. U.S.D.A. Forest Service; Gen. Tech. Rep. PNW-GTR-464. [*Juniperus occidentalis*, phytogeography, ecology, forestry].

04840f   Geng, B. Y. & J. Hilton (1999). New coniferophyte ovulate structures from the Early Permian of China. Bot. J. Linn. Soc. 129 (2): 115–138. [*Loroderma henania* gen. et sp. nov., p. 127, Coniferales, Vojnovskyales, *Cordaites* type foliage].

04841   Georgescu, C. C. (1937). Neue Beiträge zur Systematik der Schwarzföhre. Repert. Spec. Nov. Regni Veg. 41: 181–187. [*Pinus banatica* (Georgescu et Ionescu) comb. nov., *P. banatica* div. formae nov. = *P. nigra*, p. 183].

04843   Georgevitch, P. (1931). *Pinus nigra* Arn. var. *gocensis*, n. var. Österreich. Bot. Zeitschr. 80: 328–336. [ill. of bark types; *P. nigra* var. *gocensis*, var. nov. = *P. nigra* var. *nigra*, p. 336].

04850   Geraci, L. (1979). Nuovi reperti di *Abies nebrodensis* (Lojac.) Mattei rinvenuti sulle Madonie nella zona di indigenato (Coniferoxida, Pinaceae). Naturalista Sicil., n. s., 3 (1–2): 45–51.

04850b   Gernandt, D. S. & A. Liston (1999). Internal Transcribed Spacer region evolution in *Larix* and *Pseudotsuga* (Pinaceae). Amer. J. Bot. 86 (5): 711–723. [cladistic analysis of American and Eurasian species using nuclear ribosomal DNA of the ITS region].

04850c   Gernandt, D. S., A. Liston & D. Piñero (2001). Variation in the nrDNA ITS of *Pinus* subsection *Cembroides*: Implications for molecular systematic studies of pine species complexes. Mol. Phylogen. Evol. 21 (3): 449–467.

04850d Gernandt, D. S., A. Liston & D. Piñero (2003). Phylogenetics of *Pinus* subsections *Cembroides* and *Nelsoniae* inferred from cpDNA sequences. Syst. Bot. 28 (4): 657–673. [*Pinus cembroides et al.*, *P. nelsonii*, *P. krempfii*, *P. bungeana*, *Pinus* subsection *Strobi*, classification of *Pinus* subgenus *Strobus*].

04850i Geyler, T. & F. Kinkelin (1887). Oberpliocän-Flora aus den Baugruben des Klärbeckens bei Niederrad und der Schleuse bei Höchst a. M. Abh. Senckenbergischen Naturf. Ges. 15 (1): 1–49, pl. 1–4. [*Frenelites europaeus*, *Abies loehrii* sp. nov. ("*loehri*"), pp. 16–17, pl. 1, f. 13–15, *Pinus* spp., *Picea* spp.].

04850m Gibbs, L. S. (1912). On the development of the female strobilus in *Podocarpus*. Ann. Bot. (London) 26: 515–571. (ill.).

04851 Gibbs, L. S. (1917). A contribution to the phytogeography and flora of the Arfak Mountains, & c. Coniferae: pp. 78–87. London. [Coniferales, *Dacrydium novo-guineense* sp. nov., p. 78, *Libocedrus arfakensis* sp. nov. = *Papuacedrus arfakensis*, pp. 84–87, f. 6a].

04858 Gifford, D. J. (1987). An electrophoretic analysis of the seed proteins from *Pinus monticola* and eight other species of pine. Canad. J. Bot. 66: 1808–1812.

04860 Gifford, E. M., Jr. & N. T. Mirov (1960). Initiation and ontogeny of the ovulate strobilus in ponderosa pine. Forest Sci. 6: 19–25. (ill.). [*Pinus ponderosa*].

04868 Gil, L., J. Climent, N. Nanos, S. Mutke, I. Ortiz & G. Schiller (2002). Cone morphology variation in *Pinus canariensis* Sm. Pl. Syst. Evol. 235 (1–4): 35–51.

04870 Gilliland, H. B. (1938). Notes on the Flora of Rhodesian Manicaland: 1. gymnosperms. J. S. African Bot. 4 (4): 153–156, pl. 45. [*Widdringtonia nodiflora*].

04880 Gleason, H. A. (1955). Pedanticism runs amuck. Rhodora 57: 332–335. [*Pseudotsuga menziesii*, nomenclature].

04884 Glen, H. F. & J. J. Lavranos (1991). *Juniperus phoenicea*; Egypt to Morocco, Portugal, Mediterranean Europe, Cyprus, Turkey, Jordan, Saudi Arabia; Cupressaceae. Fl. Plants Africa 51 (2): pl. 2022. (3 pp., col. ill., map).

04887 Golovneva, L. B. (1988). A new genus *Microconium* (Cupressaceae) from the Late Cretaceous deposits of the North-East of the U.S.S.R. Bot. Žurn. 73: 1179–1183. (Russ., Eng. summ.). [*Microconium beringianum* sp. nov. = *Mesocyparis beringiana* (Golovneva) McIver & Aulenback, see McIver & Aulenback, 1994].

04888 Golte, W. (1993). *Araucaria*. Verbreitung und Standortsansprüche einer Coniferengattung in vergleichender Sicht. Erdwiss. Forschung Vol. 27. Stuttgart. (maps, foldout distr. map of genus, ill.). [*A. araucana*, *A. angustifolia*, *A. bidwillii*, *A. cunninghamii*, *A. hunsteinii*, *A. columnaris*, *A. heterophylla*].

04888a Golte, W. (1996). Exploitation and conservation of *Fitzroya cupressoides* in southern Chile. In: D. Hunt (ed.). Temperate trees under threat. Proceedings of an IDS symposium on the conservation status of temperate trees, University of Bonn, 30 September – 1 October 1994, pp. 133–150. International Dendrology Society. (ill., map).

04888b Golte, W. (1998). *Fitzroya cupressoides* in Zuid-Chili. Dendroflora 34 (1997): 34–42.

04888d Goncharenko, G. G., V. Y. Padutov, A. Y. Silin, A. N. Chernodubov *et al.* (1991). Genetic structure of *Pinus sylvestris* L. and *Pinus cretacea* Kalen populations and their taxonomic relationship. Dokl. Akad. Nauk S.S.S.R. 319: 1230. (Russ.).

04889 Goncharenko, G. G. & V. V. Potenko (1991). The parameters of genetic diversity and differentiation in Norway Spruce (*Picea abies* (L.) Karst.) and Siberian Spruce (*Picea obovata* Ledeb.) populations. Genetika 27: 1759–1772. (Russ., Eng. summ. p. 1772). [*P. abies* ssp. *obovata*].

04889a Goncharenko, G. G. & V. V. Potenko (1992). Genetic variability and differentiation in Norway Spruce (*Picea abies* (L.) Karst.) and Siberian Spruce (*Picea obovata* Ledeb.) populations. Soviet Genet. 28: 1235–1246.

04889b Goncharenko, G. G., V. V. Potenko, Y. N. Slobodjan & A. I. Sidor (1990). Genetic taxonomic correlation between *Picea abies* (L.) Karst., *P. montana* Schur. and *P. obovata* Ledeb. Dokl. Akad. Nauk B.S.S.R. 34: 361–364. (Russ.).

04889c Goncharenko, G. G., A. E. Silin & V. E. Padutov (1995). Intra- and interspecific genetic differentiation in closely related pines from *Pinus* subsection *Sylvestres* (Pinaceae) in the former Soviet Union. Plant Syst. Evol. 194: 39–54. [*Pinus* subsect. *Pinus*, *P. funebris*, *P. mugo*, *P. pallasiana*, *P. sylvestris*, *P. sylvestris* var. *hamata*, isozymes].

04889m Gooch, N. L. (1992). Two new species of *Pseudolarix* Gordon (Pinaceae) from the Middle Eocene of the Pacific Northwest. Paleobios 14 (1): 13–19. [*Pseudolarix wehrii* fossil sp. nov., p. 14, *P. arnoldii* fossil sp. nov., p. 14, ill.].

04889p Gondran, M., P. Woltz, R. Gajardo, J. Marguerier & J. Bernard (1999). Xylologie des conifères endemiques des Andes meridionales au MEB. II. Araucariaceae – Cupressaceae. Rev. Cytol. Biol. Veg. Bot. 21 (3–4): [11 p. following p. 42, 1998 publ. 1999]. [*Araucaria araucana*, *Austrocedrus chilensis*, *Fitzroya cupressoides*, *Pilgerodendron uviferum*; corrected version of same art. in Rev. Cytol. Biol. Veg. Bot. 20 (1–2): 3–13 (1997); SEM ill. of wood anatomy, maps].

04890 Gordon, A. G. (1952). Spruce identification by twig characteristics. Forest. Chron. 28: 45–47. [*Picea*].

04900 Gordon, A. G. (1968). Ecology of *Picea chihuahuana* Martínez. Ecology 49 (5): 880–896.

04910 Gordon, A. G. (1976). The taxonomy and genetics of *Picea rubens* and its relationship to *Picea mariana*. Canad. J. Bot. 54 (9): 781–813, 6 figs.

04915 Gordon, G. (1840). Report on new species and varieties of hardy trees and shrubs raised in the Horticultural Society's gardens since the last report made in October, 1838. Gard. Mag. & Reg. Rural Domest. Improv. 16: 1–10. [*Pinus* spp., pp. 6–9, *P. kesiya* Royle ex Gordon, sp. nov., p. 8, *Picea* spp., *Abies* spp., Cupressaceae spp., pp. 9–10).

04920 Gordon, G. (1846). New plants from the Society's garden. J. Hort. Soc. London 1: 234–239. [*Pinus orizabae* sp. nov. = *P. pseudostrobus* s.l., p. 237].

04930 Gordon, G. (1847). New plants from the Society's garden. J. Hort. Soc. London 2: 77–80, 158–160. [*Pinus grenvilleae* sp. nov., p. 77, *P. gordoniana* Hartweg ex Gord., p. 79, *P. wincesteriana* sp. nov., p. 158, ill.].

04931 Gordon, G. (1849). Notes upon some newly-introduced Conifers, collected by Mr. Hartweg in Upper California. J. Hort. Soc. London 4: 211–226, 295–296. [*Cupressus goveniana* sp. nov., p. 295, fig., *C. lambertiana* hort., pro syn. = *C. macrocarpa* Hartw. ex Gord., p. 296, fig.].

04940 Gordon, G. (1858). New plants, no. 222. Gardener's Chron., ser. 3, 1858: 358. [*Pinus bonapartea* sp. nov., *Biota* "spp. nov." = *Platycladus orientalis* cultivars].

04971 Gothan, W. (1954). Über ein Massenvorkommen von *Sciadopitys*-Nadeln in kohligen Ablagerungen des oberen Jura oder Wealden der Spanischen Ost-Pyrenäen. Svensk Bot. Tidskr. 48 (2): 337–343. [Sciadopityaceae, palaeobotany].

04972 Gottfried, G. J. & K. E. Severson (1994). Managing Pinyon-Juniper woodlands. Rangelands 16 (6): 234–236. [*Pinus edulis*, *Juniperus* sp., ecology, conservation].

04974 Gourlay, W. B. (1940). The Mexican swamp cypress (*Taxodium mucronatum* Tenore). Quart. J. Forest. 34: 53–61.

04975 Govindaraju, D. R. & C. A. Cullis (1992). Ribosomal DNA variation among populations of a *Pinus rigida* Mill. (pitch pine) ecosystem: I. Distribution of copy numbers. Heredity 69: 133–140.

04976 Govindaraju, D. R., B. P. Dancik & D. B. Wagner (1989). Novel chloroplast DNA polymorphism in a sympatric region of two pines. J. Evol. Biol. 2: 49–59. [*Pinus banksiana*, *P. contorta*].

04978 Govindaraju, D. R., P. Lewis & C. A. Cullis (1992). Phylogenetic analysis of pines using ribosomal DNA restriction fragment length polymorphisms. Plant Syst. Evol. 179: 141–153. [*Pinus* spp., molecular systematics, cladograms].

04979 Govindaraju, D. R., D. B. Wagner, G. P. Smith & B. P. Dancik (1988). Chloroplast DNA variation within individual trees of a *Pinus banksiana - P. contorta* sympatric region. Canad. J. Forest Res. 18: 1347–1350.

04980a Graham, G. C., R. J. Henry, I. D. Godwin & D. G. Nikles (1996). Phylogenetic position of Hoop pine (*Araucaria cunninghamii*). Austral. Syst. Bot. 9: 893–902. [molecular analysis of several conifers using rRNA sequencing].

04980e Grahl, N. O., O. Mjaavatten & D. O. Ovstedal (1991). A chemometric comparison between *Picea abies* and *P. obovata* (Pinaceae) in Norway. Nordic. J. Bot. 11 (6): 613–618.

04981 Gramuglio, G. (1960). Appunti sulla distribuzione geografica dell' *Abies nebrodensis*. Atti Reale Accad. Naz. Lincei, Mem. Cl. Sci. Fis., Sez. 3a, Bot., ser. 8, 29: 106–114. (ill.).

04984 Grant, M. C. & J. B. Mitton (1977). Genetic differentiation among growth form of Engelmann spruce and subalpine fir at tree line. Arctic Alpine Res. 9: 259–263. [*Abies lasiocarpa*, *Picea engelmannii*].

04986 Grauvogel-Stamm, L. (1978). La flore du Grès à *Voltzia* (Buntsandstein supérieur) des Vosges du Nord (France) – morphologie, anatomie, interprétations phylogénique et paléo-géographique. Université Louis Pasteur de Strasbourg, Institut de Géologie, Sci. Geol. Mém. 50: 1–225. [Coniferales, Voltziaceae, Cycadocarpidiaceae, *Cycadocarpidium* Nathorst, *C. pilosum* sp. nov., *Aethophyllum stipulare* Brongn., palaeobotany, ill.].

04988 Grauvogel-Stamm, L. & L. Grauvogel (1973). *Masculostrobus acuminatus* nom. nov., un nouvel organe reproducteur mâle de gymnosperme du Grès à Voltzia (Trias inférieur) des Vosges, France. Geobios 6 (2): 101–114. [male cone of the Triassic conifer *Aethophyllum* Brongn.].

04989 Grauvogel-Stamm, L. & L. Grauvogel (1975). *Aethophyllum* Brongniart, conifère (non équisétale) du Grès à Voltzia (Buntsandstein supérieur) des Vosges, France. Geobios 8 (2): 143–146. [*A. stipulare* Brongn., palaeobotany, ill.].

04991 Gray, A. M. (1968). Tasmanian Conifers. Austral. Plants 4 (34): 267–273. [*Athrotaxis cupressoides*, *A. laxifolia*, *A. selaginoides*, *Diselma archeri*, Podocarpaceae].

04991a   Gray, A. M. (1998). Pines of Tasmania. Austral. Plants 19 (156): 344–354. [*Athrotaxis cupressoides, A. laxifolia, A. selaginoides, Diselma archeri, Callitris oblonga, C. rhomboidea,* Podocarpaceae].

04992   Greene, E. L. (1882). New Western Plants. Bull. Torrey Bot. Club 9 (4): 62–65. [*Cupressus arizonica* sp. nov., p. 64].

04992a   Greene, E. L. (1891). Against the using of revertible generic names. Pittonia 2: 185–195. [*Tumion californicum* (Torr.) gen. et comb. nov. = *Torreya californica*, p. 193, *T. grande* (Fortune ex Lindl.) comb. nov. = *Torreya grandis*, p. 194, *T. nuciferum* (L.) comb. nov. = *Torreya nucifera*, p. 194, *T. taxifolium* (Arn.) comb. = *Torreya nucifera*, p. 194].

04992e   Greenwood, D. R. & J. F. Basinger (1994). The paleoecology of high-latitude Eocene swamp forest from Axel Heiberg Island, Canadian High Arctic. Rev. Palaeobot. Palynol. 81: 83–97. (ill.). [conifers, e.g. *Larix, Picea, Metasequoia,* cupressoid taxa].

04993   Gregor, H.-J. (1980). Funde von *Pinus canariensis* Ch. Smith fossilis aus dem Neogen von La Palma (Kanarische Inseln). Vieraea 9 (1–2): 57–64.

05001   Greguss, P. (1956). The phyllotaxy of *Metasequoia, Sequoia* and *Taxodium.* Acta Biol. (Szeged 1955+) 2 (1–4): 29–38. (ill.).

05002   Greguss, P. (1962). Le genre *Ducampopinus* est-il valable en vertu de sa xylotomie? Trav. Lab. Forest. Toulouse, T. 1, vol. 6, art. 13: 1–6. [*Ducampopinus = Pinus krempfii,* phot. of wood-anatomy].

05010   Greguss, P. (1970). Similar xylotomy and leaf-epidermis of the *Pseudotsuga* and the new genus *Cathaya.* Bot. Közlem. 57: 51–55. (Hung.). [*Pseudotsuga argyrophylla* (Chun et Kuang) comb. nov., p. 54].

05011   Greguss, P. (1970). Ein *Callitris*-ähnliches Holz aus dem Tertiär von Limburg (Niederlande). Senckenberg. Leth. 51 (2–3): 265–275. [*Callitris* sp., palaeobotany].

05050   Grierson, A. J. C., D. G. Long & C. N. Page (1980). Notes relating to the Flora of Bhutan: III *Pinus bhutanica:* a new 5-needle pine from Bhutan and India. Notes Roy. Bot. Gard. Edinburgh 38 (2): 297–310. [*P. bhutanica* sp. nov., p. 299].

05060   Griffin, J. R. (1964). Cone morphology in *Pinus sabiniana.* J. Arnold Arbor. 45: 260–273.

05061   Griffin, J. R. (1964). David Douglas and the Digger Pine, some questions. Madroño 17 (7): 227–230. [*Pinus sabiniana, P. coulteri,* map; see also E. L. Little, Jr., 1948].

05071   Griffin, J. R. & C. O. Stone (1967). Macnab Cypress in northern California: a geographic review. Madroño 19 (1): 19–27. [*Cupressus macnabiana,* ecology, map].

05115   Grilli Caiola, M., A. Travaglini & M. Giuliano (2000). Palynological study of *Cupressus sempervirens* L. var. *pyramidalis* and var. *horizontalis.* Plant Biosystems 134 (1): 99–109.

05120   Grisebach, A. H. R. (1846). Spicilegium florae rumelicae et bithynicae... Vol. 2. Braunschweig. [*Pinus peuce* sp. nov., p. 349, *Juniperus sabinoides* sp. nov. = *J. foetidissima,* p. 352; publ. Jan. 1846].

05121   Grisebach, A. H. R. (1862). Plantae Wrightianae e Cuba Orientali: Pars 2, Monopetalae et Monocotyledones. Mem. Amer. Acad. Arts, n. s. (or ser. 2) 8: 502–536. [*Pinus cubensis* sp. nov., p. 530].

05122   Grisebach, A. H. R. (1866). Catalogus plantarum cubensium exhibens collectionem Wrightianam aliasque minores ex insula Cuba missas. Leipzig. [*Podocarpus angustifolius* sp. nov., p. 217].

05140   Grosdemagne, C. (1895). *Sciadopitys verticillata.* Rev. Hort. 1895: 523–524.

05142   Grosfeld, J. & D. Bathélémy (2001). Dioecy in *Fitzroya cupressoides* (Molina) I. M. Johnst. and *Pilgerodendron uviferum* (D. Don) Florin (Cupressaceae). Compt. Rend. Acad. Sci. (Paris), Sciences de la vie 324: 245–250.

05147   Gross, H. (1929). Die Moorformen der Fichte. Mitt. Deutsch. Dendrol. Ges. 41: 11–23. [*Picea abies*].

05150   Grosser, D., D. Fengel & H. Schmidt (1974). Tamrit Zypresse (*Cupressus dupreziana* A. Camus); Beitrag zur Ökologie, Anatomie und Chemie. Forstwiss. Zentralbl. 93 (4): 191–207.

05155   Grotkopp, E., M. Rejmánek & T. L. Rost (2002). Toward a causal explanation of plant invasiveness: seedling growth and life-history strategies of 29 pine (*Pinus*) species. Amer. Nat. 159 (4): 396–419.

05160   Grove, A. (1938). *Taiwania.* New Fl. & Silva 10: 191–193. [phytogeography, discovery].

05160b   Gruber, F. (1989). Phänotypen der Fichte (Picea abies (L.) Karst.) I. Allg. Forst-Jagd-Zeitung 160 (8): 157–165. [ill. of habitus, crowns].

05160c   Gruber, F. (1995). Morphologie der Weisstanne (*Abies alba* Mill.) II. Wurzel-verzweigung, Architekturmodell und Kronenanalysen. Flora 190: 135–153.

05160d   Grunwald, C., G. Schiller & M. T. Conkle (1986). Isozyme variation among native stands and plantations of Aleppo Pine in Israel. Israel J. Bot. 35: 161–174. [*Pinus brutia, P. halepensis*].

05161   Guan, Z. & Y. Chen (1986). A preliminary study on the *Cathaya* mixed forest in Jinfushan, Sichuan. Acta Bot. Sin. 28 (6): 646–656. (Chin., Eng. summ.).

05162   Guédès, M. (1970). La morphologie du complexe séminifère de *Cryptomeria japonica* (L.) Don. Flora, Abt. B, 159 (1–2): 71–83. (ill.).

05163   Guédès, M. & P. Dupuy (1974). Morphology of the seed-scale complex in *Picea abies* (L.) Karst. (Pinaceae). Bot. J. Linn. Soc. 68: 127–141. (ill.). [teratology in cv. 'Acrocona' used to interpret seed scale morphology].

05171   Guillaumin, A. (1949). Contribution à la flore de la Nouvelle Calédonie. XCII. Plantes récoltées par J. Bernier (complément). Bull. Mus. Hist. Nat. (Paris), sér. 2, 21: 453–461. [*Libocedrus yateensis* sp. nov., p. 457].

05171a Gulline, H. F. (1952). The cytology of *Athrotaxis*. Pap. Proc. Roy. Soc. Tasmania 86: 131–136.

05172   Gussone, G. (1844). Flora sicula synopsis exhibens plantas vasculares in Sicilia... Vol. 2. Napoli. [*Abies pectinata* Gussone (non Lam. et DC.) = *A. nebrodensis*, p. 614, *Juniperus turbinata* sp. nov. = *J. phoenicea*, p. 634].

05181   Guyot, A.-P. & T. Mathou (1942). Contribution à l'étude du *Juniperus phoenicea* Lin. Trav. Lab. Forest. Toulouse, T. 1, vol. 3, art. 20: 1–72. [*Juniperus canariensis* "sp. nov.", p. 7, later homon. of *J. canariensis* Gord. = *J. cedrus* Webb et Berth., 1847].

05182   Haddow, W. R. (1948). Distribution and occurrence of White pine (*Pinus strobus* L.) and Red pine (*Pinus resinosa* Ait.) at the northern limit of their range in Ontario. J. Arnold Arbor. 29 (3): 217–226, pl. 1–3. (maps).

05195   Haines, R. J., N. Prakash & D. G. Nikles (1984). Pollination in *Araucaria* Juss. Austral. J. Bot. 32: 583–594.

05205   Halkett, J. & E. V. Sale (1986). The world of the Kauri. Reed Methuen, Auckland. (ill.). [*Agathis australis*].

05207   Hall, J. B. (1984). *Juniperus excelsa* in Africa: a biogeographical study of an Afromontane tree. J. Biogeogr. 11 (1): 47–61. (maps). [= *J. procera*].

05210   Hall, M. T. (1952). Variation and hybridization in *Juniperus*. Ann. Missouri Bot. Gard. 39: 1–64. (ill.).

05220   Hall, M. T. (1954). Nomenclatural note concerning *Juniperus*. Rhodora 56: 169–177. (ill.).

05230   Hall, M. T. (1971). A new species of *Juniperus* from Mexico. Fieldiana Bot. 34 (4): 45–53. [*Juniperus saltillensis* sp. nov., p. 45, ill.].

05240   Hall, M. T., J. F. McCormick & G. G. Fogg (1962). Hybridization between *Juniperus ashei* Buchholz and *Juniperus pinchotii* Sudworth in southwestern Texas. Butler Univ. Bot. Stud. 12: 9–28. (ill.).

05241   Hall, M. T., A. Mukherjee & W. R. Crowley (1979). Chromosome numbers of cultivated junipers. Bot. Gaz. (Crawfordsville) 140: 364–370. [*Juniperus*, 10 spp.: 2n = 22].

05245   Hallé, N. (1979). Analyse du réseau phyllotaxique des écussons du cône chez *Pinus*. Adansonia ser. 2, 18: 393–408.

05250   Haller, J. R. (1959). The role of hybridization in the origin and evolution of *Pinus washoensis*. Abstr. Proc. 9th Int. Bot. Congr. 2: 149.

05260   Haller, J. R. (1961). Some recent observations on Ponderosa, Jeffrey's and Washoe Pines in northeastern California. Madroño 16: 126–132. [*Pinus ponderosa, P. jeffreyi, P. washoensis*].

05270   Haller, J. R. (1962). Variation and hybridization in Ponderosa and Jeffrey pines. Univ. Calif. Publ. Bot. 34: 123–167. (ill.). [*Pinus jeffreyi, P. ponderosa*].

05280   Haller, J. R. (1965). The role of two needle fascicles in the adaptation and evolution of Ponderosa Pine. Brittonia 17: 354–382. [*Pinus ponderosa*].

05290   Haller, J. R. (1965). *Pinus washoensis* in Oregon: taxonomic and evolutionary implications. Amer. J. Bot. 52: 646.

05291   Haller, J. R. (1975). Phylogenetic and floristic antiquity in the Pines of Mexico. Abstr. Proc. 12th Int. Bot. Congr., Leningrad, p. 95. [*Pinus*].

05300   Haller, J. R. (1986). Taxonomy and relationships of the mainland and island populations of *Pinus torreyana* (Pinaceae). Syst. Bot. 11: 39–50. [*P. torreyana* ssp. *insularis*, ssp. nov., p. 45].

05304   Ham, R. W. J. M. van der, J. H. A. van Konijnenburg-van Cittert & J. van der Burgh (2001). Taxodiaceous conifers from the Maastrichtian type area (Late Cretaceous, NE Belgium, SE Netherlands). Rev. Palaeobot. Palynol. 116: 233–250. [Cupressaceae s.l., Taxodiaceae, *Elatidopsis cryptomerioides* comb. nov., *Cryptomeriopsis eluvialis* sp. nov.].

05307   Han, L. J., Y. S. Hu & J. X. Lin (1997). Anatomy of secondary phloem of *Glyptostrobus* in relation to its systematic position. Acta Phytotax. Sin. 35 (6): 527–532. (Chin., Eng. abstr., table). [bark anatomy of *Glyptostrobus pensilis, Metasequoia glyptostroboides, Taxodium distichum*].

05309   Hance, H. F. (1883). Spicilegia florae sinensis: diagnoses of new, and habitats of rare or hitherto unrecorded, Chinese plants – VIII. J. Bot., British & Foreign 21: 355–359. [*Podocarpus argotaenius* sp.nov. = *Amentotaxus argotaenia*, p. 357].

05311   Handel-Mazzetti, H. (1924). Plantae novae sinenses, diagnosibus brevibus descriptae. Akad. Wiss. Wien, Math.-Naturwiss. Kl., Anz. 61: 81–85, 107. [*Tsuga intermedia, T. leptophylla* spp. nov. = *T. dumosa*, pp. 82–83, *Juniperus wallichiana* var. *meionocarpa* var. nov. = *J. indica*, p. 107].

05320   Handel-Mazzetti, H. (1927). Das nordost-b i r m a n i s c h - w e s t - y ü n n a n i s c h e Hochgebirgsgebiet. in: G. H. H. Karsten & H. Schenck. Vegetationsbilder, Vol. 17 (7/8). Jena. [Cupressaceae, Pinaceae, Taxodiaceae, *Thuja orientalis* = *Platycladus orientalis*, pl. 38A, *Taiwania cryptomerioides*, pl. 40A].

05330   Handel-Mazzetti, H. (1931). Kleine Beiträge zur Kenntnis der Flora von China. Österreich. Bot. Zeitschr. 80: 337–338. [*Pinus fenzeliana* sp. nov.].

05350   Hanover, J. W. (1975). Genetics of blue spruce. U.S. Forest Serv. Res. Paper WO-28. (iii + 12 pp., ill., map). [*Picea pungens*].

05360   Hanover, J. W. & R. G. Wilkinson (1970). Chemical Evidence for Introgressive Hybridization in *Picea*. Silvae Genet. 19: 17–22.

05363   Hanson, L. (2001). Chromosome number, karyotype and DNA C-value of the Wollemi Pine (*Wollemia nobilis*, Araucariaceae). Bot. J. Linn. Soc. 135 (3): 271–274. [2n = 26].

05365   Hantke, R. (1973). *Keteleeria hoehneii* Kirchh., ein Zapfenrest aus der Unteren Süsswasser-molasse des Buechbergs (Kt. Schwyz). Eclogae geol. Helv. 66 (3): 739–742, pl. 1. ["*hoehnei*", fossil compressed seed cone of *Keteleeria* from Oligocene of Switzerland].

05390   Hara, H. (1977). Nomenclatural notes on some Asiatic Plants, with special reference to Kaempfer's Amoenitatum Exoticarum. Taxon 26 (5–6): 584–587. [*Larix kaempferi*, *Pseudolarix amabilis*)

05400   Hara, H. & R. K. Brummitt (1980). The problem of *Pseudolarix kaempferi*: type method or circumscription method? Taxon 29 (3): 315–317. [*Pseudolarix amabilis*, nomenclature].

05401   Hara, H. & T. T. Yü (1983). Proposal to conserve 25. *Pseudolarix* Gordon (Pinaceae) with a type specimen. Taxon 32: 485–487. [*Pseudolarix amabilis*, typification, nomen-clature].

05407   Harder, D. K. (2002). The Golden Vietnamese Cypress, *Xanthocyparis vietnamensis*: a new genus and species for science. Conifer Quarterly 19 (2): 54–57. (ill.).

05420   Harlow, W. M. (1931). The identification of the pines of the United States, native and introduced, by needle structure. N. Y. State Col. Forest., Syracuse Univ. Techn. Publ. 32. (ill.). [*Pinus*].

05430   Harlow, W. M., W. A. Côté, Jr. & A. C. Day (1964). The opening mechanism of pine cone scales. J. Forest. (Washington) 62: 538–540.

05450   Harper, R. M. (1902). *Taxodium distichum* and related species, with notes on some geological factors influencing their distribution. Bull. Torrey Bot. Club 29: 383–399. [*Taxodium imbricatum* (Nutt.) comb. nov. (as "imbricarium") = *T. distichum* var. *imbricatum* (Nutt.) Croom, p. 383, *Glyptostrobus heterophyllus* = *G. pensilis*].

05451   Harper, R. M. (1905). Further observations on *Taxodium*. Bull. Torrey Bot. Club 32: 105–115. [*Taxodium distichum*].

05452   Harper, R. M. (1912). The diverse habitats of the eastern red cedar and their interpretation. Torreya 12: 145–154. [*Juniperus virginiana*, forest fires].

05460a   Harris, S. & J. B. Kirkpatrick (1991). The phytosociology and synecology of Tasmanian vegetation with *Callitris*. in: M. R. Banks *et al.* (eds.). Aspects of Tasmanian botany – a tribute to Winifred Curtis. Roy. Soc. Tasmania, Hobart, pp. 179–189. [*C. oblonga, C. rhomboidea*].

05460b   Harris, S. & J. B. Kirkpatrick (1991). The distribution, dynamics and ecological differentiation of *Callitris* species in Tasmania. Austral. J. Bot. 39 (3): 187–202. [*C. oblonga, C. rhomboidea*].

05460d   Harris, T. M. (1943). The fossil conifer *Elatides williamsonii*. Ann. Bot. (London), n.s. 7: 325–339. [Taxodiaceae = Cupressaceae s.l.].

05469   Harrison, S. G. (1965). Note on gymnosperm nomenclature. Taxon 14: 247. [*Pinus pseudostrobus* var. *oaxacana* (Mirov) S. G. Harrison].

05484   Hart, J. A. & R. A. Price (1990). The genera of Cupressaceae (including Taxodiaceae) in the southeastern United States. J. Arnold Arbor. 71 (3): 275–322. [with extensive bibliography, ill. of *Chamaecyparis, Juniperus*].

05490   Hartesveldt, R. J., H. T. Harvey, H. S. Shellhammer & R. E. Stecker (1975). The giant Seqoia of the Sierra Nevada. U.S. Dept. of the Interior, National Park Service. Washington, D.C. [*Sequoiadendron giganteum*].

05496   Hartman, C. J. (1849–50). Handbok i Skandinaviens Flora. Ed. 5, Stockholm, Dec. 1849 – Jan. 1850. [*Pinus sylvestris* var. *lapponica* var. nov., p. 214].

05500   Hartweg, T. (1847). Journal of a mission to California in search of plants, Part III. J. Hort. Soc. London 2: 187–189. [*Cupressus macrocarpa* sp. nov., nom. subnud. (see G. Gordon, 1849), p. 187, *Pinus benthamiana* sp. nov. = *P. ponderosa*, p. 189; see also A. Murray, 1855].

05505   Havrylenko, M., P. H. A. Rosso & S. B. Fontenla (1989). *Austrocedrus chilensis*: contribución al estudio de su mortalidad en Argentina. Bosque 10 (1): 29–36. [map of Argentinian populations in Nahuel Huapi N.P.].

05510   Hayashi, Y. (1948). Preliminary observations on the natural distribution of *Thyopsis dolabrata* ("asunaro"), and *T. hondai* ("hinoki-asunaro; Hiba"). Proc. Biogeogr. Soc. Japan 2: 5–8, 20–21, map in text. [*Thujopsis dolabrata, T. dolabrata* var. *hondai*].

05570   Hayata, B. (1906). On *Taiwania*, a New Genus of Coniferae from the Island of Formosa. (comm. by M. T. Masters) J. Linn. Soc., Bot. 37: 330–331, pl. 16. [*Taiwania cryptomerioides* gen. et sp. nov., p. 330].

05580   Hayata, B. (1907). On *Taiwania* and its affinity to other genera. Bot. Mag. (Tokyo) 21: 21–27.

05620   Hayata, B. (1909). Note on *Juniperus taxifolia* Hook. & Arn. J. Linn. Soc., Bot. 39: 89–90, pl. 7.

05621   Hayata, B. (1911). Materials for a flora of Formosa. Supplementary notes to the Enumeratio plantarum formosanarum and Flora montana Formosae, ... J. Coll. Sci. Imp. Univ. Tokyo 30 (1): 1–471. [*Pinus taiwanensis* sp. nov., p. 307].

05630   Hayata, B. (1917). Some conifers from Tonkin and Yunnan. Bot. Mag. (Tokyo) 31: 113–119, 2 figs. [8 spp., *Fokienia kawaii* sp. nov., p. 116, *Cryptomeria kawaii* sp. nov. = *C. japonica*, p. 117].

05631   Hayata, B. (1917). Sur le Xuh-Peh-Muh, nouvelle espèce de *Podocarpus* du Tonkin, de concert avec quelques notes sur le Peh-Muh. Bull. Econ. Indochine 19: 435–440, 1 pl. [*Podocarpus kawaii* sp. nov. = *Dacrycarpus imbricatus* var. *patulus*, p. 439].

05650   Hayata, B. (1931). The Sciadopityaceae represented by *Sciadopitys verticillata* Sieb. et Zucc., an endemic species of Japan. Bot. Mag. (Tokyo) 45 (540): 567–569.

05660   Hayata, B. (1932). The Taxodiaceae should be divided into several distinct families, i.e. the Limnopityaceae, Cryptomeriaceae, Taiwaniaceae and the Cunninghamiaceae; and further *Tetraclinis* should represent a distinct family, the Tetraclinaceae. Bot. Mag. (Tokyo) 46 (541): 24–27.

05670   He, G. F., Z. W. Ma, W. F. Yin & M. L. Cheng (1981). On serratene components in relation to the systematic position of *Cathaya* (Pinaceae). Acta Phytotax. Sin. 19: 440–443.

05675   Hecker, U. (1991). Zur Biologie der Kiefernzapfen. Mitt. Deutsch. Dendrol. Ges. 80: 73–86. [*Pinus*].

05690   Heikinheimo, O. (1920). Über die Fichtenformen und ihren forstwirtschaftlichen Wert. Commun. Inst. Quaest. Forest. Finl. 2: 1–102. (Helsinki). [*Picea, P. abies*].

05700   Heinselman, M. L. (1957). Silvical characteristics of black spruce. U.S. Forest Serv., Lake States Forest Exp. Stat. Paper 45. [*Picea mariana*].

05701   Hemsley, W. B. (1883). The Bermuda Cedar. Gard. Chron., n. s., 19: 656–657, figs. 105–106. [*Juniperus bermudiana* L., based on "Cedrus bermudae" of Sloane and of Hermann (the latter in Hort. Acad. Lugd. Bat. Cat. pp. 345–347 of 1687), with *J. barbadensis* (= adult form?) as a probable syn.].

05702   Hemsley, W. B. (1883). Bermuda plants in the Sloane collection, British Museum. J. Bot. 21: 257–261. [*Juniperus bermudiana*, 2 figs.].

05703   Hemsley, W. B. (1885). New Chinese plants. J. Bot., British & Foreign 23: 286–287. [*Podocarpus insignis* sp. nov. = *Amentotaxus argotaenia*].

05704   Hemsley, W. B. (1896). Plantarum novarum in herbario horti regii conservatarum. Decades xxvi – xxvii. Bull. Misc. Inf. R.B.G. Kew 1896: 36–42. [*Podocarpus celebicus* sp. nov. = *Taxus sumatrana*, p. 39].

05714   Hennon, P. E., E. M. Hansen & C. G. Shaw III (1990). Dynamics of decline and mortality of *Chamaecyparis nootkatensis* in southeast Alaska, U.S.A. Canad. J. Bot. 68: 651–662. [= *Xanthocyparis nootkatensis*].

05720   Henry, A. (1906). New or noteworthy plants *Picea morindoides*. Gard. Chron., ser. 3, 39: 218–219, f. 84. [*Picea spinulosa* (Griff.) comb. nov., p. 219, publ. April 7; see also L. Beissner, 1906, p. 83, but publ. after August 7].

05730   Henry, A. (1911). A new genus of Coniferae. Gard. Chron., ser. 3, 49: 66–68, f. 32–35. [*Fokienia* gen. nov., *Fokienia hodginsii* (Dunn) A. Henry et H. H. Thomas, comb. nov. (basion. : *Cupressus hodginsii* Dunn, 1908), p. 67; protologues repeated in: Repert. Spec. Nov. Regni Veg. 12: 320, 1913].

05740   Henry, A. (1912). The giant cypress of Formosa. Gard. Chron., ser. 3, 51: 132–133, f. 53–54. [*Cupressus formosensis* = *Chamaecyparis formosensis*].

05750   Henry, A. (1915). A new species of larch. Gard. Chron., ser. 3, 57: 109, f. 31–32. [*Larix olgensis* sp. nov.].

05760   Henry, A. (1926). *Glyptostrobus pensilis*. Gard. Chron., ser. 3, 79: 309.

05770   Henry, A. (1926). The swamp cypresses of China and North America. Trans. Scott. Arbor. Soc. 40: 105–107. [*Taxodium distichum*, *Glyptostrobus heterophyllus* = *G. pensilis*].

05780   Henry, A. & M. G. Flood (1920). The Douglas Firs: a botanical and silvicultural description of the various species of *Pseudotsuga*. Proc. Roy. Irish Acad. 35, sect. B: 67–92, pl.

05790   Henry, A. & M. McIntyre (1926). The swamp cypresses, *Glyptostrobus* of China and *Taxodium* of America, with notes on allied genera. Proc. Roy. Irish Acad. 37, sect. B: 90–116, pl. 1–8. [*Taxodium distichum*, *Glyptostrobus pensilis*, monograph].

05790e   Herbert, J., P. M. Hollingsworth, M. F. Gardner, R. R. Mill, P. I. Thomas & T. Jaffré (2002). Conservation genetics and phylogenetics of New Caledonian *Retrophyllum* (Podocarpaceae) species. New Zealand J. Bot. 40 (2): 175–188. [*Retrophyllum comptonii, R.minus*, DNA, cytology/karyology].

05800   Herbin, G. A. & K. Sharma (1969). Studies on plant cuticular waxes. V. The wax coatings of pine needles: a taxonomic survey. Phytochemistry 8: 151–160. [*Pinus*].

05805   Hermann, R. K. (1985). The genus *Pseudotsuga*: ancestral history and past distribution. Special Publ. 2b. Forest Research Laboratory, Oregon State Univ., Corvallis.

05807   Hernandez-Castillo, G. R., G. W. Rothwell & G. Mapes (2001). Compound pollen cone in a Paleozoic conifer. Amer. J. Bot. 88 (6): 1139–1142. [Walchian conifer].

05808   Hernandez-Castillo, G. R., G. W. Rothwell & G. Mapes (2001). *Thucydiaceae* fam. nov., with a review and reevaluation of Palaeozoic Walchian conifers. Int. J. Plant Sci. 162: 1155–1185.

05809   Hernandez-Castillo, G. R., G. W. Rothwell, R. A. Stockey & G. Mapes (2003). Growth architecture of *Thucydia mahoningensis*, a model for primitive Walchian conifer plants. Int. J. Plant Sci. 164 (3): 443–452. [similar to *Araucaria heterophylla*].

05811   Herzfeld, S. (1909). Zur Morphologie der Fruchtschuppe von *Larix decidua* Mill. Sitzungsber. Kaiserl. Akad. Wiss., Math.-Naturwiss. Kl., Abt. 1, 118: 1345–1375. (ill.).

05811a   Herzfeld, S. (1910). Die Entwicklungsgeschichte der weiblichen Blüte von *Cryptomeria japonica* Don. Ein Beitrag zur Deutung der Fruchtschuppe der Coniferen. Sitzungsber. Kaiserl. Akad. Wiss., Math.-Naturwiss. Kl., Abt. 1, 119: 807–824.

05820   Hewes, J. J. (1981). Redwoods: the world's largest trees. London. [*Sequoia, Sequoiadendron*].

05840   Hickel, R. (1906–1908). Notes pour servir à la détermination pratique des Abietinées. Bull. Soc. Dendrol. France 2: 45–58, f. 1–7 (15-xi-1906); 3: 5–18, f. 8–30 (15-ii-1907); 4: 41–48, f. 31–46 (15-v-1907); 5: 82–86 (15-viii-1907); 7: 5–10, f. a–1 (15-ii-1908); 9: 179–185 (15-viii-1908); 10: 201–208 (15-xi-1908). [Pinaceae, *Abies, A.* sect. *Pseudopicea* sect. nov., 1906].

05851   Hickel, R. (1927). Au pays des Cèdres et des Arganiers. Bull. Soc. Dendrol. France 62: 42–55. [*Cedrus*].

05852   Hickel, R. (1929). Les Sapins de l'Himalaya. Bull. Soc. Dendrol. France 70: 37–38. [*Abies gamblei* sp. nov. = *A. pindrow* var. *brevifolia*].

05853   Hickel, R. (1930). Les conifères d'Indo-Chine. Bull. Soc. Dendrol. France 76: 73–78. [*Podocarpus fleuryi* sp. nov. = *Nageia fleuryi*, p. 75].

05865   Hida, M. (1961). Studies on the bark of *Metasequoia glyptostroboides* with special reference to giant cork cells. Bot. Mag. (Tokyo) 74: 514–518. (Japan., Engl. abstr., ill.).

05870   Hida, M. (1962). The systematic position of *Metasequoia*. Bot. Mag. (Tokyo) 75: 316–323. (Japan., Eng. summ.) [Taxodiaceae, classifications].

05870a   Hiebert, R. D. & J. L. Hamrick (1983). Patterns and levels of genetic variation in Great Basin Bristlecone pine, *Pinus longaeva*. Evolution 37: 302–310.

05870b   Hiep, Nguyen Tien & Phan Ke Loc (2002). The diversity of the flora of Vietnam 9. *Taiwania* Hayata and *T. cryptomerioides* Hayata (Taxodiaceae): new genus and species for the flora. Genetics and Applications 1: 32–40. (Vietnamese, Eng. summ.) (ill., map).

05870d   Hiitonen, I. (1964). Christian Steven. Mem. Soc. Fauna Fl. Fennica 40: 178–187. [biography, photograph of holotype of *Abies nordmanniana* (Steven) Spach, p. 184].

05871   Hildebrand, F. H. G. (1861). Die Verbreitung der Coniferen in der Jetztzeit und in den früheren geologischen Perioden. Verh. Naturhist. Vereins Preuss. Rheinl. Westphalens 18: 199–384. [*Abies concolor* (Gord. et Glend.) Lindl. ex Hildebr., comb. nov. [with basion. *Picea concolor* Gord. et Glend. in Pinet., 1858), p. 261].

05872   Hill, K. D. (1996). The Wollemi pine: discovering a living fossil. Nature & Resources 32 (1): 20–25. (ill.). [*Wollemia nobilis*].

05873   Hill, K. D. (1997). Architecture of the Wollemi pine (*Wollemia nobilis*, Araucariaceae), a unique combination of model and reiteration. Austral. J. Bot. 45: 817–826. (ill.).

05875   Hill, R. S. (1990). *Araucaria* (Araucariaceae) species from Australian Tertiary sediments – A micromorphological study. Austral. Syst. Bot. 3: 203–220. [incl. SEM examination of cuticle morphology of modern spp.].

05876b   Hill, R. S. & D. C. Christophel (2001). Two new species of *Dacrydium* (Podocarpaceae) based on vegetative fossils from Middle Eocene sediments at Nelly Creek, South Australia. Austral. Syst. Bot. 14: 193–205. [*Dacrydium mucronatum* sp. nov. ("*mucronatus*"), p. 202, *D. fimbriatum* sp. nov. ("*fimbriatus*"), p. 203, SEM ill.].

05876c   Hill, R. S., G. J. Jordan & R. J. Carpenter (1993). Taxodiaceous macrofossils from Tertiary and Quarternary sediments in Tasmania. Austral. Syst. Bot. 6: 237–249. [*Athrotaxis mesibovii* sp. nov., the <u>reliable fossil record of this genus is confined to post-Cretaceous sediments in Tasmania</u>; *Austrosequoia wintonensis, Sequoia*].

05877a Hill, R. S. & R. Paull (2003). *Fitzroya* (Cupressaceae) macrofossils from Cenozoic sediments in Tasmania, Australia. Rev. Palaeobot. Palynol. 126: 145–152. [*Fitzroya acutifolia* comb. nov. ("*acutifolius*"; basion. *Dacrycarpus acutifolius*, syn. *F. tasmanensis*, SEM ill.].

05878 Hill, R. S. & S. S. Whang (1996). A new species of *Fitzroya* (Cupressaceae) from Oligocene sediments in north-western Tasmania. Austral. Syst. Bot. 9: 867–875. [*Fitzroya tasmanensis* fossil sp. nov., fossil conifers in Cupressaceae s.l.].

05890 Hilliard, O. (1985). Trees and Shrubs of the Natal Drakensberg. Pietermaritzburg. [*Widdringtonia*].

05900 Hillier, E. L. (1941). The newer Asiatic silver firs and their characteristics described from British-grown specimens. J. Roy. Hort. Soc. (London) 66: 400–411, f. 129, 431–443, f. 137–140. [*Abies* spp., see also E. L. Hillier, 1942].

05910 Hillier, E. L. (1942). Further notes on silver firs. J. Roy. Hort. Soc. (London) 67: 162. [*Abies*, additions to E. L. Hillier, 1941].

05920 Hills, L. V. & R. T. Ogilvie (1970). *Picea banksii* n. sp. Beaufort Formation (Tertiary), northwestern Banks Island, Arctic Canada. Canad. J. Bot. 48: 457–464. [*P. banksii* fossil sp. nov.].

05925 Hirayoshi, I. & Y. Nakamura (1943). Chromosome number of *Sequoia sempervirens*. Bot. & Zool. 2: 73–75. (ill.). [first report of hexaploidy, 2n = 6x = 66].

05940 Hisauchi, K. (1935). Monophyllous pines. J. Japan. Bot. 11: 177–180, f. 1–3. [*Pinus*].

05955 Hluštik, A. (1987). Frenelopsidaceae fam. nov., a group of highly specialized *Classopollis*-producing conifers. Acta Palaeobot. 27 (2): 3–20. [Cheirolepidiaceae, *Frenelopsis* spp., palaeobotany, ill.].

05960 Ho, R. H. & J. N. Owens (1974). Microstrobilate morphology, microsporogenesis, and pollen formation in western hemlock. Canad. J. Forest Res. 4: 509–517. (ill.). [*Tsuga heterophylla*].

05960a Ho, R. H. & J. N. Owens (1974). Microstrobili of Douglas-fir. Canad. J. Forest Res. 4: 561–562. [*Pseudotsuga menziesii*].

05961 Ho, R. H. & O. Sziklai (1972). On the pollen morphology of *Picea* and *Tsuga* species. Grana 12: 31–40. [*Picea orientalis*, *P. sitchensis*, *Tsuga mertensiana*, *T. heterophylla*, SEM and light microscopy].

05962 Ho, R. H. & O. Sziklai (1973). Fine structure of pollen surface of some Taxodiaceae and Cupressaceae species. Rev. Palaeobot. Palynol. 15: 17–26. [*Chamaecyparis*, *Cryptomeria*, *Cunninghamia*, *Sciadopitys*, *Sequoia*, *Taiwania*, SEM photographs].

05970 Ho, T. H. (1949). Wood structure of *Tsuga longibracteata* and *Taxus chinensis*. Quart. J. Taiwan Mus. 2: 34–43. [*Nothotsuga longibracteata*, also *Keteleeria davidiana*, wood anatomy, phylogeny].

05980 Hoey Smith, J. R. P. van (1974). *Microbiota decussata*. Mitt. Deutsch. Dendrol. Ges. 67: 52–54.

05990 Hoey Smith, J. R. P. van (1979). *Microbiota* really dioecious? Int. Dendrol. Soc. Yearb. 1978: 101–102.

06000 Hoey Smith, J. R. P. van (1983). *Microbiota* monoecious. Int. Dendrol. Soc. Yearb. 1982: 82. [see also Zamjatnin, 1963 who observed this on wild growing plants 20 years earlier].

06010 Höfker, H. (1931). Zur Gattung *Larix*. Mitt. Deutsch. Dendrol. Ges. 43: 18–20. [taxonomy].

06011 Höfker, H. (1913). Die Zedernarten. Mitt. Deutsch. Dendrol. Ges. 22: 201–208. [*Cedrus* spp., ill. of growth habits].

06015 Hogg, K. E., A. K. Mitchell & M. R. Clayton (1996). Confirmation of cosexuality in Pacific yew (*Taxus brevifolia* Nutt.). Great Basin Naturalist 56 (4): 377–378. [SEM showing ovuliferous bud on branch with polliniferous bud of predominantly male tree].

06020 Holdridge, L. R. (1942). The pine forests of Haiti. Caribbean Forest. 4 (1): 16–21. [*Pinus occidentalis*].

06024 Holm, Y. & R. Hiltunen (1997). Variation and inheritance of monoterpenes in *Larix* species. Flavour Fragrance J. 12: 335–339. [Assessment of *Larix decidua*, *L. sibirica*, *L. kaempferi* (as *L. leptolepis*) and artificial hybrids of the first two; only quatitative variation observed].

06025 Holmboe, J. (1914). Studies on the vegetation of Cyprus; based on researches during the spring and summer 1905. Bergens Museum Skrifter, new ser. 1, vol. 2. Bergen. [gymnosperms pp. 28–30; *Juniperus oxycedrus* ssp. *rufescens* (Link) comb. nov., p. 28, *Pinus nigra* ssp. *pallasiana* (Lamb.) stat. nov., p. 29, *P. halepensis* ssp. *brutia* (Tenore) stat. nov., p. 29, *Cedrus libanotica* Link ssp. *brevifolia* (Hook. f.) comb. nov., p. 29, a floristic rather than a vegetation study].

06030 Holubcík, M. (1971). Die Veränderlichkeit der Fichte (*Picea abies* Karst.) im Gebiet der Hohen Tatra den Zapfen nach. Zborn. Tanab. (Samml. v. Studien Tatra-Nationalpark) 13: 5–47.

06040 Holubicková, B. (1965). A study of the *Pinus mugo* complex. Preslia 37: 276–288.

06040a Hong, Y. P., V. D. Hipkins & S. H. Strauss (1993). Chloroplast DNA diversity among trees, populations and species in the California closed-cone pines (*Pinus radiata*, *P. muricata* and *P. attenuata*). Genetics 135: 1187–1196.

06040b Hong Yang (1999). From fossils to molecules: the *Metasequoia* tale continues. Arnoldia 58 (4) – 59 (1): 60–71. (special issue, see K. Madsen at 08902). [*M. glyptostroboides*, palaeobotany, palaeogeography, DNA, ill., maps].

06040e Hooibrenk, D. (1853). *Cryptomeria japonica vera* (Familie Coniferen). Wien. J. Gesammte Pflanzenreich 1: 22–23. [*Cryptomeria fortunei* ("*fortunini*", nom. nud.) = *C. japonica*].

06041 Hooker, J. D. (1845). On the Huon Pine, and on *Microcachrys*, a New Genus of Coniferae from Tasmania; together with Remarks upon the Geographical Distribution of that Order in the Southern Hemisphere. London J. Bot. 4: 137–157. ["*Arthrotaxis*" = *Athrotaxis*, *Callitris gunnii* sp. nov. = *C. oblonga*, p. 147, Cupressaceae; *Microcachrys tetragona* (Hook.) gen. et comb. nov. (basion. : *Athrotaxis tetragona* Hook.), p. 149, *Podocarpus alpinus* sp. nov., p. 150, *P. lawrencei* sp. nov., p. 151, *Phyllocladus aspleniifolius* (Labill.) comb. nov., p. 151, *D. franklinii* sp. nov. = *Lagarostrobos franklinii* ("Huon Pine", pl. VI), p. 152, Podocarpaceae].

06050 Hooker, J. D. (1851). *Fitzroya patagonica*. Bot. Mag. 77: pl. 4616. [= *Fitzroya cupressoides*, Lat. diagn. of genus and sp.; comb. first publ. by J. Lindley in: J. Hort. Soc. London 6: 264, 1 Oct. 1851, also in Paxton's Fl. Gard. 2: 115, Oct. 1851, comb. ex Herb. Hook. f.; gen. descr. by Hook. f. in Bot. Mag. 77, publ. Nov. 1851].

06050a Hooker, J. D. (1852). *Phyllocladus hypophylla* Hook. f. Icon. Pl., n.s. 5 [9], t. 889. [*Phyllocladus hypophyllus* sp. nov.].

06051 Hooker, J. D. (1854). Himalayan Journals; or, notes of a naturalist in Bengal, the Sikkim and Nepal Himalayas, the Khasia mountains, ... Vol. 1–2. London. [*Cedrus libani* var. *deodara* (Roxb.) comb. et stat. nov., Vol. 1, p. 257; *Larix griffithii* Hook. f. et Thomson, sp. nov., Vol. 2, p. 44].

06053 Hooker, J. D. (1862). On the Cedars of Lebanon, Taurus, Algeria and India. Nat. Hist. Rev., n. s., 2: 11–18. [*Cedrus* spp., *C. libani* var. *atlantica* (Endl.) comb. et stat. nov., p. 15].

06054 Hooker, J. D. (1864). Handbook of the New Zealand Flora: a systematic description of the native plants of New Zealand... Part I. Order LXXV. Coniferae. (pp. 255–260). London. [*Libocedrus doniana* Endl., p. 256, *L. bidwillii* sp. nov., p. 257].

06055 Hooker, J. D. (1864). On the plants of the temperate regions of the Cameroons mountains and islands in the Bight of Benin; collected by Mr. Gustav Mann, Government Botanist. J. Proc. Linn. Soc., Bot. 7: 171–240. [*Podocarpus mannii* sp.nov. = *Afrocarpus mannii*, p. 218].

06060 Hooker, J. D. (1866). *Glyptostrobus pendulus*. Bot. Mag. 92: pl. 5603. [*G. pendulus* sensu Hook. fil., non Endl. = *Taxodium distichum* (L.) Rich. var. *imbricatum* (Nutt.) Croom].

06061 Hooker, J. D. (1880). On the discovery of a variety of the Cedar of Lebanon on the Mountains of Cyprus, with letter thereupon from Sir Samuel Baker, F. R. S. J. Linn. Soc., Bot. 17 (104–105): 517–519. [*Cedrus libani* var. *brevifolia*, var. nov.].

06062 Hooker, J. D. (1882). *Pherosphaera fitzgeraldi* F. Muell. Hooker's Icon. Pl. 14: 64, t. 1383. [*Pherosphaera fitzgeraldii* F. Muell. ex Hook. f.]

06070 Hooker, J. D. (1884). *Picea ajanensis*. Curtis's Bot. Mag. 110: pl. 6743. [*P. ajanensis* Masters = *P. brachytyla*].

06075 Hooker, J. D. (1886). *Cephalotaxus mannii*. Hooker's Icon. Pl. 16: 1 p. sub. t. 1523. [*C. mannii* sp. nov.].

06090 Hooker, W. J. (1827). *Cunninghamia lanceolata*. Curtis's Bot. Mag. 54: pl. 2743. [comb. nov., basion. : *Pinus lanceolata* Lamb.].

06105 Hooker, W. J. (1841). *Taxodium sempervirens* Lamb.? Icon. Pl. 4, t. 379. [*T. sempervirens* sensu Hooker, non D. Don (in Lambert) = *Abies bracteata* (D. Don) Poit.].

06110 Hooker, W. J. (1842). On a new species of *Thuja*, and on *Podocarpus totara* of New Zealand. London J. Bot. 1: 570–575. [*Thuja doniana* sp. nov. = *Libocedrus plumosa* (D. Don) Sargent, p. 571].

06111 Hooker, W. J. (1843). Icones plantarum; or figures with brief descriptive characters and remarks, ... Ser. 2, Vol. 6. London. [*Podocarpus biformis* sp. nov. = *Halocarpus biformis* (t. 544), *Dacrydium colensoi* sp. nov. = *Manoao colensoi* (t. 548), *Athrotaxis laxifolia* sp. nov. (t. 573), *A. selaginoides* (t. 574), *A. cupressoides* (t. 559) (Cupressaceae), *A. tetragona* sp. nov. (t. 560) = *Microcachrys tetragona* (Podocarpaceae), as "Arthrotaxis", *Podocarpus nivalis* sp. nov. (t. 582)].

06111a Hooker, W. J. (1843). Brief descriptions, with figures, of *Juniperus bermudiana*, the pencil-cedar tree; and of the *Dacrydium elatum*, Wall. London J. Bot. 2: 141–145, t. 1–2. [*D. elatum* sp. nov., p. 144, t. 2].

06112 Hooker, W. J. (1843). Figure and description of a new species of *Thuja*, from Chili. (With a Plate Tab. IV). London J. Bot. 2: 199–200, pl. 4. [*Thuja chilensis*, with syn. *Cupressus chilensis* Gillies ex Hook., but see *T. chilensis* D. Don in Lamb., 1832 = *Austrocedrus chilensis*].

06112a Hooker, W. J. (1843). Figure and description of a new species of *Araucaria*, from Moreton Bay, New Holland, detected by J. T. Bidwill, Esq. London J. Bot. 2: 498–506, pl. 18–19. [*A. bidwillii* sp. nov., p. 503].

06113 Hooker, W. J. (1844). Description, with a figure, of a new species of *Thuja*, the Alerse of Chili. London J. Bot. 3: 144–149. [*Thuja tetragona* sp. nov. = *Pilgerodendron uviferum*, p. 148, pl. 4].

06114   Hooker, W. J. (1844). *Podocarpus purdieana* Hook. Hook. Icon. Pl. 7, t. 624. [*P. purdieanus* sp. nov.].

06118   Hooker, W. J. (1850). *Cephalotaxus fortuni* – Mr. Fortune's Cephalotaxus. Curtis's Bot. Mag. 76, 2 unnumb. pp. sub t. 4499. [*C. fortunei* sp. nov.].

06120   Hooker, W. J. (1852). *Araucaria columnaris* – Pillared Araucaria. Curtis's Bot. Mag. 78, 4 unnumb. pp. sub t. 4635. [*Araucaria columnaris* (J. R. Forst.) Hook., comb. nov.].

06120a   Hooker, W. J. (1854). *Torreya myristica* – California nutmeg. Curtis's Bot. Mag. 80, 2 unnumb. Pp. sub t. 4780. [= *T. californica*].

06121   Hooker, W. J. (1855). The Big Tree (*Wellingtonia gigantea*, Lindl.). Hooker's J. Bot. Kew Gard. Misc. 7: 26–29. [on C. F. Winslow's plea to name *W. gigantea* (= *Sequoiadendron giganteum*) *Washingtonia californica*, a nom. superfl. ; see also C. F. Winslow, 1854].

06122   Hooker, W. J. & G. A. W. Arnott (1838). The botany of Captain Beechey's Voyage; comprising an account of the plants collected by Messrs. Lay and Collie, and other officers of the expedition, during the voyage to the Pacific and Bering's Strait, ... Part. 6: 241–288. London. [*Juniperus taxifolia* sp. nov., p. 271].

06123   Hopkins, M. (1938). Notes from the Herbarium of the University of Oklahoma I. The geographic Range of *Juniperus mexicana*. Rhodora 40 (479): 425–429. [map N of Mexico].

06127   Hörhammer, L. (1933). Über die Coniferen-Gattungen *Cheirolepis* Schimper und *Hirmeriella* nov. gen. aus dem Rhät-Lias von Franken. Bibl. Bot. 107: 1–33[34], pl. 1–7. [*Hirmeriella rhaetoliassica* nov. gen. et sp., p. 29; palaeobotany].

06128   Hörster, H. (1974). Vergleich der Monoter-penfraktion von *Juniperus drupacea* und *Juniperus oxycedrus*. Planta Medica 26 (2): 113–118. [*Juniperus* sect. *Oxycedrus*].

06130   Horton, K. V. (1959). Characteristics of subalpine spruce in Alberta. Canada Dept. Aff. and Nat. Res., Forestry Branch, Forest Res. Div. Techn. Note 76: 1–20. [*Picea engelmannii*].

06135   Howard, E. T. (1971). Bark structure of the Southern Pines. Wood Sci. 3 (3): 134–149. [*Pinus* spp. of SE U.S.A.].

06140   Howell, J. T. (1941). The closed-cone pines of insular California. Leafl. W. Bot. 3 (1): 1–8. [*Pinus radiata* div. f. nov., p. 3, *P. muricata* var. *cedrosensis*, var. nov., p. 7, *P. muricata* var. *remorata* (Mason) comb. not validly publ., p. 7].

06141   Howell, J. T. & E. C. Twisselmann (1968). A columnar form of *Juniperus californica*. The Four Seasons 2 (4): 16–18. [*J. californica* f. *lutheyana* f. nov., p. 16, ill.].

06150   Hruby, K. & V. Gotthard (1934). Recherches biométriques sur les feuilles et les cônes de *Larix decidua* Mill., de *L. sudetica* Dom. et de *L. polonica* Racib. Bull. Int. Acad. Sci. Bohême 1934: 1–5.

06160   Hsia, W. Y. (1936). A new species of Chinese pine. Chin. J. Bot. 1 (1): 17–18, pl. 6, f. 1. [*Pinus hwangshanensis* sp. nov.].

06180   Hsu, Y. C. (1950). A preliminary study of the forest ecology of the area about Kunming. Contr. Dudley Herb. 4 (1): 112.

06190   Hsueh, C. J. (1983). A new variety of *Keteleeria evelyniana*. Acta Phytotax. Sin. 21 (3): 253., fig. [*K. evelyniana* var. *pendula*, var. nov.].

06195   Hsueh, C. J. (1991). Reminiscences of collecting the type specimens of *Metasequoia glyptostroboides*. Arnoldia 51 (4): 17–21.

06200   Hsueh, C. J. & S. H. Hao (1981). A new species of *Keteleeria* from China *K. xerophila*. Acta Bot. Yunnan. 3 (2): 249–250.

06208   Hu, H. H. (1926). Synoptical study of Chinese Torreyas. [with observations by R. C. Ching]. Contr. Biol. Lab. Sci. Soc. China [Sect. Bot.] 3 (5): 1–9(–37). (ill. pl. 1–18). [*Torreya grandis* var. *dielsii* var. nov., p. 7, *T. grandis* var. *sargentii* var. nov., p. 7, *T. grandis* var. *chingii* var. nov., p. 8, *T. grandis* var. *merrillii* var. nov., p. 9, + two new forms under var. *grandis*].

06220   Hu, H. H. (1946). Notes on a Palaeogene species of *Metasequoia* in China. (Eng., Chin. summ.). Bull. Geol. Soc. China 26: 105–107. [fossil *M. chinensis* (Endo) comb. nov.].

06230   Hu, H. H. (1950). *Taiwania*, the monarch of Chinese conifers. J. New York Bot. Gard. 51: 63–67, 3 figs.

06240   Hu, H. H. (1951). Lecture material on the Classification of Seed Plants. (Chin.). [*Nothotsuga*, p. 64].

06250   Hu, H. H. & W. C. Cheng (1948). Some new trees from Yunnan. Bull. Fan Mem. Inst. Biol., n.s. 1: 191–198. [*Pinus wangii* sp. nov., p. 191].

06260   Hu, H. H. & W. C. Cheng (1948). On the new family Metasequoiaceae and on *Metasequoia glyptostroboides*, a living species of the genus *Metasequoia* found in Szechuan and Hupeh. Bull. Fan Mem. Inst. Biol., n.s. 1 (2): 154–161, pl. 1–2. [nom. gen. cons., see W. Greuter *et al.*, 1988), *M. glyptostroboides* sp. nov., p. 154].

06261   Hu, S. Y. (1964). Notes on the flora of China. IV. Gymnospermae. Taiwania 10: 13–62. [*Podocarpus chingianus* sp. nov., p. 32, *Cupressus chengiana* sp. nov., p. 57].

06270   Hu, S. Y. (1980). The *Metasequoia* flora and its phytogeographic significance. J. Arnold Arbor. 61 (1): 41–94.

06270a Hu, Y. S. (1986). SEM observation of the inner surface structure of needle cuticles in *Pinus*. Acta Phytotax. Sin. 24: 464–468. (ill.).

06290 Hu, Y. S. & F. H. Wang (1984). Anatomical Studies of *Cathaya* (Pinaceae). Amer. J. Bot. 71 (5): 727–735.

06300 Hu, Y. S., F. H. Wang & Y. C. Chang (1976). On the comparative morphology and systematic position of *Cathaya* (Pinaceae). Acta Phytotax. Sin. 14 (1): 73–78. (Chin., ill.).

06310 Huang, W. L., Y. L. Tu & S. Z. Fang (1984). A new species of *Abies* Mill. – *Abies fanjingshanensis*. Acta Phytotax. Sin. 22 (2): 154–155. (Chin. + Lat., ill.).

06311 Huberman, M. A. (1935). The role of Western white pine in forest succession in Northern Idaho. Ecology 16 (2): 137–151. [*Pinus monticola, Thuja plicata, Tsuga heterophylla, Abies grandis, Pseudotsuga menziesii, Larix occidentalis*, ecology].

06320 Hudson, R. H. (1960). The anatomy of the genus *Pinus* in relation to its classification. J. Inst. Wood Sci. 6: 26–46. (ill.).

06320c Hueber, F. M. & J. Watson (1988). The unusual Upper Cretaceous *Androvettia* from eastern U.S.A. J. Linn. Soc., Bot. 98: 117–133. (ill.). [Cheirolepidiaceae].

06320k Huguet del Villar, E. (1934). Apéndice a unas observaciones sobre el habitat calizo de *Pinus pinaster*. Bol. Soc. Española Hist. Nat. 33: 421–431. [*Pinus pinaster* ssp. *atlantica* nom. nov., p. 427].

06321 Huguet del Villar, E. (1938). L'aire du *Callitris articulata* en Espagne. Bull. Soc. Bot. France 85: 4–14. [*Tetraclinis articulata*, map].

06330 Huguet del Villar, E. (1947). Conifères et Fagacées de l'Afrique du Nord. annot. in: "Types de sols de l'Afrique du Nord". Fasc. 1, Rabat. [*Abies tazaotana* = *A. pinsapo* var. *tazaotana*, Lat. diagn., p. 80].

06331 Huguet del Villar, E. (1947). Quel est le nom valable du *Pinus laricio* Poir.? Ber. Schweiz. Bot. Ges. 57: 149–155. [*P. nigra*, nomenclature].

06332 Huguet del Villar, E. (1948). Les Pins de l'Afrique du nord. Pp. 235–263 in: Volume Jubilaire de la Société des Sciences Naturelles du Maroc: L'Evolution des Sciences Naturelles au Maroc de 1934 à 1947. Rabat & Paris. [*Pinus pinaster* div. var. nov., pp. 241, 244, 247, *P. clusiana* (= *P. nigra*) div. comb. et var. nov., pp. 255, 256, 257].

06340 Hultén, E. (1926). *Pinus pumila* Regel. Pflanzenareale 1 (2): 33, map 20.

06390 Hunt, D. R. (1962). Some notes on the pines of British Honduras. Empire Forest. Rev. 41: 134–145. [*Pinus caribaea, P. oocarpa*].

06400 Hunt, D. R. (1967). *Abies delavayi* var. *fabri*. J. Roy. Hort. Soc. London 92 (6): 263. [*A. delavayi* var. *fabri* (Masters) comb. et stat. nov. = *A. fabri*].

06402 Hunt, D. R. & H. J. Welch (1968). The correct varietal names of two low-growing European forms of *Juniperus communis* L. Taxon 17 (5): 545. [authors consider *J. communis* var. *saxatilis* Pall. not validly publ. and prefer *J. communis* var. *montana* Aiton].

06409 Hunt, R. S., M. D. Meagher & J. F. Manville (1990). Morphological and foliar terpene characters to distinguish between western and eastern white pine. Canad. J. Bot. 68: 2525–2530. [*Pinus monticola, P. strobus*].

06410 Hunt, R. S. & E. von Rudloff (1974). Chemosystematic studies in the genus *Abies*. I. Leaf and twig oil analysis of alpine and balsam firs. Canad. J. Bot. 52: 477–487. [*A. lasiocarpa, A. balsamea*].

06421 Hustich, I. (1966). On the Forest-Tundra and the Northern Tree-Lines. Ann. Univ. Turkuensis, ser. A. 2 (36): 11–? [*Abies, Picea, Pinus, Larix*, maps].

06430 Hutchinson, A. H. (1914). The male gametophyte of *Abies*. Bot. Gaz. (Crawfordsville) 57: 148–153. (ill.).

06440 Hutchinson, A. H. (1915). Fertilization in *Abies balsamea*. Bot. Gaz. (Crawfordsville) 60: 457–472. (ill.). [double fertilization reported].

06450 Hutchinson, A. H. (1917). Morphology of *Keteleeria fortunei*. Bot. Gaz. (Crawfordsville) 63: 124–134, pl. 7–8, f. 1–3. [embryology].

06460 Hutchinson, A. H. (1924). Embryogeny of *Abies*. Bot. Gaz. (Crawfordsville) 77: 280–288. (ill.).

06470 Hutchinson, J. (1918). *Pinus canariensis*. (I). Bull. Misc. Inform. 1918 (1): 1–3, pl. 1–2. [embryology].

06474a Hyland, B. P. M. (1978). A revision of the genus *Agathis* (Araucariaceae) in Australia. Brunnonia 1: 103–115. [*A. robusta, A. microstachya, A. atropurpurea* sp. nov., p. 109, ill.; appendix on leaf anatomy by S.G.M. Carr & D. J. Carr, ill.].

06474b Hylander, N. (1953). Taxa et nomina nova in opere meo: Nordisk kärlväxtflora I (1953) inclusa. Bot. Not. 1953 (3): 352–359. [*Picea abies* ssp. *europaea* (Tepl.) comb. nov., *Pinus mugo* var. *arborea* (Tubeuf) comb. nov., *P. nigra* var. *corsicana* (Loud.) comb. nov., p. 352].

06474c Ickert-Bond, S. M. (2000). Cuticle micromorphology of *Pinus krempfii* Lecomte (Pinaceae) and additional species from southeast Asia. Int. J. Plant Sci. 161 (2): 301–317. [*Pinus dalatensis, P. kesiya, P. krempfii, P. kwantungensis*, ill. (SEM)].

06474d Ickert-Bond, S. M. (2001). Reexamination of wood anatomical features in *Pinus krempfii* (Pinaceae). I.A.W.A. Journal 22 (4): 355–365. (ill.).

06474e Ilic, J. (1995). Distinguishing the woods of *Araucaria cunninghamii* (Hoop pine) and *Araucaria bidwillii* (Bunya pine). I.A.W.A. Journal 16 (3): 255–260.

06475 Imkhanitskaya, N. N. (1990). Taksono-micheskaya zametka o *Juniperus excelsa* (Cupressaceae). (The taxonomic note on *Juniperus excelsa* (Cupressaceae)). Bot. Žurn. 75 (3): 402–409. (Russ.). [*J. excelsa* ssp. *polycarpos* var. *pendula* (Mulk.) comb. nov., basion. *J. polycarpos* var. *pendula* Mulk., p. 407, *J. excelsa* ssp. *seravschanica* (Komarov) comb. nov., p. 407, *J. excelsa* ssp. *turcomanica* (B. A. Fedtsch.) comb. nov., p. 408].

06476 Imkhanitskaya, N. N. (1990). Kriticheskaya zametka o kavkazskikh vidakh sektsii *Juniperus* roda *Juniperus* L. (Cupressaceae). (Notula critica de speciebus caucacicis sectionis *Juniperus* generis *Juniperus* L. (Cupressaceae)). Novit. Syst. Pl. Vasc. 27: 5–16. (Russ.). [*Juniperus communis*, *J. communis* ssp. *oblonga* (M.-Bieb.) Galushko = *J. communis* var. *oblonga* (M.-Bieb.) Parl., non Loud. and other syn., *J. communis* ssp. *hemisphaerica* (C. Presl) Nym., *J. communis* ssp. *pygmaea* (K. Koch) Imkhanitskaya, comb. nov., p. 10, *J. oxycedrus*].

06477 International Union of Forestry Research Organizations (IUFRO). (1980). Forestry problems of the genus *Araucaria*. Fundacao de Pesquisas Florestais do Parana (FUPEF), Curitiba, Brasil. (382 pp.). [*Araucaria angustifolia*, *A. araucana*].

06478 Isenberg, I. H. (1943). The anatomy of redwood bark. Madroño 7: 85–91. (ill.). [*Sequoia sempervirens*].

06479 Ishii, H. & E. D. Ford (2002). Persistence of *Pseudotsuga menziesii* (Douglas Fir) in temperate coniferous forests of the Pacific Northwest coast, U.S.A. Folia Geobot. 37: 63–69. [ecology].

06480 Ishii, S. (1940–41). On the various forms of *Pinus pumila* and other northern Japanese soft pines with special reference to their distribution. J. Japan. Forest. Soc. 22: 581–586, 651–656, figs. (1940); 23: 1–7, 47–55, 107–114, figs. (1941). (Japan.).

06490 Ishii, S. (1952). A taxonomic study on the genus *Pinus*. Rep. Kôchi Univ. Nat. Sci. 2: 103–126, 1 pl. (Japan., Eng.). [ca. 80 spp. recognized].

06500 Ishii, S. (1952). On the geographical distribution of the genus *Pinus*. Res. Rep. Kôchi Univ. 1 (34): 1–6, pl. 1–4. (Japan.).

06510 Ishizuka, K. (1961). A relict stand of *Picea glehnii* Masters on Mt. Hayacine, Iwate Prefective. Ecol. Rev. 15: 155–162.

06510a Isoda, K., T. Brodribb & S. Shiraishi (2000). Hybrid origin of *Athrotaxis laxifolia* (Taxodiaceae) confirmed by random amplified polymorphic DNA analysis. Austral. J. Bot. 48: 753–758. [*Athrotaxis cupressoides*, *A. laxifolia*, *A. selaginoides*].

06510b Isoda, K., S. Shiraishi & H. Kisanuki (2000). Classifying *Abies* species (Pinaceae) based on sequence variation of tandemly repeated array found in the chloroplast DNA *trn*L and *trn*F intergeneric spacer. Silvae Genet. 49 (3): 161–165. [18 spp. analysed, phylogeny, taxonomic comparison of three classifications; see also Farjon & Rushforth, 1989 and Liu, 1971].

06510c Isoda, K., S. Shiraishi, S. Watanabe & K. Kitamura (2000). Molecular evidence of natural hybridization between *Abies veitchii* and *A. homolepis* (Pinaceae) revealed by chloroplast, mitochondrial and nuclear DNA markers. Mol. Ecol. 9: 1965–1974.

06510g Islebe, G. (1993). Will Guatemala's *Juniperus–Pinus* forests survive? Environment. Cons. 20 (2): 167–168.

06511 Ito, T. (1909). [*Abies kawakamii* (Hayata) comb. et stat. nov.] in: Encyclopaedia Japonica II, p. 167. [not seen].

06520 Ivanova, R. N. (1958). Kedr sibirskii. (The Siberian cedar). pp. 1–95, f. 1–38, map. (Russ.). [*Pinus sibirica*, ecology, phytogeography].

06531 Jack, J. G. (1893). The fructification of *Juniperus*. Bot. Gaz. (Crawfordsville) 18: 369–375. (ill.).

06540 Jackson, A. B. (1909). The Japanese Douglas fir. Gard. Chron., ser. 3, 45: 307, f. 132. [*Pseudotsuga japonica*].

06550 Jackson, A. B. (1915–21). Notes on conifers IX. Gard. Chron., ser. 3, 58: 78, (59): 58, 278–279, (60): 85, (64): 137, (69): 103. [*Cupressus pisifera* = *Chamaecyparis pisifera*, *Cupressus funebris* = *Chamaecyparis funebris*, *Cupressus obtusa* = *Chamaecyparis obtusa*, *Abies firma*, *A. mariesii*, other conifers, ill.].

06560 Jackson, A. B. (1928). *Pinus thunbergii*. Gard. Chron., ser. 3, 83: 231–232, f. 114.

06570 Jackson, A. B. (1929). The golden larch. New Fl. & Silva 2: 30–31, f. 16. [*Pseudolarix amabilis*].

06580 Jackson, A. B. (1932). The coffin juniper. New Fl. & Silva 5: 31–34, 1 pl. [*Juniperus coxii* sp. nov. = *J. recurva* var. *coxii*].

06590 Jackson, A. B. (1938). [*Pinus wallichiana* A.B. Jackson, nom. nov.; *P. wallichiana* forma *zebrina* (Croux ex Bailly) A. B. Jackson, comb. nov.] in: Bull. Misc. Inform. 1938 (2): 85. [publ. March 23, 1938, without title; comm. on syn.: *P. excelsa* Wallich ex D. Don in Lamb., *P. griffithii* McClell.].

06620   Jacobs, B. F., C. R. Werth & S. I. Guttman (1984). Genetic relationships in *Abies* (fir) of eastern United States: an electrophoretic study. Canad. J. Bot. 62: 609–616. [*A. fraseri, A. balsamea*].

06630   Jacobs, D. F., D. W. Cole & J. R. McBride (1985). Fire history and the perpetuation of natural coast redwood ecosystems. J. Forest. (Washington) 83: 494–497. [*Sequoia sempervirens*].

06640   Jaffré, T., J.-M. Veillon & J.-F. Cherrier (1987). Sur la présence de deux Cupressaceae, *Neocallitropsis pancheri* (Carrière) Laubenfels et *Libocedrus austrocaledonica* Brongn. et Griseb. dans le massif du Paeoua et localités nouvelles de gymnospermes en Nouvelle-Calédonie. Bull. Mus. Hist. Nat. (Paris), sér. 4, sect. B, Adansonia 9 (3): 273–288.

06645   Jagel, A. & T. Stützel (2000). Zypressengewächse (Cupressaceae) im Palmengarten Frankfurt – Die Vertreter Südamerikas: *Austrocedrus chilensis, Fitzroya cupressoides, Pilgerodendron uviferum*. (ill.). Palmengarten 64 (2): 138–147.

06646   Jagel, A. & T. Stützel (2001). Zur Abgrenzung von *Chamaecyparis* Spach und *Cupressus* L. (Cupressaceae) und die systematische Stellung von *Cupressus nootkatensis* D. Don [= *Chamaecyparis nootkatensis* (D. Don) Spach]. Fedde's Repert. 112 (3–4): 179–229. [*Chamaecyparis lawsoniana, C. formosensis, C. obtusa, Cupressus nootkatensis* (= *Xanthocyparis nootkatensis*), *C. funebris, C. bakeri, C. arizonica, C. duclouxiana, C. sempervirens, Fokienia hodginsii*, ovulate cone ontogeny, SEM and diagrams].

06647   Jagel, A. & T. Stützel (2001). Untersuchungen zur Morphologie und Morphogenese der Samenzapfen von *Platycladus orientalis* (L.) Franco (= *Thuja orientalis* L.) und *Microbiota decussata* Kom. (Cupressaceae). Bot. Jahrb. Syst. 123 (3): 377–404. [ovulate cone ontogeny, SEM and diagrams].

06648   Jagel, A. & T. Stützel (2003). On the occurrence of non-axillary ovules in *Tetraclinis articulata* (Vahl) Mast. (Cupressaceae s. str.). Feddes Repert. 114 (7–8): 497–507. [ovulate cone ontogeny, SEM and diagrams].

06649   Jagels, R., B. A. LePage & M. Jiang (2001). Definitive identification of *Larix* (Pinaceae) wood based on anatomy from the Middle Eocene, Axel Heiberg Island, Canadian High Arctic. I.A.W.A. Journal 22 (1): 73–83.

06650   Jährig, M. (1962). Beiträge zur Nadelanatomie und Taxonomie der Gattung *Pinus* L. Wildenowia 3 (2): 329–366.

06651   Jährig, M. (1968). Beiträge zur Kenntnis der Sippenstruktur der Gattung *Pinus* L. Arch. Forstwes. 17: 173–205.

06660   Jain, K. K. (1975). A taxonomic revision of the Himalayan firs. Indian Forester 101 (3): 199–204. [*Abies* spp.)

06670   Jain, K. K. (1976). Introgressive hybridization in the West Himalayan silver firs. Silvae Genet. 25: 107–109. [*Abies pindrow, A. spectabilis*].

06690   Jain, K. K. (1976). A taxonomic revision of the Himalayan junipers. Indian Forester 102 (2): 109–118. [*Juniperus* spp.].

06691   Jain, K. K. (1978). Morphology of [the] female strobilus in *Podocarpus neriifolius*. Phytomorphology 27: 215–233. (ill.).

06710   James, E. (1823). Account of an expedition from Pittsburgh to the Rocky Mountains, ... (known as Long's Exped.). Vol. 1–2, with atlas... Philadelphia – London. [*Pinus flexilis* sp. nov., Vol. 2, pp. 27, 35].

06720   James, R. L. (1959). Carolina hemlock wild and cultivated. Castanea 24: 112–134. (ill.). [*Tsuga caroliniana*].

06740   Jansen, A. E. & F. W. de Vries (1988). Qualitative and quantitative research on the relation between ectomycorrhiza of *Pseudotsuga menziesii*, vitality of the host and acid rain. Report 2502, LU Wageningen.

06750   Jaurès, R. & Y. de Ferré (1949). A propos des *Larix* d'Amérique du Nord. Trav. Lab. Forest. Toulouse, T. 1, vol. 4, art. 33. (16 pp., ill.). [*Larix laricina* var. *lutea, L. laricina* var. *parvistrobus* vars. nov.].

06760   Jeffers, J. N. R. & T. M. Black (1963). An analysis of variability of *Pinus contorta*. Forestry (Oxford) 36: 199–218.

06761   Jeffrey, E. C. (1903). The comparative anatomy and phylogeny of the Coniferales. Part 1. The genus *Sequoia*. Mem. Boston Soc. Nat. Hist. 5 (10): 441–459. [*Sequoia, Sequoiadendron*].

06770a   Jensen, U. & C. Lixue (1991). *Abies* seed protein divergent from other Pinaceae. Taxon 40 (3): 435–440.

06780   Jepson, W. L. (1922). A new species of Cypress. Madroño 1 (4): 75. [*Cupressus forbesii* sp. nov. = *C. guadelupensis* var. *forbesii* (Jepson) Little].

06790   Jepson, W. L. (1923). A conifer new to California. Madroño 1: 116. [*Cupressus forbesii*].

06820   Johansen, D. A. (1953). Morphological criteria for the specific validity of *Pinus jeffreyi*. Madroño 12: 92–95.

06825   Johns, R. J. (1995). *Papuacedrus papuana* var. *papuana* – Cupressaceae. Curtis's Bot. Mag. 12 (2): 66–72. [*Papuacedrus papuana* var. *arfakensis* (Gibbs) stat. nov., p. 70].

06840   Johnston, H. H. (1896). Cedar Tree of Mount Mlanje. Bull. Misc. Inform. 119: 216–217. [*Widdringtonia whytei*].

06841 Johnston, I. M. (1924). Taxonomic records concerning American spermatophytes. Contr. Gray Herb., n.s. 70: 61–91. [*Fitzroya cupressoides* (Molina) comb. nov., p. 91; basion.: *Pinus cupressoides* Molina in Sag. stor. nat. Chili, 1782].

06842 Johnston, I. M. (1943). Plants of Coahuila, eastern Chihuahua, and adjoining Zacatecas and Durango, I. J. Arnold Arbor. 24 (3): 306–339. [*Pinus* spp., *Abies coahuilensis* sp. nov. = *A. durangensis* var. *coahuilensis*, p. 332, *Cupressus arizonica*, *Juniperus* spp., *Ephedra* spp.].

06850 Johnston, M. C. (1985). Nomenclatural readjustments in Mexican *Juniperus* (Cupressaceae). Taxon 34 (3): 505–506. [*J. monticola* Martínez = *J. sabinioides* (Kunth in H. B. K.) Nees, 1847; but see *J. sabinoides* Griseb., 1846].

06851 Jones, M. E. (1891). [*Pinus monophylla* var. *edulis*) (Contrib. to Western Botany). Zoe 2: 251. [*P. monophylla* var. *edulis* (Engelm.) comb. nov.].

06853 Jones, W. G., K. D. Hill & J. M. Allen (1995). *Wollemia nobilis*, a new living Australian genus and species in the Araucariaceae. Telopea 6 (2–3): 173–176. [*Wollemia nobilis* gen. et sp. nov., p. 173, 174].

06855 Jordan, G. J. (1995). Extinct conifers and conifer diversity in the Early Pleistocene of western Tasmania. Rev. Palaeobot. Palynol. 84: 375–387. [Cupressaceae, *Athrotaxis* spp., *Phyllocladus*, Podocarpaceae].

06860 Joubert, A. & P. A. Burollet (1934). Biologie et rôle forestier des Cyprès. Rev. Eaux Forêts 72: 85–93, 165–176, 245–252. [*Cupressus*, ill., map].

06867 Jung, W. (1968). *Hirmeriella muensteri* (Schenk) Jung nov. comb., eine bedeutsame Konifere des Mesozoikums. Palaeontographica Abt. B, Paläophytol. 122: 55–93. (ill.).

06868 Jung, W. (1974). Die Konifere *Brachyphyllum nepos* Saporta aus den Solnhofener Plattenkalken (unteres Untertithon), ein Halophyt. Mitt. Bayerischen Staatssamml. Paläontol. Hist. Geol. 14: 49–58. [palaeobotany, palaeoecology, Jurassic].

06870 Jussieu, A. L. de (1789). Genera plantarum secundum ordines naturales disposita, ... Ordo V. Coniferae, Les conifères. (pp. 411–415). Paris. [*Araucaria* gen. nov., *Juniperus, Cupressus, Thuja, Pinus, Abies* (incl. *Larix*), descr.].

06883 Kaeiser, M. (1949). Embryo development of the Pond Cypress (*Taxodium ascendens* Brongn.). Trans. Illinois State Acad. Sci. 42: 63–67. (ill.). [*Taxodium distichum* var. *imbricatum*].

06884 Kaeiser, M. (1950). Microscopic anatomy of the wood of three species of *Juniperus*. Trans. Illinois State Acad. Sci. 43: 46–50. (ill.).

06885 Kaeiser, M. (1953). Microstructure of the wood of the three species of *Taxodium*. Bull. Torrey Bot. Club 80: 415–418. (ill.). [*T. distichum, T. ascendens* = *T. distichum* var. *imbricatum, T. mucronatum*].

06886 Kaeiser, M. (1953). Microstructure of the wood of *Juniperus*. Bot. Gaz. (Crawfordsville) 115: 155–162. (ill.). [*Juniperus* spp., *J.* sect. *Juniperus, J.* sect. *Sabina, J. drupacea*].

06890 Kalis, A. J. (1984). L'indigenat de l'Epicéa dans les Hautes Vosges. Revue de Paleobiologie, Vol. spécial, pp. 103–115, Genève. [*Picea abies*].

06897 Kananji, B. (1990). Determination of the origin of *Widdringtonia nodiflora* on Zomba Mountain in Malawi by the aid of allozymes. Canad. J. Forest Res. 20: 1814–1818. [*W. nodiflora, W. whytei*, plantation is of mixed provenance, see also Pauw & Linder, 1997].

06910 Kanehira, R. (1932). On *Cunninghamia lanceolata* and *C. konishii*. Trans. Nat. Hist. Soc. Taiwan 22: 233–234. (Japan.).

06931 Kaniewski, K. & O. Kucewicz (1978). Anatomical development of the *Abies alba* Mill. cone and shedding of its scales during ripening. Zesz. Nauk. Szk. Główn. Gospod. Wiejsk. Akad. Roln. w Warszawie, Lesnictwo 26: 141–158. (ill.).

06946 Karagiannakidou, V. (1996). Statistische Untersuchungen der Samen von Juniperus-Arten in Griechenland. Mitt. Deutsch. Dendrol. Ges. 82: 129–137. [*Juniperus* spp.].

06947 Karalamangala, R. R. & D. L. Nickrent (1988). An electrophoretic study of representatives of subgenus *Diploxylon* of *Pinus*. Canad. J. Bot. 67: 1750–1759. [*Pinus* subgenus *Pinus*, 14 spp. studied from SW U.S.A. and Mexico].

06950 Karavaev, M. N. (1958). *Tsuga longibracteata* Cheng, first found in a fossil condition on the territory of the USSR. Bjull. Moskovsk. Obsc. Isp. Prir., Otd. Biol. 63: 73–76. (Russ., Eng. summ.). [*Nothotsuga longibracteata* (Cheng) C.N. Page].

06960 Karavaev, M. N. (1960). Two new coniferous species (*Cathaya jacutica* M. sp. nov. and *Pinus sukaczewii* M. sp. nov.) discovered in the Tertiary deposits of central Yakutia. Bjull. Moskovsk. Obsc. Isp. Prir., Otd. Biol. (n. s.?) 3: 127–130. (Russ., Eng. summ.).

06967 Karryev, M. O. (1967). Comparative study of volatile oils in Middle Asian *Juniperus* species. News of the Acad. Sci. of Turkmenia, ser. biol. 1: 85–88.

06970   Karsten, G. H. H. & H. Schenk (1904–44). Vegetationsbilder. Vols. 1–26. Jena. [*Calocedrus decurrens*, Vol. 9, *Cunninghamia lanceolata*, Vol. 14, *Cupressus lusitanica*, Vols. 2, 21, *C. macrocarpa*, Vol. 9, *C. sempervirens*, Vols. 3, 13, 26, *Juniperus excelsa*, Vols. 17, 26, *J. oxycedrus*, Vol. 10, *J. phoenicea*, Vol. 10, *Papuacedrus papuana*, Vol. 15, *Thuja orientalis* = *Platycladus orientalis*, Vol. 17, *Sequoia sempervirens*, Vol. 9, *Sequoiadendron giganteum*, Vol. 9, *Taiwania cryptomerioides*, Vol. 17, *Taxodium distichum*, Vol. 15, *T. mucronatum*, Vol. 2, *Tetraclinis articulata*, Vol. 10].

06980   Karsten, G. K. W. H. (1881). Deutsche Flora. Pharmaceutisch-medicinische Botanik... Lieferung 2/3, pp. 129–336. Berlin. [*Picea abies* (L.) Karst., comb. nov., p. 324].

06981   Kasapligil, B. (1978). Past and present pines of Turkey. Phytologia 40 (2): 99–199. [*Pinus* spp., *P. brutia*, *P. halepensis*, *P. nigra*, *P. pinea*, *P. sylvestris*, maps, figs., palaeobotany].

06990   Katoh, T. (1948). Studies on the distribution of "Kôyamaki" (= *Sciadopitys verticillata* Sieb. et Zucc.) natural forest and a few factors of its climatic environment. Bull. Kyushu Univ. Forests 16: 55–115, f. 1–12. (Japan.).

06994   Kaundun, S. S., B. Fady & P. Lebreton (1997). Genetic differences between *Pinus halepensis*, *P. brutia* and *P. eldarica* based on needle flavonoids. Biochem. Syst. Ecol. 25 (6): 553–562. [*Pinus brutia*, *P. eldarica*, *P. halepensis*; suggests ranking *P. eldarica* as subspecies of *P. brutia* s.l.].

06995   Kaur, D., R. N. Konar & S. P. Bhatnagar (1981). Male cone development in *Agathis robusta*. Phytomorphology 31: 104–112. (ill.).

07000   Kayaçik, H. (1955). The distribution of *Picea orientalis* (L.) Carrière. Bull.Misc. Inf. R.B.G. Kew 1955 No. 3: 481–490. [phytogeography, ecology].

07010   Kayaçik, H., F. Yaltirik & G. Eliçin (1979). The floristic composition of the Italian cypress (*Cupressus sempervirens* L.) forest within the Antalya region in Turkey. Webbia 34 (1): 145–153.

07020   Kazmi, S. M. A. & A. Jehan (1975). Useful plants of Pakistan: part 1. gymnosperms. Sultania 1: 5–55. [*Abies*, *Picea*, *Cedrus*, *Pinus*, *Cupressus*, *Juniperus polycarpos* = *J. excelsa* ssp. *polycarpos*, ill.].

07030   Keast, A., ed. (1981). Ecological bio-geography of Australia. (Monogr. Biol. vol. 41). The Hague. [See C. N. Page & H. T. Clifford, 1981].

07040   Keck, D. D. (1946). Bibliographic notes on *Abies bracteata* and *Pinus coulteri*. Madroño 8 (6): 177–179. [nomenclature].

07043   Kedves, M. (1985). LM, TEM and SEM investigations on recent inaperturate Gymnospermatophyta pollen grains. Acta Biol. Szeged. 31: 129–146. (ill.). [Cupressaceae, Taxodiaceae, *Chamaecyparis*, *Cryptomeria*, *Cupressus*, *Juniperus*, *Sequoia*, *Thuja*, *Xanthocyparis nootkatensis*, in addition *Cephalotaxus*, *Taxus*].

07057   Kelch, D. G. (2002). Phylogenetic assessment of the monotypic genera *Sundacarpus* and *Manoao* (Coniferales: Podocarpaceae) utilising evidence from 18S rDNA sequences. Austral. Syst. Bot. 15 (1): 29–35.

07060   Kelley, W. A. & R. P. Adams (1977). Seasonal variation of isozymes in *Juniperus scopulorum*: systematic significance. Amer. J. Bot. 64: 1092–1096.

07061   Kelly, P. E., E. R. Cook & D. W. Larson (1992). Constrained growth, cambial mortality, and dendrochronology of ancient *Thuja occidentalis* on cliffs of the Niagara Escarpment: an eastern version of bristlecone pine? Int. J. Plant Sci. 153 (1): 117–127. (ill., map). [oldest individual 1032 years].

07065   Kendall, M. W. (1947). On five species of *Brachyphyllum* from the Jurassic of Yorkshire and Wiltshire. Ann. Mag. Nat. Hist., ser. 11, 14: 225–251. (ill.). [*Brachiophyllum* Brongn., *B. mamillare* Brongn., *B. scalbiensis* sp. nov., *B. crucis* sp. nov., *B. stemonium* sp. nov., *B. desnoyersii* (Brongn.) Saporta, palaeobotany].

07066   Kendall, M. W. (1948). On six species of *Pagiophyllum* from the Jurassic of Yorkshire and southern England. Ann. Mag. Nat. Hist., ser. 12, 1: 73–108. (ill.). [*Pagiophyllum* Heer, *P. insigne* sp. nov., *P. peregrinum* (Lindley & Hutton) Schenk, *P. connivens* sp. nov., *P. maculosum* sp. nov., *P. sewardii* sp. nov., *P. ordinatum* sp. nov.].

07067   Kendall, M. W. (1949). On a new conifer from the Scottish Lias. Ann. Mag. Nat. Hist., ser. 12, 2: 299–307. [*Brachyphyllum scotii* sp. nov., Jurassic, palaeobotany, ill.].

07068   Kendall, M. W. (1949). A Jurassic member of the Araucariaceae. Ann. Bot. (London), n.s. 13: 151–161. [*Araucarites phillipsii* Carruthers, *Brachyphyllum mammillare* Brongn. (very likely the same plant), palaeobotany, ill.].

07069   Kendall,M. W. (1949). On *Brachyphyllum expansum* (Sternberg) Seward, and its cone. Ann. Mag. Nat. Hist., ser. 12, 2: 308–320. [Jurassic, palaeobotany, ill.].

07070   Keng, H. (1957). Shui sha. [*Metasequoia*]. Forest. Bull. (Taipei) 10: i–v, 1–18, f. 1–20. (Chin.). [review of current knowledge].

07101   Kerfoot, O. (1961). *Juniperus procera* Endl. (The African Pencil Cedar) in Africa and Arabia. 1. Taxonomic affinities and Geographical distribution. E. African Agric. Forest. J. 26 (3): 170–177. (map).

07102   Kerfoot, O. (1964). The distribution and ecology of *Juniperus procera* Endl. in East Central Africa, and its relationship to the genus *Widdringtonia* Endl. Kirkia 4: 75–86. (3 pl., ill., tables, map).

07103   Kerfoot, O. (1966). Distribution of the Coniferae: the Cupressaceae in Africa. Nature 212 (5065): 961. [*Juniperus procera*, *Widdringtonia*, map].

07110   Kerfoot, O. (1975). Origin and speciation of the Cupressaceae in Sub-Saharan Africa. Boissiera 24a: 145–150.

07120   Kerfoot, O. & J. J. Lavranos (1984). Studies in the flora of Arabia: 10. *Juniperus phoenicea* L. and *J. excelsa* M.-Bieb. Notes Roy. Bot. Gard. Edinburgh 41 (3): 483–489. [*J. excelsa* in Saudi Arabia and Yemen = *J. procera*].

07130   Kermode, C. W. D. (1939). A note on the occurrence of *Taiwania cryptomerioides* in Burma and its utilisation for coffin boards in China. Indian Forester 65: 204–206.

07130f   Kerp, J. H. J. & J. A. Clement-Westerhof (1991). Aspects of Permian Palaeobotany and Palynology XII. The form-genus *Walchiostrobus* Florin reconsidered. Neues Jahrb. Geol. Paläontol. Abh. 183 (1/3): 257–268. [Coniferales, palaeobotany, ill.].

07138   Khan, N. U., W. H. Asari, J. N. Usmani, M. Ilyas & W. Rahman (1971). Biflavonoyls of the Araucariales. Phytochemistry 10: 2129–2137. [Araucariaceae].

07139   Khomentovsky, P. A. (2004). Ecology of the Siberian Dwarf Pine (*Pinus pumila* (Pallas) Regel) on Kamchatka (General Survey). Science Publishers, Inc., Enfield, N.H. (maps, ill.).

07167   Kim, K., S. S. Whang & R. S. Hill (1999). Cuticle micromorphology of leaves of *Pinus* (Pinaceae) in east and south-east Asia. Bot. J. Linn. soc. 129 (1): 55–74. [21 taxa studied (but not *P. krempfii*), SEM ill.].

07170   Kim, Y. D. (1983). Karyological relationships of the genus *Abies* in Korea. (Korean). J. Korean Forest. Soc. 62: 60–67. [*Abies koreana*, *A. nephrolepis*].

07180   Kim, Y. D. & S. S. Kim (1983). Studies on the morphological and anatomical characteristics of the genus *Abies* in Korea. (Korean). J. Korean Forest. Soc. 62: 68–75. (ill.). [*Abies koreana*, *A. nephrolepis*].

07185   Kimura, T. & J. Horiuchi (1978). *Pseudolarix nipponica* sp. nov., from the Palaeogene Noda Group, northeast Japan. Proc. Japan. Acad., ser. B, Phys. Biol. Sci. 54: 429–434. (ill.). [palaeobotany].

07190   Kindel, K.-H. (1986). Samen der Coniferae: 1. Pinaceae: *Abies* Miller. Mitt. Deutsch. Dendrol. Ges. 76: 93–98. [*Abies* spp., drawings of seeds].

07191   Kindel, K.-H. (1989). Samen der Coniferae: 2. Araucariaceae – *Agathis* Salisb., *Araucaria* Juss. Mitt. Deutsch. Dendrol. Ges. 79: 107–114.

07192   Kindel, K.-H. (1996). Samen der Coniferae: 3. Pinaceae: *Pinus* L. Mitt. Deutsch. Dendrol. Ges. 82: 117–127. [*Pinus* spp., drawings of seeds].

07193   Kindel, K.-H. (2001). Die Gattung *Araucaria*. Mitt. Deutsch. Dendrol. Ges. 86: 191–218. (ill., maps).

07195   Kinloch, B. B., M. Marosy & M. E. Huddleston (eds.) (1996). Sugar pine. Status, values, and roles in ecosystems. Proc. Symposium March 30 – April 1, 1992, Univ. California Davis. Univ. California Div. Agric. Nat. Res. Publ. 3362. [*Pinus lambertiana*, ecology, forestry, much on rust pathogens].

07220   Kirchheimer, F. (1934). Über *Tsuga*-Pollen aus dem Tertiär. Planta 22: 171–179. (ill.).

07230   Kirjasoff, A. B. (1920). Formosa the beautiful. Natl. Geogr. Mag. 37: 247–292. [ill. of *Chamaecyparis formosensis*].

07231   Kirk, T. (1878). A revised arrangement of the New Zealand species of *Dacrydium*, with descriptions of new species. Trans. & Proc. New Zealand Inst. 10: 383–391. (ill.). [*Dacrydium intermedium* sp. nov. = *Lepidothamnus intermedius*, p. 386, *D. westlandicum* sp. nov. = *Manoao colensoi*, p. 387, *D. bidwillii* Hook. f. ex Kirk = *Halocarpus bidwillii*, p. 388].

07231c   Kirk, T. (1883). Description of a new pine. Trans. & Proc. New Zealand Inst. 16: 370–371, t. 26. [*Podocarpus acutifolius* sp. nov.].

07232   Kirkwood, J. E. (1917). Bisporangiate cones in *Larix*. Bot. Gaz. (Crawfordsville) 61: 256–257.

07240   Kitagawa, M. (1979). Neo-Lineamenta florae Manshuricae, or Enumeration of the spontaneous vascular plants hitherto known from Manchuria (NE China) together with their synonymy and distribution. (Flora et Vegetatio Mundi Band 4). J. Cramer, Vaduz. [*Larix dahurica* var. *heilingensis* (Yang et Chou) comb. et stat. nov., *L. olgensis* var. *amurensis* (Dylis) comb. et stat. nov., *L. principis-rupprechtii* var. *wulingshanensis* (Liou et Wang) comb. et stat. nov., pp. 47–48; see also A. Farjon, 1990 on these taxa].

07241   Kitamura, S. (1960). Flora of Afghanistan. (Results of the Kyoto Univ. Sci. Exped. to the Karakoram and Hindukush, 1955, Vol. II). Gymnospermae, pp. 20–23. Kyoto, Japan. [*Juniperus seravschanica* Komarov, figs. 38–39].

07240a   Kitamura, K. & M. Y. B. A. Rahman (1992). Genetic diversity among natural populations of *Agathis borneensis* (Araucariaceae), a tropical rain forest conifer from Brunei Darussalam, Borneo, Southeast Asia. Canad. J. Bot. 70: 1945–1949. [*A. dammara*].

07242   Kitamura, S. (1974). Short reports of Japanese Plants. Acta Phytotax. Geobot. 26 (1–2): 1–15. [*Juniperus rigida* ssp. *conferta* (Parl.) et *J. rigida* ssp. × *pseudorigida* (Makino) comb. et stat. nov., p. 9].

07250   Klaus, W. (1972). Saccusdifferenzierungen an Pollenkörnern ostalpiner *Pinus*-Arten. Österreich. Bot. Zeitschr. 120: 93–116. (ill.).

07260   Klaus, W. (1978). On the taxonomic significance of tectum sculpture characters in alpine *Pinus* species. Grana 17: 161–166. (ill.).

07265   Klaus, W. (1979). Blaue, rote und glebgrüne männliche *Pinus*-blüten. Plant Syst. Evol. 133: 95–97. [*Pinus*, phytogeography, taxonomy].

07270   Klaus, W. (1980). Neue Beobachtungen zur Morphologie des Zapfens von *Pinus* und ihre Bedeutung für die Systematik, Fossilbestimmung, Arealgestaltung und Evolution der Gattung. Plant Syst. Evol. 134: 137–171. (ill., maps).

07280   Klaus, W. (1988). Mediterranean pines and their history. in: F. Ehrendorfer (ed.). Woody Plants Evolution and Distribution Since the Tertiary: 133–163. Spec. Ed. of Plant Syst. Evol., Vol. 162. Wien – New York. [*Pinus* spp., *Pinus* subsect. *Rzedowskiae* subsect. nov., p. 142 (nom. superfl., see Carvajal, 1986), *Pinus* subsect. *Resinosae* subsect. nov., p. 159, also *P. canariensis, P. roxburghii*, ill., maps].

07301   Klotz, G. (1966). Beiträge zur Flora und Vegetation Indiens I. Wiss. Z. Martin-Luther Univ. Halle-Wittenberg, Math.-Naturwiss. Reihe 15 (1): map 43 (*Abies densa*), map 54 (*Juniperus recurva*].

07315   Knobloch, E. (1972). *Aachenia debeyi* n.g.n.sp. – eine neue Konifere aus dem Senon von Aachen. Neues Jahrb. Geol. Paläontol., Mh., 1972/7: 400–406. (ill.). [palaeobotany].

07320   Koch, K. H. E. (1857). *Chamaecyparis thurifera* Endl. und *Cupressus benthami* Endl. Allg. Gartenzeitung 25: 310–311. [*C. thurifera* Endl. = *Juniperus flaccida* var. *poblana* Martínez, *C. benthamii* Endl. = *C. lusitanica* var. *benthamii* (Endl.) Carrière; see also M. Martínez, 1946].

07340   Koch, K. H. E. (1861). Geschichte der *Biota pendula* Endl. [*Thuja filiformis* Lodd.). Wochenschr. Vereines Beförd. Gartenbaues Königl. Preuss. Staaten 4: 191–192. [= *Cupressus pendula* Thunb., a cultivar of *Platycladus orientalis*].

07360   Koch, K. H. E. (1873). Die Cypressen der Alten und Neuen Welt. Eine monographische Szizze. (= Skizze, sic!). Wochenschr. Vereines Beförd. Gartenbaues Königl. Preuss. Staaten 16: 54–64, 109–114. [*Cupressus*].

07370   Koch, W. D. J. (1844). Synopsis florae Germanicae et Helveticae. Ed. 2, Vol. 2. Lipsiae (Leipzig). [*Pinus* classification].

07390   Koehne, B. A. E. (1905). Über Taxodien (Sumpfcypressen). Naturwiss. Wochenschr. 20 (8): 122–123. [*Taxodium distichum, T. mexicanum = T. mucronatum, T. imbricatum = T. distichum* var. *imbricatum, T. heterophyllum = Glyptostrobus pensilis*].

07401   Koidzumi, G. (1918). Contributiones ad floram Asiae orientalis. Bot. Mag. (Tokyo) 32: 53–63, 134–138, 249–259. [*Juniperus lutchuensis* sp. nov. = *J. taxifolia*, p. 138].

07402   Koidzumi, G. (1924). Contributiones ad floram Asiae orientalis. Bot. Mag. (Tokyo) 38: 87–113. [*Pinus amamiana* sp. nov., p. 113].

07410   Koidzumi, G. (1942). A new species of *Taiwania*. Acta Phytotax. Geobot. 11: 138. [*T. yunnanensis* sp. nov.].

07420   Koidzumi, G. (1942). The classification of the Coniferae. Acta Phytotax. Geobot. 11: 227–229. (Japan.). [*Taiwania fushunensis* comb. nov.; basion. : *Eotaiwania fushunensis* Yendo in Bull. Centr. Natl. Mus. Manchoukuo 3: 37, 1942, fossil taxon].

07421   Kolakovsky, A. (1970). New species of *Pinus* and the *Cathaya europaea* Sveshn. from the Tertiary flora of the Georgian S.S.R. Bot. Žurn. (Moscow & Leningrad) 55 (6): 847–861. (Russ.). [palaeobotany].

07430   Kolesnikov, B. P. (1938). High mountain silver-fir of Sikhote-Alin. Vestn. Dal'nevost. Fil. Akad. Nauk SSSR 31: 115–122 (Bull. Far East. Branch Acad. Sci. USSR); also in: Trudy Glavn. Bot. Sada 20. (Russ.). [*Abies koreana* f. *prostrata*].

07440   Kolesnikov, B. P. (1946). [*Larix komarovii, L. middendorfii, L. ochotensis* spp. nov.) in: Mater. Istorii Fl. Rastitel'n. SSSR, Fasc. 2, pp. 356, 358. (Mater. Hist. Fl. & Veg. USSR). [= *L. gmelinii*].

07441   Komarov, V. L. (1901).) Flora Manshurica I. Trudy Imp. S.-Peterburgsk. Bot. Sada 20: 1–559. (Russ., Lat. diagn.). [*Pinus funebris* sp. nov. = *P. densiflora*, p. 177, *Abies gracilis* sp. nov. = *A. sachalinensis* var. *gracilis*, p. 203, *Thuja japonica* (sensu Komarov) = *T. koraiensis*, p. 206].

07460   Komarov, V. L. (1923). De Gymnospermis nonnullis Asiaticis. I. Bot. Mater. Gerb. Glavn. Bot. Sada RSFSR 4: 177–181. [*Picea crassifolia* sp. nov., p. 177, *Microbiota decussata* gen. et sp. nov., p. 180, *Juniperus jarkendensis* sp. nov., p. 181].

07470   Komarov, V. L. (1924). De Gymnospermis nonnullis Asiaticis. II. Juniperi (Sabinae) species monospermae Asiaticae. Bot. Mater. Gerb. Glavn. Bot. Sada RSFSR 5: 25–32. [*Juniperus* spp. from Central Asia and China: *Juniperus turkestanica* sp. nov., p 26, *J. centrasiatica* sp. nov., p. 27, *J. tibetica* sp. nov., p. 27, *J. mekongensis* sp. nov., p. 28, *J. potaninii* sp. nov., p. 28, *J. przewalskii* sp. nov., p. 28, *J. zaidamensis* sp. nov., p. 29, *J. fargesii* (Rehder et Wilson) comb. nov., p. 30, *J. kansuensis* sp. nov., p. 31].

07481 Komarov, V. L. (1932). Sabinae polyspermae Asiae Mediae. Bot. Žurn. 17 (5–6): 474–482. (Russ., with Latin descr.). [*Juniperus turkestanica* var. *fruticosa* var. nov., p. 481, *J. seravschanica*, *J. schugnanica* spp. nov., pp. 481–482; see also V. L. Komarov (ed.), 1934, pp. 187, 190].

07510 Konar, R. N. (1960). The morphology and embryology of *Pinus roxburghii* Sar. with a comparison with *Pinus wallichiana* Jack. Phytomorphology 10 (3): 305–319. (ill.).

07510a Konar, R. N. (1962). Some observations on the life history of *Pinus gerardiana* Wall. Phytomorphology 12: 196–201.

07511 Konar, R. N. & S. K. Banerjee (1963). The morphology and embryology of *Cupressus funebris* Endl. Phytomorphology 13 (3): 321–338. [syn. *Chamaecyparis funebris* (Endl.) Franco, ill.].

07515 Konar, R. N. & R. Mittra (1977). Developmental anatomy of *Callitris glauca*. Phytomorphology 27: 88–92. [= *C. columellaris*].

07530 Konar, R. N. & S. Ramchandani (1958). The morphology and embryology of *Pinus wallichiana* Jack. Phytomorphology 8: 328–346. (ill.).

07534 Konijnenburg- van Cittert, J. H. A. van (1990). Taxonomical affinities of *Taxodiophyllum scoticum* and *Masculostrobus zeilleri* from the Upper Jurassic of Culgower, Sutherland, Scotland. Proc. Melbourne I.O.P. Conference 1988. [Miroviaceae, Taxodiaceae (?)].

07536 Konnov, A. A. (1990). (*Juniperus* formations of Middle Asia and neighbouring areas.). Diss. abstract, 32 pp. Akad. Nauk SSSR, Novosibirsk. (Russ.). [*Juniperus* spp., *J. seravschanica*, *J. semiglobosa*, *J. turkestanica*, *J. turcomanica*, map, vegetation].

07540 Konoe, R. (1959). Über den Impfungsversuch der endotrophen Mykorrhiza bei *Metasequoia glyptostroboides* Hu et Cheng. J. Inst. Polytech., Osaka City Univ. Ser. D, Vol. 10: 37–43, figs.

07541 Koorders, S. F. (1898). Karakter-trekken der flora van Celebes. Gymnospermae. Meded. Lands Plantentuin 19: 262–266. [*Dammara celebica* sp. nov., p. 263 = *Agathis celebica*].

07541a Koorders, S. F. & T. Valeton (1904). Bijdrage No. 10 tot de kennis der boomsoorten op Java. Meded. Lands Plantentuin 68: I–VI, 1–287. [Coniferales, pp. 257–269, *Podocarpus koordersii* Pilg. ex Koord. & Valeton = *P. rumphii*, p. 268].

07542 Köpke, E., L. J. Musselman & D. J. de Laubenfels (1981). Studies on the anatomy of *Parasitaxus ustus* and its root connections. Phytomorphology 31: 85–92. [*Parasitaxus usta*].

07543 Kormutak, A, A. Szmidt & X. R. Wang (1993). Restriction fragment length polymorphism of chloroplast DNA's in some species of fir (*Abies* sp.). Biol. Plants 35: 113–119.

07544 Kormutak, A., B. Vookova, A. Gaidosova & J. Salai (1992). Hybridological relationships between *Pinus nigra* Arn., *Pinus thunbergii* Parl. and *Pinus tabulaeformis* Carrière. Silva Genet. 41: 228–234. [*P. tabulaeformis* = *P. tabuliformis*, embryology, genetics].

07545 Korn, R. W. (2001). Analysis of shoot apical organization in six species of the Cupressaceae based on chimeric behaviour. Amer. J. Bot. 88 (11): 1945–1952. [Cupressaceae, anatomy, morphology, ill.].

07545a Korol, L., G. Shklar & G. Schiller (2002). Diversity among circum-Mediterranean populations of Aleppo Pine and Differentiation from Brutia Pine in their isozymes: additional results. Silvae Genet. 51 (1): 35–41. [*Pinus halepensis*, *P. brutia*, allozymes, genetics].

07546 Korstian, C. F. & W. D. Brush (1931). Southern white cedar. Techn. Bull. U.S.D.A. 251. (75 pp.). [*Chamaecyparis thyoides*].

07550 Korzeniewski, L. (1937). Variation de longueur des aiguilles de l'épicéa (*Picea excelsa* Link) à la limite supérieur de la forêt dans les Tatra (vallée de Sucha Woda). Sylwana 55, ser. A., 2–3: 1–13. (Polish, French summ.). [*Picea abies*].

07560 Korzeniewski, L. (1953). Variabilité de l'épicéa (*Picea excelsa* Link). Monogr. Bot. 1: 1–86, pl. I–XVII. (Polish, French summ.) [photogr. of cone forms of *Picea abies*].

07563 Kotyk, M. E. A., J. F. Basinger & E. McIver (2003). Early Tertiary *Chamaecyparis* Spach from Axel Heiberg Island, Canadian High Arctic. Canad. J. Bot. 81: 113–130. [*C. eureka* Kotyk, sp. nov., p. 115, palaeobotany, morphology of *Chamaecyparis* spp., ill.].

07565 Kovar-Eder, J. & Z. Kvaček (1995). The record of a fertile twig of *Tetraclinis brachyodon* (Brongniart) Mai et Walther from Radoboj, Croatia (Middle Miocene). Flora 190: 261–264. [palaeobotany].

07580 Krajina, V. J. (1956). A summary of the nomenclature of Douglas-fir, *Pseudotsuga menziesii*. Madroño 13 (8): 265–267.

07589 Kramer, A. (1885). Beiträge zur Kentniss der Entwicklungsgeschichte und des anatomischen Baues der Fruchtblätter der Cupressineen und der Placenten der Abietineen. Flora 68 (29): 519–528; 544–568. [Cupressaceae, Pinaceae, *Thuja occidentalis*, *T. plicata*, *Platycladus orientalis*, *Chamaecyparis lawsoniana*, *Cupressus sempervirens*, *Juniperus communis*, *Pinus sylvestris*, *P. strobus*, *P. cembra*, *Larix gmelinii*, *Abies alba*, *Picea rubra*, *Tsuga canadensis*, ill.].

07590 Kramer, K. U., L. Y. T. Westra, E. Kliphuis & T. W. J. Gadella (1972). Floristic and cytotaxonomic notes on the flora of the Maltese islands. Acta Bot. Neerl. 21 (1): 54–66. [*Tetraclinis articulata*].

07590a Krassilov, V. A. (1974). *Podocarpus* from the Upper Cretaceous of eastern Asia and its bearing on the theory of conifer evolution. Palaeontology 17: 365–370. [*Podocarpus* s.l. = *Nageia*].

07591 Krause, E. H. L. (1906). J. Sturm's Flora von Deutschland in Abbildungen nach der Natur. Zweite, umgearbeitete Auflage... Vol. 1. Stuttgart. [*Cryptomeria generalis* sp. nov. = *C. japonica*, p. 23].

07600 Kräusel, R. & B. Kubart (1929). Kritische Bemerkungen bezüglich *Glyptostrobus* und *Taxodium*. I. Von Prof. Dr. R. Kräusel. II. Von Prof. Dr. B. Kubart. Mitt. Deutsch. Dendrol. Ges. 41: 154–158, pl. 15–16. [incl. *Sequoia sempervirens*, discussions of leaf anatomy, wood anatomy, taxonomy, palaeobotany].

07601 Krishtofovich, A. N. (1959). The Oligocene flora of Mount Ashutas in Kazakhstan. Trudy Bot. Inst. Komarova Acad. Nauk SSSR, ser. 8, Paleobotanika 1: 11–166 + plates. (Russ.). [Coniferales pp. 48–59, *Taxodium*, *Metasequoia*, *Sequoia*, *Pseudolarix*, ill., map].

07605 Krupkin, A. B., A. Liston & S. H. Strauss (1996). Phylogenetic analysis of the hard pines (*Pinus* subgenus *Pinus*, Pinaceae) from chloroplast DNA restriction site analysis. Amer. J. Bot. 83 (4): 489–498. [*Pinus* subgenus *Pinus*, *P. resinosa* allied to Eurasian spp. of subsection *Sylvestres* (=subsect. *Pinus*)].

07615 Krutovskii, K., D. V. Politov & Y. P. Althukov (1994). Genetic differentiation and phylogeny of stone pine species based on isozyme loci. In: W. C. Schmidt & F.-K. Holtmeier (comps.). Proceedings – international workshop on subalpine stone pines and their environment: the status of our knowledge. U.S.D.A. Forest Service Gen. Tech. Rep. INT-GTR-309; pp. 19–30. Intermountain Research Station, Ogden, Utah [*Pinus albicaulis*, *P. cembra*, *P. koraiensis*, *P. pumila*, *P. sibirica*].

07641 Kubart, B. (1905). Die weibliche Blüte von *Juniperus communis* L. Sitzungsber. Kaiserl. Akad. Wiss., Math.-Naturwiss. Cl., Abt. 1: 1–29, 2 pl.

07650 Kuiper, L. C. (1989). The structure of natural Douglas-fir forests in western Washington and western Oregon. Agric. Univ. Wageningen Papers 88–5. [*Pseudotsuga menziesii*].

07651 Kuittinen, H., O. Mura, K. Karkkainen & Z. Bozzan (1991). Serbian Spruce, a narrow endemic with much genetic variation. Canad. J. Forest Res. 211: 363–367. [*Picea omorika*].

07660 Kullman, L. (1981). Recent tree-limit dynamics of Scots pine (*Pinus sylvestris* L.) in the southern Swedish Scandes. Wahlenbergia 8: 3–67.

07680 Kunishi, H. (1934). On the determination of some species of genus *Pinus*, by means of the anatomical characters of leaves. J. Japan. Bot. 10: 387–391, f. 1–12. (Japan., key).

07680e Kuntze, C. E. O. (1891–98). Revisio generum plantarum vascularium omnium atque cellularium multarum secundum leges nomenclaturae internationales... (3 volumes), Leipzig. [*Nageia nagi* (Thunb.) comb. nov., vol. 2: 798 (1891), *N. wallichiana* (C. Presl) comb. nov., vol. 2: 800, *N. falciformis* (Parl.) comb. nov. = *Falcatifolium falciforme*, vol. 2: 800, *N. vitiensis* (Seem.) comb. nov. = *Retrophyllum vitiense*, vol. 2: 800, *N. eurhyncha* (Miq.) comb. nov. = *Sundacarpus amarus*, vol. 2: 800, *Araucaria angustifolia* (Bertol.) comb. nov., vol. 3: 375 (1898); based on *Columbea angustifolia* Bertol., Opusc. Sci. 3: 411 (1819) n.v.].

07681 Kunze, G. (1846). Chloris Austro-Hispanica. E collectionibus Willkommianis, a.m. Majo 1844 ad finem Maji 1845 factis, ... Flora 29 (40): 625–640. [*Juniperus oophora* sp. nov. = *J. phoenicea*, p. 637].

07682 Kunze, G. (1847). Pugillus tertius plantarum adhuc ineditarum seu in hortis minus cognitarum, ... Linnaea 20: 1–64. (publ. 17–18 May 1847). [*Cupressus ehrenbergii* sp. nov. = *C. lusitanica* (var.), p. 16 (No. 68); if identical with var. *benthamii*, this name would have priority: *C. benthamii* Endl. was publ. later, i.e. May–June 1847, with 14 May as date of E.'s signature on p. IV].

07684 Kunzmann, L. (1999). Koniferen der Oberkreide und ihre Relikte im Tertiär Europas. Ein Beitrag zur Kenntnis ausgestorbener Taxodiaceae und Geinitziaceae fam. nov. (diss.). Abh. Staatl. Mus. Min. Geol. Dresden 45: 1–134. [Cupressaceae s.l., Geinitziaceae fam. nov., p. 124, Taxodiaceae, *Athrotaxis* D. Don, *Chamaecyparites* Endl., emend., *Cryptomeria* D. Don, *Cunninghamia* R. Br. ex Rich. & A. Rich., *Cupressospermum* Mai emend., *Doliostrobus* Marion (Araucariaceae), *Geinitzia* Endl. emend., *Glyptostrobus* Endl., *Metasequoia* Hu & W.C.Cheng, *Quasisequoia* Srinivasan & Friis, *Q. couttsiae* (Heer) comb. nov., p. 57, *Sequoia* Endl., *Sequoiadendron* J. Buchholz, *Taiwania* Hayata, *Taxodium* Rich., palaeobotany, ill.].

07685 Kunzmann, L. (2001). Neue Untersuchungen an *Cunninghamites oxycedrus* Presl in Sternberg 1838. Feddes Repert. 112 (7–8): 421–445. [Taxodiaceae, *Cunninghamites* spp., palaeobotany, ill.].

07686 Kunzmann, L. & E. M. Friis (1999). Zum Vorkommen von Koniferensamen mit stark gekrümmten Keimfächern in der Oberkreide. Feddes Repert. 110 (5–6): 341–347. [Cupressaceae, Taxodiaceae, fossil seeds].

07687   Kunzmann, L., H. Knoll & R. Gaipl (2003). Neue Untersuchungen an *Geinitzia* Endl. 1847 aus den Aachener Schichten von Belgien und Deutschland (Oberes Santon, Oberkreide). Feddes Repert. 114 (1–2): 1–24. [*G. schlotheimii* sp. nov., p. 5, ill.].

07690a   Kupila, S. & E. M. Gifford (1963). Shoot apex of *Pseudolarix amabilis*. Bot. Gaz. (Crawfordsville) 124 (4): 241–246.

07692   Kurmann, M. H. (1989). Pollen wall formation in *Abies concolor* and a discussion on wall layer homologies. Can. J. Bot. 67: 2489–2504. [ill., mainly light microscopy].

07693   Kurmann, M. H. (1990). Exine formation in *Cunninghamia lanceolata* (Taxodiaceae). Rev. Palaeobot. Palynol. 64: 175–179. (ill.).

07694   Kurmann, M. H. (1990). Development of the pollen wall in *Tsuga canadensis* (Pinaceae). Nord. J. Bot. 10 (1): 63–78. [SEM and TEM + light microscopy].

07695   Kurmann, M. H. (1991). Pollen ultrastructure in *Elatides williamsonii* (Taxodiaceae) from the Jurassic of North Yorkshire. Rev. Palaeobot. Palynol. 69: 291–298. [ill., aff. with *Cunninghamia* apparent].

07695a   Kurmann, M. H. (1991). Pollen morphology and ultrastructure in *Athrotaxis* D. Don (Taxodiaceae). Abstr. Symp. Assoc. Palynol. Language Fr., 12th, Caen, France.

07696   Kurmann, M. H. (1992). Exine stratification in extant gymnosperms: a review of published transmission electron micrographs. Kew Bull. 47 (1): 25–39. [Cycadales, Ginkgoales, Coniferales, Taxales, Gnetales, comparison of families: Cupressaceae and (paraphyletic) Taxodiaceae are similar].

07699   Kurz, H. & R. K. Godfrey (1962). Trees of Northern Florida. Gainsville. [gymnosperms pp. 1–23, *Taxus*, *Torreya*, *Pinus* spp., *Taxodium*, *Chamaecyparis thyoides* var. *henryae*, *Juniperus silicicola*].

07700   Kurz, S. (1873). On a few new plants from Yunnan: *Calocedrus*, nov. gen. J. Bot. 11 (2): 196, t. 133. [*Calocedrus macrolepis* sp. nov.].

07710   Kusaka, M. (1954). On the genus *Sabina*, relating with Li's classification of Cupressaceae. J. Japan. Bot. 29: 125–128, f. 1. (Japan., Eng. summ.). [*Sabina* = *Juniperus*; see also H. L. Li, 1953].

07710b   Kvaček, J. (1997). *Sphenolepis pecinovensis* sp. nov., a new taxodiaceous conifer from the Bohemian Cenomanian, Central Europe. (proceedings 4th E.P.P.C.). Med. Nederlands Inst. Toegep. Wetensch. TNO 58: 121–129. (ill.). [Cretaceous fossil conifer].

07710c   Kvaček, J. (1999). Two conifers (Taxodiaceae) of the Bohemian Cenomanian (Czech Republic, Central Europe). In: L. Stuchlik (ed.). Proceedings of the Fifth European Palaeobotanical and Palynological Conference. (June 1998, Krakow). Acta Palaeobot. Suppl. 2: 129–151. [Cupressaceae s.l., Taxodiaceae, *Quasisequoia crispa* (Velenovský) Kvaček, *Cunninghamites lignitum* (Sternberg) Kvaček, comb. nov., palaeobotany, ill.].

07710d   Kvaček, J. (2000). *Frenelopsis alata* and its microsporangiate and ovuliferous reproductive structures from the Cenomanian of Bohemia (Czech Republic, Central Europe). Rev. Palaeobot. Palynol. 112 (1–3): 51–78. [Coniferales, Cheirolepidiaceae, *Frenelopsis alata* (K. Feistmantel) E. Knobloch, *F. oligostomata* Romariz, Alvinia *bohemica* (Velenovský) J. Kvaček, p. 61, palaeobotany, ill. (incl. SEM)].

07710k   Kvaček, Z. (1971). Supplementary notes on *Doliostrobus* Marion. Palaeontographica B, 135 (3–6): 115–126. (ill.). [palaeobotany].

07710m   Kvaček, Z. (1989). The fossil *Tetraclinis* Mast. (Cupressaceae). (Fossilni *Tetraclinis* Mast. [Cupressaceae]). Časopis Nár. Muz. Praze 155 (1–2): 45–54. (ill.). [*Tetraclinis salicornioides* (Ung.) Z. Kvaček].

07710n   Kvaček, Z. (1999). An ancient *Calocedrus* (Cupressaceae) from the European Tertiary. Flora 194: 237–248. (ill.). [*Calocedrus suleticensis* (Brabenec) Z. Kvaček, p. 241].

07710p   Kvaček, Z. (2002). Novelties on *Doliostrobus* (Doliostrobaceae) an extinct conifer genus of the European Palaeogene. J. Nat. Mus., Nat. Hist. Ser. 171 (1–4): 131–175. [Araucariaceae, Doliostrobaceae fam. nov., *Doliostrobus taxiformis* Z. Kvaček, palaeobotany, ill.].

07710q   Kvaček, Z. (2002). A new juniper from the Palaeogene of Central Europe. Feddes Repert. 113 (7–8): 492–502. (ill.). [*Juniperus pauli* sp. nov., p. 493, *Cupressus*, *Chamaecyparis*, palaeobotany].

07710s   Kvaček, Z., S. R. Manchester & H. E. Schorn (2000). Cones, seeds, and foliage of *Tetraclinis salicornioides* (Cupressaceae) from the Oligocene and Miocene of western North America: a geographic extension of the European Tertiary species. Int. J. Pl. Sci. 161: 331–344. (ill.).

07710w   Kwei, Y. L. & Y. S. Hu (1974). Epidermal features of the leaves of *Taxus* in relation to taxonomy. Acta Phytotax. Sin. 12 (3): 329–334.

07711   Kwei, Y. L. & C. L. Lee (1963). Anatomical studies of the leaf structure of Chinese Pines. Acta Bot. Sin. 11: 44–66. [*Pinus*, ill.].

07712   Labillardière, J. J. H. de (1791). Icones plantarum Syriae rariorum, descriptionibus et observationibus illustratae... Decas 2. Paris. [*Juniperus drupacea* sp. nov., p. 14, t. 8].

07720   Lacassagne, M. (1929). Une nouvelle espèce d'épicéa, *Picea kamtchatkensis*. Bull. Soc. Hist. Nat. Toulouse 58: 637–638. [= *P. jezoensis*].

07730 Lacassagne, M. (1934). Étude morphologique, anatomique et systematique du genre *Picea*. Trav. Lab. Forest. Toulouse, T. 2, vol. 3, art. 1: 1–292. (ill.).

07731 Lacassagne, M. (1946). Additions et corrections à l'étude du genre *Picea*. Trav. Lab. Forest. Toulouse, T. 2, vol. 3, art. 2: 1–6.

07731c Laderman, A. D. (ed.) (1998). Coastally restricted forests. Oxford Univ. Press, New York, Oxford. [*Chamaecyparis*, *C. lawsoniana*, *C. nootkatensis*, *C. obtusa*, *C. pisifera*, *C. thyoides*, *Pinus pumila*, *Sequoia sempervirens* and some other (conifer or broadleaved) trees (shrubs) and their ecology; chapters by various authors, ill., tables, maps].

07732 Laing, E. V. (1956). The genus *Abies* and recognition of species. Scott. Forest. 10: 20–25.

07740a Lakhanpal, T. N. & S. Kumar (1995). Regeneration of Cold Desert Pine of N.W. Himalayas (India) – A preliminary study. Pp. 102–106 in: B. A. Roundy *et al.* (comp.). Proceedings: Wildland Shrub and Arid Land Restoration Symposium, Las Vegas, Nevada, Oct. 19–21, 1993. U.S.D.A. Forest Service Gen. Tech. Rep. INT-GTR-315. Intermountain Research Station, Ogden, Utah. [*Pinus gerardiana*].

07750 LaMarche, V. C., Jr. (1969). Environment in relation to age of bristlecone pines. Ecology 50: 53–59. [*Pinus aristata*, *P. longaeva*].

07758 Lamarck, J. B. A. P. M. de (1778). Flore françoise, ou description succincte de toutes les plantes qui croissent naturellement en France,... Vol. 2. Paris. [*Pinus* spp., pp. 199–202, *P. excelsa* sp. nov. ("*excelsus*") = *Picea abies* p. 202].

07759 Lamarck, J. B. A. P. M. de (1786). Encyclopédie methodique. Botanique... Vol. 2: CIC-GOR. Paris, Liège. [*Dombeya chilensis* gen. et sp. nov. = *Araucaria araucana*, p. 301].

07760 Lamarck, J. B. A. P. M. de (1804–05). Encyclopédie methodique. Botanique... Vols. 5–6. Paris, Liège. [see J. L. M. Poiret, 1804–05].

07770 Lamarck, J. B. A. P. M. de (1816–17). Encyclopédie methodique. Botanique... Supplement. Vols. 4–5. Paris. [see J. L. M. Poiret, 1816–17].

07771 Lamb, A. F. A. (1950). Pine forests of British Honduras. Empire Forest. Rev. 29: 219–226. [*Pinus caribaea*].

07780 Lamb, W. H. (1914). A conspectus of North American firs (exclusive of Mexico). Proc. Soc. Amer. Foresters 9: 528–538. [*Abies*].

07781 Lambert, A. B. (1805). On a new species of *Pinus* [*P. pungens*]. Ann. Bot. (London) 2: 188–199. [*P. pungens* sp. nov., p. 198].

07810e Lan, K. M. & F. H. Zhang (1984). A new variety of *Amentotaxus argotaenia*. Acta Phytotax. Sin. 22 (6): 492. (ill.). [*A. argotaenia* var. *brevifolia* var. nov.].

07811 Land, W. J. G. (1902). A morphological study of *Thuja*. Bot. Gaz. (Crawfordsville) 34: 249–259. (ill.). [anatomy, embryology, double fertilization reported].

07820 Landry, P. (1974). Les sous-genres et les sections du genre *Pinus*. Naturaliste Canad. 101: 769–780.

07821 Landry, P. (1976). Taxonomie tridifférentielle de *Pinus strobus* L. et de *Pinus monticola* Dougl. Bull. Soc. Bot. France 123 (1–2): 47–80.

07830 Landry, P. (1977). Taxonomie du sous-genre *Strobus* (genre *Pinus*): les sous-sections et les séries. Bull. Soc. Bot. France 124: 469–474. [*Pinus* ser. *Chylae*, ser. *Cosmopolitae*, ser. *Extremisorientaliae*, ser. nov., pp. 471–472, *Pinus* subsect. *Flexiles* (Shaw ex Rehder) comb. nov., p. 472, *Pinus* sect. *Cantona* sect. nov., p. 473].

07831 Landry, P. (1978). Réflections sur la division et la subdivision taxonomiques d'un genre: l'exemple du genre *Pinus*. Bull. Soc. Bot. France 125: 507–519. [review of classifications in *Pinus* sect. *Pinus*].

07840 Landry, P. (1984). Synopsis du genre *Abies*. Bull. Soc. Bot. France (Lettres bot.) 131 (3): 223–229. [*Abies* sect. *Illeden*, sect. nov., p. 228].

07850 Landry, P. (1988). A revised synopsis of the white pines. Phytologia 65 (6): 467–474. [*Pinus* sect. *Quinquefoliis* "sect. nov.", but see H. L. Duhamel du Monceau, 1755].

07860 Landry, P. (1988). A revised synopsis of the pines 2. The Arolla pines (*Pinus* section *Cembra* (Loud.) stat. nov.). Phytologia 65 (6): 475–481.

07861 Landry, P. (1989). A revised synopsis of the pines 3: The Parasol pine (*Pinus* section *Pinea*). Phytologia 66 (6): 477–481. [*Pinus pinea*].

07861a Landry, P. (1992). A revised synopsis of the pines 4: The Chihuahua pine (*Pinus* section *Leiophylla*). Phytologia 72 (5): 373–377. [*Pinus leiophylla*].

07861b Landry, P. (1994). A revised synopsis of the pines 5: The subgenera of *Pinus*, and their morphology and behaviour. Phytologia 76 (1): 73–79. [*Pinus* subgen. *Tamaulipasa*, subgen. nov. p. 77, ill.].

07861c Landry, P. (1995). A revised synopsis of the pines 6: supplement to the subgenera. Phytologia 78 (4): 287–290. [key to 7 "subgenera"].

07868 Langer, S. K. (1996). Singleleaf Pinyon pines in California. Fremontia 24 (3): 8–11. [*Pinus monophylla*, *P. californiarum* ssp. *californiarum*, *P. californiarum* ssp. *fallax*, photogr. of veg. and habit].

07869   Langer, S. K. (1999). Pinyon pines of the New York Mountains. Fremontia 27 (3): 24. [*Pinus californiarum, P. edulis, P. monophylla*].

07870   Langlet, O. (1936). Studier över tallens fysiologiska variabilitet och dess samband med klimatet, ett bidrag till kännedomen om tallens ekotyper. Meddeland. Statens Skogsförsökanst. 29 (4): 219–470. (German summ.). [*Pinus sylvestris*].

07880   Langlet, O. (1959). A cline or not a cline a question of Scots pine. Silvae Genet. 8: 13–22. [*Pinus sylvestris*].

07900   Lanner, R. M. (1974). A new pine from Baja California and the hybrid origin of *Pinus quadrifolia*. Southw. Naturalist 19 (1): 75–95. [*P. juarezensis* sp. nov., p. 77, ill., map].

07910   Lanner, R. M. (1974). Natural hybridization between *Pinus edulis* and *Pinus monophylla* in the American Southwest. Silvae Genet. 23: 108–116. (ill.).

07915   Lanner, R. M. (1981). The pinyon-pine: a natural and cultural history. Univ. Nevada Press, Reno. [*Pinus* subsect. *Cembroides*].

07920   Lanner, R. M. (1982). Adaptations of Whitebark Pine for seed dispersal by Clark's nutcracker. Canad. J. Forest Res. 12: 391–402. [*Pinus albicaulis*].

07930   Lanner, R. M., ed. (1984). Proceedings of the Eighth North American Forest Biology Workshop. Logan, Utah. [see W. B. Critchfield, 1984].

07931   Lanner, R. M. (1996). Whatever became of the world's oldest tree? Wildflower 12 (1): 26–27. [*Pinus longaeva*].

07932   Lanner, R. M. & A. M. Phillips III (1992). Natural hybridization and introgression of Pinyon pines in northwestern Arizona. Int. J. Plant Sci. 153 (2): 250–257. [*Pinus edulis, P. monophylla* ssp. *fallax* (Little) Zavarin in M.-F. Passini *et al.*, 1987, p. 37, comb. inval., basion. not cited].

07934   LaPasha, C. & C. N. Miller, Jr. (1981). New taxodiaceous seed cones from the Upper Cretaceous of New Jersey. Amer. J. Bot. 68: 1374–1382.

07937   Lara, A. & R. Villalba (1993). A 3,622 year temperature reconstruction from *Fitzroya cupressoides* tree rings in southern South America. Science 260: 1104–1106.

07940   La Roi, G. H. & J. R. Dugle (1968). A systematic and genecological study of *Picea glauca* and *P. engelmannii*, using paper chromatograms of needle extracts. Canad. J. Bot. 46: 649–687. (ill.).

07950   Larsen, E. (1964). A new species of pine from Mexico. Madroño 17 (6): 217–218. [*Pinus martinezii* sp. nov.].

07967   Laubenfels, D. J. de (1959). Parasitic conifer found in New Caledonia. Science 130: 97. [*Podocarpus ustus* = *Parasitaxus usta*].

07968   Laubenfels, D. J. de (1960). *Podocarpus lucienii*, a new species from New Caledonia. Brittonia 12 (1): 79–80.

07970   Laubenfels, D. J. de (1965). The relationships of *Fitzroya cupressoides* (Molina) Johnston and *Diselma archeri* J. D. Hooker based on morphological considerations. Phytomorphology 15 (4): 414–419.

07978   Laubenfels, D. J. de (1969). Diagnoses de nouvelles espèces d'Araucariacées de Nouvelle-Calédonie. Trav. Lab. Forest. Toulouse T. 1 (8) Art. 5: 1–2. [*Araucaria nemorosa, A. scopulorum, A. schmidii, Agathis montana, A. corbassonii* spp. nov.].

07991   Laubenfels, D. J. de (1972). Deux nouveaux *Podocarpus* endémiques de Madagascar. Adansonia 11 (4): 713–715. [*Podocarpus capuronii* sp. nov., p. 713, *P. humbertii* sp. nov., p. 714, *P. madagascariensis* var. *procerus* var. nov., p. 715].

07995   Laubenfels, D. J. de (1976). The genus *Dacrydium* in Malaya (Gymnospermae). Blumea 23 (1): 97–98. [*Dacrydium medium* sp. nov.].

08001   Laubenfels, D. J. de (1978). The genus *Prumnopitys* in Malesia. Blumea 24 (1): 189–190. [*Prumnopitys amara* (Blume) comb. nov. = *Sundacarpus amarus, P. andina* (Poepp. ex Endl.) comb. nov., *P. ferruginea* (G. Benn ex D. Don) comb. nov., *P. harmsiana* (Pilg.) comb. nov., *P. ladei* (F. M. Bailey) comb. nov., *P. montana* (Humb. & Bonpl. ex Willd.) comb. nov., *P. standleyi* (J. Buchholz & N. E. Gray) comb. nov., *P. taxifolia* (Banks & Sol. ex D. Don) comb. nov.].

08002   Laubenfels, D. J. de (1978). The Moluccan Dammars (*Agathis*, Araucariaceae). Blumea 24: 499–504. [*A. dammara, A. celebica*].

08003   Laubenfels, D. J. de (1979). The *Podocarpus* species of Ambon (Podocarpaceae). Blumea 24 (2): 495–497. [*Podocarpus levis* sp. nov., p. 496].

08004   Laubenfels, D. J. de (1979). The species of *Agathis* (Araucariaceae) of Borneo. Blumea 25: 531–541. [*A. borneensis, A. dammara, A. endertii, A. kinabaluensis* sp. nov., p. 535, *A. lenticula* sp. nov., p. 537, *A. orbicula* sp. nov., p. 540].

08005   Laubenfels, D. J. de (1980). The endemic species of *Podocarpus* in New Guinea. Blumea 26 (1): 139–143. [*Podocarpus crassigemmis* sp. nov., p. 141, *P. pseudobracteatus* sp. nov., p. 142].

08009   Laubenfels, D. J. de (1984). Un neovo *Podocarpus* (Podocarpaceae) de la Española. Moscosoa 3: 149–150. [*Podocarpus hispaniolensis* sp. nov., p. 149].

08010   Laubenfels, D. J. de (1985). A taxonomic revision of the genus *Podocarpus*. Blumea 30 (2): 251–278. [*Podocarpus smithii* sp. nov., p. 257, *P. borneensis* sp. nov., p. 266, *P. insularis* sp. nov., p. 266, *P. rubens* sp. nov., p. 266, *P. spathoides* sp. nov., p. 267, *P. micropedunculatus* sp. nov., p. 268, *P. globulus* sp. nov., p. 269, *P. atjehensis* (Wasscher) comb. nov., p. 271, *P. confertus* sp. nov., p. 271, *P. degeneri* (N. E. Gray) comb. nov., p. 271, *P. brassii* var. *humilis* var. nov., p. 274, *P. grayae* sp. nov., p. 275, *P. fasciculus* sp. nov., p. 277, *P. subtropicalis* sp. nov., p. 277].

08015   Laubenfels, D. J. de (1987). Revision of the genus *Nageia* (Podocarpaceae). Blumea 32: 209–211. [*N. fleuryi* (Hickel) comb. nov., p. 210, *N. maxima* (de Laub.) comb. nov., p. 210, *N. motleyi* (Parl.) comb. nov., p. 210, *N. comptonii* (J. Buchholz) comb. nov. = *Retrophyllum comptonii*, p. 211, *N. piresii* (Silba) comb. nov. = *R. piresii*, p. 211, *N. rospigliosii* (Pilg.) comb. nov. = *R. rospigliosii*, p. 211].

08021   Laubenfels, D. J. de (1991). Las Podocarpaceas del Peru. Bol. de Lima 73: 57–60. [Podocarpaceae, *Podocarpus*, *Prumnopitys*, *Retrophyllum*].

08022   Laubenfels, D. J. de (1990). The Podocarpaceae of Costa Rica. Brenesia 33: 119–121. [*Podocarpus costaricensis* sp. nov. (but see Silba, 1990 No. 13341), *P. monteverdeensis* sp. nov., p. 120].

08025   Laubenfels, D. J. de (1992). *Podocarpus acuminatus* (Podocarpaceae), a new species from South America. Novon 2 (4): 329.

08025a   Laubenfels, D. J. de (2003). A new species of *Podocarpus* from the maquis of New Caledonia. New Zealand J. Bot. 41: 715–718. [*Podocarpus beecherae* sp. nov., p. 715, ill.].

08026   Laubenfels, D. J. de & J. Silba (1987). The *Agathis* of Espiritu Santo (Araucariaceae, New Hebrides). Phytologia 61: 448–452. [*Agathis silbae* sp. nov., p. 448].

08027   Laubenfels, D. J. de & J. Silba (1988). Notes on Asian-Pacific Podocarpaceae, I (*Podocarpus*). Phytologia 64: 290–292. [*Podocarpus epiphyticus* sp. nov., p. 290, *P. palawanensis* sp. nov., p. 291, *P. indonesiensis* sp. nov., p. 292].

08028   Laubenfels, D. J. & J. Silba (1988). Notes on Asian and trans-Pacific Podocarpaceae II. Phytologia 65 (5): 329–332. [*Podocarpus aracensis* sp. nov., p. 330, *P. chinensis* var. *wardii* var. nov., p. 331].

08029   Laurent, L. (1915). Les *Podocarpus* de Madagascar. Ann. Fac. Sci. Marseille 23: 52–66. [*P. madagascariensis* var. *rotundus* var. nov., p. 59, *P. rostratus* sp. nov., p. 60].

08030   Laurent, M. & H. Gaussen (1945). Particularités anatomiques d'*Abies hickelii* Flous et Gaussen. Trav. Lab. Forest. Toulouse, T. 1, Art. div. No. 10: 1–9.

08030a   Lauria, F. (1991). Taxonomy, systematics, and phylogeny of *Pinus*, subsection *Ponderosae* Loudon (Pinaceae). Alternative concepts. Linzer Biol. Beitr. 23 (1): 129–202. [*Pinus ponderosa*, *P. jeffreyi*, *P. washoensis*, *P. engelmannii*].

08030b   Lauria, F. (1996). Typification of *Pinus benthamiana* Hartw. (Pinaceae), a taxon deserving renewed botanical examination. Ann. Naturhist. Mus. Wien 98B, Suppl.: 427–446. [*Pinus ponderosa* s.l.].

08030c   Lauria, F. (1996). The identity of *Pinus ponderosa* Douglas ex C. Lawson (Pinaceae). Linzer Biol. Beitr. 28 (2): 999–1052. [*P. ponderosa*, *P. washoensis*, typification, review of (infraspecific) names of taxa related to *P. ponderosa*].

08030d   Lauria, F. (1997). The taxonomic status of *Pinus washoensis* H. Mason & Stockw. (Pinaceae). Ann. Naturhist. Mus. Wien 99B: 655–671. [= *Pinus ponderosa*, *P. washoensis* relegated to synonymy].

08031   Lauterbach, C. (1914). Neue Pinaceae Papuasiens. (in: Beiträge zur Flora von Papuasien II). Bot. Jahrb. Syst. 50: 46–53, figs. 1–2. [*Agathis*, *Araucaria klinkii* sp. nov. = *A. hunsteinii*, p. 48, *A. cunninghamii* var. *papuana* Lauterb., p. 51, *Libocedrus papuana* = *Papuacedrus papuana*, *L. torricellensis* Schlechter ex Lauterb., sp. nov. = *P. papuana*, p. 52, f. 2].

08031a   Lavin, M. & M. Luckow (1993). Origins and relationships of tropical North America in the context of the Boreotropics hypothesis. Amer. J. Bot. 80 (1): 1–14.

08033   Lawson, A. A. (1904). The gametophytes, fertilization and embryo of *Sequoia sempervirens*. Ann. Bot. (London) 18: 1–28. (ill.).

08034   Lawson, A. A. (1904). The gametophytes, fertilization and embryo of *Cryptomeria japonica*. Ann. Bot. (London) 18: 417–444. (ill.).

08034a   Lawson, A. A. (1907). The gametophytes, fertilization and embryo of *Cephalotaxus drupacea*. Ann. Bot. (London) 21: 1–23. [= *C. harringtonii*].

08035   Lawson, A. A. (1907). The gametophytes and embryo of the Cupressineae, with special reference to *Libocedrus decurrens*. Ann. Bot. (London) 21: 281–302. [= *Calocedrus decurrens*, ill.].

08036   Lawson, A. A. (1909). The gametophytes and embryo of *Pseudotsuga douglasii*. Ann. Bot. (London) 23: 163–180. [= *P. menziesii*, ill.].

08037   Lawson, A. A. (1910). The gametophytes and embryo of *Sciadopitys verticillata*. Ann. Bot. (London) 24: 403–422. (ill.).

08040   Lawson, P. (1836). Agriculturist's Manual. Edinburgh. [*Pinus ponderosa* Dougl. ex Laws., sp. nov., p. 354].

08045 Lazzaro, M. D. (1999). Microtubule organization in germinated pollen of the conifer *Picea abies* (Norway Spruce, Pinaceae). Amer. J. Bot. 86 (6): 759–766. [micro-structure of pollen tube in Pinaceae, ill.].

08050 LeBarron, R. K. & G. M. Jemison (1953). Ecology and silviculture of the Engelmann spruce-alpine fir type. J. Forest. (Washington) 51: 349–355. [*Abies lasiocarpa, Picea engelmannii*].

08050b Lebreton, P. (1983). Nouvelles données sur la distribution au Portugal et en Espagne des sous-espèces du génévrier de Phénicie (*Juniperus phoenicea* L.). Agron. Lusit. 42 (1–2): 55–62. (map).

08050c Lebreton, P. (1990). La chimiotaxonomie des Gymnospermes. Bull. Soc. Bot. France 137 (1): 35–46. [summ. traditional classif. of gymnosperms, suggests *Pinus* as sep. fam.; chemistry of *Juniperus thurifera*].

08051 Lebreton, P. & S. Thivend (1981). Sur une sous-espèce du Genévrier de Phénicie *Juniperus phoenicea* L., définie à partir de critères biochimiques. Naturalia Monspel., Sér. Bot. 47: 1–12, maps. [*J. phoenicea* ssp. *eumediterranea* ssp. nov., p. 8].

08060 Lecomte, H. (1921). Un pin remarquable de l'Annam, *Pinus krempfii*. Bull. Mus. Hist. Nat. (Paris) 27: 191–192. [*Pinus krempfii* sp. nov., ill.].

08070 Lecomte, H. (1924). Additions au sujet de *Pinus krempfii* H. Lec. Bull. Mus. Hist. Nat. (Paris) 30: 321–325. [*P. krempfii* var. *poilanei* var. nov., p. 325, ill.].

08092 Ledig, F. T. (1987). Genetic structure and the conservation of California's endemic and near-endemic conifers. Pp. 587–594 in: T. S. Elias (ed.). Proceedings of the Symposium on Conservation and Management of Rare and Endangered Plants, November 5–9, 1986. California Native Plant Society, Sacramento. [*Abies bracteata, Pinus balfouriana*].

08092a Ledig, F. T. (1996). *Pinus torreyana* at the Torrey Pines State Reserve. In: D. A. Falk, C. I. Millar & M. Olwell (eds.). Restoring diversity – strategies for reintroduction of endangered plants. (pp. 265–271). Washington DC (with Center for Plant Conservation, Missouri Botanical Garden).

08092b Ledig, F. T. (1999). Genetic diversity, genetic structure, and biogeography of *Pinus sabiniana* Dougl. Diversity and Distributions 5: 77–90.

08092c Ledig, F. T. (2000). Locations of endangered spruce populations in Mexico and the demography of *Picea chihuahuana*. Madroño 47 (2): 71–88.

08092d Ledig, F. T., M. A. Capó-Arteaga, P. D. Hodgskiss, H. Sbay, C. Flores-López, M. T. Conkle & B. Bermejo-Velázquez (2001). Genetic diversity and the mating system of a rare Mexican Piñon, *Pinus pinceana*, and a comparison with *Pinus maximartinezii* (Pinaceae). Amer. J. Bot. 88 (11): 1977–1987.

08093 Ledig, F. T. & M. T. Conkle (1983). Gene diversity and genetic structure in a narrow endemic, Torrey Pine (*Pinus torreyana* Parry ex Carrière). Evolution 37: 79–85.

08093a Ledig, F. T., M. T. Conkle, B. Bermejo-Velázquez, T. Eguiluz-Piedra, P. D. Hodgskiss, D. R. Johnson & W. S. Dvorak (1999). Evidence for an extreme bottleneck in a rare Mexican pinyon: genetic diversity, disequilibrium, and the mating system in *Pinus maximartinezii*. Evolution 53 (1): 91–99.

08094 Ledig, F. T., V. Jacob-Cervantes, P. D. Hodgskiss & T. Eguiluz-Piedra (1997). Recent evolution and divergence among populations of a rare Mexican endemic, Chihuahua spruce, following Holocene climatic warming. Evolution 51 (6): 1815–1827. [*Picea chihuahuana*, genetics, population decline, conservation].

08097 Le Duc, A., R. P. Adams & M. Zhong (1999). Using random amplification of polymorphic DNA for a taxonomic re-evaluation of Pfitzer junipers. HortScience 34 (6): 1123–1125. [*Juniperus chinensis, J. sabina, J.* × *media = J.* × *pfitzeriana* and other cultivar hybrids of these parental species].

08098 Lee, S. C. (1962). Taiwan Red- and Yellow-cypress and their conservation. Taiwania 8: 1–13. [*Chamaecyparis formosensis, C. obtusa* var. *formosana*].

08099 Lee, Y. N. (2003). Two new varieties of *Larix sibirica* Ledeb. Bull. Korea Pl. Res. 3: 20–23. [*L. sibirica* var. *viridis, L. sibirica* var. *hybrida*, ill. (colour photographs, showing colour variation of cones in a population in Mongolia)].

08100 Lehmann, C. (1845). Plantae Preissianae, ... Vol. 1. Hamburg. [see F. A. G. Miquel, 1845].

08102 Leistner, O. A., G. F. Smith & H. E. Glen (1995). Notes on African plants – Podocarpaceae. Notes on *Podocarpus* in Southern Africa and Madagascar. Bothalia 25 (2): 233–236. [*Afrocarpus, Podocarpus, P.* sect. *Afrocarpus, P. falcatus*].

08109 Lemmon, J. G. (1888). Report of the Botanist of the California State Board of Forestry; Pines of the Pacific Slope. Calif. State Board Forest. Bienn. Rep. 2: 53–129. [*Pinus* subgen. *Strobus* subgen. nov., p. 69, 79, *P.* subgen. *Pinaster* subgen. nov. (= subgen. *Pinus*), p. 71, 85, *P. monticola* var. *minima* var. nov., p. 70, 80, *P. lambertiana* var. minor var. nov., p. 70, 83, *P. ponderosa* var. *benthamiana* (Hartweg) comb.

nov., p. 73, 97, *P. ponderosa* var. *brachyptera* (Engelm.) comb. nov., p. 73, 98, *P. jeffreyi* div. var. nov., p. 74, 100, *P. insignis* var. *radiata* (D. Don) comb. nov., p. 76, 114, *P. insignis* var. *laevigata* var. nov. (as "levigata"), p. 76, 114; a classification of Californian pines without formal naming of groupings, the subgenera named in parentheses].

08110   Lemmon, J. G. (1890). Cone-bearers of California. Third Biennial Report, Calif. State Board of Forestry 18891890 to gov. R. W. Waterman. (ill.). [Coniferales, *Hesperopeuce*, gen. nov., *H. pattoniana* (Andr. Murray) comb. nov. = *Tsuga mertensiana* ssp. *mertensiana*, pp. 100, 126, *Pseudotsuga macrocarpa* "sp. nov.", p. 134, but see H. Mayr, 1889, *Abies magnifica* var. *shastensis* var. nov., p. 145, *Juniperus utahensis* (Engelm.) comb. nov. = *J. osteosperma*, p. 183].

08111   Lemmon, J. G. (1892). Notes on the conebearers of northwest America. Mining Sci. Press 64: 45. (also publ. in: Erythaea 1, 1893). [*Pinus attenuata* nom. nov., p. 45, publ. 16 Jan. 1892; see also C. S. Sargent in Gard. & Forest 5: 65, publ. 10 Feb. 1892, the latter with descr.].

08131   Lemoine-Sebastian, C. (1965). Ecologie des Génévriers au Maroc. Bull. Soc. Sci. Phys. Nat. Maroc 45: 49–116. [*Juniperus communis, J. oxycedrus, J. phoenicea*].

08132   Lemoine-Sebastian, C. (1967). Appareil reproducteur male des *Juniperus*. Trav. Lab. Forest. Toulouse, T. 1, vol. 6, art. 29: 1–35. (ill., tables).

08140   Lemoine-Sebastian, C. (1968). L'inflorescence femelle des Juniperae: ontogenèse, structure, phylogenèse. Trav. Lab. Forest. Toulouse, T. 1, vol. 7, art. 5: 1–456. [*Juniperus* spp., ill. with figs. and plates].

08143   Lemoine-Sebastian, C. (1971). A propos de strobiles prolifères de *Sciadopitys*: l'écaille séminale et l'aiguille. Bot. Rhedonica, sér. A. 10: 227–250. (ill.).

08143a   Lemoine-Sebastian, C. (1972). Structures épidermiques chez *Sciadopitys* et interprétation des organes. Bull. Soc. Bot. France 119: 61–74. (ill., drawings + SEM).

08145   Lemoine-Sebastian, C. (1975). Structures épidermiques de l'écailles des Pins. Phytomorphology 25: 310–324. [*Pinus*, ill.].

08145a   Lemon, C. (1835). Remarks on the growth of a peculiar fir resembling the Pinaster. Trans. Hort. Soc., ser. 2, 1: 509–512, pl. [*Pinus lemoniana* Benth., sp. nov., p. 512 = *P. pinaster*].

08145b   Leng, Q., H. Yang, Q. Yang & J. P. Zhou (2001). Variation of cuticle micromorphology of *Metasequoia glyptostroboides* (Taxodiaceae). Bot. J. Linn. Soc. 136 (2): 207–219. (SEM ill.).

08145d   LePage, B. A. (2001). New species of *Picea* A. Dietrich (Pinaceae) from the middle Eocene of Axel Heiberg Island, Arctic Canada. Bot. J. Linn. Soc. 135 (2): 137–167 (ill.). [*Picea* spp., *P. sverdrupii, P. nansenii, P. palustris* fossil sp. nov., palaeobotany, evolution since middle Eocene].

08145e   LePage, B. A. (2003). A new species of *Thuja* (Cupressaceae) from the Late Cretaceous of Alaska: implications of being evergreen in a polar environment. Amer. J. Bot. 90 (2): 167–174. [*T. smileya* sp. nov., ill., map. This fossil does not belong to *Thuja*, it is evidently not "evergreen" as it bears no leaves and perhaps it is not even a conifer].

08145f   LePage, B. A. (2003). A new species of *Tsuga* (Pinaceae) from the middle Eocene of Axel Heiberg Island, Canada, and an assessment of the evolution and biogeographical history of the genus. Bot. J. Linn. Soc. 141: 257–296. [*T. swedaea* sp. nov., *Tsuga* spp. This fossil is wrongly identified and belongs to *Larix*, as is evident from the attached, curved peduncles. This was pointed out to the author by a reviewer (AF), accompanied by unequivocal evidence, but his arguments have been ignored in the published paper].

08146   LePage, B. A. & J. F. Basinger (1989). Early Tertiary *Larix* from the Canadian High Arctic. Musk-ox 37: 103–109 (*Larix* sp., ill.].

08147   LePage, B. A. & J. F. Basinger (1991). Early Tertiary *Larix* from the Buchanan Lake Formation, Canadian Arctic Archipelago, and a consideration of the phytogeography of the genus. Geol. Survey Canada, Bull. 403: 67–82. [*Larix* spp., fossil and non-fossil taxa, bibliography, ill.].

08147a   LePage, B. A. & J. F. Basinger (1991). A new species of *Larix* (Pinaceae) from the early Tertiary of Axel Heiberg Island, Arctic Canada. Rev. Palaeobot. Palynol. 70: 89–111, pl. 1–4. [*Larix altoborealis* sp. nov., p. 94, comparison to non-fossil species, phylogeny].

08147b   LePage, B. A. & J. F. Basinger (1995). The evolutionary history of the genus *Larix* (Pinaceae). In: W. C. Schmidt & K. J. McDonald (comps.). Ecology and management of *Larix* forests: a look ahead. Proceedings of an international symposium. U.S.D.A. Forest Service, Gen. Tech. Rep. INT-GTR-319: 19–29. Intermountain Research Station, Ogden, Utah. [*Larix* spp., palaeobotany, phylogeny].

08147c   LePage, B. A. & J. F. Basinger (1995). Evolutionary history of the genus *Pseudolarix* Gordon (Pinaceae). Int. J. Plant Sci. 156 (6): 910–950. [*Pseudolarix amabilis* (recent & fossil), *P. wehrii* N. L. Gooch (fossil), ill.].

08148   Leredde, C. (1957). Etude ecologique et phytogéographique du Tassili n' Ajjer. Trav. Lab. Forest. Toulouse, T. 5 (3rd. sect.), vol. 3, art. 2: 1–455. [*Cupressus dupreziana*, pp. 236–239 with ill., pp. 260–262].

08150   Lester, D. T. (1968). Variation in cone-morphology of Balsam Fir, *Abies balsamea*. Rhodora 70 (781): 83–94.

08150a   Lester, D. T. (1974). Geographic variation in leaf and twig monoterpenes of Balsam Fir. Canad. J. Forest Res. 4 (1): 55–60. [*Abies balsamea*].

08151   Léveillé, A. A. H. (1910). Decades Plantarum novarum: *Keteleeria esquirolii* spec. nov. et *Pinus* div. spec. nov. in: Repert. Spec. Nov. Regni Veg. 27/28 (8): 60–61. [for comment on Léveillé's new taxa see A. Rehder, 1929, 1937].

08160   Léveillé, A. A. H. (1914). Nouveaux conifères de Chine. Monde Plantes 2 (16): 19–20. [*Juniperus mairei* sp. nov., *Podocarpus mairei* sp. nov., *Tsuga mairei* sp. nov. (= *Taxus mairei* (Lemée & Lév.) S. Y. Hu ex Liu), attrib. to Lemée & Léveillé; for comment on L.'s new taxa see A. Rehder, 1929, 1937].

08161   Léveillé, A. A. H. (1914–15). Flore du Kouy Tchéou. [530 pp.; lithographed from handwritten MS; numerous new spp.: *Juniperus lemeeana* Léveillé et Blin = *J. squamata*, p. 111; for comment on L.'s new taxa see A. Rehder, 1929, 1937].

08162   Léveillé, A. A. H. (1916). Catalogue des plantes de Yun-Nan avec renvoi aux diagnoses originales observations et descriptions d'espèces nouvelles... Signat. No. 4: 49–64. Le Mans (par l'Auteur). [*Cupressus mairei* sp. nov., p. 56; for comment on L.'s new taxa see A. Rehder, 1929, 1937].

08162a   Lev-Yadun, S. (1992). Aggregated cones in *Pinus halepensis*. Aliso 13 (3): 475–485. (ill.).

08162b   Lev-Yadun, S. (1995). Living serotinous cones in *Cupressus sempervirens*. Int. J. Plant Sci. 156 (1): 50–54. [Israelian populations, phenology].

08162c   Lev-Yadun, S. (1996). A developmental variant of indentation in the scales of female cones of *Cupressus sempervirens* L. Bot. J. Linn. Soc. 121 (3): 263–269.

08162e   Lev-Yadun, S. & N. Liphschitz (1987). The ontogeny of gender of *Cupressus sempervirens* L. Bot. Gaz. (Crawfordsville) 148: 407–412.

08162h   Lewandowski, A., A. Boratynski & L. Mejnartowicz (2000). Allozyme investigations on the genetic differentiation between closely related pines: *Pinus sylvestris*, *P. mugo*, *P. uncinata*, and *P. uliginosa* (Pinaceae). Pl. Syst. Evol. 221 (1–2): 15–24. (maps).

08162i   Lewis, J. (1995). Proposal to conserve the name *Juniperus × media* Melle (Cupressaceae). Taxon 44 (2): 229–231. [= *J. chinensis* cultivars].

08162p   Lhote, H. (1964). L'ancienne forêt de Cyprès du Tassili-n-Ajjer (Sahara Central). J. Agric. Trop. Bot. Appl. 11 (4): 84–90. [*Cupressus dupreziana*, table with measurements of 21 trees].

08162s   Li, D. Z. (1997). A reassessment of *Pinus* subgen. *Pinus* in China. Edinburgh J. Bot. 54 (3): 337–349. [*Pinus* (10 spp. recognized), *P. thunbergiana* correct name, *P. luchuensis* ssp. *hwangshanensis* (W. Y. Hsia) comb. nov., p. 341, *P. luchuensis* ssp. *taiwanensis* (Hayata) comb. nov., p. 342, *P. merkusii* ssp. *latteri* (F. Mason) comb. nov., p. 346, neotypification of *P. kesyia*, p. 345].

08165   Li, H. L. (1952). The genus *Amentotaxus*. J. Arnold Arbor. 33: 192–198. [*A. argotaenia*, *A. cathayensis* sp. nov., p. 195, *A. formosana* sp. nov., p. 196, *A. yunnanensis* sp. nov., p. 197].

08175   Li, H. L. (1953). New species and varieties in *Cephalotaxus*. Lloydia 16 (3): 162–164. [*C. sinensis* (Rehder & E. H. Wilson) comb. nov., p. 162, *C. hainanensis* sp. nov., p. 164, *C. fortunei* Hook. var. *alpina* var. nov. ("*fortuni*"), p. 164].

08182   Li, H. L. (1962). A New Species of *Chamaecyparis*. Morris Arbor. Bull. 13 (3): 43–46. [*C. henryae* sp. nov. = *C. thyoides* var. *henryae* (H. L. Li) Little].

08191   Li, H. L. (1964). *Metasequoia*, a living fossil. Amer. Sci. 52: 93–109. (ill.).

08192   Li, H. L. (1967). Notes on the nomenclature and taxonomy of *Pseudolarix*. Taiwania 13: 147–152. (Chin., Eng. summ.).

08193   Li, H. L. (1968). The Lace-Bark Pine, *Pinus bungeana*. Morris Arbor. Bull. 19 (1): 1–7. (ill.).

08194   Li, H. L. (1968). The Golden Larch, *Pseudolarix amabilis*. Morris Arbor. Bull. 19 (2): 19–25. (ill.).

08200   Li, H. L. (1982). Contributions to botany. Taiwan. [see H.L. Li, 1982].

08220a   Li, J. H., C. C. Davis, M. J. Donoghue, S. Kelley & P. del Tredici (2001). Phylogenetic relationships of *Torreya* (Taxaceae) inferred from sequences of [the] nuclear ribosomal DNA ITS region. Harvard Pap. Bot. 6 (1): 275–281.

08220b   Li, J. H., C. C. Davis, P. del Tredici & M. J. Donoghue (2001). Phylogeny and biogeography of *Taxus* (Taxaceae) inferred from sequences of the internal transcribed spacer region of nuclear ribosomal DNA. Harvard Pap. Bot. 6 (1): 267–274.

80220c   Li, J. H., D. L. Zhang & M. J. Donoghue (2003). Phylogeny and biogeography of *Chamaecyparis* (Cupressaceae) inferred from DNA sequences of the nuclear ribosomal ITS region. Rhodora 105: 106–117.

08221   Li, L. C. (1987). A new karyotypic type in higher plants based on *Cunninghamia lanceolata*. Guihaia 7 (3): 201–204. (Chin., Eng. summ., ill.).

08222   Li, L. C. (1987). The origin of *Sequoia sempervirens* (Taxodiaceae) based on caryotype. Acta Bot. Yunnan. 9 (2): 187–192. (Chin., Eng. summ.). [ill., chrom. nos.].

08223 Li, L. C. (1988). The parents of *Sequoia sempervirens* (Taxodiaceae) based on morphology. Acta Bot. Yunnan. 10 (1): 33–37. (Chin., Eng. summ.). [anatomy and morphology].

08224 Li, L. C. (1988). The systematic position of *Sciadopitys* Sieb. et Zucc. based on cytological data. Guihaia 8 (2): 135–141. (Chin., Eng. summ.). [tables, chrom. nos.].

08225 Li, L. C. (1988). The studies on the karyotypes and cytogeography of *Taxodium* Rich. (Taxodiaceae). Acta Phytotax. Sin. 26 (5): 371–377. (Chin., Eng. summ.). [ill. (pl. 1 follows p. 408), map].

08226 Li, L. C. (1988). The comparative karyotypic studies in some species of *Tsuga* (Pinaceae). Guihaia 8 (4): 324–328. (Chin., Eng. summ.). [ill., chrom. nos.].

08226b Li, L. C. (1991). The karyotype analysis of *Tsuga longibracteata* and its taxonomic significance. Acta Bot. Yunnan. 13 (3): 309–313. (Chin., Eng. summ., ill.). [= *Nothotsuga longibracteata*].

08227a Li, L. C. (1992). Studies on the karyotype and the systematic position of *Pseudolarix amabilis* (Pinaceae). Proc. 2nd. Sino-Japanese Symposium on Plant Chromosome Research: 259–266. [suggested subfam. Pseudolaricoideae not validly published].

08227b Li, L. C. (1993). Studies on the karyotype and systematic position of *Larix* Mill. (Pinaceae). Acta Phytotax. Sinica 31: 405–412. [*Larix, L. potaninii*].

08227c Li, L. C. (1994). A cytotaxonomical study on *Pseudolarix amabilis*. Acta Bot. Yunnanica 16 (3): 248–254. (Chin., Eng. summ., ill.).

08227k Li, L. C., Y. Q. Cen, P. Xu & H. J. Wang (1996). Studies on the karyotypes of *Chamaecyparis* and the cytotaxonomy of Cupressoideae. Acta Bot. Yunnan. 18 (1): 72–76. (Chin., Eng. summ., tables). [*C. lawsoniana, C. obtusa, C. pisifera*].

08227m Li, L. C., B. Cong, G. Liu, Y. Q. Liu & R. F. Weng (1994). Karyotype analysis of 3 species of *Callitris* (Cupressaceae) in Australia and its phylogenetic significance. Acta Bot. Yunnan. 16 (4): 349–353. [*C. preissii, C. verrucosa, C. endlicheri*].

08227n Li, L. C. & Y. X. Fu (1995). Studies on the cytotaxonomy and the historical phytogeography of *Cedrus* (Pinaceae). Acta Bot. Yunnan. 17 (1): 41–47. (Chin., Eng. summ., tables). [*Cedrus atlantica, C. brevifolia, C. deodara, C. libani*].

08227o Li, L. C. & Y. X. Fu (1996). Studies on the karyotypes and the cytogeography of *Cupressus* (Cupressaceae). Acta Phytotax. Sin. 34 (2): 117–123.

08228 Li,. L. C. & P. S. Hu (1984). Karyotype analyses in *Platycladus orientalis* and *Fokienia hodginsii*. Acta Bot. Yunnan. 6: 447–451. (Chin., Eng. summ.). [both with 2n = 22].

08228e Li, L. C., A. Xu & G. Wang (1997). A karyotypic study of three species in *Abies* (Pinaceae) from East Asia. Acta Bot. Yunnan. 19 (3): 289–291. (Chin., Eng. summ.). [*Abies georgii* var. *smithii* = *A. forrestii* var. *smithii, A. nephrolepis, A. holophylla*, 2n =24].

08229a Li, N. (1993). Phytogeographic studies on the genus *Pseudotsuga*. Bull. Bot. Res. (China) 13 (4): 404–411. (Chin., Eng. summ.). (maps).

08229b Li, N. & L. K. Fu (1997). Notes on gymnosperms I. Taxonomic treatments of some Chinese conifers. Novon 7 (3): 261–264. [*Abies ferreana* var. *longibracteata* var. nov., *A. beshanzuensis* var. *ziyuanensis* (L. K. Fu & S. L. Mo) comb. nov., *Keteleeria fortunei* var. *oblonga* (Cheng & L. K. Fu) comb. nov. (p. 261), *Larix potaninii* var. *chinensis* (Beissn.) comb. nov., *Picea asperata* var. *heterolepis* (Rehder & E. H. Wilson) comb. nov. superfl. vide 11941, *Pinus fenzeliana* var. *dabeshanensis* (Cheng & Y. W. Law) comb. nov., *P. kwangtungensis* var. *varifolia* var. nov. (p. 262), *Pseudotsuga sinensis* var. *wilsoniana* (Hayata) comb. nov., *Tsuga oblongisquamata* (Cheng & L. K. Fu) comb. nov., *T. chinensis* var. *patens* (Downie) comb. nov., *Cephalotaxus chinensis* var. *wilsoniana* (Hayata) comb. nov., *Taxus wallichiana* var. *mairei* (Lemée & Lév.) comb. nov., *T. fuana* Nan Li & R. R. Mill sp. nov. (p. 263), new forms in *Larix*, very short list of ref. ignores Farjon and many other authors on these taxa since 1975–78].

08230 Li, S. J. (1972). The female reproductive organs of *Chamaecyparis*. Taiwania 17 (1): 27–39. [*C. formosensis, C. obtusa* var. *formosana*, ill.].

08232 Li, S. Y. & K. T. Adair (1994). New taxa and new combinations in Chinese plants. Sida 16 (1): 183–190. (ill., maps). [*Larix gmelinii* var. *genhensis* var. nov., p. 183, *Pinus sylvestris* var. *manguiensis* var. nov., p. 184].

08237 Li, X. W. (1992). A new series and a new species of *Pinus* from Yunnan. Acta Bot. Yunnan. 14: 259–260. [*P. squamata* sp. nov., p. 259, *P.* ser. *Squamatae* X. W. Li & Hsueh ser. nov., p. 259].

08240 Li, Y. H. (1948). Anatomical study of the wood of "shui sha" (*Metasequoia glyptostroboides* Hu et Cheng). Trop. Woods 94: 28–29, 1 pl.

08242 Li, Z. Y., Z. C. Tang & N. Kang (1995). Studies on the taxonomy of the genus *Torreya*. (Chin., Eng. abstr.). Bull. Bot. Res. (Harbin) 15 (3): 349–362. [*Torreya fargesii* var. *yunnanensis* (W. C. Cheng & L. K. Fu) N. Kang, comb. nov., p. 353, *T. grandis* var. *jiulongshanensis* var. nov., p. 356].

08245   Libby, W. J., M. H. Bannister & Y. B. Linhart (1968). The Pines of Cedros and Guadalupe Islands. J. Forest. 66: 846–853. [*Pinus radiata* var. *binata*].

08246   Lickey, E. B., F. D. Watson & G. L. Walker (2002). Differences in bark thickness among populations of baldcypress (*Taxodium distichum* (L.) Rich. var. *distichum*) and pondcypress (*T. distichum* var. *imbricarium* (Nuttall) Croom). Castanea 67 (1): 33–41.[*T. distichum* var. *imbricatum*].

08250   Lin, D. Y. (T. Y.) (1934). *Cunninghamia lanceolata* as a possible paper-making wood. Proc. 5th Pacific Sci. Congr. (Canada) 5: 3997–3999.

08250a   Lin, J. X., Y. S. Hu, X. Q. He & R. Ceulemans (2002). Systematic survey of resin canals in Pinaceae. Belg. J. Bot. 135 (1–2): 3–14. [anatomy of wood and leaves, phylogeny, ill.].

08252   Lin, J. X., Y. S. Hu, X. P. Wang & L. B. Wei (1995). The biology and conservation of *Tsuga longibracteata*. Chin. Biodiv. 3 (3): 147–152. (Chin., Eng. summ.). [= *Nothotsuga longibracteata*, endangered].

08253   Lin, J. X., Y. S. Hu & F. H. Wang (1995). Wood and bark anatomy of *Nothotsuga* (Pinaceae). Ann. Missouri Bot. Gard. 82: 603–609. [*Nothotsuga longibracteata*, *Tsuga*].

08253b   Lin, J. X., E. Liang & A. Farjon (2000). The occurrence of vertical resin canals in *Keteleeria*, with reference to its systematic position in Pinaceae. Bot. J. Linn. Soc. 134 (4): 567–574.

08254   Lin, T. P., C. T. Wang & J. C. Yang (1998). Comparison of genetic diversity between *Cunninghamia konishii* and *C. lanceolata*. J. Hered. 89 (4): 370–373. (maps).

08260   Lindley, J. (1833). [Botany in] The Penny Cyclopaedia of the Society for the diffusion of useful knowledge. London. [*Abies*, Vol. 1: 29–34, ill., *A. grandis* (Dougl. ex D. Don) comb. nov., p. 30, *A. nobilis* (D. Don) comb. nov. = *A. procera*, p. 30, *A. webbiana* sp. nov. = *A. spectabilis*, p. 30, *A. hirtella* sp. nov. = *A. religiosa*, p. 31, *A. kaempferi* (Lamb.) comb. nov. = *Pseudolarix amabilis* (p. p.), p. 34, *A. thunbergii* sp. nov. = *A. firma*, p. 34.].

08270   Lindley, J. (1836). Natural System of Botany. Ed. 2. London. [Pinaceae, fam. nov., p. 313].

08280   Lindley, J. (1839). Miscellaneous notices: Mexican pines. Edwards's Bot. Reg. 25: 62–64. (repr. in: Allg. Gartenzeitung 7: 324–325, 1839). [*Pinus hartwegii* sp. nov., *P. devoniana* sp. nov., *P. russeliana* sp. nov., *P. macrophylla* sp. nov., *P. pseudostrobus* sp. nov., *P. apulcensis* sp. nov., *Cupressus thurifera*, = *Juniperus flaccida* var. *poblana* (see M. Martínez, 1946), *J. tetragona*, *J. flaccida*, *J. mexicana*].

08290   Lindley, J. (1840). Miscellaneous notices: Mexican pines. Edwards's Bot. Reg. 26: 61. [*Pinus filifolia* sp. nov.].

08291   Lindley, J. (1851). (*Juniperus sphaerica* sp. nov.) in: Paxton's Fl. Gard. 1: 58, f. 35. [= *J. chinensis*, see also P. J. van Melle, 1946].

08292   Lindley, J. (1851). Notices of certain ornamental plants lately introduced into England. J. Hort. Soc. London 6: 258–273. [*Saxegothaea conspicua* gen. et sp. nov., p. 258, *Fitzroya patagonica* Hooker f. ex Lindl., gen. et sp. nov., p. 264; publ. 1 Oct. 1851, see also J. D. Hooker, 1851; = *F. cupressoides*, *Podocarpus nubigenus* sp. nov., p. 264, *Araucaria cookii* R. Br. ex Lindl. = *A. columnaris*, p. 267, *Dammara obtusa* sp. nov. = *Agathis macrophylla*, p. 270, *D. macrophylla* sp. nov. = *A. macrophylla*, p. 271, *D. moorei* sp. nov. = *A. moorei*, p. 271].

08300   Lindley, J. (1853). New Plants. 33 *Wellingtonia gigantea*. Gard. Chron. 1853: (819–820) 823. [*Wellingtonia gigantea* gen. et sp. nov., nom. illeg. as *Wellingtonia* Lindl. is a later homonym of *Wellingtonia* Meisn. (1840) in Sabiaceae; unsigned but ed. by J. Lindley, see also H. St. John & R. W. Krauss, 1954; = *Sequoiadendron giganteum*].

08301   Lindley, J. (1855). New Plants. 132. *Juniperus pyriformis* (sp. nov.), 133. *Cupressus macnabiana* A. Murray,.. Gard. Chron. 1855, p. 420. [as co-editor Lindley probably wrote this unsigned note; *J. pyriformis* = *J. californica*].

08309   Lindley, J. (1857). New plants 216. *Torreya grandis*, Fortune. Gard. Chron. 1857: 788–789. [*T. grandis* Fortune ex Lindl., p. 788].

08310   Lindley, J. (1857). *Thujopsis dolabrata*. Gard. Chron. 1857: 379–380, 1 fig.

08311   Lindley, J. (1861). New Plants 1–10. Gard. Chron. 1861: 22–23. [*Abies microsperma* sp. nov. = *Picea jezoensis*, p. 22, *A. veitchii* sp. nov., p. 23, *A. alcockiana* Veitch ex Lindl., sp. nov. ("alcoquiana") = *Picea alcockiana*, p. 23].

08312   Lindley, J. (1861). New Plants. 16. *Veitchia japonica*: Lindley, n. g. Gard. Chron. 1861: 265. [= *Picea?*, sp. dubia].

08320   Lindley, J. (1861). *Sciadopitys verticillata*. Gard. Chron. 1861: 359–360.

08321   Lindley, J. (1861). *Araucaria rulei*. Gard. Chron. 1861: 868, ill.

08340   Lindquist, B. (1948). The main varieties of *Picea abies* (L.) Karst. in Europe, with a contribution to the theory of a forest vegetation in Scandinavia during the last Pleistocene glaciation. Acta Horti Berg. 14: 249–342. [*P. abies* var. *obovata* (Ledeb.) comb. nov., p. 307, *P. abies* var. *germanica*, *P. abies* var. *arctica*, vars. nov., pp. 308–309, ill., maps, tables].

08341　Lindquist, B. (1956). Provenances and type variation in natural stands of Japanese Larch. Acta Horti Gothob. 20 (1–2): 1–34. [*Larix kaempferi*, ill., map].

08346　Linhart, Y. B., B. Burr & M. T. Conkle (1967). The closed-cone pines of the northern Channel Islands. In: R. N. Philbrick (ed.). Proceedings of the Symposium on the Biology of the Californian Islands, pp. 151–177. Santa Barbara Bot. Gard., Santa Barbara, California. [*Pinus radiata*, *P. muricata*, *P. remorata*, cone morphology, chemistry].

08349　Link, J. H. F. (1821–22). Enumeratio plantarum hortii regii botanici berolinensis altera. 2 vols. Berlin. [*Dammara loranthifolia* (Salisb.) Link = *Agathis dammara*, vol. 2: 411 (1822)].

08350　Link, J. H. F. (1829). Handbuch zur Erkennung der nutzbarsten und am häufigsten vorkommenden Gewächse... Vol. 2. Berlin. [*Cedrus libanotica* Link = *C. libani* A. Richard, p. 480].

08351　Link, J. H. F. (1830). Über die Familie *Pinus* und die Europäischen Arten derselben. Abh. Königl. Akad. Wiss. Berlin 1827: 157–192. [*Pinus rotundata* "sp. nov." (first. publ. in Flora 10: 218, 1827) = *P. mugo* s. l., p. 168].

08361　Link, J. H. F. (1847). Revisio abietinarum horti Regii botanici Berolinensis. Linnaea 20: 283–298. [Pinaceae, *Pinus rotundata* = *P. mugo* s. l., p. 285 (see also J. H. F. Link, 1830), *Picea orientalis* (L.) comb. nov., p. 294 (but see for earlier comb. by Petermann: Mabberley, 1998)].

08370　Linnaeus, C. (1737–54). Genera Plantarum. Ed. 1–5. Lugduni Batavorum (Leiden) & Holmiae (Stockholm). [see esp. Ed. 5, 1754, with generic names as in Sp. Pl. Ed. 1, 1753, descr: *Pinus* gen. nov. (p. 434), *Thuja* gen. nov. (p. 435 as "Thuya"), *Cupressus* gen. nov. (p. 435), *Juniperus* gen. nov. (p. 461), their dates of publ. being 1 May 1753 (in Sp. Pl. Ed. 1) according to ICBN Art. 13.4].

08380　Linnaeus, C. (1753). Species Plantarum, exhibentes plantas rite cognitas, ... Vol. II.; Monoecia Monadelphia, pp. 1000–1003. Holmiae (Stockholm). [Coniferales: *Pinus* gen. nov., *P. taeda*, *P. pinea*, *P. cembra* spp. nov., p. 1000, *P. sylvestris*, *P. strobus* spp. nov., p. 1001, *P. cedrus*, *P. larix*, *P. picea*, spp. nov. (= *Cedrus libani*, *Larix decidua*, *Abies alba*, resp.), p. 1001, *P. balsamea*, *P. abies* spp. nov. (= *Abies balsamea*, *Picea abies*, resp.), p. 1002, *Thuja* gen. nov., p. 1002, *T. occidentalis* sp. nov., p. 1002, *T. orientalis* (= *Platycladus orientalis*), p. 1002, *Cupressus* gen. nov., p. 1002, *C. sempervirens* sp. nov., p. 1002, *C. disticha* sp. nov. (= *Taxodium distichum*), p. 1003, *C. thyoides* sp. nov. (= *Chamaecyparis thyoides*), p. 1003].

08381　Linnaeus, C. (1753). Species Plantarum, exhibentes plantas rite cognitas, ... Vol. II.; Dioecia Monadelphia, pp. 1038–1040. Holmiae (Stockholm). [Coniferales: *Juniperus* gen. nov. (et in Gen. Pl. Ed. 5, 1754), *J. oxycedrus* sp. nov., p. 1038, *J. thurifera*, *J. lycia* (= *J. phoenicea*), *J. barbadensis*, *J. bermudiana*, *J. sabina*, *J. virginiana* spp. nov., p. 1039, *J. communis*, *J. phoenicea* spp. nov., p. 1040, *Taxus* gen. nov., p. 1040, *T. baccata* sp. nov., p. 1040, *T. nucifera* sp. nov. = *Torreya nucifera*, p. 1040; Ephedrales: *Ephedra* gen. et sp. nov.].

08390　Linnaeus, C. (1763). Species Plantarum, ... Vol. II.; Monoecia Monadelphia, pp. 1418–1423. Ed. 2. Holmiae (Stockholm). [Coniferales: *Pinus* as in Ed. 1, + *P. canadensis* sp. nov. = *Tsuga canadensis*, p. 1421, *P. orientalis* sp. nov. = *Picea orientalis*, p. 1421, (*Thuja aphylla* = *Tamarix articulata*, angiosperm), *Cupressus juniperoides* sp. nov. = *incertae sedis*, p. 1422].

08391　Linnaeus, C. (1767). Mantissa plantarum. Generum editionis VI. et Specierum editionis II. No. 1. Holmiae (Stockholm). [*Thuja cupressoides* sp. nov. = *Widdringtonia nodiflora*, p. 125, *Juniperus chinensis* sp. nov., p. 127].

08392　Linné, C. von, filius (1782). Supplementum plantarum systematis vegetabilium editionis decimae tertiae, generum plantarum editionis sextae, et specierum plantarum editionis secundae. Braunschweig. (1781 on t.p.). [*Thuja dolabrata* Thunb. ex L. f., sp. nov. = *Thujopsis dolabrata*, p. 420, *Cupressus japonica* Thunb. ex L. f., sp. nov. = *Cryptomeria japonica*, p. 421].

08400　Liou, T. N. (1958). Illustrated Flora of Ligneous Plants of NE China. (Chin., Lat. diagn. p. 547). [*Larix wulingshanensis* Liou et Wang, sp. nov. = *L. gmelinii* var. *principis-rupprechtii*].

08402　Liphschitz, N., S. Lev-Yadun & Y. Waisel (1981). The annual rhytm of activity of the lateral meristems (cambium and phellogen) in *Cupressus sempervirens* L. Ann. Bot. (London), n.s. 47 (4): 485–496 (ill.). [anatomy and phenology of wood].

08403　Lipsky, V. I. (1912). Explorations botaniques. in: Travaux d'expédition pour exploration des regions de Colonisation Russe d'Asie, II, Fasc. 6. Leningrad. [*Juniperus talassica* sp. nov., p. 185 + pl.; see also V. L. Komarov, 1932, p. 482].

08406　Liston, A., W. A. Robinson, D. Piñero & E. R. Alvarez-Buylla (1999). Phylogenetics of *Pinus* (Pinaceae) based on nuclear ribosomal DNA Internal Transcribed Spacer region sequences. Mol. Phylogen. Evol. 11 (1): 95–109. [cladistic analysis of 47 species of *Pinus* using ITS region sequences of DNA; the data on *P. krempfii* appear to have been erroneously taken from *P. armandii*].

08410 Little, E. L., Jr. (1944). Nomina conservanda proposals for ten genera of trees and shrubs. Madroño 7 (8): 240–251. [Pinaceae: *Abies*, *Cedrus*].

08420 Little, E. L., Jr. (1944). Notes on nomenclature in Pinaceae. Amer. J. Bot. 31 (9): 587–596. [*Juniperus ashei* Buchholz, *J. mexicana* Spreng. on p. 593].

08430 Little, E. L., Jr. (1948). Older names for two western species of *Juniperus*. Leafl. W. Bot. 5: 125–132. [*Juniperus deppeana* Steud., *J. osteosperma* (Torrey) Little, comb. et stat. nov., p. 125].

08480 Little, E. L., Jr. (1952). The genus *Pseudotsuga* (Douglas-fir) in North America. Leafl. W. Bot. 6: 181–198. [taxonomy, nomenclature, uses *P. taxifolia* (Lamb.) Britton for. *P. menziesii* (Mirb.) Franco, 1950].

08490 Little, E. L., Jr. (1953). A natural hybrid spruce in Alaska. J. Forest. (Washington) 51: 745–746. [*Picea × lutzii*, nothosp. nov].

08520 Little, E. L., Jr. (1962). Key to Mexican species of pines. Caribbean Forest. 23: 72–77. [*Pinus* spp.].

08530 Little, E. L., Jr. (1966). A new pinyon variety from Texas. Wrightia 3: 181–187. [*Pinus cembroides* var. *remota*].

08540 Little, E. L., Jr. (1966). Varietal transfers in *Cupressus* and *Chamaecyparis*. Madroño 18 (6): 161–167. [*Cupressus arizonica* var. *glabra* (Sudw.), p. 162, var. *montana* (Wiggins), p. 163, var. *nevadensis* (Abrams), p. 164 et var. *stephensonii* (C. B. Wolf), comb. et stat. nov., p. 164, *Chamaecyparis thyoides* var. *henryae* (H. L. Li) comb. et stat. nov., p. 165].

08550 Little, E. L., Jr. (1968). Two new pinyon varieties from Arizona. Phytologia 17: 329–342. [*Pinus edulis* var. *fallax*, p. 331, *P. cembroides* var. *bicolor*, p. 336].

08555 Little, E. L., Jr. (1968). *Pinus hartwegii* in Honduras. Phytologia 17 (6): 439–440.

08560 Little, E. L., Jr. (1970). Names of New World Cypresses (*Cupressus*). Phytologia 20 (7): 429–445. [*Cupressus goveniana* var. *abramsiana* (C. B. Wolf) comb. et stat. nov., p. 435, *C. guadalupensis* var. *forbesii* (Jepson) comb. et stat. nov., p. 435; nomenclature].

08590 Little, E. L., Jr. (1979). Four varietal transfers of United States trees. Phytologia 42 (3): 219–222. [*Pinus aristata* var. *longaeva* (D. K. Bailey) comb. et stat nov., p. 221].

08620 Little, E. L., Jr. (1981). Bald Cypress (*Taxodium distichum*) in Oklahoma, U.S.A. Proc. Oklahoma Acad. Sci. 60: 105–107.

08630 Little, E. L., Jr. & W. B. Critchfield (1969). Subdivisions of the Genus *Pinus*. (Pines). U.S. Forest Serv. Misc. Publ. 1144. Washington, D.C. [classification, synonymy with references,

table, maps, *Pinus* subgen. *Ducampopinus* (A. Chevalier) Y. de Ferré ex Little & Critchfield, p. 7, *P.* subsect. *Pineae*, subsect. nov., p. 12].

08640 Little, E. L., Jr. & K. W. Dorman (1952). Geographic differences in cone-opening in sand pine. J. Forest. (Washington) 50: 204–205. [*Pinus clausa*].

08650 Little, E. L., Jr. & K. W. Dorman (1952). Slash pine (*Pinus elliottii*), its nomenclature and varieties. J. Forest. (Washington) 50: 918–923. [*Pinus caribaea, P. elliottii* var. *densa* var. nov., p. 921, ill.].

08660 Little, E. L., Jr. & K. W. Dorman (1954). Slash pine (*Pinus elliottii*) including South Florida slash pine. Nomenclature and description. U.S. Forest Serv. Southeast Forest Exp. Stat. Pap. 36. (ii + 82 pp., ill.). [*Pinus elliottii* + var. *densa*].

08680 Little, E. L., Jr., S. Little & W. T. Doolittle (1967). Natural hybrids among pond, loblolly, and pitch pines. U.S. Forest Serv. Res. Pap. NE-67. (22 pp.). [*Pinus rigida, P. serotina, P. taeda*].

08690 Little, E. L., Jr. & S. S. Pauley (1958). A natural hybrid between black and white spruce in Minnesota. Amer. Midl. Naturalist 60: 202–211. (ill.). [*Picea mariana × P. glauca*].

08695 Little, S. (1950). Ecology and silviculture of white cedar and associated hardwoods in southern New Jersey. Bull. Yale Univ. School Forest. 56: 1–103. [*Chamaecyparis thyoides*].

08696 Little, S. (1951). Observations on the minor vegetation of the pine barren swamps in southern New Jersey. Bull. Torrey Bot. Club 78: 153–160. [*Chamaecyparis thyoides*].

08699 Litvinov, D. I. (1905). (*Pinus sylvestris* var. *mongolica* var. nov.) in: Spisok Rast. Gerb. Russk. Fl. Bot. Muz. Imp. Akad. Nauk (Schedae Herb. Fl. Ross. ...) 5: 160 (156–162). [not seen, diagn. repr. in Repert. Spec. Nov. Regni Veg. 4: 11, 1907].

08700 Litvinov, D. I. (1913). *Pinus coronans* sp. n. Gornyi sibirskii kedr (Mountain Siberian pine). Trudy Bot. Muz. Imp. Akad. Nauk 11: 20–26. (Russ.). [= *Pinus sibirica*].

08701 Liu, Q. X. (1988). [*Abies dayuanensis* sp. nov.) in: Bull. Bot. Res. North-East. Forest. Inst. 8 (3): 85. (Chin., Lat.).

08710 Liu, T. S. (1971). A Monograph of the Genus *Abies*. Nat. Taiwan Univ., Taipeh. (608 pp., ill., maps, tables). [*A. chensiensis* var. *ernestii* (Rehder), p. 135, *A. delavayi* var. *smithii* (Viguié et Gaussen), p. 143, *A. fargesii* var. *faxoniana* (Rehder et Wilson), p. 151, *A. sibirica* var. *semenovii* (B.A. Fedtsch.), p. 188, *A. cephalonica* var. *graeca* (Fraas), p. 222, *A. vejarii* var. *mexicana* (Martínez), p. 261, comb. et stat. nov.; hybrids: *A. × borisii-regis* Mattf. emend. Liu, *A. × phanerolepis* (Fern.) Liu, *A. × shastensis* Lemmon emend. Liu, *A. × umbellata* Mayr emend. Liu].

08720 Liu, T. S. (1982). A New Proposal for the Classification of the Genus *Picea*. Acta Phytotax. Geobot. 33: 227–244.

08730 Liu, T. S. & H. J. Su (1983). Biosystematic studies on *Taiwania* and numerical evaluations of the systematics of Taxodiaceae. Taipei, Taiwan Mus. Special Publ. Ser. 2. (Chin., ill.).

08741 Liu, Y. J. & C. S. Li (2000). On *Metasequoia* in Eocene Age from Liaoning Province of Northeast China. Acta Bot. Sin. 42 (8): 873–878. (Chin., Eng. summ., ill.).

08742 Liu, Y. J., C. S. Li & Y. F. Wang (1999). Studies on fossil *Metasequoia* from north-east China and their taxonomic implications. Bot. J. Linn. Soc. 130 (3): 267–297. [*Metasequoia glyptostroboides*, number of fossil taxa reduced to two recognized species: *M. milleri* Rothwell & Basinger and *M. occidentalis* (Newberry) Chaney, ill.].

08743 Liu, Y. S. & J. F. Basinger (2000). Fossil *Cathaya* (Pinaceae) pollen from the Canadian High Arctic. Int. J. Plant Sci. 161 (5): 829–847. [*Cathaya gaussenii* Sivak, *Pityosporites microalatus* (Potonié) Thomson & Pflug (form-taxon), Eocene, review of fossil *Cathaya* from Cretaceous to Pleistocene, ill. of SEM-TEM].

08744 Liu, Z. L., D. Zhang, X. Q. Wang, X. F. Ma & X. R. Wang (2003). Intragenomic and interspecific 5S rDNA sequence variation in five Asian pines. Amer. J. Bot. 90 (1): 17–24. [*Pinus densata* is a hybrid between *P. tabuliformis* and *P. yunnanensis*].

08745 Llorens, L. (1984).00 Notas floristicas Baleáricas. Fol. Bot. Misc. (Barcelona) 4: 55–58. [*Pinus ceciliae* A. & L. Llorens sp. nov., p. 55 = *P. halepensis*].

08750 Loder, E. G. (1919). Notes on *Taxodium* and *Glyptostrobus*. Gard. Chron., ser. 3, 66: 259, f. 118–124.

08760 Loder, E. G. (1920). *Picea jezoensis*. Gard. Chron., ser. 3, 67: 139, f. 56–57.

08770 Long, D. G. (1980). Notes relating to the Flora of Bhutan: IV, The weeping cypress, *Cupressus corneyana* Carrière. Notes Roy. Bot. Gard. Edinburgh 38 (2): 311–314. [= *C. cashmeriana* Carrière].

08771 Long, Y. H. & Y. Wu (1984). [*Metasequoia glyptostroboides* var. *caespitosa*, var. nov.] in: Bull. Bot. Res. North-East. Forest. Inst. 4 (1): 149.

08780 Looby, W. J. & J. Doyle (1937). Fertilization and proembryo formation in *Sequoia*. Sci. Proc. Roy. Dublin Soc. 21 (44): 457–476, 14 pl. [*Sequoia*, *Sequoiadendron*].

08790 Looby, W. J. & J. Doyle (1940). New observations on the life-history of *Callitris*. Sci. Proc. Roy. Dublin Soc. 22 (24): 241–255, ill.

08800 Looby, W. J. & J. Doyle (1942). Formation of gynospore, female gametophyte, and archegonia in *Sequoia*. Sci. Proc. Roy. Dublin Soc. 23 (5): 35–54, ill. [*Sequoia*, *Sequoiadendron*].

08810 Loock, E. E. M. (1950). The pines of Mexico and British Honduras. S. Africa Dept. Agric. Forest. Bull. 35: 1–244. [*Pinus* spp., *P. hondurensis* "sp. nov.", but see A. Sénéclauze, 1868].

08810c López Almirall, A. (1978). Valor taxonómico del número de agujas por fascículo en los pinos cubanos. Ciencias Biol. 2: 49–57. [*Pinus caribaea* var. *caribaea*, P. *cubensis*, P. *maestrensis*, P. *tropicalis*].

08811 Loret, H. & A. Barrandon (1876). Flore de Montpellier, comprenant l'analyse descriptive des Plantes Vasculaires de l'Hérault. 2 Vols. Montpellier – Paris. [*Juniperus heterocarpa* Timb.-Lagr. ex Loret & Barrandon, sp. nov., p. 610 = *J. oxycedrus* var. *brachyphylla* Loret in Billot: Annot. fl. France et Allem., p. 282, 1862].

08821 Lotsy, J. P. (1893). The formation of the so-called Cypress-knees on the roots of *Taxodium distichum*. Studies Biol. Lab., Baltimore 5: 269–277. (ill.).

08841 Lovric, A. (1971). Nouveautés de la flore halophile du Littorale croate. Österreich. Bot. Zeitschr. 119 (4–5): 567–571. [*Pinus nigra* ssp. *croatica* ssp. nov., p. 569].

08841c Lu, S. Y., Y. F. Li, Z. K. Chen & J. X. Lin (2003). Pollen development in *Picea asperata* Mast. Flora 198 (2): 112–117.

08842 Luchnik, Z. I. (1976). Izmenchivost' eli Siberskoi (*Picea obovata* Ledeb.) na Altae (Variabilitas Piceae obovatae Ledeb. in montibus Altaicis). Novosti Sist. Vyssh. Rast. (Novit. Syst. Pl. Vasc.) 13: 4–8. Inst. Bot. V. L. Komarov, Leningrad. (Russ., Lat.). [numerous var. nov. in *P. obovata*].

08850 Lückhoff, H. A. (1964). The natural distribution, growth and botanical variation of *Pinus caribaea* and its cultivation in South Africa. Ann. Univ. Stellenbosch 39, ser. A, 1: 1–153.

08860 Lückhoff, H. A. (1971). The Clanwilliam cedar (*Widdringtonia cedarbergensis* Marsh): its past history and present status. J. Bot. Soc. South Africa 57: 17–23.

08869 Lundell, C. L. (1937). Studies of Mexican and Central American plants II. Phytologia 1: 212–222. [*Podocarpus matudae* sp. nov. ("*matudai*"), p. 212].

08870 Lundell, C. L. (1940). Two new trees from the mountains of Mexico. Amer. Midl. Naturalist 23: 175–176. [*Abies tacanensis* sp. nov. = *A. guatemalensis* var. *tacanensis*].

08871a Ma, Q. W. & C. S. Li (2002). Epidermal structures of *Sequoia sempervirens* (D. Don) Endl. (Taxodiaceae). Taiwania 47 (3): 194–202. (ill.).

08872 Mabberley, D. J. (1982). William Roxburgh's "Botanical Description of a new species of Swietenia (Mahogany)" and other overlooked binomials in 36 vascular plant families. (Appendix). Taxon 31 (1): 72. [*Pinus pindrow* Royle ex D. Don in Brewster, Taylor & Phillips: Phil. Mag. & J. Sci. 8: 255, March 1836 = *Abies pindrow* (D. Don) Royle].

08874 Mabberley, D. J. (1998). Wilhelm Petermann and the Oriental Spruce. Thaiszia 8: 111–114. [*Picea orientalis* (L.) Peterm., Pflanzenreich: 235 (1838–45); Link in Linnaea 20: 294 (1847)].

08875 Mabberley, D. J. (2002). The *Agathis brownii* case (Araucariaceae). Telopea 9 (4): 743–754. [*Agathis* spp., nomenclature, typifications].

08876 Mabberley, D. J. (2002). The coming of the Kauris. Curtis's Bot. Mag. 19 (4): 252–264. [*Agathis* spp., discovery, early introductions to Europe, iconography].

08880 MacDonald, R. D. & M. E. MacDonald (1966). A new method for maintaining the cones of *Abies* and *Cedrus* intact for study and storage. Rhodora 68 (776): 516–517.

08886 Macphail, M., K. Hill, A. Partridge, E. Truswell & C. Foster (1995). 'Wollemi pine': old pollen records for a newly discovered genus of gymnosperm. Geology Today 11 (2): 48–50. (ill.) [Araucariaceae, *Wollemia nobilis*, palynology, *Dilwynites* fossil pollen type].

08900 Madrigal S., X. & M. Caballero D. (1969). Una nueva especie Mexicana de *Pinus*. Bol. Técn. Inst. Nac. Invest. Forest. México 26: 1–11. [*Pinus rzedowskii* sp. nov., p. 1].

08902 Madsen, K. (ed.). (1999). *Metasequoia* after fifty years. Arnoldia 58 (4) – 59 (1) (1998–99). [volume dedicated to *Metasequoia glyptostroboides*; reissues of papers and new papers, e.g. J. H. Li. *Metasequoia*: An overview of its phylogeny, reproductive biology, and ecotypic variation. pp. 54–59].

08910 Maeda, T. (1951). Sociological study of *Chamaecyparis obtusa* forest and its Japan Sea elements. Misc. Inform. Tokyo Univ. Forests 8: 21–44, pl. 1–3, f. 1–2. (Japan., Eng. summ.).

08920 Maekawa, F. (1948). Phyllotaxy of *Metasequoia*. J. Japan. Bot. 22 (3–4): 58–59. (Japan.).

08937 Magistris, A. A. de & M. A. Castro (1999). Anatomía de la plántula de tres especies de *Cupressus* (Cupressaceae). Darwiniana 37 (3–4): 199–207. [*Cupressus lusitanica*, *C. macrocarpa*, *C. sempervirens*, ill., seedlings].

08938 Magistris, A. A. de & M. A. Castro (2001). Bark anatomy of southern South American Cupressaceae. I.A.W.A. Journal 22 (4): 367–383. (ill.). [*Austrocedrus chilensis*, *Fitzroya cupressoides*, *Pilgerodendron uviferum*].

08940 Maheshwari, P. & C. Biswas (1970). *Cedrus*. Botanical Monograph 5. Council of Sci. & Ind. Res., New Delhi. [anatomy, embryology].

08950 Maheshwari, P. & R. N. Konar (1971). *Pinus*. Botanical Monograph 7. Council of Sci. & Ind. Res., New Delhi. [anatomy, embryology].

08962 Mai, D. H. (1986). Über Typen und Originale tertiärer Arten von *Pinus* L. (Pinaceae) in mitteleuropäischen Sammlungen: ein Beitrag zur Geschichte der Gattung in Europa. Fedde's Repert. 97 (9–10): 571–605. [palaeobotany, anatomy, morphology, ill., keys].

08962a Mai, D. H. & E. Velitzelos (1992). Über fossile Pinaceen-Reste im Jungtertiär von Griechenland. Fedde's Repert. 103 (1–2): 1–18. [*Abies resinosa*, *Cathaya bergeri*, *Cedrus vivariensis*, *Pinus* spp., *P. vegorae* sp. nov., p. 10, ill.].

08962d Mainieri, C. & J. M. Pires (1973). O gênero *Podocarpus* no Brasil. Silvicultura en S. Paulo. Revista Inst. Florestal 8. (report).

08963 Makino, T. (1901). Observations on the flora of Japan. Bot. Mag. (Tokyo) 15: 102–114. [*Thujopsis dolabrata* var. *hondai* var. nov., p. 104].

08970 Makino, T. (1926). *Chamaecyparis obtusa* Endl. var. *filicoides* Masters. J. Japan. Bot. 3 (9): 1 pl. [photogr. with brief Japan. text].

08980 Makino, T. (1928). *Cryptomeria japonica* D. Don. J. Japan. Bot. 5 (9): 1 pl. [photogr. with brief Japan. text].

08990 Malejeff, W. (1928). Schema einer natürlichen Klassifikation der *Cupressus*-Arten. Mitt. Deutsch. Dendrol. Ges. 40: 57–61.

09000 Malejeff, W. (1929). *Pinus pithyusa* (Stev.) und *Pinus eldarica* (Medw.), zwei relikt-Kiefern der taurisch-kaukasischen Flora. Mitt. Deutsch. Dendrol. Ges. 1929: 138–150.

09000a Malusa, J. (1992). Phylogeny and biogeography of the pinyon pines (*Pinus* subsect. *Cembroides*). Syst. Bot. 17 (1): 42–66.

09000b Mamaev, S. A. & P. P. Popov (1989). El' sibirskaja na Urale. Moskva. [*Picea obovata*].

09000c Manders, P. T. (1986). An assessment of the current status of the Clanwilliam Cedar (*Widdringtonia cedarbergensis*) and the reasons for its decline. S. African Forest. J. 139: 48–53.

09000d Manders, P. T., S. A. Botha, W. J. Bond & M. E. Meadows (1990). The enigmatic Clanwilliam Cedar. Veld & Flora 76: 8–11. [*Widdringtonia cedarbergensis*].

09001 Manko, J. I. (1976). On the northern border of distribution of *Abies sachalinensis* Fr. Schm. on the Sachalin. Bot. Žurn. (Moscow & Leningrad) 61 (3): 393–395. (Russ., Eng. summ.).

09010 Manley, S. A. M. (1972). The occurrence of hybrid swarms of red and black spruces in central New Brunswick. Canad. J. Forest Res. 2: 381–391. [*Picea rubens, P. mariana*].

09012 Manum, S. B. (1987). Mesozoic *Sciadopitys*-like leaves with observations on four species from the Jurassic of Andøya, northern Norway and emendation of *Sciadopityoides* Sveshnikova. Rev. Palaeobot. Palynol. 51: 145–168.

09013 Manum, S. B., J. H. A. van Konijnenburg-van Cittert & V. Wilde (2000). *Tritaenia* Mägdefrau et Rudolf, Mesozoic "*Sciadopitys*-like" leaves in mass accumulations. Rev. Palaeobot. Palynol. 109: 255–269. [Miroviaceae, Sciadopityaceae, Taxodiaceae (?), *Tritaenia crassa, T. linkii, T. scotica* comb. nov. (*Taxodiophyllum scoticum*)].

09014 Mao, Z. Z. (1989). *Cathaya argyrophylla*, an endemic tree of China, its resources, distribution and environment. Guihaia 9 (1): 1–11. (Chin., Eng. summ., map).

09016 Mapes, G. & G. W. Rothwell (1984). Permineralized ovulate cones of *Lebachia* from Late Paleozoic limestones of Kansas. Palaeontology 27 (1): 69–94. [*Lebachia lockardii* sp. nov. (= *Emporia lockardii* (Mapes & Rothwell) Mapes & Rothwell, 1991), walchian conifers, ill.].

09016a Mapes, G. & G. W. Rothwell (1988). Diversity among Hamilton conifers. In: G. Mapes & R. H. Mapes (eds.). Regional geology and palaeontology of [the] Upper Paleozoic Hamilton quarry area in southeastern Kansas. Guidebook 6, Kansas Geological Survey, Lawrence, Kansas, pp. 225–244. [*Walchia* (fossil "morpho-genus"), *Lebachia* (= *Utrechtia*), *Gomphostrobus* Marion, walchian conifers, ill.].

09018 Mapes, G. & G. W. Rothwell (1998). Primitive pollen cone structure in Upper Pennsylvanian (Stephanian) walchian conifers. J. Palaeontol. 72 (3): 571–576. [*Emporia, Walchia* (fossil "morpho-genus"), ill.].

09019 Mapes, G., G. W. Rothwell & M. T. Haworth (1989). Evolution of seed dormancy. Nature 337: 645–646. [seeds containing embryos in a walchian conifer cone of Permo-Carboniferous age are evidence of embryogeny similar to modern conifers].

09020 Marco, H. F. (1939). The anatomy of spruce needles. J. Agric. Res. 58: 357–368, pl. 1–7. [*Picea*].

09021 Marie-Victorin, Frère (1927). Les Gymnospermes du Quebec. Contr. Lab. Bot. Univ. Montréal 10: (i–xii) 1–147. [*Abies, Larix, Picea, Pinus, Tsuga, Juniperus, Thuya = Platycladus, Taxus*].

09021a Marion, A. F. (1884). Sur les chractères d'une conifère tertiaire voisine des Dammarées (*Doliostrobus sternbergii*). Compte Rendu Acad. Sci. 99: 821–823. [palaeobotany].

09022 Markgraf, F. (1931). Die Panzerkiefer. Mitt. Deutsch. Dendrol. Ges. 43: 250–254. [*Pinus heldreichii*, ill., map].

09030 Marloth, R. (1905). Eine neue Kap-Cypresse. Bot. Jahrb. Syst. 36: 206. [*Callitris schwarzii* sp. nov. = *Widdringtonia schwarzii*].

09031 Marschall von Bieberstein, F. A. (1800). Beschreibung der Länder zwischen den Flüssen Terek und Kur am Caspischen Meere. Mit einem botanischen Anhang. Frankfurt am Main. (211 pp.). [*Juniperus excelsa* sp. nov., p. 204; also in his Flora taurico-caucasica... Vol. 2: 425, 1808; *J. excelsa* Willd. (non M.-Bieb.) in Sp. pl. 4 (2): 852 = 1806].

09032 Marschall von Bieberstein, F. A. (1808). Flora taurico-caucasica exhibens stirpes phaenogamas, ... Vol. 2. Charkoviae (Kharkov). [*Juniperus oblonga* sp. nov., = *J. communis* f. *oblonga* (M.-Bieb.) B. K. Boom, p. 426].

09040 Marsh, J. A. (1966). Notes on *Widdringtonia* (Cupressaceae), in: New and Interesting Records of African Flowering Plants. Bothalia 9 (1): 124–126. [*Widdringtonia cedarbergensis* sp. nov., p. 125].

09050 Marsh, J. A. (1966). Cupressaceae. in: L. E. Codd, B. de Winter & H. B. Rycroft (eds.), Flora of Southern Africa Vol. 1, pp. 43–48. [*Widdringtonia* spp.].

09057 Martin, P. C. (1950). A morphological comparison of *Biota* and *Thuja*. Proc. Pennsylvania Acad. Sci. 24: 65–112. [*Biota orientalis = Platycladus orientalis, Thuja occidentalis*, ill.].

09059 Martínez M., F. J. & L. M. Aísa (1995). Estudio comparado de la posición de los canales resiníneros y de la biometría de las hojas de los pinsapos bético-rifeños (*Abies pinsapo* Boiss.). Transfretana Monografía 2. Estudios sobre el media natural de Ceuta y su entorno. pp. 115–130. Revista del Instituto de Estudios Ceutíes, Ceuta. [found no significant diff. between Spanish and Moroccan populations in leaf morphology and anatomy].

09060 Martínez, M. (1939). Nueva especie de *Abies*. Prot. Nat. 3 (9): 10–11. [descr. and fig. of *A. hickelii* Flous et Gaussen, new for Mexico].

09070   Martínez, M. (1940). Pinaceas Méxicanas: Descripción de algunas especies y variedades nuevas. Anales Inst. Biol. Univ. Nac. México 11 (1): 57–84, figs. [*P. oocarpa* var. *ochoterenai* et var. *manzanoi*, var. nov., pp. 65, 70, *Pinus herrerae* sp. nov. ("*herrerai*"), p. 76, *P. strobus* var. *chiapensis* var. nov., p. 81].

09080   Martínez, M. (1942). Una nueva pinacea Méxicana. Anales Inst. Biol. Univ. Nac. México 13 (1): 23–29. [*Pinus durangensis* sp. nov., p. 23, f. 1–4].

09090   Martínez, M. (1942). Una nueva Pinacea Méxicana. *Picea chihuahuana* sp. nov. Anales Inst. Biol. Univ. Nac. México 13 (1): 31–34.

09100   Martínez, M. (1942). Tres especies nuevas méxicanas del género *Abies*. Anales Inst. Biol. Univ. Nac. México 13 (2): 621–634. [*A. durangensis*, p. 621, *A. mexicana*, p. 626, *A. vejarii*, p. 629, spp. nov.].

09110   Martínez, M. (1943). Una nueva especie de pinus Méxicano. Madroño 7: 5–8. [*Pinus douglasiana* sp. nov.].

09120   Martínez, M. (1944). Una nueva especie del genero *Pinus*, *Pinus michoacana*. Anales Inst. Biol. Univ. Nac. México 15 (1): 1–6. [*Pinus michoacana* sp. nov., *P. michoacana* var. *cornuta* et var. *quevedoi*, var. nov. [= *P. devoniana* Lindl.), p. 1, f. 1–4].

09130   Martínez, M. (1944). Nuevas especies de *Juniperus*. Anales Inst. Biol. Univ. Nac. México 15 (1): 7–15. [*J. gamboana* sp. nov., p. 7, *Juniperus comitana* sp. nov., p. 12].

09135   Martínez, M. (1944). El *Pinus macrophylla* Engelm. y su variedad *blancoi*. Anales Inst. Biol. Univ. Nac. México 15 (2): 341–348. [*Pinus macrophylla* ("*macrofila*") var. *blancoi*, var. nov. = *P. engelmannii* var. *blancoi*, p. 345].

09150   Martínez, M. (1946). Los *Juniperus* méxicanos. Anales Inst. Biol. Univ. Nac. México 17 (1): 3–128. [*Juniperus deppeana* var. *pachyphlaea* (Torrey) comb. nov., p. 53, *J. patoniana* sp. nov. (see also T. A. Zanoni, 1978), p. 62, *J. jaliscana* sp. nov., p. 69, *J. blancoi* sp. nov., p. 73, *J. monticola* sp. nov., p. 79, *J. monticola* f. *compacta* f. nov., p. 87, *J. monticola* f. *orizbensis* f. nov., p. 91, *J. durangensis* sp. nov., pp. 94–95, *J. monosperma* var. *gracilis* var. nov., pp. 111–112].

09160   Martínez, M. (1946). Sobre la no existencia del ciprés, *Cupressus thurifera* [Kunth in] H. B. K. Ciencia 7 (4–6): 135. [= *Juniperus flaccida* var. *poblana*, var. nov.].

09170   Martínez, M. (1947). Los *Cupressus* de México. Anales Inst. Biol. Univ. Nac. México 18 (2): 71–149. [descr. repr. in: Bol. Soc. Bot. Méx. 6: 1–6, 1948].

09180   Martínez, M. (1948). Las *Abies* méxicanas. Anales Inst. Biol. Univ. Nac. México 19 (1/2): 11–104. [*A. oaxacana* sp. nov. = *A. hickelii* var. *oaxacana*, p. 39; div. var. et comb. nov., e.g. *A. guatemalensis* var. *jaliscana*, *A. vejarii* var. *macrocarpa*].

09190   Martínez, M. (1948). *Picea chihuahuana*. Anales Inst. Biol. Univ. Nac. México. 19 (2): 393–405. (ill.).

09200   Martínez, M. (1948). Los Pinos méxicanos. Ed. 2, Univ. of Mexico, Mexico-city. [*Pinus* spp., many ill., *P. pseudostrobus* var. *coatepecensis* var. nov., p. 194, *P. michoacana* var. *cornuta* "var. nov.", p. 260, *P. michoacana* var. *quevedoi* "var. nov.", p. 267 (see M. Martinez, 1944), *P. engelmannii* var. *blancoi*, comb. nov., p. 288, *P. arizonica* var. *stormiae* var. nov., p. 295, *P. oocarpa* var. *trifoliata* var. nov., p. 308–309, *P. patula* var. *longipedunculata* Loock ex Martínez, var. nov., p. 334].

09210   Martínez, M. (1949). Los Pseudotsugas de México. Anales Inst. Biol. Univ. Nac. México 20 (2): 129–184. [*Pseudotsuga*].

09220   Martínez, M. (1950). El Ahuehuete. (*Taxodium mucronatum* Ten.). Anales Inst. Biol. Univ. Nac. México 21 (1): 25–82.

09230   Martínez, M. (1961). Una nueva especie de *Picea* en México. Anales Inst. Biol. Univ. Nac. México 32 (2): 137–142. [*P. mexicana* sp. nov. = *P. engelmannii* ssp. *mexicana* (Martínez) P. Schmidt, p. 137].

09241   Masamune, G. (1930). A contribution of the phytogeography of the Island of Yakushima [Kagoshima-ken]. I. Bot. Mag. (Tokyo) 44: 43–52. (Japan.). [*Juniperus tsukusiensis* sp. nov., p. 50].

09250   Masamune, G. (1949). On *Chamaecyparis taiwanensis* Masamune et Suzuki. Bull. Tokyo Univ. Forest 37: 135–140, pl. 1–2. (Japan., Eng. summ.). [= *C. obtusa* var. *formosana*, *C. taiwanensis* not mentioned in Index Kewensis].

09251   Masamune, G. & S. Suzuki (1933). [*Chamaecyparis taiwanensis* sp. nov.) in: Sylvia 4: 57, ill. [= *C. obtusa* var. *formosana*, not seen, *C. taiwanensis* not in Index Kewensis].

09260   Masamune, G. & S. Suzuki (1934). Vergleichende Anatomie der Blätter der Kiefer in Taiwan. Trans. Nat. Hist. Soc. Taiwan 24 (135): 391–396. (Japan., Germ. summ., figs.). [*Pinus armandii*, *P. formosana*, *P. massoniana*, *P. taiwanensis*].

09261   Mason, F. (1849). The Pine tree of the Tenasserim provinces. J. Asiat. Soc. Bengal Sci. 18: 73–75. [*Pinus latteri* sp. nov., p. 74].

09280   Mason, H. L. (1930). The Santa Cruz Island pine. Madroño 2: 8–10. [*Pinus remorata* sp. nov., p. 9; see also J. T. Howell, 1941].

09290   Mason, H. L. (1932). A phylogenetic series of the California closed-cone pines suggested by the fossil record. Madroño 2: 49–55. [*Pinus masonii* (fossil), *P. linguiformis* (fossil), *P. muricata*, *P. remorata*, *P. radiata*, *P. attenuata*].

09300   Mason, H. L. & W. P. Stockwell (1945). A new pine from Mount Rose, Nevada. Madroño 8: 61–62. [*Pinus washoensis* sp. nov.].

09310   Masters, M. T. (1878). *Thuja standishii*. Gard. Chron., n. s., 10: 397. [*Thuja standishii* descr.].

09330   Masters, M. T. (1880). *Picea ajanensis*. Gard. Chron., n. s., 14: 427–428. (ill.). [= *P. brachytyla*].

09350   Masters, M. T. (1882). Proliferous cones. Gard. Chron., n. s., 17: 112–113. [ill., e.g. *Sciadopitys*].

09360   Masters, M. T. (1882). *Pinus bungeana*. Gard. Chron., n. s., 18: 8. (ill.).

09370   Masters, M. T. (1882). *Pinus ayacahuite*. Gard. Chron., n. s., 18: 492–493. (ill.).

09380   Masters, M. T. (1882). *Thuja* (§ *Thuiopsis*) *dolabrata*. Gard. Chron., n. s., 18: 556. [*Thujopsis dolabrata*].

09390   Masters, M. T. (1883). The Larches. Gard. Chron., n. s., 19: 88. (ill.). [*Larix kaempferi*, *L. leptolepis*].

09400   Masters, M. T. (1884). On the comparative morphology of *Sciadopitys*. J. Bot. 22: 97–105.

09410   Masters, M. T. (1884). The Arboretum *Picea omorika*. Gard. Chron., n. s., 21: 308–309, figs. 56–58.

09420   Masters, M. T. (1884). *Abies fortunei*. Gard. Chron., n. s., 21: 348. (ill.). [= *Keteleeria fortunei*].

09430   Masters, M. T. (1884). *Pseudolarix kaempferi* (The golden larch). Gard. Chron., n. s., 21: 584. (ill.); 22: 238. (ill.). [= *Pseudolarix amabilis*].

09450   Masters, M. T. (1887). *Abies davidiana*. Gard. Chron., ser. 3, 1: 481. [= *Keteleeria davidiana*].

09480   Masters, M. T. (1895). The Guadeloupe (Guadelupe) cypress. Gard. Chron., ser. 3, 18: 62. [*Cupressus macrocarpa* var. *guadeloupensis* (for "*guadalupensis*") (S. Watson) Masters, comb. et stat. nov.].

09490   Masters, M. T. (1896). A general view of the genus *Cupressus*. J. Linn. Soc., Bot. 31 (216): 312–363, f. 1–29.

09500   Masters, M. T. (1897). The species of *Thuja*. Gard. Chron., ser. 3, 21: 213–214, 258. (ill.). [*Thuja* spp., nomenclature].

09510   Masters, M. T. (1898). De coniferis quibusdam sinicis vel japonicis adnotationes quedam porrigit. Bull. Herb. Boissier 6: 269–274. [*Pinus koraiensis et al.*, *Pinus scipioniformis* sp. nov., p. 270, *Cephalotaxus oliveri* sp. nov., p. 270; coll. conif. A. Faurie in Japan].

09515   Masters, M. T. (1899). Bermuda juniper and its allies. J. Bot. 37: 1–11. [*Juniperus bermudiana*, *J. virginiana*, *J. barbadensis*].

09520   Masters, M. T. (1900). *Taxodium* and *Glyptostrobus*. J. Bot. 38: 37–40.

09560   Masters, M. T. (1904). A General view of the genus *Pinus*. J. Linn. Soc., Bot. 35: 560–659, pl. 20–23.

09570   Masters, M. T. (1905). Notes on the genus *Widdringtonia*. J. Linn. Soc., Bot. 37: 267–274. [*W. schwarzii* (Marloth) comb. nov., p. 269, *W. mahoni* sp. nov. = *W. nodiflora*, p. 271, *W. equisetiformis* sp. nov. = *Callitris preissii*, p. 271, see also M. T. Masters, 1906].

09590   Masters, M. T. (1906). Correction of *Widdringtonia equisetiformis* to *Callitris robusta*. J. Linn. Soc., Bot. 37: 332. [= *C. columellaris*].

09610   Masters, M. T. (1906). *Abies mariesii*. Bot. Mag. 132: pl. 8098.

09612   Masters, M. T. (1907). *Abies magnifica* var. *xanthocarpa*. Gard. Chron., ser. 3, 41: 114–115.

09630   Mastrogiuseppe, R. J. & J. D. Mastrogiuseppe (1980). A study of *Pinus balfouriana* Grev. & Balf. (Pinaceae). Syst. Bot. 5: 86–104. [*P. balfouriana* ssp. *austrina*, ssp. nov.].

09636   Matos, J. A. (1995). *Pinus hartwegii* and *P. rudis*: A critical assessment. Syst. Bot. 20 (1): 6–21. [tables, figs., concludes synonymy].

09641   Mattei, G.E. (1908). (*Abies nebrodensis*) in: Boll. Reale Orto Bot. Giardino Colon. Palermo 7: 59–69. [*Abies nebrodensis* (Lojac.) comb. et stat. nov. [basion. : *A. pectinata* var. *nebrodensis* Lojac. in Flora Sicula, Vol. II, part 2: 401, 1907), p. 64].

09650   Mattfeld, J. (1925). Die in Europa und dem Mittelmeergebiet wildwachsenden Tannen. Mitt. Deutsch. Dendrol. Ges. 35: 1–37, f. 1–10. [*Abies* spp., *A. bornmuelleriana* "sp. nov." = *A. nordmanniana* ssp. *equi-trojani*, p. 24, *A. borisii-regis* "sp. nov.", p. 26, *A. equi-trojani* (Aschers. et Sint. ex Boiss.) comb. et stat. nov., p. 29; the first 2 spp. were earlier publ. in Notizbl. Bot. Gart. Berlin-Dahlem 9, 1925, p. 239, resp. p. 235].

09651   Mattfeld, J. (1926). Das Areal der Weiss-Tanne. Mitt. Deutsch. Dendrol. Ges. 37: 15–35. [*Abies alba*, map].

09660   Mattfeld, J. (1926). Die europäischen und mediterranen *Abies*-Arten. Pflanzenareale 1 (2): 22–29, maps 14–16. [fossils in map 16].

09661   Mattfeld, J. (1930). Über hybridogene Sippen der Tannen, nachgewiesen an den Formen der Balkanhalbinsel, zugleich ein Beitrag zur Waldgeschichte der Balkanhalbinsel. Stuttgart. [*Abies cephalonica*, *A. borisii-regis*, ill. of leaf apices, forms named].

09670   Mattoon, W. R. (1915). Life history of the shortleaf pine. U.S.D.A. Bull. (1915–23) 244: 1–46, pl. 1–10. [*Pinus echinata*].

09672   Mattos, J. R. de (1997). Uma variedade nova de *Araucaria angustifolia*. Loefgrenia 111: 1. [*A. angustifolia* var. *vinacea*].

09680   Matzenko, A. E. (1957). Abieties geronotogeae clavis analytica. Bot. Mater. Gerb. Glavn. Bot. Sada SSSR 18: 311–315. [*Abies*, key].

09690 Matzenko, A. E. (1963). Conspectus generis *Abies*; Observations on the genus *Abies* Miller. Bot. Mater. Gerb. Bot. Inst. Komarova Akad. Nauk SSSR 22: 33–42. (Russ.).

09700 Matzenko, A. E. (1964). The firs of the Eastern Hemisphere. Trudy Bot. Inst. Akad. Nauk. SSSR, ser. 1, 13: 1–103. Inst. Bot. V. L. Komarov, Leningrad. (Russ., Lat.). [*Abies*].

09710 Matzenko, A. E. (1968). Series novae generis *Abies* Mill. Novosti Sist. Vysših. Rast. (Novit. Syst. Pl. Vasc.) 6: 9–12. Inst. Bot. V. L. Komarov, Leningrad. (Russ., Lat.).

09720 Matzenko, A. E. (1971). Lectotypi specierum nonnulis Juniperi L. ex Asia centrali. Novosti Sist. Vysših. Rast. (Novit. Syst. Pl. Vasc.) 7: 41–42. Inst. Bot. V. L. Komarov, Leningrad. (Russ., Lat.). [*Juniperus tibetica* Komarov, *J. przewalskii* Komarov].

09721 Maximowicz, C. J. (1859). Primitiae florae Amurensis. Mem. Acad. Imp. Sci. Saint.-Pétersbourg (Sav. Etr.) 9. [*Abies sibirica* var. *nephrolepis* Trautv. ex Maxim., var. nov. = *A. nephrolepis*, p. 206; see also C. J. Maximowicz, 1866].

09730 Maximowicz, C. J. (1866). Diagnoses breves plantarum novarum Japoniae et Mandschuriae. Bull. Acad. Imp. Sci. Saint.-Pétersbourg 10: 485–490. [*Abies nephrolepis* (Trautv. ex Maxim.) comb. et stat. nov., p. 486, *A. holophylla* sp. nov., p. 487, *A. brachyphylla* sp. nov. = *A. homolepis*, p. 488, *A. bicolor* sp. nov., *Picea japonica* pro syn. = *Picea alcockiana*, p. 488., *Thuja japonica* sp. nov. = *T. standhisii*, p. 490].

09731 Maximowicz, C. J. (1868). Diagnoses breves plantarum novarum Japoniae et Mandschuriae. Bull. Acad. Imp. Sci. Saint.-Pétersbourg 12: 225–231. [*Abies diversifolia* sp. nov. = *Tsuga diversifolia*, p. 229, *Juniperus nipponica* sp. nov. = *J. rigida* ssp. *nipponica*, p. 230].

09760 Mayr, H. (1894). Die Unterschiede zwischen der Hondo-Fichte (*Picea hondoensis*) und der Ajans-Fichte (*Picea ajanensis*). Mitt. Deutsch. Dendrol. Ges. 3: 30–32. [*P. jezoensis* + ssp. *hondoensis*].

09770 Mayr, H. (1894). Über die Kiefern des japanischen Reiches. Bot. Centralbl. 58: 148–151. [*Pinus sinensis* = *P. tabuliformis*, *P. thunbergii*, *P. luchuensis* sp. nov., p. 149].

09771 Mayr, H. (1901). Kleinere Mitteilungen über Coniferen. Mitt. Deutsch. Dendrol. Ges. 10: 318–319 (pag. in Ed. 2). [*Pseudotsuga glauca* (Beissn.) comb. et stat. nov. = *P. menziesii* var. *glauca*, *P. macrocarpa*, nomenclature].

09790 Mayr, H. (1910). Neue Arten aus: Fremdländische Wald- und Parkbäume Europas. Repert. Spec. Nov. Regni Veg. 8: 90–92. [diagnoses, first publ. in H. Mayr, 1906].

09792 McCarter, P. S. & J. S. Birks (1985). *Pinus patula* subspecies *tecunumanii*: the application of numerical techniques to some problems of its taxonomy. Commonw. Forest. Rev. 64 (2): 117–132.

09800 McCune, B. (1988). Ecological diversity in North American pines. Amer. J. Bot. 75: 353–368. (ill.). [*Pinus* spp.].

09800a McIver, E. E. (1992). Fossil *Fokienia* (Cupressaceae) from the Paleocene of Alberta, Canada. Can. J. Bot. 70: 742–749. (ill.).

09800c McIver, E. E. (1994). An early *Chamaecyparis* (Cupressaceae) from the Late Cretaceous of Vancouver Island, British Columbia, Canada. Canad. J. Bot. 72: 1787–1796. (ill.). [*Chamaecyparis corpulenta* (Bell) comb. nov., *Cupressinocladus interruptus*].

09800d McIver, E. E. (2001). Cretaceous *Widdringtonia* Endl. (Cupressaceae) from North America. Int. J. Plant Sci. 162 (4): 937–961. (ill.). [Cheirolepidiaceae, Cupressaceae s.s., *Widdringtonia* Endl., *W. americana* McIver, sp. nov., p. 941, *Widdringtonites* Endl., *W. subtilis* Heer sensu Berry (1919), *Elatocladus* Halle, *Brachyphyllum* Brongn., evolution, palaeobotany, western Laurasian origin of genus].

09800g McIver, E. E. & K. R. Aulenback (1994). Morphology and relationships of *Mesocyparis umbonata* sp. nov.: fossil Cupressaceae from the Late Cretaceous of Alberta, Canada. Canad. J. Bot. 72 (2): 273–295. (ill.: LM, SEM). [*Chamaecyparis*, *Thuja*, *Mesocyparis borealis*, *M. umbonata* sp. nov., p. 275, *M. beringiana* (Golovneva) comb. nov., p. 276].

09800k McIver, E. E. & J. F. Basinger (1987). *Mesocyparis borealis* gen. et sp. nov.: fossil Cupressaceae from the early Tertiary of Saskatchewan, Canada. Canad. J. Bot. 65 (11): 2338–2351. [comparison with *Chamaecyparis*, *Thuja*, ill.: LM, SEM].

09800l McIver, E. E. & J. F. Basinger (1989). The morphology and relationships of *Thuja polaris* sp. nov. (Cupressaceae) from the early Tertiary, Ellesmere Island, Arctic Canada. Canad. J. Bot. 67: 1903–1915. (ill.).

09800m McIver, E. E. & J. F. Basinger (1990). Fossil seed cones of *Fokienia* (Cupressaceae) from the Paleocene Ravenscrag formation of Saskatchewan, Canada. Canad. J. Bot. 68: 1609–1618. [*F. ravenscragensis* fossil sp. nov., p. 1610].

09800n McIver, E. E. & J. F. Basinger (1999). Early Tertiary floral evolution in the Canadian High Arctic. Ann. Missouri Bot. Gard. 86: 523–545. (ill.). [Coniferales, angiosperms, palaeoflora of Palaeocene-Eocene, *Chamaecyparis*, *Glyptostrobus*, *Larix*, *Metasequoia*, *Picea*, *Pinus*, *Pseudolarix*, *Taiwania*, *Tsuga*].

09800p McKown, A. D., R. A. Stockey & C. E. Schweger (2002). A new species of *Pinus* subgenus *Pinus* subsection *Contortae* from Pliocene sediments of Ch'ijee's Bluff, Yukon Territory, Canada. Int. J. Plant Sci. 163 (4): 687–697. (ill.). [*Pinus contorta, P. matthewsii* sp. nov. (fossil), palaeobotany].

09801 McNab, W. R. (1876). Notes on the synonymy of certain species of *Abies*. Trans. Bot. Soc. Edinburgh 12: 503–506.

09802 McNab, W. R. (1877). Remarks on the structure of the leaves of certain Coniferae. Proc. Roy. Irish Acad., ser. 2, 2: 209–213. [e.g. *Abies hookeriana, A. pattoniana = Tsuga mertensiana* s.l.)

09810 McNab, W. R. (1877). A revision of the species of *Abies*. Proc. Roy. Irish Acad., ser. 2, 2: 673–704.

09811 McNab, W. R. (1882). Note on *Abies pattonii*, Jeffrey MSS, 1851. J. Linn. Soc., Bot. 19 (120): 208–212. [*Tsuga mertensiana, T. mertensiana* ssp. *grandicona*, as *A. pattoniana, A. hookeriana*].

09820 McWilliam, J. R. (1958). The role of the micropyle in the pollination mechanism of *Pinus*. Bot. Gaz. (Crawfordsville) 120: 109–117. (ill.).

09830 McWilliam, J. R. (1959). Interspecific incompatibility in *Pinus*. Amer. J. Bot. 46: 425–433.

09840 Medley, M. E. & B. E. Wofford (1980). *Thuja occidentalis* L. and other noteworthy collections from the Big South Fork of the Cumberland River in McCreary County, Kentucky. Castanea 45 (3): 213–215.

09841 Medwedew, Y. S. (1903). [*Pinus eldarica* sp. nov.] in: Trudy Tiflissk. Bot. Sada (Acta Hort. Tiflis) 6 (2): 21, fig.

09842 Medwedew, Y. S. (1903). (*Juniperus foetidissima* Willd. var. *squarrosa*, var. nov.) in: Trudy Bot. Sada Imp. Jur'evsk. Univ. 3: 229. (Acta Hort. Bot. Jurjev. 3: 229 (whole art. pp. 227–330) 1903). [diagn. repr. in Repert. Spec. Nov. Regni Veg. 2 (14–15): 136, 1906].

09850 Mehra, P. N. (1976). Conifers of the Himalayas with particular reference to the *Abies* and *Juniperus* complexes. Nucleus 19 (2): 123–139.

09858 Mehra, P. N. & P. D. Dogra (1965). Normal and abnormal pollen grains in *Abies pindrow* (Royle) Spach. Palynol. Bull. 1: 16–23. (ill.).

09867 Meijer, J. J. F. (2000). Fossil woods from the Late Cretaceous Aachen Formation. Rev. Palaeobot. Palynol. 112: 297–336. (ill.). [Taxodiaceae, *Taxodioxylon gypsaceum, T. albertense, Dammaroxylon aachenense* sp. nov., *Pinuxylon* sp., palaeobotany, wood anatomy].

09875 Mejnartowicz, L. & F. Bergmann (1975). Genetic studies on European larch (*Larix decidua* Mill.) employing isoenzyme polymorphisms. Genet. Polon. 16: 29–34.

09890 Melle, P. J. van (1946). The junipers commonly included in *Juniperus chinensis*. Phytologia 2 (6): 185–195. [*J. sphaerica = J. chinensis, J. sheppardii* (Veitch) comb. nov., p. 187, × *J. media* hybr. nov., p. 189, nomenclature, taxonomy based on cultivated plants, see also P. J. van Melle, 1947].

09900 Melle, P. J. van (1947). Review of *Juniperus chinensis et al.* New York Bot. Gard., pp. 1–108, pl. 1–12. New York. [deals primarily with cultivated plants, see e.g. P. J. van Melle, 1946].

09910 Melle, P. J. van (1952). *Juniperus texensis* sp. nov. West-Texas juniper in relation to *J. monosperma, J. ashei et al.* Phytologia 4: 26–35. [*J. texensis* sp. nov., p. 26].

09920 Melville, R. (1948). *Abies koreana*. Bot. Mag. 165: pl. 40, 1 f.

09930 Melville, R. (1949). *Abies mariesii*. Bot. Mag. 166: pl. 45, 1 f.

09940 Melville, R. (1950). *Picea wilsonii*. Bot. Mag. 167: pl. 107, 1 f.

09950 Melville, R. (1951). *Larix gmelini* var. *japonica*. Bot. Mag. 168: pl. 159, 1 f.

09960 Melville, R. (1959). Notes on gymnosperm nomenclature. Bull. Misc. Inf. R.B.G. Kew 1958: 531–535. [*Abies delavayi* var. *georgei* (Orr) comb. et stat. nov., p. 533, *A. grandis, A. amabilis, Juniperus recurva* var. *coxii* (A. B. Jackson) comb. et stat. nov., p. 533, *Pinus nigra* var. *maritima* (Ait.) comb. nov., p. 534].

09965 Meng, X. Y., F. Chen & S. H. Deng (1988). Fossil plant *Cunninghamia asiatica* (Krassilov) comb. nov. Acta Bot. Sinica 30 (6): 649–654. (Chinese, Eng. summ.). [*Elatides asiatica* Krassilov, Early Cretaceous, compared with *Cunninghamia lanceolata*].

09970 Mergen, F. (1963). Ecotypic variation in *Pinus strobus* L. Ecology 44: 716–727, (ill.).

09975 Mergen, F. (1976). Microsporogenesis and macrosporogenesis in *Pseudolarix amabilis*. Silvae Genet. 25: 183–188.

09980 Mergen, R. & J. Burley (1964). *Abies* karyotype analysis. Silvae Genet. 13: 63–68. [*A. alba, A. cephalonica, A. × borisii-regis, A. lasiocarpa, A. firma, A. procera, A. guatemalensis*, all n = 12].

09990 Merkle, M. & K. Napp-Zinn (1977). Anatomische Untersuchungen an Pinaceen-Deckschuppen I. *Abies koreana* E. H. Wilson. Bot. Jahrb. Syst. 97 (4): 475–502.

10000 Merkle, M. & K. Napp-Zinn (1977). Anatomische Untersuchungen an Pinaceen-Deckschuppen II. *Tsuga canadensis* (L.) Carr. var. *aurea* Beissn. und var. *parvifolia* (P. Smith) Beissn. Bot. Jahrb. Syst. 98 (4): 549–572.

10001 Merriam, C. H. (1896). A new Fir from Arizona, *Abies arizonica*. Proc. Biol. Soc. Washington 10: 115–118, figs. 24–25. [= *A. lasiocarpa* var. *arizonica*].

10001a   Merrill, E. D. (1907). The flora of Mount Halcon, Mindoro. Philipp. J. Sci., C. Botany 2: 251–309. [*Podocarpus glaucus* Foxw. sp. nov., p. 258, *P. pilgeri* Foxw. sp. nov., p. 259].

10002   Merrill, E. D. (1922). Notes on the flora of southeastern China. Philipp. J. Sci. 21: 491–512. [*Fokienia maclurei* sp. nov. = *F. hodginsii*, p. 492].

10011   Merrill, E. D. (1948). *Metasequoia*, another "living fossil". Arnoldia 8 (1): 1–8. (ill.).

10021   Messeri, A. (1958). Nuovi dati sulla sistematica dell' *Abies nebrodensis* (Lojac.) Mattei. Atti Reale Accad. Naz. Lincei, Mem. Cl. Sci. Fis., Sez. 3a, Bot., ser. 8, 25: 547–556.

10030   Metcalf, F. P. (1934). "Shan" or Foochow pine. Lingnan Agric. J. 1: 71–77, f. 1–2. (Chin., Eng.). [*Cunninghamia lanceolata*].

10050   Metcalfe, C. R. (1931). The wood structure of *Fokienia hodginsii* and certain related Coniferae. Bull. Misc. Inform. Roy. Bot. Gard. Kew 1931: 420–424, f. 1–6. [*Calocedrus decurrens, C. macrolepis, Cupressus sempervirens, Fokienia hodginsii, Thuja plicata, Thujopsis dolabrata*].

10085   Miehe, S., G. Miehe, J. Huang, T. Otsu, T. Tuntsu & Y. Tu (2000). Sacred forests of South-Central Xizang and their importance for restauration of forest resources. (Contributions to ecology, phytogeography and environmental history of High Asia. 2). Marburg. Geogr. Schr. 135: 228–249. [*Juniperus convallium, J. tibetica, Cupressus gigantea*, ecology and phytogeography in S Tibet].

10090   Miki, S. (1941). On the change of flora in Eastern Asia since Tertiary period (I). The clay or lignite beds' flora of Japan, with special reference to the *Pinus trifolia* beds in Central Hondo. J. Japan. Bot. 11: 237–303. [with fossil *Metasequoia disticha* (Heer) Miki, p. 262].

10091   Miki, S. (1950). Taxodiaceae of Japan, with special reference to its remains. J. Inst. Polytechn. Osaka City Univ., ser. D., Biol. 1: 63–77, fig. 1–4. [*Cryptomeria, Sciadopitys*, map, mainly palaeobotany].

10092   Miki, S. (1953). On *Metasequoia*, Fossil and Living. Osaka Univ. Press, Osaka. (pp. 1–142). (Japan.). [*Metasequoia, Sequoia*, ill., maps].

10100   Miki, S. (1954). The occurrence of the remains of *Taiwania* and *Palaeotsuga* (n. subg.) from Pliocene beds in Japan. Proc. Imp. Acad. Japan. 30: 976–981. [palaeobotany, *Palaeotsuga* subg. nov., p. 977 = *Nothotsuga longibracteata* (Cheng) C. N. Page].

10120   Miki, S. (1958). Gymnosperms in Japan, with special reference to the remains. J. Inst. Polytechn. Osaka City Univ., ser. D., Biol. 9: 125–156, pl. 1–3, f. 1–6. [Coniferales, *Thuja protojaponica*, fossil sp. nov., map, mainly palaeobotany].

10121   Miki, S. (1969). *Protosequoia* (n. g.) in Taxodiaceae from *Pinus trifolia* beds in Central Honshu, Japan. Proc. Imp. Acad. Japan. 45 (8): 727–732. [palaeobotany]

10122   Mildenhall, D. C. & M. R. Johnston (1971). A megastrobilus belonging to the genus *Araucarites* from the Upper Motuan (Upper Albian), Wairavapa, North Island, New Zealand. New Zealand J. Bot. 9: 67–79. [Araucariaceae, *Araucarites* Presl, palaeobotnany].

10124   Mill., R. R. (1999). A new species of *Larix* (Pinaceae) from southeast Tibet and other nomenclatural notes on Chinese *Larix*. Novon 9 (1): 79–82. [*Larix kongboensis* sp. nov.].

10124b   Mill, R. R., M. Möller, F. Christie, S. M. Glidewell, D. Masson & B. Williamson (2001). Morphology, anatomy and ontogeny of female cones in *Acmopyle pancheri* (Brogn. & Gris) Pilg. (Podocarpaceae). Ann. Bot. (London) 88: 55–67. (SEM and other ill.).

10124c   Mill, R. R., M. Möller, S. M. Glidewell, D. Masson & B. Williamson (2004). Comparative anatomy and morphology of fertile complexes of *Prumnopitys* and *Afrocarpus* (Podocarpaceae) as revealed by histology and NMR imaging, and their relevance to systematics. Bot. J. Linn. Soc. 145 (3): 295–316. (colour ill. and drawings).

10125   Mill, R. R. & C. J. Quinn (2002). *Prumnopitys andina* reinstated as the correct name for "lleuque", the Chilean conifer recently renamed *P. spicata* (Podocarpaceae). Taxon 50 (4): 1143–1154.

10126   Mill, R. R. & P. Thomas (1999). *Falcatifolium taxoides* (Podocarpaceae). Curtis's Bot. Mag. 16 (3): 199–211. (ill.).

10128   Millar, C. I. (1983). A steep cline in *Pinus muricata*. Evolution 37: 311–319.

10130   Millar, C. I. (1986). Bishop pine (*Pinus muricata*) of inland Marin County, California. Madroño 33 (2): 123–129.

10140   Millar, C. I. (1986). The Californian closed cone pines (subsection *Oocarpae* Little and Critchfield): a taxonomic history and review. Taxon 35: 657–670. [*Pinus radiata, P. muricata*, (syn. *P. remorata*), *P. attenuata*].

10145   Millar, C. I. (1993). Impact of the Eocene on the evolution of *Pinus*. Ann. Missouri Bot. Gard. 80: 471–498.

10146   Millar, C. I. (1998). Early evolution of pines. In: D. M. Richardson (ed.). Ecology and biogeography of *Pinus*. (pp. 69–91, maps). Cambridge University Press, Cambridge, UK. [*Pinus*, palaeobotanical records from Cretaceous and Tertiary, theory of Eocene refugia at extreme palaeolatitudes].

10147   Millar, C. I. (1998). Reconsidering the conservation of Monterey pine. Fremontia 26 (3): 12–16. [*Pinus radiata*].

10148 Millar, C. I. (1999). Evolution and biogeography of *Pinus radiata*, with a proposed revision of its Quarternary history. New Zealand J. Forest. Sci. 29 (3): 335–365.

10150 Millar, C. I. & W. B. Critchfield (1988). Crossability and relationships of *Pinus muricata* (Pinaceae). Madroño 35 (1): 39–53, f. 1–2.

10160 Millar, C. I., S. H. Strauss, M. T. Conkle & R. D. Westfall (1988). Allozyme differentiation and biosystematics of the Californian closed-cone pines (subsect. Oocarpae). Syst. Bot. 13: 351–370.

10170 Miller, C. N., Jr. (1970). *Picea diettertiana*, a new species of petrified cones from the Oligocene of western Montana. Amer. J. Bot. 57: 579–585. (ill.).

10175 Miller, C. N., Jr. (1975). Petrified cones and needle bearing twigs of a new taxodiaceous conifer from the Early Cretaceous of California. Amer. J. Bot. 62: 706–713. (ill.).

10210 Miller, C. N., Jr. (1985). *Pityostrobus pubescens*, a new species of pinaceous cones from the Late Cretaceous of New Jersey. Amer. J. Bot. 72: 520–529. [Pinaceae, palaeobotany].

10212 Miller, C. N., Jr. (1986). Seed cones of *Pinus* from the Late Cretaceous of New Jersey. Rev. Palaeobot. Palynol. 46: 257–272. (ill.).

10212a Miller, C. N., Jr. (1989). A new species of *Picea* based on silicified seed cones from the Oligocene of Washington. Amer. J. Bot. 76: 747–754. [*Picea eichornii* fossil sp. nov.].

10212b Miller, C. N., Jr. (1990). Stems and leaves of *Cunninghamiostrobus goedertii* from the Oligocene of Washington. Amer. J. Bot. 77: 963–971. [Cupressaceae/Taxodiaceae, palaeobotany].

10213 Miller, C. N., Jr. (1992). Silicified *Pinus* remains from the Miocene of Washington. Amer. J. Bot. 79 (7): 754–760. [*Pinus foisyi* fossil sp. nov., p. 755].

10216 Miller, C. N., Jr. & D. R. Crabtree (1989). A new taxodiaceous seed cone from the Oligocene of Washinton. Amer. J. Bot. 76: 133–142. [*Cunninghamia, Cunninghamiostrobus goedertii* sp. nov., Taxodiaceae].

10217 Miller, C. N., Jr. & C. A. LaPasha (1983). Structure and affinities of *Athrotaxites berryi* Bell, an Early Cretaceous conifer. Amer. J. Bot. 70: 772–779. [ill., resembles *Athrotaxis cupressoides*].

10217a Miller, C. N., Jr. & C. A. LaPasha (1984). Flora of the Early Cretaceous Kootenai Formation in Montana, conifers. Palaeontographica, B 193: 1–17. [Coniferales, *Athrotaxites berryi, Elatides curvifolia, Conites* sp., *Elatocladus dunnii* sp. nov., *E. montanensis* sp. nov., *Masculostrobus montanensis* sp. nov., ill.].

10218 Miller, C. N., Jr. & J. M. Malinky (1986). Seed cones of *Pinus* from the late Cretaceous of New Jersey, U.S.A. Rev. Palaeobot. Palynol. 46: 257–272. (ill.).

10220 Miller, Ph. (1754). The Gardener's Dictionary. Abr. Ed. 4, Vol. 1. London. [*Abies* gen. nov., *Larix* gen. nov., *Sabina* gen. nov. = *Juniperus sabina* L.].

10230 Miller, Ph. (1768). The Gardener's Dictionary. Ed. 8, London. [*Abies alba* sp. nov., *A. balsamea* (L.) comb. nov., *A. mariana* sp. nov. = *Picea mariana, Cupressus horizontalis* sp. nov. = *C. sempervirens, C. lusitanica* sp. nov., *Juniperus caroliana* sp. nov. = *J. virginiana, J. hispanica* sp. nov. = *J. thurifera, J. lusitanica* sp. nov. = *J. sabina, J. suecica* sp. nov. = *J. communis, Larix decidua* sp. nov., *Pinus echinata, P. halepensis, P. maritima* = *P. pinaster, P. montana* = *P. mugo, P. palustris, P. rigida, P. virginiana* spp. nov.; pp. without numbers, spp. with numbers in margin of text; Ed. 8 is the first ed. in which M. consistently employs binary nomenclature (ICBN Art. 23)].

10231 Miller, P. M., L. E. Eddleman & J. M. Miller (1995). *Juniperus occidentalis* juvenile foliage: advantages and disadvantages for a stess-tolerant, invasive conifer. Canad. J. Forest. Res. 25: 470–479.

10237 Minghetti, P. & E. Nardi (1999). Lecto-typification of *Pinus mugho* Turra (Pinaceae). Taxon 48 (3): 465–469. (ill.).

10240 Minnich, R. A. (1982). *Pseudotsuga macrocarpa* in Baja California? Madroño 29 (1): 22–31. [the species does not occur there].

10260 Miquel, F. A. G. (1845). Cupressinae Richard. in: C. Lehmann: Plantae Preissianae, ... Vol. 1, pp. 643–645. Hamburg. [*Callitris preissii*, sp. nov, p. 643, *Actinostrobus*, gen. nov., *A. pyramidalis*, sp. nov., p. 644].

10261 Miquel, F. A. G. (1856). Stirpes novo-hollandas a Ferd. Müllero collectas. Ned. Kruidk. Arch. 4 (2): 97–150. [*Frenela crassivalvis* sp. nov. = *Callitris preissii*, p. 97].

10262 Mirbel, C. F. B. (1812). Sur l'Abies, genre de la famille des Conifères. Nouv. Bull. Sci. Soc. Philom. Paris 3: 73–76. [*Abies*].

10263 Mirbel, C. F. B. (1812). [*Schubertia* gen. nov.) in: Nouv. Bull. Sci. Soc. Philom. Paris 3: 123. [*Schubertia* Mirb., nom. rej. (vs. *Schubertia* Mart., 1824) = *Taxodium* Rich.].

10270 Mirbel, C. F. B. (1825). Essai sur la distribution géographique des conifères. Mem. Mus. Hist. Nat. 13: 28–76. [*Phyllocladus* Rich. ex Mirb., p. 48, *Abies menziesii* nom. nov. = *Pseudotsuga menziesii* (Mirb.) Franco, pp. 63, 70, *Cedrus libanensis* Juss. ex Mirb. = *C. libani*, p. 71, *Frenela* spp. = *Callitris* spp., p. 74, all nom. nud., the basion. sub *Callitris* were either on labels or in ms. given by A. Cunningham or R. Brown, but never published, e.g. *Frenela australis* Mirb. non Endl. = *Callitris oblonga*; *Podocarpus latifolius* (Thunb.) comb. nov., p. 75, *P. spinulosus* (Sm.) R. Br. ex Mirb., p. 75].

10280   Mirov, N. T. (1938). Phylogenetic relations of *Pinus jeffreyi* and *Pinus ponderosa*. Madroño 4: 169–171.

10290   Mirov, N. T. (1952). Mr. Pince's Mexican pine. Madroño 11 (7): 270–273. [*Pinus pinceana*].

10300   Mirov, N. T. (1953). Taxonomy and chemistry of the white pines. Madroño 12: 81–89. [*Pinus* sect. *Strobus*].

10301   Mirov, N. T. (1954). Lodgepole Pine discovered and misnamed. Madroño 12 (5): 156–157. [*Pinus contorta*, nomenclature].

10310   Mirov, N. T. (1954). Composition of turpentines of Mexican pines. Unasylva 8 (4): 167–173. [chemistry + phytogeography, map].

10320   Mirov, N. T. (1955). Relationship between *Pinus halepensis* and other *Insignes* pines of the Mediterranean region. Bull. Res. Council Israel, sect. D, Bot. 5D: 65–72.

10330   Mirov, N. T. (1956). Photoperiod and flowering of pines. Forest Sci. 2: 328–332. [*Pinus*].

10340   Mirov, N. T. (1958). *Pinus oaxacana*, a new species from Mexico. Madroño 14: 145–150. [*P. oaxacana* sp. nov., p. 145; see also M. Martínez, 1945].

10350   Mirov, N. T. (1961). Composition of gum turpentines of pines. Techn. Bull. U.S.D.A. 1239. [*Pinus*, maps].

10360   Mirov, N. T. (1962). Phenology of tropical pines. J. Arnold Arbor. 43: 218–219. [*Pinus*].

10370   Mirov, N. T. (1967). The Genus *Pinus*. (viii + 602 pp.) New York. [palaeobotany, phytogeography, chemistry etc., but not a taxonomic treatment, ill., maps, tables].

10380   Mirov, N. T., E. Frank & E. Zavarin (1965). Chemical composition of *Pinus elliottii* var. *elliottii* turpentine and its possible relation to taxonomy of several pine species. Phytochemistry 4: 563–568.

10385   Mirov, N. T. & J. Hasbrouck (1976). The story of pines. London. (ill.).

10390   Mirov, N. T. & R. G. Stanley (1959). The pine tree. Annual Rev. Plant Physiol. 10: 223–238. [*Pinus*].

10391   Mirov, N. T., E. Zavarin & K. Snajberk (1966). Chemical composition of the turpentines of some eastern Mediterranean Pines in relation to their classification. Phytochemistry 5: 97–102. [*Pinus brutia*, *P. halepensis*].

10392   Mirov, N. T., E. Zavarin, K. Snajberk & K. Costello (1966). Further studies of turpentine composition of *Pinus muricata* in relation to its taxonomy. Phytochemistry 5: 343–355.

10399   Mitchell, A. F. (1970). A note on two new hybrid Cypresses. J. Roy. Hort. Soc. 95 (10): 453–454. [× *Cupressocyparis notabilis* hybr. nov., × *C. ovensii* hybr. nov. = *Chamaecyparis nootkatensis* × *Cupressus lusitanica* from Silk Wood, Gloucestershire, U.K.].

10420   Mitchell, A. J. (1918). Incense-cedar. U.S.D.A. Bull. (1915–23) 604: 1–40. [*Calocedrus decurrens*].

10430   Mitsopoulos, D. J. & C. P. Panetsos (1987). Origin of Variation in Fir forests of Greece. Silvae Genet. 36 (1): 1–15. [*Abies alba*-*A. cephalonica* complex].

10440   Mittak, W. L. & J. P. Perry, Jr. (1979). *Pinus maximinoi*: its taxonomic status and distribution. J. Arnold Arbor. 60: 386–395.

10444   Mitton, J. B. & R. Andalora (1981). Genetic and morphological relationships between blue spruce, *Picea pungens*, and Engelmann spruce, *Picea engelmannii*, in the Colorado Front Range. Can. J. Bot. 59 (11): 2088–2094.

10451   Miyake, K. (1911). The development of the gametophytes and embryogeny in *Cunninghamia sinensis*. Bot. Centralbl. Beih. 27: 1–25. (ill.). [= *C. lanceolata*].

10452   Miyake, K. & K. Yasui (1911). On the gametophytes and embryo of *Pseudolarix*. Ann. Bot. (London) 25: 639–648. (ill.).

10454   Miyoshi, M. (1925). Bericht über die neuerdings gesetzlich geschützten botanischen Naturdenkmäler. Bot. Mag. (Tokyo) 39: 235–238. [*Torreya unda* sp. nov. = *T. nucifera*, p. 236].

10460   Moe, D. (1970). The Post-Glacial Immigration of *Picea abies* into Fennoscandia. Bot. Not. 123: 61–66.

10461   Moench, C. (1794). Methodus plantas horti botanici et agri marburgensis, ... Marburg. [*Thuja obtusa* sp. nov. = *T. occidentalis*, p. 691, *T. acuta* sp. nov. = *Platycladus orientalis*, p. 692, *Juniperus horizontalis*, *J. oppositifolia* = *J. bermudiana*, *J. tetragona* = *J. phoenicea*, spp. nov., p. 699].

10462   Mohl, H. von (1871). Morphologische Betrachtung der Blätter von *Sciadopitys*. Bot. Zeitung (Berlin) 29: 1–14, 17–23, figs.

10470   Mohr, C. (1897). Timber pines of the southern United States, together with a discussion of the structure of their wood, by F. Roth. U.S.D.A. Bull. (1895–1901) 13. (176 pp., 27 pl., maps). [*Pinus echinata*, *P. elliottii*, *P. glabra*, *P. palustris*, *P. taeda*, *P. serotina*].

10470a   Mohr, C. (1901). Notes on the Red Cedar. U.S.D.A. Division of Forestry Bull. No. 31. Washington, D.C. [*Juniperus virginiana*, ill.].

10470e   Molina, G. I. (1782). Saggio sulla storia naturale del Chili del signor Abate Giovanni Ignazio Molina. Bologna. [*Pinus araucana* Molina, p. 182].

10471   Molina, R. A. (1964). Coniferas de Honduras. Ceiba 10 (1): 5–21. [*Abies guatemalensis, Pinus ayacahuite, P. caribaea, P. oocarpa, P. pseudostrobus, Podocarpus oleifolius, Taxus globosa, Cupressus lusitanica,* ill.].

10480   Moll, E. (1981). Trees of Natal. Cape Town. [*Widdringtonia*].

10480a   Möller, M., R. R. Mill, S. M. Glidewell, D. Masson, D. Williamson & R. M. Bateman (1999). Comparative biology and taxonomic significance of the pollination mechanisms in *Acmopyle pancheri* and *Phyllocladus hypophyllus.* Ann. Bot. (London) 86: 149–158. [Podocarpaceae, reproductive biology].

10480e   Molloy, B. P. J. (1995). *Manoao* (Podocarpaceae), a new monotypic conifer genus endemic to New Zealand. New Zealand J. Bot. 33: 183–201. [*Dacrydium, Halocarpus, Lagarostrobos, Lepidothamnus, Manoao* gen. nov., p. 196, *M. colensoi* (Hook.) comb. nov., p. 196, ill.].

10480f   Molloy, B. P. J. (1996). A new species name in *Phyllocladus* (Phyllocladaceae) from New Zealand. New Zealand J. Bot. 34: 287–297. [*Phyllocladus glaucus* = *P. aspleniifolius, P. toatoa* sp. nov., p. 290, lectotype chosen of *Podocarpus aspleniifolius* Labill., p. 289, ill.].

10480g   Molloy, B. P. J. & K. R. Markham (1999). A contribution to the taxonomy of *Phyllocladus* (Phyllocladaceae) from the distribution of key flavonoids. New Zealand J. Bot. 37 (3): 375–382. (ill.).

10480h   Molloy, B. P. J. & M. Muñoz-Schick (1999). The correct name for the Chilean conifer Lleuque (Podocarpaceae). New Zealand J. Bot. 37: 189–193. [*Prumnopitys andina* (Poepp.) de Laub., *P. spicata* (Poepp.) Molloy & Muñoz-Schick, comb. nov. & superfl., based on erroneous notion that *Podocarpus spicatus* R. Br. is a nomen nudum: it is a nomen novum].

10481   Montacchini, F. & R. Caramiello (1968). Il *Pinus mugo* Turra ed il *Pinus uncinata* Miller in Piemonte. Giorn. Bot. Ital. 102 (6): 529–535. [*P. uncinata* possible hybr. betw. *P. sylvestris* and *P. mugo,* ill., map].

10490   Moore, H. E., Jr. (1965). *Chrysolarix,* a New Name for the Golden Larch. Baileya 13 (3): 131–134. [*C. amabilis* (Nelson) comb. nov. = *Pseudolarix amabilis* (nom. cons.), p. 133].

10500   Moore, H. E., Jr. (1966). Nomenclatural Notes on the Cultivated Conifers. Baileya 14 (1): 1–11. [*Austrocedrus, Callitris, Chamaecyparis, Calocedrus, Platycladus, Juniperus, Abies, Picea, Pinus, Pseudotsuga, Pinus tenuifolia* Benth. = *P. maximinoi* nom. nov., p. 8].

10505   Moore, H. E., Jr. (1966). In defense of *Chrysolarix.* Taxon 15: 258–264. [*Psudolarix amabilis,* nomenclature].

10510   Moore, H. E., Jr. (1967). Further Notes on Conifer Nomenclature. Baileya 15: 26. [*Callitris columellaris, C. preissii* ssp. *verrucosa,* syn.].

10520   Moore, H. E., Jr. (1973). *Chrysolarix* renounced – A comedy of restoration. Taxon 22 (5/6): 587–589. [*Pseudolarix* Gord., nom. cons.].

10530   Morandini, R. (1969). *Abies nebrodensis* (Lojac.) Mattei inventario 1968. Pubbl. Istituto Sperimentale Selvicoltura Arezzo 18: I–VI + 1–93. (map, photogr.).

10540   Morelet, A. (1851). Description de deux nouvelles espèces de pins. Rev. Hort. Côte d'Or 1: 105–107. [*Pinus tropicalis* sp. nov. p. 106, *P. caribaea* sp. nov. p. 107; again publ. in: Bull. Soc. Hist. Nat. Dép. Moselle 7: 97–101, 1855].

10548   Morgan, C. S. (1999). *Platycladus orientalis* (Cupressaceae). Curtis's Bot. Mag. 16 (3): 185–192, pl. 368.

10550   Morgenstern, E. K. & J. L. Farrar (1964). Introgressive hybridization in red spruce and black spruce. Univ. Toronto Fac. Forest., Techn. Rep. 4. (46 pp., ill.). [*Picea rubens, P. mariana*].

10560e   Morikawa, K. (1928). *Torreya igaensis,* a new species of the genus *Torreya,* and *Torreya macrosperma.* Bot. Mag. (Tokyo) 42: 533–536. (ill.). [*Torreya igaensis* Doi & Morikawa sp. nov. = *T. nucifera,* p. 536, *T. macrosperma* Miyoshi = *T. nucifera*].

10561   Morley, B. D. & H. R. Toelken, eds. (1983). Flowering plants in Australia. Adelaide. [see J. Venning, 1983].

10562   Morley, T. (1948). On leaf arrangement in *Metasequoia glyptostroboides.* Proc. Nation. Acad. Sci. U.S.A. 34 (12): 574–578.

10570   Morren, C. (1848). Notice sur les thuyas et particulièrement sur celui de Tartarie. Ann. Soc. Roy. Agric. Gand 4: 462–468. [*Thuja, Platycladus orientalis*)

10580   Morren, C. (1853). Le cyprès funèbre ou pleureur, ou *Cupressus funebris* des botanistes. Hort. Belge 3: 126–127, pl. 20.

10590   Morton, C. V. (1941). Notes on *Juniperus.* Rhodora 43: 344–348. [nomenclature].

10600   Morton, F. von (1933). Der Sandarakbaum (*Callitris quadrivalvis* Vent.). Mitt. Deutsch. Dendrol. Ges. 45: 369–370, 1 pl. [*Tetraclinis articulata*].

10610   Moseley, M. F., Jr. (1943). Contributions to the Life History, Morphology and Phylogeny of *Widdringtonia cupressoides.* Lloydia 6 (2): 109–132. [= *W. nodiflora*].

10630   Mottet, S. (1903). *Picea ajanensis* et *Picea alcoquiana.* Rev. Hort. 1903: 339–342, f. 137–140. [*Picea jezoensis, P. alcockiana*].

10640   Mottet, S. (1903). *Cunninghamia sinensis*. Rev. Hort. 1903: 549–552, f. 232–234.

10650   Mottet, S. (1904). Les *Keteleeria*. Rev. Hort. 1904: 129–131, f. 52–53. [4 spp.].

10660   Mottet, S. (1910). *Pinus armandii*. Rev. Hort. 1910: 423–426, f. 177–179.

10670   Mottet, S. (1919). Un nouveau *Chamaecyparis* (*C. formosensis*). Rev. Hort. 1918–19: 342–344, f. 105.

10680   Mottet, S. (1929). Le pin de la Corée (*Pinus koraiensis*). Rev. Hort. 101: 479–480, f. 192.

10681   Mueller, F. J. H. von (1858). On the *Octoclinis macleayana* a new Australian pine. Trans. & Proc. Philos. Inst. Victoria 2: 20–22, 1 pl. [*Octoclinis macleayana* gen. et sp. nov. = *Callitris macleayana*, p. 22; see also F. J. H. von Mueller, 1860].

10681a   Mueller, F. J. H. von (1860). Description of the Australian Kauri pine. Quart. J. Trans. Pharm. Soc. Victoria 2 (8): 173–175. [*Dammara robusta* sp. nov. = *Agathis robusta*, p. 174].

10682   Mueller, F. J. H. von (1860). Essay on the plants collected by Mr. Eugene Fitzalan, during Lieut. Smith's expedition to the estuary of the Burdekin. Rep. govt. Printer, Melbourne. [*Callitris actinostrobus* nom. nov. et illeg. = *Actinostrobus pyramidalis*, *C. macleayana* (F. Mueller) comb. nov., p. 19, *C. verrucosa* (Endl.) comb. nov., p. 19].

10682a   Mueller, F. J. H. von (1863–64). Fragmenta phytographiae Australiae, ... Coniferae. Vol. 4 (26): 86–87. Melbourne. [*Podocarpus drouynianus* sp. nov., p. 86, t. 31].

10683   Mueller, F. J. H. von (1866). Fragmenta phytographiae Australiae, ... Coniferae. Vol. 5 (40): 191–208. Melbourne. [*Callitris parlatorei* sp. nov., p. 186 = *C. macleayaya*, *C. columellaris* sp. nov., p. 198].

10684   Mueller, F. J. H. von (1866). *Callitris* (*Frenela*) *parlatorei*, F. Muell. J. Bot. 4: 267–268. [*Callitris parlatorei* sp. nov. = *C. macleayana*; first publ. in 10683].

10685   Mueller, F. J. H. von (1882). Systematic census of Australian plants, with chronologic, literary and geographic annotations. Part 1. Vasculares. Melbourne. [*Callitris acuminata* (Parl.) comb. nov. = *Actinostrobus acuminatus*, *C. drummondii* (Parl.), *C. muelleri* (Parl.), *C. roei* (Endl.), comb. nov., p. 109].

10685a   Mueller, F. J. H. von (1888). Key to the system of Victorian plants... Vol. 1, with map. Melbourne. [*Callitris cupressiformis* sp. nov. = *C. rhomboidea*, p. 402].

10686   Mueller, F. J. H. von (1889). Records of observations on Sir William MacGregor's Highland plants from New Guinea. Trans. Roy. Soc. Victoria 1 (2): 1–45. [*Libocedrus papuana* sp. nov. = *Papuacedrus papuana*, p. 32].

10687   Mueller, F. J. H. von (1891). Descriptions of new Australian plants, with occasional other annotations. Victorian Nat. 8: 45–46. [*Dammara palmerstonii* sp. nov. = *Agathis robusta*].

10689   Muir, J. (1876). On the post-glacial history of *Sequoia gigantea*. Amer. Ass. Adv. Sci. Proc. 25: 242–253. [*Sequoiadendron giganteum*].

10691   Muir, N. (1992). Some notes on nut pines. The Plantsman 14 (2): 80–98. [*Pinus cembroides*, *P. culminicola*, *P. discolor*, *P. edulis*, *P. johannis*, *P. maximartinezii*, *P. monophylla*, *P. nelsonii*, *P. pinceana*, *P. quadrifolia*, *P. remota*, ill.].

10700   Müller-Using, B. & R. Lassig (1986). Zur Verbreitung der Chihuahua-Fichte (*Picea chihuahuana* Martínez) in Mexico. Mitt. Deutsch. Dendrol. Ges. 76: 157–169.

10700a   Mullin, L. J. (2001). Conifers in Zimbabwe. Kirkia 17 (2): 199–217. [*Afrocarpus falcatus*, *Juniperus procera*, *Widdringtonia nodiflora*].

10701a   Murbeck, S. S. (1900). Conributions à la connaissance de la Flore du Nord-Ouest de l'Afrique, et plus spécialement de la Tunisie. IV. Graminaceae-Polypodiaceae. (entire series 1897–1905). Lund. [*Callitris articulata* (Vahl) comb. nov. = *Tetraclinis articulata*, p. 29–30].

10702   Murray, A. (1850). Botanical expedition to Oregon. No. 2. Oregon Committee, Edinburgh. (pamphlet, 2 pp., with figs.; enum. coll. of J. Jeffrey). [*Thuja craigiana* sp. nov. (as "*Craigana*") = *Calocedrus decurrens*, p. 2, t. 5].

10710   Murray, A. (1853). Botanical expedition to Oregon. No. 6; No. 8. Oregon Committee, Edinburgh. (pamphlets, 3 pp.; 2 pp. + 4 pl.; enum. coll. of J. Jeffrey). [*Abies pattoniana* Jeffrey sp. nov. (sine descr.), No. 6, p. 1 (No. 430 in: "Oregon Botanical Expedition": publ. letter signed John Jeffrey, with preceding handwritten minute of meeting by J.H. Balfour, orig. at E); *Abies pattoniana* Jeffrey ex Balf. sp. nov. = *Tsuga mertensiana* ssp. *mertensiana*, No. 8, p. 1 (No. 430), *Pinus balfouriana* Balf., sp. nov., No. 8, p. 1, *P. jeffreyi* Balf., sp. nov., No. 8, p. 2, *P. murrayana* Balf., sp. nov. = *P. contorta*, No. 8, p. 2, *Thuja craigiana* "sp. nov." (see A. Murray, 1850) = *Calocedrus decurrens*, No. 8, p. 2. Although M. stated in the preceding note of pamphlet No. 8, that these spp. were described by "Professor Balfour" (and named by him and other members of the committee), their authority in the list is given as "Oregon Committee", while M. signed the note. Following Murray's indication, Balfour is cited as the author of these spp.].

10720 Murray, A. (1855). Description of new coniferous trees from California. Edinburgh New Philos. J. 1: 284–295. [*Pinus beardsleyi* sp. nov., p. 286, t. 6, *Pinus craigiana* sp. nov. (as "Craigana") = *P. ponderosa*, p. 288, t. 7, *Abies hookeriana* sp. nov. = *Tsuga mertensiana* ssp. *grandicona*, p. 289, t. 9 (t. 9 has also a fig. of *A. pattoniana* = *T. mertensiana* ssp. *mertensiana*), *Cupressus lawsoniana* sp. nov. = *Chamaecyparis lawsoniana*, p. 292, t. 10, *Cupressus macnabiana* sp. nov. (as "M'Nabiana"), p. 293, t. 11, *Pinus benthamiana* Hartw. (t. 8), *Taxus lindleyana* sp. nov. = *T. brevifolia*, p. 294].

10721 Murray, A. (1860). Notes on Californian Trees. Part 2. Trans. Bot. Soc. Edinburgh 6: 330–353. [*Sequoiadendron giganteum* (syn.: *Wellingtonia gigantea*), *Pinus* spp., nomenclature].

10760 Murray, A. (1867). Description of a new conifer from Arctic America. J. Bot. 5: 253–254, pl. 69. [*Abies arctica* Andr. Murray, non Gord. = *Picea glauca*].

10763 Murray, B. G., N. Friesen & J. S. Heslop Harrison (2002). Molecular cytogenetic analysis of *Podocarpus* and comparison with other gymnosperm species. Ann. Bot. (London) 89 (4): 483–489.

10770 Murray, E. (1982). Notae Spermatophytae No. 1. Unum minutum monographum generis *Pinus*. Kalmia 12: 18–27. [see No. 10790].

10780 Murray, E. (1983). Notae Spermatophytae No. 2. Unum minutum monographum generis *Pinus*. Kalmia 13: 3–24. [see No. 10790; *Juniperus virginiana* var. *silicicola* (Small) comb. nov., p. 8].

10790 Murray, E. (1984). Notae Spermatophytae No. 4. Unum minutum monographum generis *Abietarum*. Kalmia 14: 2–8. [other "small monographs" in Pinaceae: *Larix*, *Picea*, *Pseudotsuga*, *Tsuga*; Kalmia 14: 9–19; numerous unnessesary (new) names in a home-made "journal", but validly publ. under ICBN].

10791 Murray, J. A. (1784). Caroli à Linné equitis Systema vegetabilium secundum classes ordines genera species cum characteribus et differentiis. Ed. 14, Göttingen. [*Cupressus pendula* "sp. nov.", p. 861 (publ. May–June 1784, but preceded by C.P. Thunberg in Nova Acta Regiae Soc. Sci. Upsal. 4: 40, 1783) = cultivar of *Platycladus orientalis*, *Myrica nagi* sp. nov. = *Nageia nagi*, p. 884, *Taxus macrophylla* sp. nov. = *Podocarpus macrophyllus*, p. 895, *T. verticillata* sp. nov. = *Sciadopitys verticillata*, p. 895 (new taxa by C. P. Thunberg)].

10795 Mustart, P. (1993). What is the Cedarberg without the cedar? Veld & Flora (Dec. 1993) 79: 114–117. [*Widdringtonia cedarbergensis*, ill. conservation].

10796 Mustart, P., J. Juritz, C. Makua, S. W. van der Merwe & N. Wessels (1995). Restoration of the Clanwilliam Cedar *Widdringtonia cedarbergensis*: the importance of monitoring seedlings planted in the Cederberg, South Africa. Biol. Cons. 72: 73–76.

10798 Myburg, H. & S. A. Harris (1997). Genetic variation across the natural distribution of the South East Asian pine, *Pinus kesiya* Royle ex Gordon (Pinaceae). Silvae Genet. 46 (5): 295–301.

10800 Myers, O., Jr. & F. H. Borman (1963). Phenotypic variation in *Abies balsamea* in response to altitudinal and geographic gradients. Ecology 44: 429–436. (ill.).

10803 Nadolny, C. & J. Benson (1993). The biology and management of the Pigmy Cypress Pine (*Callitris oblonga*) in NSW. New South Wales National Parks and Wildlife Service, Species management report No. 7. [with detailed distribution maps of populations].

10805 Nahal, I. (1962). Le Pin d'Alep: Etude taxonomique, phytogéographique, écologique et silvicole. Ann. Ecole Nation. Eaux Fôrets Nancy 19: 475–686. [*Pinus brutia*, *P. halepensis*, *P. eldarica*, *P. brutia* div. ssp. comb. nov., p. 521].

10806 Nahal, I. (1983). Le Pin Brutia (*Pinus brutia* Ten. subsp. *brutia*). Fôrêt Méditerr. 5: 164–172. [*P. brutia* div. ssp., *P. halepensis*].

10810 Nair, P. K. K., ed. (1980–84). Glimpses in plant research, 5, 6. New Delhi. [see P. D. Dogra, 1980, 1984, P. D. Dogra & S. Tandon, 1984].

10820 Nakai, T. (1919). Taxaceae et Coniferae in flora Coreano-Manshurica novae. J. Japan. Bot. 2: 9–12, 1 fig. [*Picea koraiensis* sp. nov., *Pinus mukdensis* sp. nov., *Thuja koraiensis* sp. nov. [see also E. H. Wilson, 1920); diagn. (also) publ. in: Bot. Mag. (Tokyo) 33: 195–196, 1919].

10821 Nakai, T. (1919). Notulae ad plantas Japoniae et Koreae XXI. Bot. Mag. (Tokyo) 33: 193–216. [*Torreya nucifera* var. *radicans* var. nov., p. 194, *Picea koraiensis* sp. nov., p. 195, *Pinus mukdensis* sp. nov., p. 195, *Thuja koraiensis* sp. nov., p. 196; this publ. is to be accepted as the place of valid publ. of these taxa].

10821a Nakai, T. (1926). Notulae ad plantas Japoniae et Koreae XXXI. Bot. Mag. (Tokyo) 40: 161–171. [*Juniperus coreana* sp. nov., p. 161].

10822 Nakai, T. (1930). Notulae ad plantas Japoniae et Koreae XXXIX. Bot. Mag. (Tokyo) 44: 507–537. [*Cephalotaxus koreana* sp. nov., p. 508, *Juniperus sargentii* (Henry) Takeda ex Nakai = *J. chinensis* var. *sargentii*, p. 511].

10830 Nakai, T. (1935). Species nova generis Pini in Jehol. Rep. First Sci. Exped. Manchoukuo Sect. IV, Pt. 2: 164–166, pl. 19, f. 24–25. (in: T. Nakai, M. Honda & M. Kitagawa (eds.). Contr. Cogn. Fl. Mansh., No. 10). [*Pinus tokunagai* sp. nov.].

10831 Nakai, T. (1937). Notulae ad plantas Asiae Orientalis. I. J. Japan. Bot. 13: 393–406. [*Cryptomeria mairei* (Léveillé) comb. nov. = *C. japonica*, p. 395].

10840 Nakai, T. (1938–39). Indigenous species of conifers and taxads of Korea and Manchuria, and their distribution. I–IV. Bull. Forest. Soc. Korea, Tokyo. (Japan., Lat. descr.; repr. from Chôsen Sanrin Kaihô, 158–167: 1–29. May 1938). [Coniferales, Taxales, *Larix koreana* (Nakai) comb. nov. in Vol. I (158), p. 6, 1938, basion. not cited (in Vol. III (165), p. 32, 1938 with syn. *L. olgenis* var. *koreana*), *Taxus cuspidata* var. *latifolia* (Pilg.) comb. nov. in Vol. I (158), p. 19, 1938, *T. caespitosa* sp. nov. = *T. cuspidata* in Vol. I (158), p. 20, 1938, *Torreyaceae* fam. nov. in Vol. I, p. 23, 1938, *Torreya fruticosa* sp. nov. = *T. nucifera* in Vol. I, p. 26, 1938, *Juniperus chekiangensis* sp. nov. in Vol. II (163), p. 26, 1938, *Thuja kongoensis* Doi ex Nakai, sp. nov. (with syn. *T. koraiensis*, but see T. Nakai, 1919 and E. H. Wilson, 1920). Vol. III (165), p. 23, 1938, *Pinus takahasii*, *P. sosnowskyi*, spp. nov. in Vol. IV (167): 32–33, 1939].

10850 Nakai, T. (1941). Notulae ad plantas Asiae Orientalis (XV). J. Japan. Bot. 17 (1): 1–17. [*Picea tonaiensis* sp. nov., p. 1, *P. pungsanensis* Uyeki ex Nakai, sp. nov., p. 3, *P. intercedens* sp. nov., p. 4].

10860 Nakai, T. (1943). Notulae ad plantas Asiae Orientalis (XXVI). J. Japan. Bot. 19: 245–251. [*Picea manshurica* sp. nov. = *P. jezoensis*, p. 251].

10870 Nakayama, T. (1957). On the *Chamaecyparis obtusa* forest in Shikoku, Southern Japan. Japan. J. Ecol. 6: 149–152. (Japan., Eng. summ.).

10915 Nardi, E. & P. Minghetti (1999). Proposal to conserve the name *Pinus mugo* (Pinaceae) with a conserved type. Taxon 48 (3): 571–572.

10920 Naumenko, Z. M. (1964). *Picea obovata* Ledeb. at the northeastern-most limit of its range. Bot. Žurn. (Moscow & Leningrad) 49: 1008–1013. (Russ.).

10925 Neale, D. B. & R. R. sederoff (1989). Paternal inheritance of chloroplast DNA and maternal inheritance of mitochondrial DNA in Loblolly pine. Theor. Appl. Genet. 77: 212–216. [*Pinus taeda*].

10930 Necker, N. J. de (1790). Elementa botanica genera genuina, species naturales... Vol. 3. (456 p.). Paris. [*Apinus* gen. nov., with *A. cembra* as type = *Pinus cembra*, p. 369].

10930c Neeman, G. & L. Trabaud (eds.) (2000). Ecology, biogeography and management of *Pinus halepensis* and *P. brutia* forest ecosystems in the Mediterranean Basin. Backhuys, Leiden.

10931 Nees von Esenbeck, C. G. D. & S. Schauer (1847). Enumeratio et descriptiones generum novorum specierumque plantarum in terris mexicanis crescentium. Linnaea 19 (6): 681–734. [*Juniperus sabinioides* (Kunth in H. B. K.) Nees, "comb. nov." (but see A. H. R. Grisebach, 1846) = *J. monticola*, p. 706].

10947 Nelson, C. D., W. L. Nance & D. B. Wagner (1994). Chloroplast DNA variation among and within taxonomic varieties of *Pinus caribaea* and *Pinus elliottii*. Canad. J. Forest. Res. 24: 424–426. (ill.).

10949 Nelson, E. C. (1991). *Araucaria bidwillii*: Andrew Petrie's Pine. Kew Mag. 8 (4): 175–185. (colour ill.).

10960 Neubauer, H. F. (1976). Über Zapfen und Zapfenmissbildungen bei *Metasequoia*. Bot. Jahrb. Syst. 95 (3): 321–326.

10970 Newcomb, G. B. (1959). The relationships of the pines of insular Baja California. Proc. 9th Int. Bot. Congr. 2: 281. [*Pinus muricata* var. *cedrosensis*, *P. radiata* var. *binata*].

10971 Nichols, G. E. (1910). A morphological study of *Juniperus communis* var. *depressa*. Bot. Centralbl. Beih. 25 (1): 201–241. (ill.). [anatomy, embryology].

10972 Nichols, G. E. (1935). Hemlock – White Pine – Northern Hardwood Region of Eastern North America. Ecology 16 (3) 403–422. [*Abies*, *Picea*, *Pinus*, *Tsuga*, maps].

10978 Nicholson, D. H. (1980). Point of view on *Pseudolarix*. Taxon 29: 318. [*Pseudolarix amabilis*, nomenclature].

10980 Nicholson, H. B. (1971). Notes on the Mlanje cedar and some other trees of the Mlanje Mountains in Malawi. Trees in South Africa 23 (3): 58–68. [*Widdringtonia whytei*, ill.].

10990 Nicholson, R. G. (1986). Collecting rare conifers in North Africa. Arnoldia 46 (1): 20–29. [*Abies*].

10996 Niebling, C. R. & M. T. Conkle (1990). Diversity of Washoe Pine and comparisons with allozymes of Ponderosa Pine races. Canad. J. Forest. Res. 20: 298–308. [*Pinus ponderosa*, *P. washoensis* = *P. ponderosa* "North Plateau race"].

11002 Nienstaedt, H. (1983). *Picea chihuahuana* og *Picea mexicana*. To sjaeldene mexicanske granarter. Dansk Dendrol. Årsskrift 6: 53–66.

11004 Nimsch, H. & J. C. Liu (1990). *Cathaya* – Eine wenig bekannte Pinaceen-Gattung aus China. Freiburg. [24 pp., colour photographs, map, privately publ. by the first author; also in Palmengarten 55 (2): 47–53, 1991].

11008 Nishida, M., T. Ohsawa & H. Nishida (1992). Structure and affinities of the petrified plants from the Cretaceous of northern Japan and Saghalien VIII. *Parataiwania nihongii* gen. et sp. nov., a taxodiaceous cone from the Upper Cretaceous of Hokkaido. J. Japan. Bot. 67: 1–9. (ill.).

11010   Nitzelius, T. G. (1969). *Abies*, a review of the firs in the Mediterranean. Lustgården 49 (1968), reprint pp. 146–189, Uppsala. [*A. numidica*, *A. pinsapo*, *A. cephalonica*, *A. cilicica*, *A. nordmanniana*, *A. nordmanniana* ssp. *equitrojani*, *A. alba*, *A. alba* ssp. *nebrodensis* (Lojac.) comb. et stat. nov., p. 178, and cult. spp., figs., key].

11011   Nitzenko, A. A. (1959). The Larch (*Larix sukaczewii* Dyl.) grove at Lindulovo (Karelian isthmus). Bot. Žurn. (Moscow & Leningrad) 44 (9): 1249–1260. (Russ., Eng. summ.). [= *L. sibirica*].

11011e   Nkongolo, K. K., L. Deverno & P. Michael (2003). Genetic validation and characterization of RAPD markers differentiating black and red spruces: molecular certification of spruce trees and hybrids. Plant Syst. Evol. 236: 151–163. [*Picea mariana*, *P. rubens*, DNA, genetics].

11013   Noren, C. O. (1907). Zur Entwicklungsgeschichte des *Juniperus communis*. Årsskr. Univ. (Uppsala) 1: 1–64. (ill.).

11015   Noss, R. F. (ed.) (2000). The Redwood forest: history, ecology and conservation of the Coast Redwoods. Washington, D.C.; Covelo, California. [*Sequoia sempervirens*, ill.].

11019   Novák, F. A. (1927). Zur fünfzigjährigen Entdeckung der *Picea omorika*. Mitt. Deutsch. Dendrol. Ges. 38: 47–56.

11020   Novák, F. A. (1934). *Pinus pindica* et *Pinus magellensis*. Bull. Int. Acad. Sci. Bohême 1934: 1–4. [= *P. nigra* et *P. mugo*].

11020a   Nunes, E., T. Quilhó & H. Pereira (1996). Anatomy and chemical composition of *Pinus pinaster* bark. I.A.W.A. Journal 17 (2): 141–149.

11021   Nuttall, T. (1818). The genera of North American plants, and catalogue of the species, to the year 1817... Vol. 2. Philadelphia. [*Cupressus disticha* var. *imbricata* (as "imbricaria") = *Taxodium distichum* var. *imbricatum*, p. 224].

11023   Nuttall, T. & N. B. Wyeth (1834). A catalogue of a collection of plants made chiefly in the valleys of the Rocky Mountains or Northern Andes, towards the sources of the Columbia River. J. Acad. Nat. Sci. Philadelphia 7: 5–60. [*Thuja gigantea* Nutt., sp. nov. = *T. plicata*, p. 52].

11023c   Offord, C. A., C. L. Porter, P. F. Meagher & G. Errington (1999). Sexual reproduction and early plant growth of the Wollemi pine (*Wollemia nobilis*), a rare and threatened Australian conifer. Ann. Bot. 84: 1–9.

11023d   Ohana, T. & T. Kimura (1995). Further observations of *Cunninghamiostrobus yubariensis* Stopes and Fujii from the Upper Yezo Group (Upper Cretaceous), Hokkaido, Japan. Trans. Proc. Palaeontol. Soc. Japan, n.s. 178: 122–141. (ill.).

11023e   Ohsawa, T. (1994). Anatomy and relationships of petrified seed cones of the Cupressaceae, Taxodiaceae, and Sciadopityaceae. J. Plant Res. 107: 503–512. [palaeobotany, fossil genera].

11023g   Ohsawa, T., M. Nishida & H. Nishida (1991). Structure and affinities of the petrified plants from the Cretaceous of northern Japan and Saghalien IX: a petrified cone of *Sciadopitys* from the Upper Cretaceous of Hokkaido. J. Phytogeogr. Taxon. 39: 97–105.

11023h   Ohsawa, T., H. Nishida & M. Nishida (1992). Structure and affinities of the petrified plants from the Cretaceous of northern Japan and Saghalien XI: a cupressoid seed cone from the Upper Cretaceous of Hokkaido. Bot. Mag. (Tokyo) 105: 125–133. [*Archicupressus nihongii* sp. nov., p. 126].

11023i   Ohsawa, T., H. Nishida & M. Nishida (1993). Structure and affinities of the petrified plants from the Cretaceous of northern Japan and Saghalien XIII: *Yubaristrobus* gen. nov., a new taxodiaceous cone from the Upper Cretaceous of Hokkaido. J. Plant Res. 106: 1–9. [*Yubaristrobus nakajimae* sp. nov., p. 1].

11025   Okubo, A. & T. Kimura (1991). *Cupressinocladus obatae* sp. nov., from the Lower Cretaceous Chosi Group, in the Outer Zone of Japan. Bull. Nat. Sci. Mus. 17 (3): 91–109. (ill.). [Cupressaceae (?)].

11030   Oosting, H. J. & W. D. Billings (1943). The red fir forest of the Sierra Nevada: Abietum magnificae. Ecol. Monogr. 13: 259–274. (ill.). [*Abies magnifica* and other conifers].

11040   Oosting, H. J. & W. D. Billings (1951). A comparison of virgin spruce-fir forests in the northern and southern Appalachian system. Ecology 32: 84–103. [*Abies balsamea*, *A. fraseri*, *Picea rubens*.)

11041   Opiz, P. M. (1854). Nachtrag zu meinem Seznam rostlin kveteny ceské. Lotos 1854 (April): 94. [*Strobus* gen. nov. (1853), with basion.: *Pinus* sect. XII *Strobus* Endl., *Strobus weymouthiana* Opiz 1853 sp. nov. = *Pinus strobus* L.; new comb. under this genus e.g. by H. N. Moldenke in Revista Sudamer. Bot. 6 (1–2): 30, 1939 and in Phytologia 4: 128–129, 1952; in the first e.g. *Strobus griffithii* to replace *S. weymouthiana* Opiz and *P. wallichiana* A. B. Jackson in Bull. Misc. Inform. 1938: 85, but the latter epithet would have priority under this genus].

11049   Orlova, L. V. (2001). On the diagnostic features of the vegetative organs in the genus *Pinus* (Pinaceae). Bot. Zhurn. 86 (9): 33–44. (Russ.). (ill.).

11049a   Orlova, L. V. & K. I. Christensen (2002). Typification of *Pinus pallasiana*, *P. salzmannii* and *P. pityusa* (Pinaceae). Nordic J. Bot. 22 (2): 171–175. (ill.).

11050   Orlova, S. Ya. & Yu. B. Kerimov (1982). Morfologo-anatomicheskie issledovaniya i tyazhelopakhuchego. *Juniperus polycarpos, J. foetidissima.* Izv. Akad. Nauk Azerbajdzansk. SSR, Ser. Biol. Med. Nauk 4: 12–19 (Russ., ill.].

11051   Orr, M. Y. (1923). On the resin ducts in the leaves of *Picea brachytyla* Pritzel. Notes Roy. Bot. Gard. Edinburgh 14: 21–24.

11060   Orr, M. Y. (1933). A New Chinese Silver Fir. Notes Roy. Bot. Gard. Edinburgh 18 (86): 1–5, t. 236. [*Abies georgei*, sp. nov. = *A. forrestii* var. *georgei* (see A. Farjon, 1990), p. 1].

11070   Orr, M. Y. (1933). *Taiwania* in Upper Burma – a new record. Notes Roy. Bot. Gard. Edinburgh 18 (86): 6.

11080   Orr, M. Y. (1933). Plantae Chinenses Forrestianae: Coniferae. Notes Roy. Bot. Gard. Edinburgh 18 (86): 119–157, t. 240–242. [*Abies georgei* "sp. nov.", but see M. Y. Orr on p. 1].

11090   Orr, M. Y. (1935). Plantae Chinenses Forrestianae: Coniferae. Two additions and a correction. Notes Roy. Bot. Gard. Edinburgh 18: 275–276. [*Abies georgei, Juniperus squamata, Pinus yunnanensis*].

11110   Ortgies, E. (1876). Beiträge zur Kenntnis der Weisstanne-Arten. Gartenflora 25: 131–136, pl. 866. [*Abies*].

11115   Ortiz, P. L., M. Arista & S. Talavera (1998). Low reproductive success in two subspecies of *Juniperus oxycedrus* L. Int. J. Plant Sci. 159 (5): 843–847. [*J. oxycedrus* ssp. *macrocarpa, J. oxycedrus* ssp. *oxycedrus*].

11120   Osborn, A. (1931). The Keteleerias. Gard. Chron., ser. 3, 90: 327, f. 134. [*Keteleeria davidiana, K. fortunei*].

11130   Ostenfeld, C. H. & C. Syrach Larsen (1930). *Larix.* Pflanzenareale 2 (7): 59–63, maps 62–64. [*L. gmelinii* var. *olgensis* (Henry) comb. et stat. nov., p. 62, *L. decidua* var. *polonica* (Racib.) comb. et stat. nov., p. 63; see also II.7.63 for correction].

11140   Ostenfeld, C. H. & C. Syrach Larsen (1930). The species of the genus *Larix* and their geographical distribution. Biol. Meddel. Kongel. Danske Vidensk. Selsk. 9 (2): 1–106. [see review + key in: Bull. Soc. Dendrol. France 1930: 63–72, 1930].

11140a   Ottley, A. M. (1909). The development of the gametophytes and fertilization of *Juniperus communis* and *Juniperus virginiana.* Bot. Gaz. (Crawfordsville) 48: 31–46.

11140b   Otto, A., J. Kvaček & K. Goth (1999). Biomarkers from the taxodiaceous conifer *Sphenolepis pecinovensis* Kvaček and resin from [the] Bohemian Cenomanian. Acta Palaeobot. Suppl. 2: 153–157. [palaeobotany, chemistry].

11150   Ovsîannikov, V. F. (1929). Our cedar pines. Zap. Vladivostoksk. Otd. Gosud. Russk. Geogr. Obsc. Izuc. Amursk. Kraja 20 (2): 71–116, f. 1–18. (Russ.). [*Pinus cembra* = *P. sibirica, P. koraiensis, P. pumila*].

11160   Owens, J. N. & M. D. Blake (1983). Pollen morphology and development of the pollination mechanism in *Tsuga heterophylla* and *T. mertensiana.* Canad. J. Bot. 61: 3041–3048. (ill.).

11160a   Owens, J. N., G. L. Catalano, S. J. Morris & J. Aitken-Christie (1995). The reproductive biology of Kauri (*Agathis australis*) 1. Pollination and prefertilization development. Int. J. Plant Sci. 156 (3): 257–269.

11160d   Owens, J. N. & M. Molder (1974). Cone initiation and development before dormancy in yellow cedar (*Chamaecyparis nootkatensis*). Canad. J. Bot. 52: 2075–2084.

11160e   Owens, J. N. & M. Molder (1975). Pollination, female gametophyte, and embryo and seed development in Yellow Cedar (*Chamaecyparis nootkatensis*). Canad. J. Bot. 53: 186–199.

11160f   Owens, J. N. & M. Molder (1977). Vegetative bud development and cone differentiation in *Abies amabilis.* Canad. J. Bot. 55: 992–1008.

11160g   Owens, J. N. & M. Molder (1979). Sexual reproduction in western red cedar (*Thuja plicata*). Canad. J. Bot. 58: 1376–1392.

11160h   Owens, J. N. & R. P. Pharis (1971). Initiation and development of western red cedar cones in response to gibberellin induction and under natural conditions. Canad. J. Bot. 49: 1165–1175. [*Thuja plicata*].

11161   Owens, J. N. & F. H. Smith (1965). Development of the seed cone of douglas-fir following dormancy. Canad. J. Bot. 43: 317–332. [*Pseudotsuga menziesii*, ill.].

11170   Page, C. N. (1974). Morphology and affinities of Pinus canariensis. Notes Roy. Bot. Gard. Edinburgh 33: 317–324.

11180   Page, C. N. (1979). The earliest known find of living *Taiwania* (Taxodiaceae). Kew Bull. 34 (3): 527–528.

11195   Page, C. N. (1980). Leaf micromorphology in *Agathis* and its taxonomic implications. Plant Syst. Evol. 135: 71–79. (SEM photographs of leaf epidermis in *Agathis* spp.].

11220   Page, C. N. & R. C. Hollands (1987). The taxonomic and biogeographic position of Sitka spruce. Proc. Roy. Soc. Edinburgh 93B: 13–24. [*Picea sitchensis*].

11230   Page, C. N. & K. D. Rushforth (1980). *Picea farreri*, a new temperate conifer from Upper Burma. Notes Roy. Bot. Gard. Edinburgh 38 (1): 129–136. [*P. farreri* sp. nov., p. 130].

11233 Page, V. M. (1973). A new conifer from the Upper Cretaceous of central California. Amer. J. Bot. 60 (6): 570–575. [*Margeriella cretacea* gen. et sp. nov., p. 572, of uncertain taxodiaceous aff., Taxodiaceae].

11240 Pallas, P. S. (1784–89). Flora rossica seu stirpium Imperii rossici par Europam et Asiam indigenarum descriptiones et icones. Vol. 1. St. Petersburg. [*Pinus cembra* var. *pumila* var. nov. = *P. pumila*, No. 1, p. 5, t. 2, 1784; *Juniperus communis* var. *saxatilis* var. nov., No. 2, p. 12, *J. davurica* sp. nov., No. 2, p. 13, t. 55, *J. phoenicea* Pall., non L., No. 2, p. 16, t. 57, 1789 (t. p. 1788, but predates W. Aiton, 1789) = *J. foetidissima*].

11260 Pančić, J. (1876). Eine neue Conifere in den östlichen Alpen... Belgrad. (pp. 1–8). [Discovery of *Picea omorika* (as *Pinus omorika* sp. nov., p. 4), in Tara Mts. of Yugoslavia (sic!); diagn. repr. in Gard. Chron., n. s., 7: 620, 19 May 1877].

11270 Panetsos, C. P. (1975). Natural hybridization between *Pinus halepensis* and *Pinus brutia* in Greece. Silvae Genet. 24: 163–168.

11271 Panetsos, C. P. (1981). Monograph of *Pinus halepensis* Mill. and *P. brutia* Ten. An. Sumarstvo (Ann. Forest.) 9/2: 39–77. (ill., maps).

11272 Panetsos, C. P. (1987). Monograph of *Abies cephalonica* Loudon. An. Sumarstvo (Ann. Forest.) 7 (1): 1–22. (ill., maps).

11273 Panetsos, C. P., A. Christou & A. Scaltsoyiannes (1992). First analysis of allozyme variation in Cedar species (*Cedrus* sp.). Silvae Genet. 41: 339–342.

11274 Panetsos, K., A. Scaltsoyiannes, F. A. Aravanopoulos, K. Dounavi & A. Demetrakopoulos (1997). Identification of *Pinus brutia* Ten., *P. halepensis* Mill. and their putative hybrids. Silvae Genet. 46 (5): 253–257. [Panetsos, K. in this paper is presumably identical with Panetsos, C. P. in previous papers].

11280 Pant, D. D. & N. Basu (1978). A comparative study of the leaves of *Cathaya argyrophylla* Chun et Kuang and three species of *Keteleeria* Carrière. Bot. J. Linn. Soc. 75 (3): 271–282.

11280a Papageorgiou, A. C., K. P. Panetsos & H. H. Hattemer (1994). Genetic differentiation of natural Mediterranean cypress (*Cupressus sempervirens* L.) populations in Greece. Forest Genet. 1: 1–12.

11281 Papajoannou, J. (1936). Eine neue Varietät von *Pinus Brutia* Ten. : var. *agraphistii*. Praktika Akad. Athen. 11: 14–24. (ill.).

11300 Pardé, L. (1924). Trois espèces du genre *Picea* à réunir en une seule. Bull. Soc. Dendrol. France 50: 22–23. [*P. ascendens*, *P. sargentiana*, (*P. brachytyla*) reduced to syn. of *P. complanata*].

11303 Pardos, J. A. & M. Pardos (1997). *Tetraclinis articulata* Masters, 1893. Enzyklopädie der Holzgewächse III–1 (9): 1–6. (ill., map).

11309 Parker, E. L. (1963). The geographic overlap of Noble Fir and Red Fir. Forest Sci. 9: 207–216. [*Abies procera*, *A. magnifica*, hybridization].

11310 Parker, E. L. (1988). Those Amazing Siskiyou Firs and a New Discovery. Four Seasons (J. Region. Parks Bot. Gard.) 8 (1): (4)5–16. [*Abies amabilis*, *A. magnifica* var. *shastensis*, *A. procera*, ill.].

11320 Parker, S. P., ed. (1982). Synopsis and Classification of Living Organisms. New York. [see T. A. Zanoni, 1982].

11330 Parker, W. H., J. Maze & G. E. Bradfield (1981). Implications of morphological and anatomical variation in *Abies balsamea* and *A. lasiocarpa* (Pinaceae) from western Canada. Amer. J. Bot. 68: 843–854.

11331 Parlatore, F. (1860). Description de trois espèces nouvelles de Cypres (*C. globulifera*, *C. sphaerocarpa*, et *C. umbilicata*). Ann. Sci. Nat. Bot., sér. 4, 13: 377–379. [= *Cupressus sempervirens*].

11331a Parlatore, F. (1861). Note sur l'*Araucaria brasiliensis* et sur une nouvelle espèce d'*Araucaria*. Bull. Soc. Bot. France 8: 84–91. [*A. brasiliana* A. Rich. = *A. angustifolia*, *A. saviana* sp. nov. = *A. angustifolia*].

11332 Parlatore, F. (1862). Index seminum in horto botanico regii musei florentini: 1862. Florence. [*Frenela sulcata* sp. nov. = *Callitris sulcata*, p. 23, *F. subcordata* sp. nov. = *C. roei*, p. 24, *Actinostrobus acuminatus* sp. nov., p. 25].

11340 Parlatore, F. (1863). Coniferas novas nonnullas descripsit. pp. 1–4, publ. Jan. 1863, Florence. [*Juniperus conferta* sp. nov., p. 1, *Larix lyallii* sp. nov., p. 3, see also in: B. Seemann, 1863 for diagn. of these and other spp.].

11340a Parlatore, F. (1863). Enumeratio seminum in horto botanico regii musaei florentini...Anno 1862. Florence. [*Dammara motleyi* sp. nov. = *Nageia motleyi* (Podocarpaceae), p. 26].

11341 Parlatore, F. (1864). Studi organografici sui fiori e sui frutti delle Conifere. Ann. Mus. Imp. Fis. Firenze 1: 155–181. [*Chamaecyparis lawsoniana* comb. nov., p. 181].

11355 Parry, C. C. & G. Engelmann (1862). Supplements to the enumeration of plants of Dr. Parry's collection in the Rocky Mountains. Supplement I. Coniferae. Amer. J. Sci. Arts, ser. 2, 34: 330–332. [*Pinus aristata* sp. nov., p. 331, *P. flexilis*, mention of seed wing of *P. cembroides* in footnote, p. 332; date publ. Nov 1862].

11360 Parsons, D. J. (1972). The southern extensions of *Tsuga mertensiana* (Mountain Hemlock) in the Sierra Nevada. Madroño 21 (8): 536–539. [= *T. mertensiana* ssp. *grandicona*].

11370   Parfenov, V. I. (1971). De systematica intraspecifica *Picea abies* (L.) Karst. Novosti Sist. Vyssh. Rast. (Novit. Syst. Pl. Vasc.) 8: 4–11. Inst. Bot. V. L. Komarov, Leningrad. (Russ., Lat.).

11378   Passini, M.-F. (1982). Les forêts de *Pinus cembroides* au Mexique. Etudes Mésoaméricaines II–5. Editions recherche sur les civilisations, Paris.

11379   Passini, M.-F. (1985). Structure et régéneration des formations ligneuses à *Pinus maximartinezii* Rzed., Mexique. Bull. Soc. Bot. France 132, Lettres bot. (4/5): 327–339.

11380   Passini, M.-F. (1987). The endemic pinyon of Lower California. Phytologia 63: 337–338. [*Pinus lagunae* (M.-F. Robert-Passini) comb. et stat. nov. = *P. cembroides* var. *lagunae*].

11382   Passini, M.-F. (1994). Synonymie entre *Pinus discolor* Bailey & Hawksworth et *Pinus johannis* M.-F. Robert. Acta Bot. Gallica 141 (3): 387–388. [= *P. cembroides* var. *bicolor*].

11385   Passini, M.-F., D. Cibrian Tovar & T. Eguiluz Piedra, comp. (1988). II Simposio nacional sobre pinos piñoneros, 6.7.8. de agosto de 1987. C.E.M.C.A., Univ. Autónoma Chapingo, Centro de Genética Forestal A.C., Chapingo, México, D.F. [*Pinus* subsect. *Cembroides*, *Pinus* spp.; proceedings with contrib. in Spanish and English, see also D. K. Bailey & F. G. Hawksworth, 1988 and E. Zavarin, 1988].

11390   Passini, M.-F. & N. Pinel (1987). Morphology and Phenology of *Pinus lagunae*. Phytologia 63: 331–336. [= *P. cembroides* var. *lagunae*].

11390a   Passini, M.-F. & N. Pinel (1989). Ecology and distribution of *Pinus lagunae*, in the Sierra de Laguna, Baja California Sur, Mexico. Madroño 36 (2): 84–92. [= *P. cembroides* var. *lagunae*, ill., tables].

11391   Patel, R. N. (1968). Wood anatomy of Cupressaceae and Araucariaceae indigenous to New Zealand. New Zealand J. Bot. 6 (1): 9–18. [*Agathis australis*, *Libocedrus bidwillii*, *L. plumosa*, ill.].

11408   Patten, A. M. & S. J. Brunsfeld (2002). Evidence of a novel lineage within the *Ponderosae*. Madroño 49 (3): 189–192. [*Pinus ponderosa*, *P. jeffreyi*, *P. arizonica*, *P. engelmannii*, *P. washoensis* = *P. ponderosa*, DNA, phylogenetic analysis, taxonomy].

11410   Patterson, T. F. (1988). A new species of *Picea* (Pinaceae) from Nuevo Léon, Mexico. Sida 13 (2): 131–135. [*P. martinezii* sp. nov. = *P. chihuahuana*].

11410a   Pauly, G., A. Yani, L. Piovetti & C. Bernard-Dagan (1983). Volatile constituents of the leaves of *Cupressus dupreziana* and *Cupressus sempervirens*. Phytochemistry 22: 957–959.

11410b   Pauw, C. A. & H. P. Linder (1997). Tropical African cedars (*Widdringtonia*, Cupressaceae): systematics, ecology and conservation status. Bot. J. Linn. Soc. 123 (4): 297–319. [*Widdringtonia nodiflora*, *W. whytei*].

11411   Pavari, A. (1954). Cenni botanici sulle Conifere (Abete bianco, Pino domestico, Pinastro, Pino d'Aleppo, Cipresso, Tasso). Monti & Boschi 5 (11–12): 485–585. [*Abies alba*, *Pinus pinea*, *P. pinaster*, *P. halepensis*, *Cupressus sempervirens*, *Taxus baccata* in Italy, ill.].

11411a   Pavon y Jiménez, J. A. (1797). Disertacion botánica sobre los generos *Tovaria*, *Actinophyllum*, *Araucaria* y *Salmia*, con la reunion de algunos que Linneo publicó como distintos. Mém. Real. Acad. Méd. Madrid 1: 191–204. [*Araucaria imbricata* sp. nov. = *A. araucana*, p. 199].

11412a   Pearson, H. L. (1988). New records of fossil conifers from Suffolk. Trans. Suffolk Nat. Soc. 24: 84–87. (ill. in appendix). [fossil wood of *Cedroxylon* and Taxodiaceae from Upper Cretaceous to Pliocene].

11414   Peirce, A. S. (1935). Anatomy of the xylem of *Sciadopitys*. Amer. J. Bot. 22: 895–902. (ill.).

11431   Peirce, G. J. (1901). Studies on the Coast Redwood, *Sequoia sempervirens* Endl. Proc. Calif. Acad. Sci., ser. 3, Bot., 2: 83–106. (ill.).

11437   Peng, Z. H. & Z. H. Jiang (1999). *Pinus dabeshanensis* and its origin. (Chin., Eng., ill.). China Forestry Publishing House, Beijing.

11441a   Penny, J. S. (1947). Studies on the conifers of the Magothy Flora. Amer. J. Bot. 34: 281–296. (ill.). [Cupressaceae, *Cupressinostrobus*, palaeobotany].

11442   Pénzes, A. (1970). *Picea* und *Juniperus* Studien. Bot. Közlem. 57 (1): 45–50. (Hung., Germ. summ.). [*Juniperus communis* L. ssp. *brevifolia* (Sanio) comb. nov., *J. communis* L. ssp. *cupressiformis* (Marie-Victorin et Sennen) comb. nov., *J. albanica* sp. nov., p. 49, ill.].

11450   Pérez de la Rosa, J. A. (1983). Una nueva especie de pino de Jalisco, México. Phytologia 54: 289–298. [*Pinus jaliscana* sp. nov., p. 290].

11460   Pérez de la Rosa, J. A. (1985). Una nueva especie de *Juniperus* de México. Phytologia 57 (2): 81–86. [*J. martinezii*, sp. nov., p. 81].

11462   Pérez de la Rosa, J. A. (1991). Identificacion de los pinos silvestres de Jalisco, atendiendo la morfologia de las aciculas (1). Bol. Inst. Bot. I.B.U.G. 1 (1): 23–32. [*Pinus* spp.].

11465   Pérez de la Rosa, J. A. (1994). Nota de *Pinus ayacahuite* (Pinaceae). Bol. Inst. Bot. I.B.U.G. 2 (3–4): 103–104. [note on type locality].

11466   Pérez de la Rosa, J. A., S. A. Harris & A. Farjon (1995). Noncoding chloroplast DNA variation in Mexican pines. Theor. Appl. Genet. 91: 1101–1106. [*Pinus* spp.].

11470 Perry, J. P., Jr. (1982). The taxonomy and chemistry of *Pinus estevezii*. J. Arnold Arbor. 63: 187–198. [*P. estevezi* (Martínez) comb. et stat. nov., p. 187].

11480 Perry, J. P., Jr. (1987). A new species of *Pinus* from Mexico and Central America. J. Arnold Arbor. 68: 447–459. [*P. nubicola*, sp. nov., p. 447].

11490 Perry, R. S. (1954). Yellow cedar; its characteristics, properties and uses. Canada Dept. Aff. and Nat. Res., Forestry Branch Bull. 114: 1–19. [*Chamaecyparis nootkatensis*].

11491 Persoon, C. H. (1807). Synopsis plantarum, seu enchiridium botanicum complectens... Vol. 2, part 2. Paris. [*Podocarpus* l'Hér. ex Pers., p. 580, *P. elongatus* (Aiton) L'Hér. ex Pers., p. 580, *Juniperus prostrata* sp. nov. = *J. horizontalis* Moench, p. 632].

11500 Peters, E. J. (1902). Die Sonnencypresse (*Chamaecyparis obtusa* S. et Z.). Ill. Gart.-Zeit. (Wien) 27: 14–16. [ill., descr. hortic.].

11504 Peters, M. D. & D. C. Christophel (1978). *Austrosequoia wintonensis*, a new taxodiaceous cone from Queensland, Australia. Canad. J. Bot. 56: 3119–3128. [*Austrosequoia wintonensis* gen. et sp. nov., p. 3124, ill.].

11509 Philippi, R. A. (1861). Zwei neue Gattungen der Taxineen aus Chile. Linnaea 30: 730–735. [*Lepidothamnus* gen. nov.,p. 730, *L. fonkii* sp. nov., p. 731, *Prumnopitys* gen. nov., p. 731, *P. elegans* sp. nov. = *P. andina*, p. 732]

11510a Phillips, J. F. (1927). Fossil *Widdringtonia* in lignite of the Knysa series with a note on fossil leaves of several other species. South African J. Sci. 24: 188–197. [Cupressaceae, *Widdringtonia* in the Tertiary of Table Mountain, Cape Province, South Africa].

11510d Phipps, C. J., J. M. Osborn & R. A. Stockey (1995). *Pinus* pollen cones from the Middle Eocene Princeton chert (Allenby Formation) of British Columbia, Canada. Int. J. Plant Sci. 156 (1): 117–124. [*Pinus* spp., palaeobotany].

11511 Pichi Sermolli, R. E. G. & M. P. Bizzarri (1978). The botanical collections (Pteridophyta and Spermatophyta) of the AMF Mares-G. R. S. T. S. Expedition to Patagonia, Tierra del Fuego and Antarctica. Webbia 32 (2): 455–534. [*Austrocedrus chilensis* (D. Don) Pichi-Serm. et Bizzarri, comb. nov., p. 482].

11511a Piemenov, G. M. (1986). On the mass burial of larches in the Miocene Primor'ye. Palaeontol. J. 20: 114–118. [*Larix primoriensis* sp. nov.].

11511b Pilger, R. (1905). Ein neuer andiner *Podocarpus*. Fedde's Repert. Sp. Nov. Regni Veg. 1 (12): 189–190. [*P. utilior* sp. nov. = *Prumnopitys harmsiana*].

11511c Pilger, R. (1913). Ein neuer *Podocarpus*. Notizbl. Bot. Gart. Berlin-Dahlem 5: 299. [*P. roraimae* sp. nov.].

11512 Pilger, R. (1913). Juniperi species antillanae. in: I. Urban. Symbolae Antillanae... Vol. 7: 478–481. Berlin. [*Juniperus australis* (Endl.) comb. nov., p. 479, *J. gracilior* sp. nov., p. 481].

11513 Pilger, R. (1926). Pinaceae. in: I. Urban. Plantae Haitienses... III. Ark. Bot. 20 (4): A15: 9–10. [*Juniperus urbaniana* Pilger et Ekman, sp. nov.].

11530 Pilger, R. (1931). Die Gattung *Juniperus* L. Mitt. Deutsch. Dendrol. Ges. 43: 255–269, pl. 29–32, f. 1–5. [informal classification].

11531 Pilger, R. (1934). Taxaceae. In: J. Mildbraed (ed.). Neue und seltene Arten aus Ostafrika (Tanganyika-Territ. Mandat) leg. H. J. Schlieben, VI. [Taxaceae = *Podocarpus uluguren sis* sp. nov. = *P. milanjianus*, p. 82].

11532 Pilger, R. (1937). Die Podocarpaceae der Sammlung L. J. Brass aus Südost-Neuguinea. In: C. Lauterbach (ed.). Beiträge zur Flora von Papuasien XXII. Bot. Jahrb. Syst. 68: 244–247. [*Podocarpus brassii* sp. nov., p. 246].

11533 Pilger, R. (1938). Neue Podocarpaceae aus Neuguinea. In: L. Diels (ed.). Beiträge zur Flora von Papuasien XXIII. Bot. Jahrb. Syst. 69: 252–253. [*Dacrydium xanthandrum* sp. nov., *Podocarpus cinctus* sp. nov. = *Dacrycarpus cinctus*].

11542 Pillai, S. K. (1963). Structure and seasonal study of the shoot apex of some *Cupressus* species. New Phytol. 62: 335–341.

11550 Pinto da Silva, A. R. (1947). Sobre a sistemática dos pinheiros bravos portugueses. Brotéria Ci. Nat., Vol. 16 (43) fasc. 1–2: 61–76. (French summ.). [*Pinus maritima*, *P. mesogeensis* = *P. pinaster*].

11555 Piovesan, G., C. Pelosi, A. Schirone & B. Schirone (1993). Taxonomic evaluations of the genus *Pinus* (Pinaceae) based on electrophoretic data of salt soluble and insoluble seed storage proteins. Plant Syst. Evol. 186: 57–68.

11560 Pirotta, R. (1887). Sul genere *Keteleeria* di Carrière (*Abies fortunei* Murr.). Bull. Soc. Tosc. Ortic. 12: 269–274. [key Abietineae = Pinaceae, *Keteleeria fortunei*].

11570 Plancarte, B. (1990). Variacion e longitud de cono y peso de semilla en *Pinus greggii* Engelm. de tres procedencias de Hidalgo y Queretaro. Centro de Genetica Forestal A.C., Nota técnica 4: 1–6. Chapingo, Mexico.

11571 Plavsic, S. (1936). Anatomische Untersuchungen über *Picea omorica*. Bot. Centralbl. Beih. A, 54 (3): 429–493. [*Picea omorika*, ill.].

11572 Plavsic, S. (1938). Die Standorte von *Picea omorica* im südlichen Drina-Gebiet. Österreich. Bot. Zeitschr. 87: 140–145. [*Picea omorika*, ecology, map].

11580 Plavsic, S. (1938). Phylogenetische Untersuchungen über die Gattung *Picea* auf Grund der Blattanatomie. Planta 28: 453–463.

11580c Pocknall, D. T. (1981). Pollen morphology of the New Zealand species of *Libocedrus* Endlicher (Cupressaceae) and *Agathis* Salisbury (Araucariaceae). New Zealand J. Bot. 19 (3): 267–272. [*Libocedrus bidwillii*, *L. plumosa*, *Agathis australis*].

11580d Podogas, A. V., A. V. Shurkhal, V. L. Semerikov & L. A. Zhyvotovsky (1991). Evaluation of genetic differentiation between two pine species *Pinus sibirica*, subgenus *Strobus*, and *P. sylvestris*, subgenus *Pinus*, in specimens from the botanical garden and from natural populations. Genetika 7: 758–762. (Russ., Eng. summ.).

11581 Poeppig, E. F. & S. L. Endlicher (1841). Nova genera ac species plantarum quas in regno chilensi peruviano et in terra amazonica... Vol. 3, part 3–4, pp. 17–32, plates 220–240. Lipsiae (Leipzig). [*Thuja andina* Poeppig et Endl., sp. nov. = *Austrocedrus chilensis*, p. 17, pl. 220, *Podocarpus spicatus* Poepp. = *Prumnopitys andina*, p. 18].

11590 Poiret, J. L. M. (1804). Pin – *Pinus*. in: J. B. A. P. M. de Lamarck. Encyclopédie methodique. Botanique... Vol. 5, pp. 329–342. Paris – Liège. [*Pinus laricio* sp. nov. = *P. nigra* Arnold (1785), p. 339].

11600 Poiret, J. L. M. (1805). Sapin – *Abies*. in: J. B. A. P. M. de Lamarck. Encyclopédie methodique. Botanique... Vol. 6, pp. 509–524. Paris – Liège 1804 (on t. p., but in fact 1805, see F. A. Stafleu & R. S. Cowan, 1979: TL-2 Vol. II: 732). [*Abies*, *Cedrus*, *Larix*, *Picea*, *A. vulgaris* sp. nov. = *A. alba*, p. 514, *A. taxifolia* (Lamb.) Poir. (non Desf.) = *Pseudotsuga menziesii*, p. 523, *A. araucana* (Molina) Poir. = *Araucaria araucana*, p. 524].

11610 Poiret, J. L. M. (1816). Pin – *Pinus*. in: M. Lamarck (= J. B. A. P. M. de Lamarck). Encyclopédie methodique. Botanique... Supplément Vol. 4, pp. 415–418. Paris.

11620 Poiret, J. L. M. (1817). Sapin – *Abies*. in: M. Lamarck (= J. B. A. P. M. de Lamarck). Encyclopédie methodique. Botanique... Supplément Vol. 5, p. 35. Paris. [*Abies fraseri* (Pursh) comb. nov.].

11621 Poiret, J. L. M. (1817). *Thuya*. Suite des espèces: 8. *Thuya lineata*; var. ? *Thuya (lavandulaefolia)*. in: M. Lamarck (= J. B. A. P. M. de Lamarck). Encyclopédie methodique. Botanique... Supplément Vol. 5, pp. 302–303. Paris. [*Thuja australis* Bosc ex Poir., sp. nov. = *Callitris rhomboidea*, p. 302, *T. lineata* sp. nov. = cultivar, probably of *Taxodium distichum*, p. 303: "Elles sont toutes deux cultivées dans le jardin de M. Noisette."].

11630 Poiteau, A. (1845). Des Conifères. Rev. Hort., sér. 2, 4: 4–13. [*Abies bracteata* comb. nov., p. 7, with basion. *Pinus bracteata* D. Don in Trans. Linn. Soc. London 17: 442, 1836, publ. before July 9].

11630a Pole, M. S. (1995). Late Cretaceous macrofloras of eastern Otago, New Zealand: gymnosperms. Austral. Syst. Bot. 8: 1067–1106. (ill.). [palaeobotany, *Araucaria desmondii*, *A. taieriensis* fossil spp. nov., *A. owenii* (Ettingsh.) comb. nov., *Araucarioides falcata* fossil sp. nov., *A. taenioides* (Cantrill) comb. nov., *Kaia minuta*, *Katikia inordinata* fossil spp. nov. (Podocarpaceae), *Otakauia lanceolata* fossil sp. nov. (Taxodiaceae), *Sequoiadendron novaezeelandiae* (Ettingsh.) comb. nov.].

11635 Ponchet, J. (1997). *Cupressus dupreziana* A. Camus, 1926. Enzyklopädie der Holzgewächse III–1 (?): 1–8. (ill., map).

11366 Ponchet, J. (1997). *Cupressus funebris* Endl., 1847. Enzyklopädie der Holzgewächse III–1 (8): 1–6. (ill., map).

11640 Posey, C. E. & J. F. Goggans (1967). Observations on species of cypress indigenous to the United States. Auburn Univ. Agric. Exp. Stat. Circ. 153. [*Cupressus* spp., 190 pp., ill.)

11650 Posey, C. E. & J. F. Goggans (1968). Variation in seeds and ovulate cones of some species and varieties of *Cupressus*. Auburn Univ. Agric. Exp. Stat. Circ. 160. (23 pp., ill.).

11651 Pourtet, J. & P. Turpin (1954). Catalogue des espèces cultivées dans l'Arboretum des Barres. Ann. École Natl. Eaux 9 (1): 97–120, t. 1–3. [*Abies pinsapo* var. *tazaotana* (Cozar ex Hug. del Vill.) Pourtet, comb. et stat. nov., p. 100].

11660 Powrie, E. (1972). The typification of *Brunia nodiflora* L. J. S. African Bot. 38 (4): 301–304. [*Widdringtonia nodiflora* (L.) Powrie, comb. nov., p. 303].

11680 Pravdin, L. F. (1969). Scots Pine Variation, infraspecific taxonomy and selection. Jerusalem. (transl. from Russ. orig. 1964). [*Pinus sylvestris* L. et var.)

11690 Pravdin, L. F., G. A. Abaturova & O. P. Shershukova (1976). Karyological analysis of European and Siberian spruce and their hybrids in the U.S.S.R. Silvae Genet. 25: 89–95. [*Picea abies*, *P. obovata*].

11690a Pravdin, L. F. & A. I. Irishnikov (1982). Genetics of *Pinus sibirica* Du Tour, *P. koraiensis* Sieb. et Zucc. and *P. pumila* Regel. An. Sumarstvo (Ann. Forest.) 9/3: 79–123.

11690b Premoli, A. C. (1994). South American temperate conifer species: a larger list. Biodiv. Conserv. 3: 295–297. [*Prumnopitys andina*, *Podocarpus salignus*, *P. nubigenus*, *Saxegothaea conspicua*, conservation].

11690c Premoli, A. C., C. P. Souto, A. E. Rovere, T. R. Allnut & A. C. Newton (2002). Patterns of isozyme variation as indicators of bio-geographic history in *Pilgerodendron uviferum* (D. Don) Florin. Div. Distrib. 8: 57–66. [genetics].

11690e Presl, J. S. & K. B. Presl (1822). Deliciae Pragenses, historiam naturalem spectantes. Prague. [*Juniperus hemisphaerica* sp. nov. = *J. communis* var. *communis*, p. 142].

11690f Presl, K. B. (1846). Botanische Bemerkungen. Abh. Königl. Böhm. Ges. Wiss., ser. 5, 3: 431–584. [*Podocarpus wallichianus* sp. nov. = *Nageia wallichiana*, p. 540].

11690g Presl, K. B. (1851). Epimeliae botanicae... Prague. [*Podocarpus costalis* sp. nov., p. 236].

11721 Price, R. A., A. Liston & S. H. Strauss (1998). Phylogeny and systematics of *Pinus*. In: D. M. Richardson (ed.). Ecology and biogeography of *Pinus*. (pp. 49–68). Cambridge University Press, Cambridge, UK. [*Pinus*: overview of major classifications and species, new classification of 111 species; for nearly simultaneously published revision of Mexican pines see Farjon & Styles, 1997].

11725 Price, W. R. (1931). On the distribution of *Pseudolarix fortunei*, the golden larch. Kew Bull. 2: 67–68. [= *Pseudolarix amabilis*].

11730 Priehäusser, G. (1958). Die Fichtenvariationen und –kombinationen des Bayerischen Waldes nach phänotypischen Merkmalen mit Bestimmungsschlüssel. Forstwiss. Centralbl. 77: 151–171. [*Picea abies* var., key].

11730a Primack, R. B. (2003). Genetic piracy: a newly discovered marvel of the plant world. Arnoldia 62 (2): 27–30. [*Cupressus dupreziana*, genetics, DNA, embryology (claiming "male apomixis" in this conifer)].

11750 Prus-Glowacki, W., J. Szweykowski & R. Nowak (1985). Serotaxonomical investigation of the European pine species. Silvae Genet. 34: 162–170. [*Pinus* spp.].

11751 Purkyne, E. von (1877). Eine asiatische Conifere in den Balkanländern. Österreich. Monatsschr. Forstwesen 27: 446. [*Picea omorika* (Pančić) comb. nov.].

11752 Pursh, F. T. (1814). Flora americae septemtrionalis... Vol. 1–2. London. [*Pinus fraseri* sp. nov. = *Abies fraseri*, Vol. 2, p. 639, *Juniperus communis* var. *depressa* var. nov., Vol. 2, p. 646].

11752a Pye, M. G., P. A. Gadek & K. J. Edwards (2003). Divergence, diversity and species of the [sic!] Australasian *Callitris* (Cupressaceae) and allied genera: evidence from ITS sequence data. Australian Syst. Bot. 16: 505–514. [*Actinostrobus*, *Callitris*, *Neocallitropsis*, *Widdringtonia*, DNA, phylogeny, cladistics, cladograms; based on ITS, an unreliable ribosomal gene for phylogenetic inference, see Álvarez & Wendel in Mol. Phylogen. Evol. 29: 417–434 (2003), the authors propose that *Callitris* is paraphyletic].

11752e Qian, T., R. A. Ennos & T. Helgason (1995). Genetic relationships among Larch species based on analysis of restriction fragment variation for chloroplast DNA. Canad. J. Forest Res. 25: 1197–1202. [*Larix decidua*, *L. gmelinii*, *L. griffithii*, *L. kaempferi*, *L. laricina*, *L. occidentalis*, *L. potaninii*, *L. sibirica*, no support of classif. based on morphology].

11753 Qu, S. Z., K. Y. Wang & T. C. Cui (1988). A new species of *Pseudotsuga* Carr. from Shaanxi and a new subspecies of *Fraxinus* Linn. Acta Bot. Bor.-Occid. Sin. 8 (2): 129. (Chin., Lat.). [*Pseudotsuga shaanxiensis* S. Z. Qu & K. Y. Wang, p. 129].

11754 Quinn, C. J. (1982). Taxonomy of *Dacrydium* Sol. ex Lamb. emend. de Laub. (Podocarpaceae). Austral. J. Bot. 30 (3): 311–320. [Podocarpaceae, *Dacrydium* amended, p. 313, *Lagarostrobos* gen. nov., p. 316, *L. franklinii* (Hook. f.) comb. nov., p. 316, *Lepidothamnus* Phil. reinstated, p. 316, *L. intermedius* (Kirk) comb. nov., p. 316, *L. laxifolius* (Hook. f.) comb. nov., p. 316, *Lagarostrobos colensoi* (Hook.) comb. nov. = *Manoao colensoi*, p. 317, *Halocarpus* gen. nov., p. 317, *H. bidwillii* (Hook. f. ex Kirk) comb. nov., p. 317, *H. biformis* (Hook.) comb. nov., p. 318, *H. kirkii* (F. Muell. ex Parl.) comb. nov., p. 318, generic key to fam., p. 319].

11755 Quinn, C. J. & P. A. Gadek (1988). Sequence of xylem differentiation in leaves of Cupressaceae. Amer. J. Bot. 75 (9): 1344–1351. [*Callitris muelleri*, *Chamaecyparis lawsoniana*, *Cupressus sempervirens*, *Juniperus conferta*, *Widdringtonia cedarbergensis*, ill.].

11755e Quijada, A., G. Mendez Cardenas, S. Ortiz Garcia & E. Alvarez Buylla (1997). La region de los ITS del AND ribosomal del nucleo (nrADN), fuente de caracteres moleculares en la sistematica de las gimnospermas. Bol. Soc. Bot. Mexico 60: 159–168. [*Pinus* spp., rnDNA].

11756 Rac, M. & A. Z. Lovric (1990). Taxonomische Variationsübersicht von Tannensippen West-balkans. VI. IUFRO – Tannensymposium, Zagreb, 24–27.09.1990. Zusammenfassungen der Referate. Forstwissenschaftliche Fakultät der Universität Zagreb. [with other contrib. on *Abies alba* and *A. cephalonica* complexes].

11760 Raciborski, M. & W. Szafer, eds. (1919). Flora polska rosliny maczyniowe polski i ziem osciennych... Vol. 1. [*Larix polonica* Racib. = *L. decidua* var. *polonica*, p. 51; descr. earlier in Kosmos (Lvov) ..: 494 (1890), and by Wóycicki in Obraz. Rośl. Król. Polsk. 2: 15–16, tab. 1 (1912)].

11772   Raddi, S. & S. Sümer (1999). Genetic diversity in natural *Cupressus sempervirens* L. populations in Turkey. Biochem. Syst. Ecol. 27: 799–814.

11790   Ramond, L. F. E. de C. (1805). (*Pinus uncinata*) in: A. P. de Candolle & J. B. A. P. M. de Lamarck. Flore française Ed. 3, Vol. 3: 726. Paris. [often ascribed to Ramond, e.g. in Christensen, 1987, p. 392, but validly published by Candolle in Lamarck & Candolle].

11795   Ratzel, S. R., G. W. Rothwell, G. Mapes, R. H. Mapes & L. A. Doguzhaeva (2001). *Pityostrobus hokodzensis*, a new species of pinaceous cone from the Cretaceous of Russia. J. Paleontol. 75 (4): 895–900.

11800   Raup, H. M. (1946). Phytogeographic studies in the Athabaska – Great Slave Lake region. II. J. Arnold Arbor. 27: 1–85. [*Picea mariana, P. glauca, Larix laricina, Pinus banksiana*].

11810   Raup, H. M. (1947). The Botany of southwestern Mackenzie. Sargentia 6: 1–275. [Pinaceae, Cupressaceae pp. 102–107, *Picea glauca* var. *porsildii* var. nov., p. 102, *Larix laricina* var. *alaskensis* (Wight) comb. nov., p. 105].

11820   Ray, J. (1688). Historia Plantarum. Tomus Secundus, "De Pino Arbore". Vol. II. (pp. 1398–1401). London. [Pinaceae, *Pinus*].

11830   Read, R. A. (1980). Genetic variation in seedling progeny of Ponderosa Pine provenances. Forest Sci. Monogr. 23. [*Pinus ponderosa*].

11835   Rees, A. (1819). The Cyclopaedia; or, universal dictionary of arts, sciences, and literature by Abraham Rees, ...Vol. 35 (T-TOL). London. [*Taxus spinulosa* (*Taxus* No. 7) Sm. sp. nov. = *Podocarpus spinulosus*].

11840   Regal, P. (1979). Australia's own pines. Austral. Nat. Hist. 19 (12): 386–391. [Cupressaceae, *Actinostrobus, Callitris*].

11850   Regel, E. A. von (1857). Neue oder interessante Pflanzen des Botanischen Gartens zu St. Petersburg. Gartenflora 6: 342–346. [*Cupressus karwinskiana* sp. nov. = *C. lusitanica, Juniperus caesia* sp. nov. (non Carrière) = *J. communis* var. ?].

11860   Regel, E. A. von (1859). Index Semina Hortus Petropolitanus 1858. St. Petersburg. [*Pinus pumila* (Pall.) comb. et stat. nov., p. 23].

11870   Regel, E. A. von (1871). Bemerkungen und Untersuchungen über die Arten der Gattung *Larix*. Gartenflora 20: 99–107, pl. 684–685. [*Larix dahurica* var. *japonica* Maxim. ex Regel = *L. gmelinii* var. *japonica*, p. 105, morphology; also publ. in Trudy Imp. S.-Petersburgsk. Bot. Sada 1, 1871].

11880   Regel, E. A. von (1872). Observations sur les espèces du genre *Larix* ou mélèze. Hort. Belge 22: 96–106, pl. 7–10. [morphology, transl. of E.A. von Regel, 1871].

11881   Regel, E. A. von (1879). Descriptiones plantarum novarum in regionibus Turkestanicis a cl. viris Fedjenko, Korolkow, Kuschakewicz et Krause collectis... No. 7. Trudy Imp. S.-Petersburgsk. Bot. Sada 6 (2): 287–538. [*Juniperus semiglobosa* sp. nov., p. 487].

11890   Regel, E. A. von (1888). *Sciadopitys verticillata* Sieb. et Zucc. Gartenflora 37: 437–438, f. 101.

11891   Rehder, A. (1919). New species, varieties and combinations from the Herbarium and the Collections of the Arnold Arboretum. J. Arnold Arbor. 1: 44–60. [*Taxus chinensis* (Pilg.) comb. et stat. nov., p. 51, *Pseudolarix amabilis* (Nelson) comb. nov., p. 53, *Abies homolepis* var. *tomomi* (Bobb. et Atk.) comb. et stat. nov., p. 53].

11892   Rehder, A. (1922). New species, varieties and combinations from the Herbarium and the Collections of the Arnold Arboretum. J. Arnold Arbor. 3: 207–224. [*Juniperus squamata* var. *meyeri* var. nov., p. 207, *J. lucayana* var. *bedfordiana* var. nov. = *J. virginiana*, p. 208, *Pinus nigra* var. *cebennensis* (Gren. et Godron) comb. nov. = *P. nigra* ssp. *salzmanii* (Dunal) Franco, p. 208].

11900   Rehder, A. (1923). New species, varieties and combinations from the Herbarium and the Collections of the Arnold Arboretum. J. Arnold Arbor. 4: 107–116. [Coniferales, *Cupressus*].

11900a   Rehder, A. (1923). Enumeration of the ligneous plants of northern China. Ginkgoaceae to Ranunculaceae. J. Arnold Arbor. 4: 117–192. [Coniferales pp. 118–128, *Abies, Pinus, Juniperus, Platycladus orientalis* etc.].

11900b   Rehder, A. (1926). New species, varieties and combinations from the Herbarium and the Collections of the Arnold Arboretum. J. Arnold Arbor. 7: 22–23. [*Taxodium ascendens* f. *nutans* (Sweet) comb. nov., p. 22, *Pinus tabuliformis* var. *densata* (Masters) comb. nov., p. 23].

11902   Rehder, A. (1927). Neuere und seltene Gehölze. Mitt. Deutsch. Dendrol. Ges. 35: 34–47. [*Thuja koraiensis, Juniperus squamata* var. *meyeri, Abies koreana, Picea glauca* var. *conica* (Rehder) comb. nov., p. 37].

11910   Rehder, A. (1929). Notes on the ligneous plants described by H. Léveillé from Eastern Asia. J. Arnold Arbor. 10: 108–132. [*Keteleeria esquirolii* Léveillé and other Pinaceae; full reference to Rehder's comment on Léveillé-taxa is given in F. A. Stafleu & R. S. Cowan, TL-2 Vol. II, p. 863].

11920   Rehder, A. (1937). Notes on the ligneous plants described by H. Léveillé from Eastern Asia: additions and corrections. J. Arnold Arbor. 18: 206–257. [*K. esquirolii* Léveillé, p. 254].

11930 Rehder, A. (1939). New species, varieties and combinations from the collections of the Arnold Arboretum. J. Arnold Arbor. 20 (1): 85–86. [*Abies ernestii* sp. nov. = *A. recurvata* var. *ernestii*, p. 85, *Picea* × *notha* Rehder].

11940 Rehder, A. (1939). The firs of Mexico and Guatemala. J. Arnold Arbor. 20 (3): 281–287. [*Abies* spp., *A guatemalensis* sp. nov., p. 285].

11942 Rehder, A. (1940). *Abies procera*, a new name for *Abies nobilis* Lindl. Rhodora 42: 522–524. [*Abies procera* nom. nov., p. 522].

11943 Rehder, A. (1941). New species, varieties and combinations from the collections of the Arnold Arboretum. J. Arnold Arbor. 22: 569–579. [*Cephalotaxus harringtonii* var. *nana* var. nov., p. 571].

11987 Rehfeldt, G. E. (1997). Quantitative analyses of the genetic structure of closely related conifers with disparate distributions and demographics: the *Cupressus arizonica* (Cupressaceae) complex. Amer. J. Bot. 84 (2): 190–200. [*Cupressus arizonica, C. glabra, C. momtana, C. nevadensis, C. sargentii, C. stephensonii* (= var. under *C. arizonica*); (population) genetics inferred from growth performance traits in common garden trials of 0–3 year seedlings].

11987a Rehfeldt, G. E. (1999). Systematics and genetic structure of *Ponderosae* taxa (Pinaceae) inhabiting the mountain islands of the southwest. Amer. J. Bot. 86 (5): 741–752. [*Pinus ponderosa, P. washoensis*].

11988 Rehfeldt, G. E., B. C. Wilson, S. P. Wells & R. M. Jeffers (1996). Phytogeographic, taxonomic, and genetic implications of phenotypic variation in the *Ponderosae* of the Southwest. Southwestern Naturalist 41 (4): 409–418. [*Pinus ponderosa, P. durangensis, P. arizonica, P. engelmannii*; phenotypic study of foliage and cone characteristics in SW U.S.A.].

11990 Rendle, A. B. (1894). Gymnospermae. in: The Plants of Milanji, Nyasa-land, collected by Mr. Alexander Whyte, ... Trans. Linn. Soc. London, Bot., ser. 2, 4 (1): 60–62, pl. 19. [*Widdringtonia whytei* sp. nov., p. 60, *Podocarpus milanjianus* sp. nov., p. 61].

12000a Renner, O. (1904). Über Zwitterblüten bei *Juniperus communis*. Flora 93: 297–300. (ill.). [Ovuliferous cones subtended by whorls of microsporophylls alternating with sterile scale leaves].

12001 Renner, O. (1907). Über die weibliche Blüte von *Juniperus communis*. Flora 97: 421–430. (ill.).

12004 Rentería-Arrieta, L. I. & A. Garcia-Arévalo (1997). Las coníferas de la Reserva de la Biosfera "La Michilía", Durango, México. Madera y Bosques 3 (1): 53–70. [*Cupressus lusitanica, Juniperus* spp., *Pinus* spp., *Pseudotsuga menziesii*].

12010 Reshetnyak, T. A. (1977). Classification of *Chamaecyparis* Spach by morphological differences of seeds, cones and seedlings. Ukrajins'k. Bot. Žurn. 34 (6): 649–651. (Russ., Eng. summ.).

12012 Reveal, J. L. & A. B. Doweld (2002). Proposal to conserve the family name Microcachrydaceae (Pinophyta). Taxon 51 (3): 573. [Podocarpaceae, *Microcachrys*, nomenclature].

12014 Rezzi, S., C. Cavaleiro, A. Bighelli, L. Salgueiro, A. Proença da Cunha & J. Casanova (2001). Intraspecific chemical variability of the leaf essential oil of *Juniperus phoenicea* subsp. *turbinata* from Corsica. Biochem. Syst. Ecol. 29: 179–188.

12019 Richard, A. (1822). *Araucaria brasiliana* sp. nov., *A. dombeyi* sp. nov. (Coniferae). in: J. B. G. M. Bory de Saint-Vincent *et al.* Dictionnaire classique d'histoire naturelle. Vol. 1, p. 512. [= *A. angustifolia* and *A. araucana* respectively].

12020 Richard, A. (1823). *Cedrus libani* sp. nov. (Coniferae). in: J. B. G. M. Bory de Saint-Vincent *et al.* Dictionnaire classique d'histoire naturelle. Vol. 3, p. 299.

12020a Richard, A. (1832). Botanique, part 1. Essai d'une Flore de la Nouvelle Zélande. In: A. Lesson & A. Richard. Voyage découvertes de l'Astrolabe: exécuté par l'ordre du Roi pendant les années 1826–1827–1828–1829 sous le commendement de M. J. Dumont d'Urville. Paris. [*Podocarpus dacrydioides* sp. nov. = *Dacrycarpus dacrydioides*, p. 358].

12021 Richard, L. C. M. (1810). Note sur les plantes dites Conifères. Ann. Mus. Natl. Hist. Nat. (Paris) 16: 296–299. [*Taxodium* gen. nov., p. 298, *T. distichum* (L.) comb. nov., p. 298].

12031 Richardson, D. M. (ed.) (1998). Ecology and biogeography of *Pinus*. Cambridge University Press, Cambridge, UK. [*Pinus* all spp.; Part 1: Introduction (D. M. Richardson & P. W. Rundel), Part 2: Evolution (C. I. Millar), phylogeny and systematics (R. A. Price, A. Liston & S. H. Strauss), Part 3: Historical biogeography, Part 4: Macroecology and recent biogeography, Part 5: Ecological themes, Part 6: Pines and humans; extensive bibliography, ill., maps].

12033 Richardson, D. M., P. A. Williams & R. J. Hobbs (1994). Pine invasions in the Southern Hemisphere: determinants of spread and invadability. J. Biogeogr. 21: 511–527. [*Pinus radiata, P. caribaea, P. patula, P. oocarpa, P. elliottii, P. taeda, P. kesiya, P. halepensis*, other species].

12040 Rickett, H. W. (1950). The botanical name of the big tree. J. New York Bot. Gard. 51: 15. [*Sequoidendron giganteum*].

12041   Rigual, A. & F. Esteve (1952). Algunas anotaciones sobre los últimos ejemplares de *Callitris quadrivalvis* Vent. en la Sierra de Cartagena. Anales Inst. Bot. Cavanilles 11 (1): 437–476. [= *Tetraclinis articulata*, ecology, distribution, habitus photogr.].

12050   Rikli, M. (1909). Die Arve in der Schweiz. Ein Beitrag zur Waldgeschichte und Waldwirtschaft der Schweizer Alpen. Schweiz. Naturf. Ges. 44 (I–XL): 1–455. [*Pinus cembra*].

12060   Rikli, M. (1926). *Pinus pinea* L. Pflanzenareale 1 (1): 12–13, map 9.

12070   Riskind, D. H. & T. F. Patterson (1975). Distributional and Ecological Notes on *Pinus culminicola*. Madroño 23: 159–161.

12071   Risso, J. A. (1827). Histoire naturelle des principales productions de l'Europe méridionale... Vol. 2. Paris. [*Juniperus prostrata* et *racemosa*, spp. nov., p. 459: species dubiae cf. Parlatore (in DC.), 1867].

12088   Robert, M.-F. (1977). Aspect phytogéo-graphiques et écologiques des forêts de *Pinus cembroides*. I. Les forêts de l'est et du nord-est du Mexique. Bull. Soc. Bot. France 124: 197–216.

12090   Robert, M.-F. (1978). Un nouveau pin pignon Mexicain: *Pinus johannis*. Adansonia, sér. 2, 18: 365–373. [*P. johannis* sp. nov., p. 366].

12100   Robert-Passini, M.-F. (1981). Deux nouveaux pins pignons du Mexique. Bull. Mus. Hist. Nat. (Paris), sér. 4, sect. B., Adansonia 1: 61–73. [*Pinus cembroides* var. *lagunae* var. nov., p. 64, *Pinus catarinae* sp. nov., p. 70].

12101   Robinson, B. L. & M. L. Fernald (1894). New plants collected by Messrs. C. V. Hartman and C. E. Lloyd upon an archeological expedition to northwestern Mexico under the direction of Dr. Carl Lumholtz. Proc. Amer. Acad. Arts 30: 114–123. [*Pinus lumholtzii* sp. nov., p. 122].

12101a Robinson, C. R. (1977). *Prepinus parlinensis* sp. nov., from the Late Cretaceous of New Jersey. Bot. Gaz. (Crawfordsville) 138: 252–256. (ill.). [palaeobotany].

12101b Robinson, C. R. & C. N. Miller, Jr. (1977). Anatomically preserved seed cones of the Pinaceae from the Early Cretaceous of Virginia. Amer. J. Bot. 64: 770–779. (ill.). [palaeobotany].

12102   Robinson, J. F. & E. Thor (1969). Natural variation in *Abies* of the southern Appalachians. Forest Sci. 15 (3): 238–246. [*Abies fraseri*].

12104   Robson, K. A., J. Maze, R. K. Scagel & S. Banerjee (1993). Ontogeny, phylogeny and intraspecific variation in North American *Abies* Mill. (Pinaceae): an empirical approach to organization and evolution. Taxon 42 (1): 17–34. [*Abies* spp., cladistic anlysis based on data by Farjon, 1990].

12105   Robyns, W. (1935). Sur les espèces de *Podocarpus* du Congo belge et du Ruanda-Urundi. Bull. Séances Inst. Roy. Colon. Belge 6 (1): 226–241, t. 1–5. [*Afrocarpus, Podocarpus*].

12106   Robyns, W. (1946). Sur l'existence du *Juniperus procera* Hochst. au Congo Belge. Bull. Jard. Bot. de l'Etat, Bruxelles 18 (1/2): 125–131. (map).

12110   Roche, L. (1964). A further taxonomic distinction between White and Engelmann Spruce. Canad. Dept. Lands Forests, British Columbia Forest Serv. Forest. Res. Rev. 1964: 58. [*Picea glauca, P. engelmannii*].

12120   Roche, L. (1969). A genecological study of the genus *Picea* in British Columbia. New Phytologist 68: 505–554, pl. 1.

12130   Rodin, L. E. (1934). Materialen zur Kenntnis der Wälder des Tian-Schan. Die Fichtenwälder des Nordabhanges im Dshungarischen Alatau. Trudy Bot. Inst. Akad. Nauk SSSR, Ser. 3, Geobot. 1: 272–300, f. 1–3. (Russ., Germ. summ.). [*Picea schrenkiana*].

12136   Rodriguez R., R., O. Matthei S. & M. Quezada M. (1983). Flora Arbórea de Chile. Univ. de Concepción, Chile. [*Araucaria araucana* pp. 64–68, *Austrocedrus chilensis* pp. 73–75, *Fitzroya cupressoides* pp. 155–157, *Pilgerodendron uviferum* pp. 267–269, *Podocarpus nubigenus* pp. 273–275, *P. salignus* pp. 277–279, *Prumnopitys andina* pp. 293–295, *Saxegothaea conspicua* pp. 317–319, table of general distribution].

12140   Roezl, B., et Cie. (1857). Catalogue des graines des Conifères Mexicaines qui se trouvent chez B. Roezl et Cie, Horticulteurs à Napoles près Mexico. Pour Automne 1857 et Printemps 1858. Impr. de M. Murguia, Mexico City. [*Pinus*, spp. nov.; see also D. F. L. von Schlechtendal, 1857–58, for Latin descr. of these spp., and G. R. Shaw, 1909, for comment].

12150   Roezl, B., et Cie. (1858). Catalogue des graines et plantes Mexicaines. En vente chez B. Roezl et Cie, Horticulteurs à Napoles près Mexico. Pour Automne 1858 et Printemps 1859. Impr. Félix Malteste et Cie, Paris. [*Pinus* spp., nomina nuda; see G. R. Shaw, 1909, for comment].

12151   Rohrig, E. (1957). Ueber die Schwarzkiefer (*Pinus nigra* Arnold) und ihre Formen. Silvae Genet. 6 (1): 39–53. (ill.).

12153   Roig, F. A. (1992). Comparative wood anatomy of southern South American Cupressaceae. I.A.W.A. Bull., n.s. 13 (2): 151–162. [*Austrocedrus chilensis, Fitzroya cupressoides, Pilgerodendron uviferum*, dist. map., ill.].

12160   Roldan, A. & M. Martínez (1923–24). Nuestros árboles forestales. México Forest. 1 (9–10): 5–6, ill., (11–12): 7–9, ill. (1923); 2 (13–14): 11–14, ill. (1924). [*Taxodium mucronatum, Abies religiosa, Cupressus thurifera* Schltdl. = *C. lusitanica* var. *benthamii*].

12170   Roller, K. J. (1966). Resin Canal Position in the needles of Balsam, Alpine and Fraser firs. Forest Sci. 12 (3): 348–355. [*Abies balsamea, A. lasiocarpa, A. fraseri*].

12170a   Romanova, E. V. (1975). Representatives of Cupressaceae family from the upper Cretaceous deposits of Mount Juvankara (Zaissan Depression). Bot. Žurn. 60: 1191–1194. (Russ.). [*Thuja cretaceae* (Heer) Newberry = *Cupressinocladus interruptus* (Newberry) Schweitzer, *Libocedrus catenulata* (Bell) Krysht.].

12170b   Romero, A., E. Garcia & M.-F. Passini (1996). *Pinus cembroides* s.l. y *Pinus johannis* del Altiplano Mexicano. Acta Bot. Gallica 143 (7): 681–693. [*Pinus cembroides, P. johannis* (= *P. cembroides* var. *bicolor*), ecology].

12170c   Romero, A., M. Luna, E. García & M.-F. Passini (2000). Phenetic analysis of the Mexican midland pinyon pines, *Pinus cembroides* and *Pinus johannis*. Bot. J. Linn. Soc. 133 (2): 181–194. (ill.). [*Pinus cembroides, P. culminicola, P. johannis* = *P. cembroides* var. *bicolor*].

12170d   Romero, J. L., T. K. Stanger & W. S. Dvorak (2001). Collection and conservation of *Pinus maximartinezii*, a rare and endangered pine species. S. African Forest. J. 190: 95–98.

12170e   Roof, J. (1978). Two-needled pinyon in California. The Four Seasons 5 (3): 2–8. [*Pinus edulis, P. monophylla*].

12171   Ross, J. G. & R. E. Duncan (1949). Cytological evidences of hybridization between *Juniperus virginiana* and *J. horizontalis*. Bull. Torrey Bot. Club 76 (6): 414–429. [hybridization in Wisconsin].

12180   Ross, R. (1956). Homonyms, nomina nuda, and the Douglas-fir. Taxon 5: 43–46. [*Pseudotsuga menziesii*].

12190   Roth, I. (1962). Histogenese und morphologische Deutung der Doppelnadeln von *Sciadopitys*. Flora 152: 1–23.

12191   Rothmaler, W. (1941). Nomenklatorisches, meist aus dem westlichen Mittelmeergebiet. III. d) Abraham D. Juslenius, Centuria I Plantarum. 19-II-1755. Repert. Spec. Nov. Regni Veg. 50 (1): 72. [*Tetraclinis aphylla* (Jusl.) Rothm., comb. nov., with basion. *Thuja aphylla* Jusl. = *T. aphylla* L., 1763, p. 1422].

12191d   Rothwell, G. W. & J. F. Basinger (1979). *Metasequoia milleri* n. sp., anatomically preserved pollen cones from the Middle Eocene (Allenby Formation) of British Columbia. Canad. J. Bot. 57 (8): 958–970. [leaves decussate, microsporophylls helically arranged, as in extant species].

12191e   Rothwell, G. W., L. Grauvogel-Stamm & G. Mapes (2000). An herbaceous fossil conifer: Gymnospermous ruderals in the evolution of Mesozoic vegetation. Palaeo 156: 139–145. (ill.). [*Aethophyllum stipulare* Brongn., Triassic ruderal conifer].

12191h   Rothwell, G. W. & G. Mapes (2001). *Barthelia furcata* gen. et sp. nov., with a review of Palaeozoic coniferophytes and a discussion of coniferophyte systematics. Int. J. Plant Sci. 162 (3): 637–667. (ill.). [Walchian conifers, cordaiteans, ginkgophytes, coniferophytes, palaeobotany].

12192   Rouane, P. (1962). Croissance et ramification de quelques Cupressacées. Trav. Lab. Forest. Toulouse, T. 1, vol. 6, art. 11: 1–15. [*Chamaecyparis lawsoniana, Thuja plicata*, ill.].

12202   Rouse, R. J., P. R. Fantz & T. E. Bilderback (1997). Problems identifying Japanese cedar cultivated in the United States. (Review). HorTechnology 7 (2): 129–133. [*Cryptomeria japonica*, nomenclature, cultivars].

12210   Rowlee, W. W. (1903). Notes on Antillean Pines with description of a new species from the Isle of Pines. Bull. Torrey Bot. Club 30: 106–108. [*Pinus recurvata* sp. nov. = *P. caribaea* Morelet].

12211   Roxburgh, W. (1814). Hortus bengalensis, or a catalogue of the plants growing in the honourable East India Company's Botanic Garden at Calcutta. Serampore. [*Pinus longifolia* "sp. nov." (non Salisb.) = *P. roxburghii*, p. 68, *P. deodara* sp. nov. = *Cedrus deodara*, p. 69, *Juniperus dimorpha* sp. nov. = *J. chinensis* L., p. 73, *J. chinensis* sp. nov. (non L.) = *Podocarpus chinensis, J. elata* sp. nov. = *Dacrydium elatum, J. patens* sp. nov. = nom. nud., p. 73].

12212   Royle, J. F. (1833–40). Illustrations of the Botany and other branches of the Natural History of the Himalaya Mountains and of the Flora of Cashmere. Vol. 1–2. London. [*Abies pindrow* (D. Don) comb. nov. [as "sp. nov.", but see D.J. Mabberley, 1982), vol. 1, p. 350, t. 86, 1836, *Cupressus sempervirens* var. *indica*, nom. nud.].

12216   Rubner, K. (1936–41). Beitrag zur Kenntnis der Fichtenformen und Fichtenrassen. Tharandt Forstl. Jb. 87 (1936): 101–176; 90 (1939): 883–915; 92 (1941): 462–472, 526–545. [*Picea abies*, ill.].

12220   Ruby, J. L. & J. W. Wright (1976). A revised classification of geographic varieties in Scots Pine. Silvae Genet. 25: 169–175. [*Pinus sylvestris*].

12230   Rudloff, E. von (1967). Chemosystematic studies in the genus *Picea* (Pinaceae). I. Introduction. Canad. J. Bot. 45: 891–901. [*Picea glauca, P. mariana*].

12240   Rudloff, E. von (1967). Chemosystematic studies in the genus *Picea* (Pinaceae). II. The leaf oil of *Picea glauca* and *P. mariana*. Canad. J. Bot. 45: 1703–1714.

12245   Rudloff, E. von (1972). Chemosystematic studies of the genus *Pseudotsuga*. I. Leaf oil analysis of the coastal and Rocky Mountain varieties of Douglas Fir. Canad. J. Bot. 50: 1025–1040. [*Pseudotsuga menziesii* var. *menziesii*, *P. menziesii* var. *glauca*].

12253   Rudloff, E. von & M. Granat (1982). Seasonal variation of the terpenes of the leaves, buds and twigs of Balsam Fir (*Abies balsamea*). Canad. J. Bot. 60 (12): 2682–2685.

12255   Rüffle, L. & H. Süss (2001). Beitrag zur systematischen Stellung der ausgestorbenen Koniferengattung *Doliostrobus* Marion nach holzanatomischen Gesichtspunkten. Feddes Repert. 112 (7–8): 413–419. [Araucariaceae, Taxodiaceae, palaeobotany].

12260   Rundel, P. W. (1971). Community structure and stability in the Giant Sequoia Groves of the Sierra Nevada, California. Amer. Midl. Naturalist 85 (2): 478–492. [*Sequoiadendron giganteum*].

12270   Rundel, P. W. (1972). An annotated check list of the groves of *Sequoiadendron giganteum* in the Sierra Nevada, California. Madroño 21 (5): 319–328.

12274   Runions, C. J. & J. N. Owens (1999). Sexual reproduction of interior spruce (Pinaceae). I. Pollen germination to archegonial maturation. Int. J. Plant Sci. 160 (4): 631–640. [*Picea engelmannii*, *P. glauca*, embryology, development].

12275   Runions, C. J. & J. N. Owens (1999). Sexual reproduction of interior spruce (Pinaceae). II. Fertilization to early embryo formation. Int. J. Plant Sci. 160 (4): 641–652. [*Picea engelmannii*, *P. glauca*, embryology, development].

12280   Ruprecht, F. J. (1845). Flores samojedorum cisuralensium. in: Beitr. Pflanzenk. Russ. Reiches 2: 1–67. (St. Petersburg). [Coniferales pp. 55–57, *Juniperus communis*, *Pinus sylvestris*, *Abies obovata* = *Picea obovata*, *A. ledebourii* sp. nov. = *Larix sibirica*, p. 56, *A. gmelinii* sp. nov. = *Larix gmelinii*, p. 56].

12290   Ruprecht, F. J. (1854). Flora boreali-uralensis. Über die Verbreitung der Pflanzen im nördlichen Ural... St. Petersburg. (50 pp., ill.). [*Larix gmelinii* comb. nov., p. 48].

12300   Ruprecht, F. J. (1869). Sertum tianschanicum. Botanische Ergebnisse einer Reise im mittleren Tian-Schan. Mém. Acad. Imp. Sci. Saint-Pétersbourg, Sér. 7, Vol. 14, No. 4. [*Picea tianschanica* sp. nov., p. 72].

12310   Rushforth, K. D. (1972). Chinese Silver Firs. Honours thesis, Dept. of Forest., Univ. Aberdeen. [*Abies* spp.)

12320   Rushforth, K. D. (1983). *Abies chengii*, A previously overlooked Chinese Silver Fir. Notes Roy. Bot. Gard. Edinburgh 41 (2): 333–338. [*A. chengii* sp. nov., p. 333].

12330   Rushforth, K. D. (1984). Notes on Chinese Silver Firs 2. Notes Roy. Bot. Gard. Edinburgh 41 (3): 535–540. [*Abies* spp., *A. recurvata* var. *ernestii* (Rehder) "comb. et stat nov.", p. 536, (but see C. T. Kuan, 1983, p. 48), *A. chensiensis* ssp. *salouenensis* (Bord.-Rey et Gaussen) comb. et stat. nov. et ssp. *yulongxueshanensis* ssp. nov., p. 539].

12340   Rushforth, K. D. (1984). *Abies delavayi* and *A. fabri*. Int. Dendrol. Soc. Yearb. 1983: 118–120.

12350   Rushforth, K. D. (1986). Notes on Chinese Silver Firs 3. Notes Roy. Bot. Gard. Edinburgh 43 (2): 269–275. [*Abies* spp., *A. fabri* ssp. *minensis* (Bord.-Rey et Gaussen) comb. et stat. nov., p. 273].

12360   Rushforth, K. D. (1986). Mexico's Spruces – rare members of an important genus. The Kew Magazine 3 (3): 119–124. [*Picea chihuahuana*, *P. mexicana*].

12380   Rushforth, K. D. (1989). Two new species of *Abies* (Pinaceae) from western Mexico. Notes Roy. Bot. Gard. Edinburgh 46 (1): 101–109. [*A. flinckii* sp. nov., p. 101, *A. colimensis* sp. nov., p. 105, ill.)

12380a   Rushforth, K. D. (1999). *Abies recurvata* (Pinaceae). Curtis's Bot. Mag. 16 (3): 193–198, pl. 369.

12380b   Rushforth, K. D. (1999). Taxonomic notes on some Sino-Himalayan conifers. Int. Dendrol. Soc. Yearb. 1998: 60–63. [*Abies delavayi* ssp. *fansipanensis* comb. nov., *A. pindrow* ssp. *gamblei* comb. nov.].

12380c   Rushforth, K. D., R. P. Adams, M. Zhong, X. Q. Ma & R. N. Pandey (2003). Variation among *Cupressus* species from the eastern hemisphere based on random amplified polymorphic DNAs (RAPDs). Biochem. Syst. Ecol. 31 (1): 17–24.

12381   Rushton, W. (1915). Structure of the wood of Himalayan Junipers. J. Linn. Soc., Bot. 43 (288): 1–13, pl. 1. [*Juniperus recurva*, *J. wallichiana* = *J. indica*, *J. macropoda* = *J. excelsa* ssp. *polycarpos*, *J. communis*, table].

12390   Ruth, R. H. (1974). *Tsuga* (Endl.) Carr. in: Seeds of woody plants in the United States. U.S.D.A. Agric. Handb. 450, pp. 819–827. Washington, D.C. (ill.).

12399   Rydberg, P. A. (1905). Studies on the Rocky Mountain flora XV. Bull. Torrey Bot. Club 32: 597–610. [*Caryopitys monophylla* (Torr. & Frém.) comb. nov., p. 597, *Apinus* Necker, p. 597, *A. flexilis* (James) comb. nov., p. 598, *A. albicaulis* (Engelm.) comb. nov., p. 598, *Sabina* div. comb. nov., p. 598; *Caryopitys* & *Apinus* = *Pinus*, *Sabina* = *Juniperus*].

12400 Rydberg, P. A. (1912). Studies on the Rocky Mountain flora XXVI. Bull. Torrey Bot. Club 39 (3): 99–111. [*Hesperopeuce mertensiana* (Bong.) comb. nov. = *Tsuga mertensiana*, p. 100, *Sabina horizontalis* (Moench) comb. nov. = *Juniperus horizontalis*, p. 100].

12410 Rzedowski, J. (1964). Una especie nueva de pino piñonero del estado de Zacatecas (México). Ciencia 23: 17–20, t. 2. [*Pinus maximartinezii* sp. nov., p. 17].

12420 Rzedowski, J. & L. Vela (1966). *Pinus strobus* var. *chiapensis* en la Sierra Madre del Sur de México. Ciencia 24: 211–216.

12430 Safford, L. O. (1974). *Picea* (L.) Karst. in: Seeds of woody plants in the United States. U.S.D.A. Agric. Handb. 450, pp. 587–597. Washington, D.C. (ill.). [= *Picea* A. Dietr.].

12430e Sahni, B. (1920). On the structure and affinities of *Acmopyle pancheri* Pilger. Philos. Trans. Roy. Soc. London, ser. B, 210: 253–310. (ill.).

12431 Sahni, B. & T. C. N. Singh (1931). Notes on the vegetative anatomy and female cones of *Fitzroya patagonica* Hook. f. J. Indian Bot. Soc. 10: 1–20. [= *F. cupressoides*, ill.].

12440 Said, M. (1959). *Pinus gerardiana* (Chilgoza) in the Zhob District. Pakistan J. Forest. 9 (2): 118–123.

12450a Saiki, K. (1992). A new Sciadopityaceous seed cone from the Upper Cretaceous of Hokkaido, Japan. Amer. J. Bot. 79 (9): 989–995. [Sciadopityaceae, Taxodiaceae, *Sciadopityostrobus kerae* gen. et sp. nov., p. 992, ill.].

12450b Saiki, K. & T. Kimura (1993). Permineralized taxodiaceous seed cone from the Upper Cretaceous of Hokkaido, Japan. Rev. Palaeobot. Palynol. 76: 83–96.

12451 Salisbury, R. A. (1796). Prodromus stirpium in horto ad Chapel Allerton vigentium... London. [*Juniperus humilis* sp. nov. = *J. sabina* L., p. 397, *Callitris humilis* sp. nov. = *Widdringtonia cedarbergensis*, p. 398, *Pinus longifolia* sp. nov. = *P. palustris* Mill., p. 398, *Thuja decora* sp. nov. = *Platycladus orientalis*, p. 398, *Pinus taxifolia* sp. nov. = *Abies balsamea*, p. 399, *P. tenuifolia* sp. nov. = *P. strobus*, p. 399].

12460 Salisbury, R. A. (1807). The characters of several genera in the natural order of Coniferae: with remarks on their stigmata, and cotyledons. Trans. Linn. Soc. London 8: 308–317, pl. 15. [*Agathis* gen. nov., p. 311 (*nom. cons.*), *A. loranthifolia* sp. nov. = *A. dammara*, p. 312, *Belis jaculifolia* gen. et sp. nov. = *Cunninghamia lanceolata*, p. 316, *Eutassa heterophylla* gen. et sp. nov. = *Araucaria heterophylla*, p. 316, *Columbea quadrifaria* gen. et sp. nov = *Araucaria araucana*, *Abies*, *Pinus*, *Larix*, morphological discussion, taxonomy].

12470 Salisbury, R. A. (1817). On the coniferous plants of Kaempfer. Quart. J. Sci. Lit. Arts (London) 2: 309–314. [comm. on spp. in E. Kaempfer, 1712, *Dolophyllum* gen. nov. = *Thujopsis* nom. cons. prop. (see A. Farjon & D. R. Hunt, 1994) p. 313].

12476 Sánchez, X. M. (1967). Algunas aspectos ecológicas de las bosques de coniferas mexicanas. Revista México y sus bosques. Epoca 3 (16): 11–19. [Cupressaceae, Pinaceae, *Pinus*].

12480 Santamour, F. S., Jr. (1960). New chromosome counts in *Pinus* and *Picea*. Silvae Genet. 9: 87–88.

12480a Santiago V., T., S. Ochoa G. & T. Alemán S. (1997). Guía para identificar pinos de la Meseta Central de Chiapas, México. Guías Scientificas ECOSUR. (ill.). [*Pinus*, 8 spp. + 3 var. recognized].

12480e Santisuk, T. (1997). Geographical and ecological distributions of the two tropical pines, *Pinus kesiya* and *Pinus merkusii*, in Southeast Asia. Thai Forest. Bull. (Bot.) 25: 102–123. [*Pinus merkusii* = *P. latteri*].

12481a Sargent, C. S. (1889). Notes upon some North American trees – XIV. Garden & Forest 2: 496–497, f. 135. [*Pinus latifolia* sp. nov., p. 496 = *P. engelmannii*].

12491 Sargent, C. S. (1897). Notes on cultivated conifers. Gard. & Forest 10: 420–421, f. 54. [*Juniperus scopulorum* nom. prov. (valid. publ. in Sargent, Silva N. Amer. 14: 93, t. 739. 1902), id. p. 481, *Picea parryana* (Barrow) comb. nov. = *P. pungens*, p. 421].

12520 Sargent, C. S. (1914). Plantae Wilsonianae II. The Univ. Press, Cambridge, Mass. [see A. Rehder & E. H. Wilson, 1914].

12550 Sasaki, S. (1916). On *Cunninghamia sinensis*, *C. konishii*, *Pseudotsuga japonica* and *P. wilsoniana*. (Japan.). Trans. Nat. Hist. Soc. Taiwan 6: 255–260.

12560 Sasaki, S. (1925). On the nomenclature of *Cunninghamia lanceolata* Hook. Trans. Nat. Hist. Soc. Taiwan 15: 184–200. (Japan., Eng. summ.).

12605 Savi, P. (1846). Descrizione di una nuova specie d'Araucaria. Giorn. Bot. Ital. 2 (1): 52–59. [*Araucaria ridolfiana* sp. nov., p. 52 = *A. angustifolia*].

12619 Saxton, W. T. (1909). Preliminary account of the development of the ovule, gametophytes and embryo of *Widdringtonia cupressoides*. Bot. Gaz. (Crawfordsville) 48: 161–178. (ill.). [= *W. nodiflora*, embryology, ontogeny, morphology].

12620 Saxton, W. T. (1910). Contributions to the life-history of *Widdringtonia cupressoides*. Bot. Gaz. (Crawfordsville) 50: 31–48. [= *W. nodiflora*].

12630 Saxton, W. T. (1910). Contributions to the life-history of *Callitris*. Ann. Bot. (London) 48: 429–431.

12640 Saxton, W. T. (1913). Contributions to the life-history of *Actinostrobus pyramidalis*, Miq. Ann. Bot. (London) 27: 321–346.

12650 Saxton, W. T. (1913). Contributions to the life-history of *Tetraclinis articulata*, Masters, with some notes on the phylogeny of the Callitroideae. Ann. Bot. (London) 27: 577–606.

12661 Saxton, W. T. (1930). Notes on Conifers. IV. Some points in the leaf anatomy of *Fokienia hodginsii* and *Libocedrus macrolepis*. Ann. Bot. (London) 44: 167–171. [*L. macrolepis* = *Calocedrus macrolepis*, ill.].

12661a Saxton, W. T. (1934). Notes on Conifers. VII. The morphology of *Austrotaxus spicata* Compton. Ann. Bot. (London) 48: 411–427.

12662 Saxton, W. T. (1934). Notes on Conifers. IX. The ovule and embryogeny of *Widdringtonia*. Ann. Bot. (London) 48: 429–431. (ill.). [*W. juniperoides* sensu Endl. = *W. cedarbergensis*].

12663 Saxton, W. T. & J. C. Doyle (1929). The ovule and gametophytes of *Athrotaxis selaginoides* Don. Ann. Bot. (London) 43: 834–840. (ill.).

12670 Saylor, L. C. (1972). Karyotype analysis of the genus *Pinus* subgenus *Pinus*. Silvae Genet. 21: 155–163.

12680 Saylor, L. C. (1983). Karyotype analysis of the genus *Pinus* subgenus *Strobus*. Silvae Genet. 32: 119–124.

12682 Saylor, L. C. & H. A. Simons (1970). Karyology of *Sequoia sempervirens*: karyotype and accessory chromosomes. Cytologia 35: 294–303.

12686 Scaltsoyiannes, A. (1999). Allozyme differentiation and phylogeny of Cedar species. Silvae Genet. 48 (2): 61–68. [confirms currently recognized taxa: *Cedrus atlantica*, *C. brevifolia*, *C. libani* ssp. *libani*, *C. libani* ssp. *stenocoma*].

12687 Scaltsoyiannes, A., M. Tsaktsira & A. D. Drouzas (1999). Allozyme differentiation in the Mediterranean firs (*Abies*, Pinaceae): a first comparative study with phylogenetic implications. Plant Syst. Evol. 216 (3–4): 289–307. [*Abies alba*, *A. cephalonica*, *A. cilicica*, *A. nebrodensis*, *A. nordmanniana*, *A. numidica*, *A. pinsapo*].

12690 Schaarschmidt, F. (1975). Missbildungen an Lärchenzapfen von stammesgeschichtlicher Bedeutung. Natur & Mus. (Frankfurt) 105 (3): 65–71. [*Larix*].

12700 Schantz, M. von & S. Juvonen (1966). Chemotaxonomische Untersuchungen in der Gattung *Picea*. Acta Bot. Fenn. 73: 1–51.

12707 Schirone, B., G. Piovesan, R. Bellarosa & C. Pelosi (1991). A taxonomic analysis of seed proteins in *Pinus* spp. (Pinaceae). Plant Syst. Evol. 178: 43–53.

12720 Schiskin, I. K. (1933). Zur Kenntnis der Olga-Lärche (*Larix olgensis* A. Henry). Bot. Žurn. SSSR 18: 162–210, 1 map. (Russ., Germ. summ.). [= *L. gmelinii* var. *olgensis*].

12730 Schiskin, I. K. (1935). *Microbiota decussata* Kom. as an element of the vegetation cover of the Ussuri region. Trudy Dal'nevost. Fil. Akad. Nauka SSSR, Ser. Bot. 1: 227–243, f. 1–2. (Russ., Eng. summ.). [ecology].

12740 Schlarbaum, S. E., L. C. Johnson & T. Tsuchiya (1983). Chromosome studies of *Metasequoia glyptostroboides* and *Taxodium distichum*. Bot. Gaz. (Crawfordsville) 144 (4): 559–565.

12750 Schlarbaum, S. E. & T. Tsuchiya (1984). The chromosomes of *Cunninghamia konishii*, *C. lanceolata* and *Taiwania cryptomerioides* (Taxodiaceae). Plant Syst. Evol. 145 (3–4): 169–181.

12760 Schlarbaum, S. E. & T. Tsuchiya (1985). Karyological derivation of *Sciadopitys verticillata* Sieb. et Zucc. from a pro- taxodiaceous ancestor. Bot. Gaz. (Crawfordsville) 146: 264–267.

12770 Schlarbaum, S. E., T. Tsuchiya & L. C. Johnson (1984). The chromosomes and relationships of *Metasequoia* and *Sequoia* (Taxodiaceae): an update. J. Arnold Arbor. 65 (2): 251–254.

12780 Schlechtendal, D. F. L. von (1838). Vorläufige Nachricht über die Mexicanischen Coniferen. Linnaea 12: 486–496. [*Pinus oocarpa* Schiede ex Schltdl., sp. nov., p. 491, *P. ayacahuite* Ehrenb. ex Schltdl., sp. nov., p. 492, *Cupressus thurifera* "sp. nov." = *C. lusitanica* var. *benthamii*, p. 493, *Juniperus flaccida* sp. nov., p. 495, *J. tetragona* "sp. nov." (non Moench, 1794) = *J. ashei* Buchholz, p. 495, *Taxus globosa* sp. nov., p. 496].

12790 Schlechtendal, D. F. L. von (1857). De pinastris germaniae et helvetiae. Linnaea 29: 357–384. [*Pinus* L.].

12800 Schlechtendal, D. F. L. von (1857–58). Coniferae Mexicanae. Linnaea 29: 326–356 (1857); Ad coniferas mexicanas Roezlianas. Linnaea 29: 699–704 (1858). ["ex catalogo clar. Roezl translatae"; see B. Roezl et Cie., 1857, 1858; *Cupressus*, *Taxodium*, *Juniperus*, *Tsuga*, *Abies*, *Pinus*: many Roezl-names, here annot. as "new", with Lat. descr. 1857, only names in 1858, see also G.R. Shaw, 1909, for comment, and Farjon & Styles, 1997, for typification of these synonyms].

12820 Schlechtendal, D. F. L. von & A. von Chamisso (1830). Plantarum Mexicanarum a cel. viris Schiede et Deppe collectarum (recensio brevis). Coniferae. Linnaea 5: 72–77. [Coniferales, *Pinus teocote* Schiede ex Schltdl. et Cham., sp. nov., p. 76, *Abies religiosa* (Kunth in H. B. K.) comb. nov., p. 77,

Juniperus mexicana Schiede ex Schltdl. et Cham., "sp. nov." (non Sprengel, 1826) = *J. deppeana*, p. 77].

12830   Schlechtendal, D. F. L. von & A. von Chamisso (1831). Plantarum mexicanarum a cel. viris Schiede et Deppe collectarum (recensio brevis). Addenda (1). Linnaea 6: 352–384. [*Pinus patula* Schiede ex Schltdl. et Cham., sp. nov., *P. leiophylla* Schiede ex Schltdl. et Cham., sp. nov., p. 354].

12831   Schlechter, R. (1906). Beiträge zur Kenntnis der Flora von Neu-Kaledonien. Bot. Jahrb. Syst. 39 (1): 1–274. [Coniferales: p. 16, *Callitris subumbellata* (Parl.) comb. nov., *C. sulcata* (Parl.) comb. nov., *C. balansae* (Brongn. et Gris) comb. nov. = *C. sulcata*].

12839   Schmid, R. & M. J. Schmid (1975). Living links with the past. Nat. Hist. 84 (3): 38–45. [*Pinus aristata, P. longaeva*, dendrochronology].

12840   Schmid, W. (1937). Ergebnisse der Reise von Dr. A. U. Däniker nach Neu-Kaledonien und den Loyalty-Inseln (1924/26). Beitrag zur Kenntnis von *Callitropsis araucarioides* Compton. Ber. Schweiz. Bot. Ges. 47: 124–159. (ill.). [= *Neocallitropsis pancheri*].

12841   Schmidt, F. (1868). Reisen in Amur-lande und auf der Insel Sachalin. In Aufträge der kaiserlich-russischen geographischen Gesellschaft. Botanischer Teil. Mem. Acad. Imp. Sci. Saint-Pétersbourg, Sér. 7, 12 (2): 1–227, t. 1–8. [*Abies veitchii* var. *sachalinensis* var. nov., p. 175; F. Schmidt mentioned this taxon earlier (1868) as a species, but without descr.].

12850   Schmidt, P. A. (1984). *Pinus mugo* Turra s. l. Berg-kiefer, Latschen-Kiefer. Gärtn.-Bot. Briefe 77: 34–39.

12860   Schmidt, P. A. (1986). Systematische Übersicht der Gattung *Picea* A. Dietr. Index seminum et plantarum 1986 – Forstbot. Gart. Tharandt, pp. 14–23. [nomenclatural validation in: Haussknechtia 4: 37–38, 1988].

12870   Schmidt, P. A. (1987). Untersuchungen zur Morphologie, Systematik, Verbreitung und Introduktion der in der DDR in Kultur befindlichen Fichten-Arten – ein Beitrag zur botanischen Monographie der Gattung *Picea* A. Dietr. – Diss. B, Techn. Univ. Dresden, Sekt. Forstw. Tharandt (unpubl. mscr.)].

12875   Schmidt, P. A. (1987). Übersicht der Fichten-Arten und ihrer infraspezifischen Sippen (Gattung *Picea* A. Dietr.) in der DDR. Beitr. Gehölzk. 1987: 21–36.

12880   Schmidt. P. A. (1988). Taxonomisch-nomenklatorische Notiz zur Gattung *Picea* A. Dietr. Haussknechtia 4: 37–38. [*Picea* subgen. *Casicta* (Mayr) stat. nov., sect. *Pungentes* (Bobrov) stat nov., with *P. engelmannii* ssp. *mexicana* (Martínez) comb. et stat. nov., p. 38].

12890   Schmidt, P. A. (1989). Beitrag zur Systematik und Evolution der Gattung *Picea* A. Dietr. Flora 182: 435–461. [classification, phylogeny, ill.].

12891   Schmidt, P. A. (1991). Beitrag zur Kenntnis der in Deutschland anbaufähigen Fichten (Gattung *Picea* A. Dietr.). Mitt. Deutsch. Dendrol. Ges. 80: 7–72. [*Picea* spp., classification, key, ill.].

12892   Schmidt, P. A. (1998). *Picea* A. Dietr., 1824. Enzyklopädie der Holzgewächse III–1, 14: 1–14. Ecomed Verlag, Landsberg am Lech. [*Picea* spp., distrib., classification, ecology, ill.].

12893   Schmidt, P. A. (2002). *Picea abies* (L.) H. Karst., 1881. Enzyklopädie der Holzgewächse III–1, 28: 1–18. Ecomed Verlag, Landsberg am Lech. (ill., map). [distrib., ecology, forestry].

12895   Schmidt, P. A. (2002). *Picea obovata* Ledeb., 1833. Enzyklopädie der Holzgewächse III–1, 30: 1–14. Ecomed Verlag, Landsberg am Lech. (ill., map). [distrib., ecology, forestry].

12900   Schmidt, R. L. (1957). The silvics and plant geography of the genus *Abies* in the coastal forests of British Columbia. Canad. Dept. Lands Forests, British Columbia Forest Serv. Techn. Publ. T 46, 31 pp. [*Abies amabilis, A. grandis*].

12904   Schmidt, W. C. & K. J. McDonald (comps.) (1995). Ecology and management of *Larix* forests: a look ahead. Proceedings of an international symposium October 5–9, 1992, Whitefish, Montana, U.S.A. U.S.D.A. Forest Service Gen. Tech. Rep. INT-GTR-319, Intermountain Research Station, Ogden, Utah. [*Larix* spp.].

12905   Schmidt, W. C. & F.-K. Holtmeier (comps.) (1994). Proceedings – international workshop on subalpine stone pines and their environment: the status of our knowledge. St. Moritz, Switzerland, September 5–11, 1992. U.S.D.A. Forest Service Gen. Tech. Rep. INT-GTR-309, Intermountain Research Station, Ogden, Utah. [*Pinus albicaulis, P. cembra, P. koraiensis, P. pumila, P. sibirica*, taxonomy, genetics (isozymes, DNA), co-evolution, physiology, other subjects].

12908   Schmidtling, R. C. & V. Hipkins (1998). Genetic diversity in longleaf pine (*Pinus palustris*): influence of historical and prehistorical events. Canad. J. Forest. Res. 28: 1135–1145.

12909   Schmidtling, R. C. & V. Hipkins (2001). Evolutionary relationships of Slash Pine (*Pinus elliottii*) with its temperate and tropical relatives. S. African Forest. J. 190: 73–78. [*P. elliottii, P. elliottii* var. *densa, P. caribaea* var. *bahamensis*, allozymes, monoterpenes].

12910   Schmidt-Vogt, H. (1972). Studien zur morphologischen Variabilität der Fichte (*Picea abies* (L.) Karst.). Allg. Forst-Jagd-Zeitung 143 (7,9,11): 133–240 (Sonderdr.].

12920 Schmidt-Vogt, H. (1974). Die systematische Stellung der gemeine Fichte (*Picea abies* (L.) Karst.) und der siberischen Fichte (*Picea obovata* Ledeb.) in der Gattung *Picea*. Allg. Forst-Jagd-Zeitung 145 (3,4): 45–60 (Sonderdr.].

12930 Schmidt-Vogt, H. (1974). Das natürliche Verbreitungs-gebiet der Fichte (*Picea abies* [L.] Karst.) in Eurasien. Allg. Forst-Jagd-Zeitung 145: 185–197.

12940 Schmidt-Vogt, H. (1975). The taxonomic status of Norway spruce (*Picea abies*) and Siberian spruce (*P. obovata*) in the genus *Picea*. Proc. 12th Int. Bot. Congr., Leningrad. [see also H. Schmidt-Vogt, 1974].

12950 Schmidt-Vogt, H. (1977). Die Fichte; Ein Handbuch in zwei Banden. Band I: Taxonomie, Verbreitung, Morphologie, Ökologie, Waldgesellschaften. Hamburg – Berlin. [*Picea*, *P.* spp.; a very comprehensive monograph on esp. *P. abies*, with extensive bibliography, ill. and maps; the second volume deals with forestry aspects of *P. abies*].

12950a Schmidt-Vogt, H. (1986-90). Die Fichte; Ein Handbuch in zwei Bänden. Band I, 2. Aufl. (1987). Band II/1: Wachstum, Züchtung, Boden, Umwelt, Holz. (1986). Band II/2: Krankheiten, Fichtensterben, Ernährung und Düngung, Waldbau. (1990). Hamburg – Berlin. [*Picea* spp., *P. abies*, reprint of vol. I, vol. II on forestry aspects of mainly *P. abies*, extensive bibliography].

12950b Schmithüsen, J. (1960). Die Nadelhölzer in den Waldgesellschaften der südlichen Anden. Vegetatio 10: 313–327. [Cupressaceae, Araucariaceae, *Austrocedrus*, *Fitzroya*, *Pilgerodendron*, *Araucaria araucana*].

12950c Schmithüsen, J. (1978). Konkurrenz als begrenzender Faktor bei Restarealen alter Koniferentaxa mit einem Ausblick auf ökologische Konsequenzen für die Forstwirtschaft. In: C. Troll & W. Lauer (eds.). Geoökologische Beziehungen zwischen der Temperierten Zone der Südhalbkugel und den Tropengebirgen. Erdwiss. Forsch. 11: 124–134. Wiesbaden.

12958 Schneider, F. (1857). *Cryptomeria* Don und *Lobbii* Hort. angl. Berlin. Allg. Gartenzeit. 1857 (5): 37–38. [*Cryptomeria lobbii* = *C. japonica*].

12960 Schopf, J. M. (1943). The embryology of *Larix*. Illinois Biol. Monogr. 19 (4): 1–97. (ill.).

12970 Schopf, J. M. (1948). Should there be a living *Metasequoia*? Science 107: 344–345. [nomenclature].

12980 Schopf, J. M. (1948). Precedence of Modern plant names over names based on fossils. Science 108: 483. [nomenclature of *Metasequoia glyptostroboides*, nom. gen. et typ. cons.: see W. Greuter *et al.*, ICBN 1988: 160, No. 32a].

12982 Schorn, H. E. (1986). *Abies milleri*, sp. nov., from the Middle Eocene Klondike Mountain Formation, Republic, Ferry County, Washington. Burke Mus. Contr. Anthrop. Nat. Hist. 1: 1–7. [palaeobotany, ill.].

12983 Schorn, H. E. (1994). A preliminary discussion of fossil larches (*Larix*, Pinaceae) from the Arctic. Quarternary International 22/23: 173–183. [*Larix* "omolica-altoborealis" complex, Pliocene, classification].

12990 Schroeder, F.-G. & R. Bayer (1973). Zur Nomenklatur von *Picea polita* und *Pinus thunbergii*. Mitt. Deutsch. Dendrol. Ges. 66: 6168.

12985 Schreber, J. D. C. von (1791). Caroli a Linné ... Genera plantarum, ... Varrentrapp & Wenner, Frankfurt am Main.

13000 Schröter, C. (1898). Über die Vielgestaltigkeit der Fichte. Vierteljahrsschr. Naturf. Ges. Zürich 43: 1–130. [*Picea abies* variability].

13010 Schulman, E. (1958). Bristlecone pine, oldest known living thing. Natl. Geogr. Mag. 113 (March): 354–372. (ill.). [*Pinus longaeva*, *Pinus aristata*].

13011 Schulz, C., A. Jagel & T. Stützel (2003). Cone morphology in *Juniperus* in the light of cone evolution in Cupressaceae s.l. Flora 198: 161–177. [*J. communis*, *J. oxycedrus*, *J. rigida*, *J. squamata*, *J. chinensis*, *J. phoenicea*, ontogeny, SEM ill., diagrams].

13013 Schumann, K. M. & U. M. Hollrung (1889). Die Flora von Kaiser Wilhelms Land. Beiheft zu den Nachrichten über Kaiser Wilhelms Land und den Bismarck-Archipel. Berlin. [*Araucaria hunsteinii* K. Schum., p. 11].

13014 Schur, F. (1851). Beiträge zur Kenntnis der Flora von Siebenbürgen. Verhandl. Mitteil. Siebenbürgischen Ver. Naturwiss. Hermannstadt 2: 167–171. [*Juniperus intermedia* Schur, p. 169 = *J. communis*].

13015 Schurkhal, A. V., A. V. Podogas & L. A. Zhyvotovsky (1991). Phylogenetic analysis of [the] genus *Pinus* by allozyme loci; genetical differentiation of subgenera. Genetika 7: 1193–1205. (Russ., Eng. summ. p. 1205).

13017 Schütt, P. (1991). Tannenarten Europas und Kleinasiens. Basel – Boston – Berlin. [*Abies alba*, *A. borisii-regis*, *A. bornmuelleriana*, *A. cephalonica*, *A. cilicica*, *A. equi-trojani*, *A. marocana*, *A. nebrodensis*, *A. nordmanniana*, *A. numidica*, *A. pinsapo*, many ill.].

13020 Schwarz, O. (1936). Über die Systematik und Nomenklatur der europäischen Schwarzkiefern. Notizbl. Bot. Gart. Berlin-Dahlem 13: 226–243. [*Pinus nigra* s. l.].

13021 Schwarz, O. (1938). Nachträgliche Notiz zur Nomenklatur der Europäischen Schwarzkiefer. Notizbl. Bot. Gart. Berlin-Dahlem 14: 135. [*Pinus nigra*].

13021a   Schwarz, O. (1938). Notiz zur Nomenklatur von *Pinus excelsa* Wall. Fedde's Repert. Sp. Nov. Regni Veg. 44: 160. [*Pinus excelsa, P. griffithii, P. wallichiana*].

13022   Schwarz, O. (1939). Zweiter Nachtrag zur Systematik und Nomenklatur der Schwarz-kiefern. Notizbl. Bot. Gart. Berlin-Dahlem 14: 381. [*Pinus nigra*].

13023   Schwarz, O. (1944). Anatolica. I. Feddes Repert. Spec. Nov. Regni Veg. 54 (1): 26–34. [*Cedrus libanitica* Trew ssp. *stenocoma*, ssp. *libani*, ssp. *atlantica* et ssp. *brevifolia*, ssp. nov. et comb. nov. (nom. illeg., vide *C. libani* A. Rich.), key].

13030   Schwarz, O. & H. Weide (1962). Systematische Revision der Gattung *Sequoia* Endl. Fedde's Repert. 66 (3): 159–192. [brings *Metasequoia* and *Sequoiadendron* under one genus: *Sequoia* Endl., with fossil spp.; *Sequoia glyptostroboides* (Hu et Cheng) H. Weide, comb. nov. = *M. glyptostroboides*, p. 185].

13031   Schweinfurth, U. (1988). *Pinus* in Southeast Asia. Beitr. Biol. Pflanzen 63 (1–2): 253–268.

13038   Schweitzer, H.-J. (1963). Der weibliche Zapfen von *Pseudovoltzia liebeana* und seine Bedeutung für die Phylogenie der Koniferen. Palaeonto-graphica B, 113 (1–4): 1–29, pl. 1–9. [Voltziaceae, palaeobotany].

13041   Schweitzer, H.-J. (1996). *Voltzia hexagona* (Bischoff) Geinitz aus dem mittleren Perm Westdeutschlands. Palaeontographica B, 239 (1–3): 1–22. [Voltziaceae, palaeobotany].

13050   Schwerdtfeger, F. (1953). Informe al govierno de Guatemala sobre la entomologia forestal de Guatemala. Vol. I. Los pinos de Guatemala. UN FAO/ETAP Info. No. 202, Rome. [*Pinus tecumumani* sp. nov., p. 39 (52), nom. inval. publ. (ICBN Art. 36.1), see T. Eguiluz Piedra & J. P. Perry, Jr., 1983 and B. T. Styles, 1984].

13054   Scott, L. J., M. Shepherd & H. J. Henry (2003). Characterization of highly conserved microsatellite loci in *Araucaria cunninghamii* and related species. Pl. Syst. Evol. 236 (3–4): 115–123. [*Agathis, Araucaria*, DNA, molecular systematics].

13056   Sebastian, C. (1958). Essais de germination de quatre espèces du genre *Juniperus*. Sci. Nat. Phys. Maroc 38: 115–122. [*Juniperus communis, J. oxycedrus, J. phoenicea, J. thurifera*, seed dormancy, germination treatments].

13057   Sebert, H. (1874). Notice sur les bois de la Nouvelle-Calédonie… Paris. [*Dammara lanceolata* Sebert & Pancher sp. nov. = *Agathis lanceolata*, p. 169; name also publ. in Rev. Marit. Col. 40: 555 (Feb 1874)].

13060   Seemann, B. (1855). On the identity of *Pinus hirtella* and *Pinus religiosa* of Humboldt, Bonpland, and Kunth. Proc. Linn. Soc. London 2: 351–352. [*Pinus (Abies) religiosa*; see also D. F. L. von Schlechtendal & A. von Chamisso, 1830].

13070   Seemann, B. (1858). Der Mammuth-Baum Ober-Californiens (*Sequoia wellingtonia* Seem.). Bonplandia 6: 343–354, 1 fig. [= *Sequoiadendron giganteum*].

13071   Seemann, B. (1863). *Podocarpus vitiensis*, a new coniferous tree, from the Viti Islands. J. Bot. 1: 33–36, pl. 2. [*Podocarpus vitiensis* Seemann = *Retrophyllum vitiense*; includes the diagn., by F. Parlatore, of *Juniperus conferta* (see also F. Parlatore, 1863), *Frenela sulcata* = *Callitris sulcata, F. subcordata* = *C. roei, F. drummondii* = *C. drummondii, Actinostrobus acuminatus, Larix lyallii* (see also F. Parlatore, 1863), *Dammara motleyi* = *Nageia motleyi; A. acuminatus* and *F. sulcata* + *subcordata* were first publ. in Index Sem. Hort. Florent., 1862].

13072   Selik, M. (1959). *Pinus brutia* in der Türkei. Forstwiss. Centralbl. 78: 43–58. (ill.).

13073   Selik, M. (1962). Eine neue Varietät von *Pinus brutia* Ten.: var. *pyramidalis*. Mitt. Deutsch. Dendrol. Ges. 62: 98–101. [map, habitus photographs, growth form].

13077   Semerikov, V. L., H. Q. Zhang, M. Sun & M. Lascoux (2003). Conflicting phylogenies of *Larix* (Pinaceae) based on cytoplasmic and nuclear DNA. Mol. Phylogen. Evol. 27: 173–184. [*Larix* spp., molecular phylogenetics, introgression].

13080a   Serbet, R. & R. A. Stockey (1991). Taxodiaceous pollen cones from the Upper Cretaceous (Horseshoe Canyon Formation) of Drumheller, Alberta, Canada. Rev. Palaeobot. Palynol. 70: 67–76.

13080b   Setoguchi, H., T. A. Osawa, J.-C. Pintaud, T. Jaffré & J.-M. Veillon (1998). Phylogenetic relationships within Araucariaceae based on *rbc*L gene sequences. Amer. J. Bot. 85 (11): 1507–1516. [Araucariaceae, *Agathis, Araucaria, Wollemia*, DNA, phylogeny].

13080d   Shang, H., J.Z. Cui & C.S. Li (2001). *Pityostrobus yixiangensis* sp. nov., a pinaceous cone from the Lower Cretaceous of north-east China. Bot. J. Linn. Soc. 136 (4): 427–437. (ill.). [fossil cone quite similar to *Picea*].

13080e   Sharma, B. D. & D. R. Bohra (1977). Petrified araucarian megastrobili from the Jurassic of the Rajmahal Hills, India. Acta Palaeobot. 18: 31–36. [Araucariaceae, palaeobotany].

13081   Sharp, A. J. (1946). *Pinus strobus* south of the United States. J. Elisha Mitchell Sci. Soc. 62: 629–630. [*Pinus strobus* var. *chiapensis*].

13087   Shaw, D. W., E. F. Aldon & C. LoSapio (tech. coord.) (1995). Desired future conditions of Piñon-Juniper ecosystems. U.S.D.A. Forest Service Gen. Tech. Rep. GTR RM-258, Rocky Mountain Research Station, Fort Collins, Colorado. [*Juniperus* spp., *Pinus* subsect. *Cembroides* spp., ecology, management].

13088   Shaw, G. R. (1904). The pines of Cuba. Gard. Chron., ser. 3, 35: 179, f. 74. [*Pinus bahamensis* Griseb. (Fl. Brit. W. I.: 503, 1862) = *P. caribaea* var. *bahamensis*, *P. cubensis*, *P. terthrocarpa* = *P. tropicalis*].

13090   Shaw, G. R. (1904). *Pinus leiophylla*. Gard. Chron., ser. 3, 36: 175. (ill.).

13100   Shaw, G. R. (1904–05). *Pinus nelsoni*. Gard. Chron., ser. 3, 36: 122 (1904); 37: 306–307 (1905). [*P. nelsonii* sp. nov., 1904; ill.].

13110   Shaw, G. R. (1905). *Pinus pinceana*. Gard. Chron., ser. 3, 36: 122. (ill.).

13120   Shaw, G. R. (1909). The Pines of Mexico. Publ. Arnold Arbor. 1. (Harvard Univ.) Cambridge, Mass. [*Pinus* spp., *P. ayacahuite* var. *brachyptera* et var. *veitchii*, var. nov., p. 10, *P. leiophylla* var. *chihuahuana* (Engelm.) comb. et stat. nov., p. 14, *P. montezumae* var. *hartwegii* (Lindl.) comb. et stat. nov. (ascr. to Engelmann, 1880), p. 23, *P. ponderosa* var. *arizonica* (Engelm.) comb. et stat. nov., p. 24, *P. oocarpa* var. *microphylla* var. nov., p. 27; ill.].

13130   Shaw, G. R. (1914). The Genus *Pinus*. Publ. Arnold Arbor. 5. (Harvard Univ.) Cambridge, Mass. (ill.). [early and still valuable classification; for ref. see e.g. N. T. Mirov, 1967 and E. L. Little, Jr. & W. B. Critchfield, 1969].

13140   Shaw, G. R. (1924). Notes on the genus *Pinus*. The oblique cone. J. Arnold Arbor. 5: 225–227.

13152   Shi, D. X. & M. L. Wang (1994). Karyomorphological studies of 6 species in genus *Picea*. Acta Bot. Yunnanica 16 (2): 157–164. (Chin., Eng. abstr.). [*Picea asperata*, *P. likiangensis*, *P. likiangensis* var. *balfouriana*, *P. purpurea*, *P. retroflexa*, *P. wilsonii*].

13153   Shi, F. C., S. Kazuo & K. Hiromitsu (1998). The study on relationship of larches in northeast China by RAPD. Bull. Bot. Res. (China) 18 (1): 55–62. (Chin., Eng. summ., ill.). [*Larix gmelinii* s.l.].

13154   Shi, F. C., J. C. Nie & Z. X. Wang (1998). Analysis of taxonomic characters of *Larix* in northeast of China: 2. Geographic variation of seed scales and needle leaves. Bull. Bot. Res. (China) 18 (4): 468–471. (Chin., Eng. summ., ill.). [*Larix gmelinii* s.l.].

13155   Shi, F. C., J. C. Nie, B. Zhao & T. Cheng (1998). Analysis on taxonomic characters of *Larix* in northeast of China: geographic variation of cones morphology. Bull. Bot. Res. (China) 18 (2): 173–176. (Chin., Eng. summ., ill.). [*Larix gmelinii* s.l.].

13158   Shilkina, I. A. & M. P. Doludenko (1985). *Frenelopsis* and *Cryptomeria* as dominants of the Late Albian Ukraine Flora. Bot. Žurn. 70 (8): 1019–1030. (Russ., Eng. summ.). [Cupressaceae s.l., Taxodiaceae, *Cryptomeria pimenovae* sp. nov., palaeobotany].

13160   Shimakura, M. (1937). Anatomy of the wood of *Taiwania*. Bot. Mag. (Tokyo) 51: 694–700, f. 1–12.

13163   Shimizu, Y. (1991). Forest types and vegetation zones of Yunnan, China. J. Fac. Sci. Univ. Tokyo, Sect. 3, Bot. 15 (1): 1–71. [Western Yunnan, vegetation zonation, (mixed) conifer forest, *Cupressus duclouxiana*, *Abies delavayi*, *A. forrestii* var. *georgei*, *Keteleeria evelyniana*, *Larix potaninii*, *Picea likiangensis*, *Pinus* spp.].

13170   Shinners, L. H. (1956). *Pseudotsuga* and pseudo-science. Taxon 5: 43–46. [*Pseudotsuga menziesii*, nomenclature].

13180   Shirasawa, H. (1895). Eine neue Coniferenart in Japan. Bot. Mag. (Tokyo) 9: 41–43, 84–86, pl. 3. [*Tsuga japonica* sp. nov. = *Pseudotsuga japonica*; see L. Beissner, 1896].

13190   Shirasawa, H. (1914). Neue und wenig bekannten *Picea*- und *Abies*-Arten in Japan. Mitt. Deutsch. Dendrol. Ges. 23: 254–256, 1 pl. [*Picea koyamae* "sp. nov.", *Picea bicolor* var. *acicularis* et *reflexa*, *Picea maximowiczii*, *Abies veitchii* var. *olivacea*; see also H. Shirasawa & M. Koyama, 1913].

13200   Shirasawa, H. & M. Koyama (1913). Some new species of *Picea* and *Abies* in Japan. Bot. Mag. (Tokyo) 27: 127–132, pl. 2. [*Picea koyamae* sp. nov. (as "*P. koyamai*"), p. 127, *P. bicolor* var. *acicularis* var. nov. = *P. alcockiana* var. *acicularis*, *P. bicolor* var. *reflexa* var. nov. = *P. alcockiana* var. *reflexa*, p. 130, *Abies veitchii* var. *olivacea* var. nov.; new taxa by H. S. or by both authors].

13210   Shirley, J. C. (1947). The redwoods of coast and Sierra. Ed. 4, 84 pp. Berkeley – Los Angeles. [*Sequoia*, *Sequoiadendron*].

13220   Shishkin, B. K. & A. I. Vvedensky, eds. (1950–62). Flora Kirghizskoj SSR... Vol. 1–11 (ill.). Akad. Nauka Kirghizskoj SSR. (Russ., Lat. diagn.). [*Picea prostrata* Isakov, sp. nov. = *P. schrenkiana*, Vol. 10: 374].

13225   Sholars, R. E. (1982). The Pygmy Forest and associated plant communities of coastal Mendocino County, California. Genesis – Vegetation – Soils. Mendocino, California. [*Cupressus goveniana* var. *pygmaea*, *Pinus contorta* ssp. *bolanderi*, distr. map of *C. goveniana* var. *pygmaea*].

13230   Shrestha, T. B. (1974). gymnosperms of Nepal. Cahiers nepalais No. 3. Paris. [*Cupressus torulosa*, *Juniperus communis*, *J. indica*, *J. recurva*, *J. squamata*, *Pinus wallichiana*, *P. roxburghii*, *Cedrus deodara*, *Larix griffithii*, *Tsuga dumosa*, *Picea smithiana*, *Abies spectabilis*, *A. pindrow*, distr. maps].

13230a   Shurkhal, A. V., A. V. Podogas & L. A. Zhyvotovsky (1991). Genetic differentiation of 18 pine species in allozyme loci (Genus *Pinus*, subgenus *Strobus*, subgenus *Pinus*). Dokl. Akad. Nauk S.S.S.R. 316: 484–488. (Russ., Eng. summ.).

13230b   Shurkhal, A. V., A. V. Podogas & L. A. Zhyvotovsky (1992). Phylogenetic analysis of allozyme loci in the genus *Pinus*: genetic differentiation of subgenera. Soviet Genet. 28: 837–846.

13230c   Shurkhal, A. V., A. V. Podogas & L. A. Zhyvotovsky (1992). Allozyme differentiation in the genus *Pinus*. Silvae Genet. 41: 105–109.

13231   Sibthorp, J. & J. E. Smith (1816). Flora graeca prodromus: sive plantarum omnium enumeratio, ... Vol. 2 (2): 211–422. London. [*Juniperus macrocarpa* sp. nov. = *J. oxycedrus* ssp. *macrocarpa*, p. 263].

13256   Sigurgeirsson, A. & A. E. Szmidt (1988). Chloroplast DNA variation among North American *Picea* species, and its phylogenetic implications. In: J. E. Hällgren, ed. Molecular genetics of forest trees; Proc. Frans Kempe Symp., Rep. No. 9, pp. 49–65. Swedish Univ. Agric. Sci, Dept. Forest Genetics and Plant Physiology, Umeå.

13257   Sigurgeirsson, A. & A. E. Szmidt (1993). Phylogenetic and biogeographic implications of chloroplast DNA variation in *Picea*. Nord. J. Bot. 13 (3): 233–246. [*Picea* spp., 31 species analyzed with phenetic and cladistic methods].

13259   Silba, J. (1981). A long-misunderstood Asiatic Cypress: *Cupressus chengiana* Hu (Cupressaceae). Baileya 21 (3): 143–144.

13260   Silba, J. (1981). Revised generic concepts of *Cupressus* L. (Cupressaceae). Phytologia 49 (4): 390–399. [*C. chengiana* var. *jiangeensis* (N. Zhao) comb. nov., p. 394, *C. sempervirens* var. *atlantica* (Gaussen) et *dupreziana* (A. Camus) comb. et stat. nov., p. 398].

13270   Silba, J. (1982). Distribution of *Chamaecyparis funebris* (Endl.) Carr. and *Cupressus chengiana* Hu (Cupressaceae). Phytologia 51 (2): 157–160.

13280   Silba, J. (1983). Addendum to a revision of *Cupressus* L. (Cupressaceae). Phytologia 52 (5): 349–361.

13300   Silba, J. (1985). The infraspecific taxonomy of *Pinus culminicola* (Pinaceae). Phytologia 56: 489–491. [*P. culminicola* var. *discolor*, comb. nov., p. 490].

13330   Silba, J. (1987). Nomenclature of the Weeping Himalayan Cypress (*Cupressus*, Cupressaceae). Phytologia 64 (1): 78–80. [*Cupressus himalaica* sp. nov., p. 80, in syn.: *C. pendula* Griff., non Thunb., *C. corneyana* Knight et Perry ex Carrière, *C. cashmeriana* Royle ex Carrière, the latter two as insertae sedis].

13340   Silba, J. (1988). A new species of *Cupressus* L. from Tibet (Cupressaceae). Phytologia 65 (5): 333–336. [*C. austro-tibetica* sp. nov., p. 334, *C. gigantea* (= *C. chengiana* ?), *C. duclouxiana*].

13342   Silba, J. (1990). The genus *Cupressus* in Assam (Northeast India). Newsletter Conifer Soc. Austral. 8: 4–5. [*Cupressus himalaica* = *C. cashmeriana*, map].

13343   Silba, J. (1991). Geographic variation in Likiang spruce. Amer. Conifer Soc. Bull. 8 (5): 198–202. [*Picea likiangensis* var. *bhutanica*, var. *forrestii*, var. *linzhiensis*].

13344   Silba, J. (1994). The trans-pacific relationship of *Cupressus* in India and North America. J. Int. Conifer Preserv. Soc. 1: 1–28. [a home-made "journal" by Silba, presenting numerous new taxa in Asian *Cupressus* based on minimal evidence; names not listed here, see synonyms in Farjon, A. (2005). A monograph of Cupressaceae and Sciadopitys].

13344a   Silba, J. (1996). A new species of *Pseudotaxus* Cheng (Taxaceae) from China. Phytologia 81 (4): 322–328. (ill.). [*P. liana* sp. nov. ("*liiana*") = *P. chienii*, p. 327].

13345   Silba, J. (1998). A monograph of the genus *Cupressus* L. J. Int. Conifer Preserv. Soc. 5 (2): 1–98. [a home-made "journal" by Silba, of unknown distribution (but to NYBG and RBG Kew), photocopied texts from his earlier writings pasted together, see above; *Cupressus dupreziana* var. *atlantica* (Gaussen) Silba, comb. nov., p. 29].

13346   Silba, J. (1999). A review of *Cupressus* L. in Sikkim and Bhutan, and adjoining areas. J. Int. Conifer Preserv. Soc. 6 (1): 16–25. [reiteration of earlier accounts and taxonomies by this author, see above].

13347   Sillett, S. C. & J. C. Spickler (2000). Crown structure of the world's second largest tree. Madroño 47 (2): 127–133. [*Sequoiadendron giganteum*, tree architecture].

13360   Simak, M. (1971). De amerikanska lärkarterna; *Larix occidentalis*, *Larix lyallii*, *Larix laricina*. Sveriges Skogsvårdsforbunds Tidskr. 69 (1): 59–80.

13370   Simmon, I. G. & T. R. Vale (1975). Conservation of the California coast redwood and its environment. Environment. Conservation 2 (1): 29–38. [*Sequoia sempervirens*, ill., maps].

13375   Sinclair, W. T., R. R. Mill, M. F. Gardner, P. Woltz, T. Jaffré, J. Preston, M. L. Hollingsworth, A. Ponge & M. Möller (2002). Evolutionary relationships of the New Caledonian heterotrophic conifer, *Parasitaxus usta* (Podocarpaceae), inferred from chloroplast *trn*L-F intron/spacer and nuclear rDNA ITS2 sequences. Pl. Syst. Evol. 233 (1–2): 79–104.

13381   Singh, H. & J. Chatterjee (1963). A contribution to the life history of *Cryptomeria japonica* D. Don. Phytomorphology 13: 429–445.

13382   Singh, H. & Y. P. Oberoi (1962). A contribution to the life history of *Biota orientalis* Endl. Phytomorphology 12: 373–393. [*Platycladus orientalis*].

13383   Singh, K. J., G. W. Rothwell, G. Mapes & S. Chandra (2003). Reinvestigation of the coniferophyte morphospecies *Buriadia heterophylla* Seward and Sahni, with reinterpretation of vegetative diversity and putative seed attachments. Rev. Palaeobot. Palynol. 127: 25–43. (ill.). [Palaeozoic morphospecies of which reproductive structures remain unknown].

13390   Sivak, J. (1973). Observations nouvelles sur les grains de pollen de *Tsuga*. Pollen Spores 15: 397–457. (ill.).

13401   Sivak, J. (1976). Nouvelles espèces du genre *Cathaya* d'après leurs grains de pollen dans le Tertiaire du Sud de la France. Pollen Spores 18 (2): 243–288. (ill.). [palaeobotany].

13402   Sivak, J. (1978). Histoire de genre *Tsuga* en Europe d'après l'étude des grains de pollen actuels et fossiles. Paleobiol. Continent. 9: 1–226. (ill.).

13415   Skottsberg, C. (1916). Botanische Ergebnisse der schwedischen Expedition nach Patagonien und dem Feuerlande 1907–1909. V. Die Vegetationsverhältnisse längs der Cordillera de los Andes S. von 41° S.Br., ein Beitrag zur Kenntnis der Vegetation in Chiloé, West-Patagonien, dem andienen Patagonien und Feuerland. Kongl. Svenska Vetenskapsakad. Handl. 56 (5): 1–360, pl. 1–23. [*Libocedrus chilensis* = *Austrocedrus chilensis*, *L. tetragona* = *Pilgerodendron uviferum*, p. 170, pl. 2, f. 1].

13424   Slavin, A. D. (1933). Our deciduous conifers. Amer. Hort. 12: 48–53. [*Larix laricina*, *L. occidentalis*, *L. lyallii*, *Taxodium distichum*].

13450   Smith, H. (1927). Plantae sinenses. Acta Horti Gothob. 3. [see R. Florin, 1927].

13451   Smith, J. (1979). The Mexican White Pine. Arnoldia 39: 278–285. [*Pinus ayacahuite*].

13453   Smith, J. D. (1891). Undescribed plants from Guatemala. IX. Bot. Gaz. (Crawfordsville) 16 (6): 190–200. [*Pinus donnell-smithii* Masters, sp. nov., p. 199: species descr. by M. T. Masters].

13460   Smith, R. H. (1981). Variation in immature cone color of Ponderosa Pine (Pinaceae) in northern California and southern Oregon. Madroño 28: 272–274. [*Pinus ponderosa*].

13465   Smith, S. Y. & R. A. Stockey (2001). A new species of *Pityostrobus* from the Lower Cretaceous of California and its bearing on the evolution of Pinaceae. Int. J. Plant Sci. 162 (3): 669–681. [*Pityostrobus californiensis* sp. nov., p. 670, palaeobotany of Pinaceae, evolution, ill.].

13466   Smith, S. Y. & R. A. Stockey (2002). Permineralized pine cones from the Cretaceous of Vancouver Island, British Columbia. Int. J. Plant Sci. 163 (1): 185–196. [*Pinus*, *Pityostrobus beardii* sp. nov., Pinaceae, palaeobotany, ill.].

13468   Smith, W. W. (1920). Diagnoses specierum novarum in herbario Horti Regii Botanici Edinburgensis cognitarum. CCCCLI-D Species Asiaticae. Notes Roy. Bot. Gard. Edinburgh 12: 191–230. [*Podocarpus forrestii* Craib & W. W. Sm. sp. nov. = *P. macrophyllus*, p. 219].

13470   Smouse, P. E. & L. C. Saylor (1973). Studies of the *Pinus rigida-serotina* complex: 1. A study of geographic variation. Ann. Missouri Bot. Gard. 60 (2): 174–191. (ill.).

13480   Smouse, P. E. & L. C. Saylor (1973). Studies of the *Pinus rigida-serotina* complex: 2. Natural hybridization among the *Pinus rigida-serotina* complex, *P. taeda* and *P. echinata*. Ann. Missouri Bot. Gard. 60 (2): 192–203.

13490   Sniezko, R. A. & L. J. Mullin (1987). Taxonomic implications of Bush Pig damage and basal shoots in *Pinus tecunumanii*. Commonw. Forest. Rev. 66: 313–316. [+ *P. oocarpa*].

13500   Sochava, V. B. (1944). On the genesis and phytocenology of *Picea ajanensis*. Bot. Žurn. SSSR 29: 205–218. (Russ., Eng. summ.). [*Picea jezoensis*, *Abies nephrolepis*, *A. sachalinensis*].

13513   Soljan, D. & Solic, M. E. (1987). Prilog poznavanju jele na Biokovu. (A contribution to the study of fir on the mountain of Biokovo). Glasn. Zemaljsk. Muz. u Zerajevu, n.s. 25/26: 53–69. (Croatian, Eng. summ., ill.). [*Abies alba*, *A. cephalonica*, *A.* × *borisii-regis*].

13515   Soltis, P. S., D. E. Soltis & C. J. Smiley (1992). An *rbc*L sequence from a Miocene *Taxodium* (bald cypress). Proc. Nation. Acad. Sci. U.S.A. 89: 449–451. [sequence of 1320 base-pairs from the *rbc*L gene of Clarkia beds *Taxodium distichum*, i.e. 92% of extant taxon].

13517   Song, Z. Q., Z. Q. Lang & X. Liu (1998). Chemical characteristics of oleoresin from *Pinus bungeana* Zucc. and its taxonomic implications. Acta Phytotax. Sin. 36 (6): 511–514. (Chin., Eng. summ.).

13520   Sorger, O. (1925). Die systematische Stellung von *Taiwania cryptomerioides* Hayata. Österreich. Bot. Zeitschr. 74: 81–102, f. 1–3. [see also G. Koidzumi in Acta Phytotax. Geobot. 11: 138, 1942].

13530   Sosa, A. H. (1935). Algunas consideraciones sobre el *Juniperus* de México. México Forest. 13: 117–122.

13540   Sowerby, A. D. C. (1937). The white-barked pine (*Pinus bungeana* Zuccarini) in North China. J. Roy. Hort. Soc. London 62: 443–445.

13550   Spach, E. (1841). Révision des *Juniperus*. Ann. Sci. Nat. Bot., sér. 2, 16: 282–305. [*Juniperus* sect. *Oxycedrus*, sect. nov., p. 288, *J.* sect. *Sabina* (Mill.) comb. et stat. nov., p. 291].

13581   Sprague, T. A. & M. L. Green (1938). The botanical name of the Douglas Fir. Bull. Misc. Inform. 1938 (2): 79–80. [proposed *Pseudotsuga taxifolia* as legit. name, but see J. do A. Franco, 1950: *P. menziesii* (Mirb.) Franco].

13582   Sprengel, K. P. J. (1826). Caroli Linnaei ... Systema vegetabilium. Editio decima sexta. Vol. 3. Göttingen. [*Juniperus mexicana* sp. nov. (nom. superfl.) =*J. ashei*, p. 909].

13590   Squillace, A. E. (1966). Geographic variation in slash pine. Forest Sci. Monogr. 10: 1–56. (ill.). [*Pinus elliottii*].

13591   Squillace, A. E. & J. P. Perry, Jr. (1992). Classification of *Pinus patula*, *P. tecunumanii*, *P. oocarpa*, *P. caribaea* var. *hondurensis*, and related taxonomic entities. U.S.D.A. Forest Service Res. Paper SE-285. [chemical analysis of terpenes].

13601   Srivastava, S. K. (1976). The fossil pollen genus *Classopollis*. Lethaia 9: 437–457. [Cheirolepidiaceae, *Classopollis* Pflug, palynology, palaeobotany, ill.].

13610   Srodon, A. (1936). La répartition et la protection de l'Arole dans les Carpathes Polonaises. Nadbitka Z "Ochrony Przyrody", Rocznik 16, 1936. (Polish, French summ., ill.). [*Pinus cembra*].

13620   Srodon, A. (1937). Mélèze Polonais à Maniawa dans les monts Gorgany. Nadbitka Z "Ochrony Przyrody", Rocznik 17, 1937. (Polish, French summ., ill. of cones). [*Larix decidua* var. *polonica*].

13630   Stafleu, F. A. (1956). *Pseudotsuga menziesii* versus *Pseudotsuga taxifolia*. Taxon 5: 19, 38–39.

13651a   Standley, P. C. (1924). Nine new species of plants from Central America. Proc. Biol. Soc. Washington 37: 49–53. [*Podocarpus guatemalensis* sp. nov., p. 49].

13652   Standley, P. C. (1950). El cipres centroamericano. Ceiba 1 (3): 180–185. [*Cupressus lusitanica*, syn. *C. lindleyi*].

13653   Standley, P. C. & J. A. Steyermark (1943). Studies of Central American plants III. Publ. Field. Mus. Nat. Hist., Bot. Ser. 23 (1): (1–)3–28. [*Juniperus standleyi* Steyerm., sp. nov., p. 3].

13656   Stankiewicz, B. A., M. Mastalerz, M. A. Kruge, P. F. van Bergen & A. Sadowska (1997). A comparative study of modern and fossil cone scales and seeds of conifers: a geochemical approach. New Phytol. 135 (2): 375–393. [*Pinus strobus*, *Sequoia sempervirens*, fossil *Pinus leitzii*, *Sequoia langsdorfii*, chemosystematics].

13660   Stapf, O. (1910). *Pinus armandii*. Bot. Mag. 136: pl. 8347.

13670   Stapf, O. (1918). *Widdringtonia dracomontana* Stapf, sp. nov. Bull. Misc. Inform. 6: 206. [= *W. nodiflora*].

13680   Stapf, O. (1923). *Picea brachytyla*. Bot. Mag. 148: pl. 8969.

13690   Stapf, O. (1924). *Picea glehnii*. Bot. Mag. 149: pl. 9020.

13700   Stapf, O. (1925). *Cupressus duclouxiana*. Bot. Mag. 150: pl. 9049.

13701   Stapf, O. (1930). *Widdringtonia stipitata* sp. nov. (Tabula 3126). Hooker's Icon. Pl. Ser. 5, Vol. 32, t. 3126. [= *W. nodiflora*, form with smooth cones].

13710   Stapf, O. (1930). *Tsuga chinensis*. Bot. Mag. 153: pl. 9193.

13720   Stapf, O. (1930). *Abies faberi* et *Abies forrestii*, China. Bot. Mag. 153: pl. 9201. [*A. fabri* as "A. faberi"].

13740   Staszkiewicz, J. (1961). Biometric studies on the cones of *Pinus silvestris* L., growing in Hungary. Acta Bot. Acad. Sci. Hung. 8 (3–4): 451–466. [*P. sylvestris*].

13741   Staszkiewicz, J. (1961). Variation in recent and fossil cones of *Pinus silvestris* L. Fragm. Florist. Geobot. 7 (1): 97–160 (map p. 109). [*P. sylvestris*].

13750   Staszkiewicz, J. (1966). Preliminary studies on the variability of cones of *Picea abies* (L.) Karst. ssp. *abies* in Poland. Fragm. Florist. Geobot. 12 (4): 349–371.

13750a   Staszkiewicz, J. (1992). Variability of the cones of *Picea jezoensis* and *P. koraiensis* (Pinaceae) in North Korea. Fragm. Flor. Geobot. Ann. 37: 241–249.

13750b   Staszkiewicz, J. (1992). Variability of the cones of *Larix olgensis* (Pinaceae) in the massif of Paekdu-san in North Korea. Fragm. Flor. Geobot. Ann. 37: 487–497. [= *Larix gmelinii* var. *olgensis*].

13750c   Staszkiewicz, J. (1993). Morphological differentiation of the cones of selected *Picea* (Pinaceae) species from Europe and North Asia. Polish Bot. Stud. 5: 25–31. (maps. graphs). [*Picea abies*, *P. obovata*, *P. jezoensis*, *P. omorika*, *P. orientalis*, *P. koraiensis*].

13751   Staunton, G. L. (1797). An authentic account of an embassy from the King of Great Britain to the emperor of China; including cursory observations made, ... Vol. 1. London. [*Thuja pensilis* = *Glyptostrobus pensilis*, p. 436; validly published by D. Don in Lambert, Descr. *Pinus* ed. 2, 2: 115 (1828)].

13760   Stead, J. W. (1983). A study of variation and taxonomy of the *Pinus pseudostrobus* complex. Commonw. Forest. Rev. 62: 25–35.

13770   Stead, J. W. (1983). Studies of variation in Central American pines V: a numerical study of variation in the *Pseudostrobus* group. Silvae Genet. 32: 101–115.

13780   Stead, J. W. & B. T. Styles (1984). Studies of Central American pines: a revision of the "pseudostrobus" group (Pinaceae). Bot. J. Linn. Soc. 89: 249–275. [*Pinus* subsect. *Pseudostrobi*].

13781   Stearn, W. T. (1970). Ventenat's "Decas generum novorum" (1808). in: P. Smit & R. J. C. V. ter Laage (eds.). Essays in biohistory. Regnum Veg. Vol. 71: 342–352. I.A.P.T., Utrecht.

13790   Stebbins, G. L., Jr. (1948). The chromosomes and relationships of *Metasequoia* and *Sequoia*. Science 108: 95–98.

13796   Stein, B. (1887). *Picea alpestris* Brügger. Eine neue Fichte der Schweizer Alpen. Gaertenflora 37: 346–350. [*Picea alpestris* (Brügger) Stein, comb. nov., p. 347, *P. excelsa* = *P. abies*, basion. cit. in ed. footnote: *Abies alpestris* Bruegger, Jahresber. Naturf. Ges. Graubünden [ser. 2] 17: 154. [1874], but see Brügger, 1886].

13798   Steinhoff, R. J. (1972). White pines of western North America and Central America. In: Biology of rust resistance in forest trees: proceedings of a NATO-IUFRO advanced study institute, August 1969, pp. 215–232. USDA Misc. Publ. 1221, Washington, DC. [*Pinus albicaulis, P. aristata, P. ayacahuite, P. balfouriana, P. flexilis, P. lambertiana, P. monticola, P. strobiformis*, ill.].

13800   Steinhoff, R. J. & J. W. Andresen (1971). Geographic variation in *Pinus flexilis* and *Pinus strobiformis* and its bearing on their taxonomic status. Silvae Genet. 20: 159–167.

13803   Stellfeld, C. (1944). Vellozoa seccao de botanica No. 7. Trib. Farm. Brasil 12: 179–183. [*Araucaria dioica* (Vell.) comb. nov. = *A. angustifolia*, p. 181].

13806   Stephenson, N. L. (2000). Estimated ages of some large Giant Sequoias: General Sherman keeps getting younger. Madroño 47 (1): 61–67. [*Sequoiadendron giganteum*, max. estim. age of living tree 2890 years, oldest known (cut!) tree 3266 years].

13807   Stephenson, N. L. & A. Demetry (1995). Estimating ages of Giant Sequoias. Canad. J. Forest Res. 25: 223–233. [*Sequoiadendron giganteum*].

13810   Sterling, C. (1949). Some features in the morphology of *Metasequoia*. Amer. J. Bot. 36: 461–470.

13821   Steudel, E. G. von (1840). Nomenclator botanicus enumerans ordine alphabetico nomina atque synonyma... Ed. 2, Vol. 1 (sect. 5–6). Stuttgart – Tübingen. [*Fresnelia* nom.

gen. nov. = *Callitris* Vent., p. 648, *Juniperus deppeana* sp. nov., p. 835; *Fresnelia* Steud. is a nom. superfl. for *Frenela* Mirb. (1825), which is in turn a nom. illeg. proposed as a substitute for *Callitris* Vent. (1808) as it is not a homonym of *Calytrix* in Myrtaceae].

13822   Steven, C. von (1838). De Pinubus Taurico-Caucasicis. Bull. Soc. Imp. Naturalistes Moscou 11: 43–53. [*Pinus nordmanniana* sp. nov. = *Abies nordmanniana*, p. 45, t. 2, *P. pityusa* sp. nov. = *P. brutia* var. *pityusa*, p. 49, *P. sylvestris* var. *hamata* var. nov., p. 52].

13823   Steven, C. von (1857). Verzeichnis der auf der Taurischen Halbinsel wildwachsenden Pflanzen. Bull. Soc. Imp. Naturalistes Moscou 30: 325–398. [Pinus spp., *Juniperus marschalliana* sp. nov., *J. rhodocarpa* sp. nov. (in text) = *J. oxycedrus*, p. 397, *J. depressa* Rafin. = *J. communis*, p. 398].

13824   Steven, H. M. & A. Carlisle (1959). The Native Pinewoods of Scotland. Edinburgh – London. [*Pinus sylvestris*].

13825   Stewart, P. J. (1969). *Cupressus dupreziana*, threatened Conifer of the Sahara. Biol. Conservation (Barking) 2 (1): 10–12.

13830   St. John, H. & R. W. Krauss (1954). The taxonomic position and scientific name of the big tree known as *Sequoia gigantea*. Pacific Sci. 8: 341–358. [*Sequoiadendron giganteum, Sequoia sempervirens*].

13831   St. Paul, U. von (1901). *Pseudotsuga douglasii*, Carrière. Mitt. Deutsch. Dendrol. Ges. 10: 263–270 (pag. in Ed. 2). [= *P. menziesii*].

13832   Stockey, R. A. (1975). Seeds and embryos of *Araucaria mirabilis*. Amer. J. Bot. 73: 1079–1081. [palaeobotany].

13832a   Stockey, R. A. (1977). Reproductive biology of the Cerro Cuadrado (Jurassic) conifers: *Pararaucaria patagonica*. Amer. J. Bot. 64: 733–744.

13832b   Stockey, R. A. (1978). Reproductive biology of Cerro Cuadrado fossil conifers: ontogeny and reproductive strategies in *Araucaria mirabilis* (Spegazzini) Windhausen. Palaeontographica Abt. B, Paläophytol. 166: 1–15. [Araucariaceae, palaeobotany].

13832c   Stockey, R. A. (1980). Anatomy and morphology of *Araucaria sphaerocarpa* Carruthers from the Jurassic inferior oolite of Bruton, Somerset. Bot. Gaz. 141: 116–124. (ill.).

13832d   Stockey, R. A. (1980). Jurassic araucarian cone from southern England. Palaeontology 23: 657–666. [Araucariaceae, *Araucaria*, palaeobotany].

13833   Stockey, R. A. (1981). *Pityostrobus mcmurrayensis* sp. nov., a permineralized pinaceous cone from the Cretaceous of Alberta. Canad. J. Bot. 59: 75–82. [Pinaceae, palaeobotany].

13834 Stockey, R. A. & I. J. Atkinson (1993). Cuticle micromorphology of *Agathis* Salisbury. Int. J. Plant Sci. 154 (1): 187–224. [SEM photography, 21 species].

13834a Stockey, R. A. & B. J. Frevel (1997). Cuticle micromorphology of *Prumnopitys* Philippi (Podocarpaceae). Int. J. Plant Sci. 158 (2): 198–221. [10 spp. investigated (incl. *Sundacarpus* as section of *Prumnopitys* which perhaps merits recogn. at gen. rank on these characters), SEM ill.].

13834b Stockey, R. A., B. J. Frevel & P. Woltz (1998). Cuticle micromorphology of *Podocarpus*, subgenus *Podocarpus*, section *Scytopodium* (Podocarpaceae) of Madagascar and South Africa. Int. J. Plant Sci. 159 (6): 923–940. [7 spp. of *Podocarpus*, detailed stomatal anatomy, SEM ill.).

13834c Stockey, R. A. & H. Ko (1986). Cuticle morphology of *Araucaria* de Jussieu. Bot. Gaz. (Crawfordsville) 147 (4): 508–548. [detailed stomatal anatomy of 20 spp. in the genus, SEM ill.].

13834d Stockey, R. A. & H. Ko (1988). Cuticle micromorphology of some New Caledonian podocarps. Bot. Gaz. (Crawfordsville) 149 (2): 240–252. [*Acmopyle pancheri, Retrophyllum comptonii, R. minor, Falcatifolium taxoides, Prumnopitys ferruginoides*, SEM ill.].

13834e Stockey, R. A. & H. Ko (1990). Cuticle micromorphology of *Dacrydium* (Podocarpaceae) from New Caledonia. Bot. Gaz. (Crawfordsville) 151 (1): 138–149. [detailed stomatal anatomy of 4 spp., SEM ill.].

13834f Stockey, R. A., H. Ko & P. Woltz (1992). Cuticle micromorphology of *Falcatifolium* de Laubenfels (Podocarpaceae). Int. J. Plant Sci. 153 (4): 589–601. [detailed stomatal anatomy of 5 spp., SEM ill.].

13834g Stockey, R. A., H. Ko & P. Woltz (1995). Cuticle micromorphology of *Parasitaxus* de Laubenfels (Podocarpaceae). Int. J. Plant Sci. 156 (5): 723–730. [*P. usta*, SEM ill.].

13834h Stockey, R. A., G. W. Rothwell, H. D. Addy & R. S. Currah (2001). Mycorrhizal association of the extinct conifer *Metasequoia milleri*. Mycol. Res. 105 (2): 202–205. (ill.).

13834i Stockey, R. A., G. W. Rothwell & A. B. Falder (2001). Diversity among taxodioid conifers: *Metasequoia foxii* sp. nov. from the Paleocene of central Alberta, Canada. Int. J. Plant Sci. 162 (1): 221–234 (ill.).

13835 Stockey, R. A. & T. N. Taylor (1978). Scanning electron microscopy of epidermal patterns and cuticular structure in the genus *Araucaria*. Scanning Electron Microscopy 2: 223–228.

13835a Stockey, R. A. & T. N. Taylor (1978). On the structure and evolutionary relationships of the Cerro Cuadrado fossil conifer seedlings. Bot. J. Linn. Soc. 76: 161–176. [Araucariaceae, *Araucaria mirabilis* (Spegazzini) Windhausen, palaeobotany].

13836 Stockey, R. A. & T. N. Taylor (1978). Cuticular features and epidermal patterns in the genus *Araucaria* de Jussieu. Bot. Gaz. (Crawfordsville) 139: 490–498.

13837 Stockey, R. A. & T. N. Taylor (1981). Scanning electron microscopy of epidermal patterns and cuticular structure in the genus *Agathis*. Scanning Electron Microscopy 3: 207–212.

13840 Stockwell, W. P. (1939). Cone variation in Digger pine. Madroño 5: 72–73. [*Pinus sabiniana*].

13848 Stoffberg, E. (1991). Morphological and ontogenetic studies on the southern African podocarps. Shoot apex morphology and ovuliferous cone initiation. Bot. J. Linn. Soc. 105 (1): 1–19. [*Afrocarpus, Podocarpus*, SEM ill.].

13849 Stoffberg, E. (1991). Morphological and ontogenetic studies on the southern African podocarps. Initiation of the seed scale complex and early development of integument, nucellus and epimatium. Bot. J. Linn. Soc. 105 (1): 21–35. [*Afrocarpus, Podocarpus*, SEM ill.].

13860 Stone, C. O. (1965). Modoc Cypress, *Cupressus bakeri* Jeps., does occur in Modoc County. Aliso 6 (1): 77–87.

13870 Stones, M. & W. Curtis (with appendix by T. de Malahide) (1969). No. 70. *Athrotaxis cupressoides*. Endemic Flora of Tasmania 2: 118–119 + 131, pl. 42.

13881 Strauss, S. H. & A. H. Doerksen (1990). Restriction fragment analysis of pine phylogeny. Evolution 44 (4): 1081–1096. [*Pinus* spp, 19 spp. DNA tested in broad agreement with Little & Critchfield, 1969, but with *P. leiophylla* in a clade with *P. taeda*].

13882 Strauss, S. H., A. H. Doerksen & J. R. Byrne (1990). Evolutionary relationships of Douglas-fir and its relatives (genus *Pseudotsuga*) from DNA restriction fragment analysis. Canad. J. Bot. 68: 1502–1510. [*Pseudotsuga, P. menziesii, Larix*].

13883 Strauss, S. H., Y. P. Hong & V. D. Hipkins (1993). High levels of population differentiation for mitochondrial DNA haplotypes in *Pinus radiata, P. muricata*, and *P. attenuata*. Theor. Appl. Genet. 86: 605–611.

13884 Strauss, S. H. & F. T. Ledig (1985). Seedling architecture and life history evolution in pines. Amer. Nat. 125: 702–715. [*Pinus*].

13885   Stuart, J. D. (1987), Fire history of an old-growth forest of *Sequoia sempervirens* (Taxodiaceae) forest in Humboldt Redwoods State Park, California. Madroño 34 (2): 128–141.

13900   Sturgeon, K. B. & J. B. Mitton (1980). Cone color polymorphism associated with elevation in white fir, *Abies concolor*, in southern Colorado. Amer. J. Bot. 67: 1040–1045.

13910   Styles, B. T. (1976). Studies of variation in Central American pines I. The identity of *Pinus oocarpa* var. *ochoterenai* Martínez. Silvae Genet. 25: 109–118. [*P. patula* p. p.].

13920   Styles, B. T. (1984). The identity of Schwerdtfeger's Central American pine. F.A.O. Forest. Genet. Resources. Inform. 13: 47–51. [*Pinus patula* ssp. *tecunumanii* (Eguiluz et Perry) comb et stat. nov., p. 50].

13921   Styles, B. T. (1988). The genus *Pinus*: a Mexican review. in: Simposio sobre diversidad biologica de México.

13922   Styles, B. T. (1993). Genus *Pinus*: A Mexican purview. In: T. P. Ramamoorthy, R. E. Bye, A. Lot & J. Fa (eds). Biological diversity of Mexico: Origin and distribution. Chapter 13: 397–420. (ill.). Oxford Univ. Press, Oxford – New York.

13923   Styles, B. T. (1993). Pine kernels. In: R. Macrae, R. K. Robinson & M. J. Sadler (eds.). Encyclopaedia of food science, food technology and nutrition. Vol. 6: pH – Soya Milk. (Vols. 1–8). Academic Press, London. [*Pinus* spp. with edible seeds].

13930   Styles, B. T. & J. Burley (1972). The botanical name of the Khasi Pine (*Pinus kesiya* Royle ex Gordon). Commonw. Forest. Rev. 51 (3), No. 149: 241–245. [= *P. insularis* Endl.].

13940   Styles, B. T. & C. E. Hughes (1983). Studies of variation in Central American pines III. Notes on the taxonomy and nomenclature of the pines and related gymnosperms in Honduras and adjacent Latin America republics. Brenesia 21: 269–291. [Coniferales, Pinaceae, *Pinus* spp., *Abies guatemalensis*, Cupressaceae, *Cupressus lusitanica*].

13941   Styles, B. T. & P. S. McCarter (1988). The botany, ecology, distribution and conservation status of *Pinus patula* ssp. *tecunumanii* in the Republic of Honduras. Ceiba 29 (1): 3–30. [ill. maps. tables].

13943   Styles, B. T. & R. McVaugh (1990). A Mexican pine promoted to specific status: *Pinus praetermissa*. Contr. Univ. Mich. Herb. 17: 307–312. [*P. praetermissa* sp. nov., p. 310, ill., *P. oocarpa*].

13950   Styles, B. T., J. W. Stead & K. J. Rolph (1982). Studies of variation in Central American pines II. Putative hybridization between *Pinus caribaea* var. *hondurensis* and *P. oocarpa*. Turrialba 32: 229–242. [*Pinus caribaea* var. *hondurensis* × *P. oocarpa*].

13960   Sudo, S. (1968). Anatomical studies on the wood of *Picea* with some considerations on their geographical distribution and taxonomy. Bull. Gov. Forest Exp. Sta. (Tokyo) 215: 39–130. (ill.).

13968   Sudworth, G. B. (1893). On legitimate authorship of certain binomials with other notes on nomenclature. Bull. Torrey Bot. Club 20 (1): 40–46. [*Abies campylocarpa*, *A. concolor*, *A. magnifica*, *Pinus* spp. (names coined by D. Douglas), *P. banksiana* (*P. sylvestris* [var.] *divaricata* Aiton), *P. cubensis*, *P. ponderosa* (as to type), *P. heterophylla* (Elliott) comb. nov., p. 45 (*P. taeda* var. *heterophylla*, *Thuja plicata*].

13970   Sudworth, G. B. (1905). A new species of Juniper for Texas. Forestry & Irrig. 11: 203–206, f. 1–4. [*Juniperus pinchotii* sp. nov., p. 204].

13971   Sudworth, G. B. (1907). A new tree Juniper for New Mexico. Forestry & Irrig. 13: 307–310, f. 1–2. [*Juniperus megalocarpa* sp. nov. = *J. osteosperma*, p. 307].

13973   Sudworth, G. B. (1910). A new cypress for Arizona. Amer. Forestry 16: 88–90, with plate. [*Cupressus glabra* sp. nov. = *C. arizonica* var. *glabra*].

13980   Sudworth, G. B. (1913). Forest atlas: Geographic distribution of North American trees. Part I. Pines. Govt. Printing Office, Washington, D.C. [*Pinus* spp.].

13990   Sudworth, G. B. (1915). The cypress and juniper trees of the Rocky Mountain region. U.S.D.A. Bull. (1915–23) 207: 1–36. (ill., maps). [*Cupressus*, *Juniperus*].

14000   Sudworth, G. B. (1916). The spruce and balsam firs of the Rocky Mountain region. U.S.D.A. Bull. (1915–23) 327: 1–43. [*Picea engelmannii*, *P. pungens*, *Abies lasiocarpa*].

14010   Sudworth, G. B. (1917). The pine trees of the Rocky Mountain region. U.S.D.A. Bull. (1915–23) 460: 1–46. (with 28 pl. + 14 maps). [*Pinus* spp.].

14040a   Suguhara, Y. (1938). Fertilization and early embryogeny of *Chamaecyparis pisifera* Siebold et Zuccarini. Sci. Rep. Tohoku Imp. Univ., Ser. 4, Biol. 13: 9–14.

14040b   Sugihara, Y. (1943). Embryological observations on *Keteleeria davidiana* Beissner var. *formosana* Hayata. Sci. Rep. Tohoku Imp. Univ., Ser. 4, Biol. 17: 215–221.

14041   Sugihara, Y. (1947). The embryogeny of *Cryptomeria japonica* D. Don. Bot. Mag. (Tokyo) 60: 47–52. (ill.). [comparison with *Cunninghamia* and *Taiwania*].

14041a   Sugihara, Y. (1947). The embryology of *Abies firma* Siebold et Zuccarini. Bot. Mag. (Tokyo) 60: 58–62.

14042   Sugihara, Y. (1947). The embryogeny of *Cryptomeria japonica.* Bot. Mag. (Tokyo) 60: 703–714. (ill.).

14043   Sugihara, Y. (1956). The embryogeny of *Cupressus funebris.* Bot. Mag. (Tokyo) 69: 439–441. [*syn. Chamaecyparis funebris,* ill.].

14044   Sugihara, Y. (1969). On the embryo of *Cryptomeria japonica.* Phytomorphology 19: 110–111.

14049   Sukachev, V. N. (1906). Über eine für die Krim neue Kiefer. Bot. Žurn. [Trav. Soc. Imp. Naturalistes St. Pétersbourg, sect. Bot. 23 (3)] 1: 34–38. (Russ., Germ. summ.). [*Pinus pityusa* var. *stankewiczii* var. nov., p. 37].

14050   Sukachev, V. N. (1928). Forest species; their systematics, geography and phytosociology. Part 1 (Coniferae); pp. 1–84, f. 1–24. Akad. Nauk SSSR, Moscow – Leningrad. (Russ.). [*Picea obovata, P. abies, P. jezoensis, Pseudotsuga, Taxus*].

14070   Sukachev, V. N. (1931). Über zwei neue wirtschaftlich wertvolle Holzarten. Trudy Issl. Lesn. Khoz. Lesn. Promysl. (Mitt. Staatsinst. Wiss. Forsch.) 10: 3–20, 11 figs. (Russ., Germ. summ.). [*Larix maritima* sp. nov., *L. lubarskii* sp. nov., p. 9, *L. stenophylla* sp. nov. = *L. gmelinii* et var. *olgensis*; new names not in Index Kewensis !].

14071   Sumnevicz, G. P. (1946). Species novae generis *Juniperus* L. ex Asia media. Bot. Mater. Gerb. Inst. Bot. Zool. Akad. Nauk Uzbeksk. SSR 8: 22–26. (Russ., Lat. diagn.). [*Juniperus drobovii* sp. nov., p. 22, *J. tianschanica* sp. nov., p. 24].

14076   Surova, T. D. & V. Kvavadze (1988). Sporoderm ultrastructure in some gymnosperms (*Metasequoia, Cunninghamia, Sciadopitys*). Bot. Žurn. 73: 34–44. [LM, SEM, TEM ill., *Sciadopitys* differs].

14078   Süss, H. & E. Velitzelos (1997). Fossile Hölzer der Familie Taxodiaceae aus tertiären Schichten des Versteinerten Waldes von Lesbos, Griechenland. Fedde's Repert. 108 (1–2): 1–30. [*Taxodioxylon, Glyptostroboxylon,* anatomy of Upper Oligocene – Lower Miocene coniferous wood].

14080   Sutherland, M. (1934). A microscopical study of the structure of the leaves of the genus *Pinus.* Trans. & Proc. New Zealand Inst. 63: 517–568. (ill.).

14080a   Sutton, B. C., D. J. Flanagan & Y. A. El-Kassaby (1991). A simple and rapid method for estimating representation of species in spruce seedlots using chloroplast DNA restriction fragment length polymorphism. Silvae Genet. 40: 119–123. [*Picea engelmannii, P. glauca, P. sitchensis*].

14088   Suyama, Y., H. Yoshimaru & Y. Tsumura (2000). Molecular phylogenetic position of Japanese *Abies* (Pinaceae) based on chloroplast DNA sequences. Mol. Phylogen. Evol. 16 (2): 271–277. [*Abies firma, A. homolepis, A. mariesii, A. sachalinensis, A. veitchii*].

14089   Suzuki, K. (1985). *Larix* remains from Pleistocene strata of northeast Japan with special reference to the distribution of *Larix* in the latter half of the last glacial age. Trans. Proc. Palaeontol. Soc. Japan 137: 64–74.

14090   Suzuki, S. (1930). Species and their distribution of *Pinus* in Japan. Sylvia 1 (4): 20–21. [12 spp. recognized].

14100   Svechnikova, I. N. (1957). The Eocene Flora of SW Ukraine. Sbornik pamjati Afrikana Niklajevica Krischtofovica, Isd. Akad. Nauk SSSR. Moscow – Leningrad. (Russ.). [*Taiwania araucarioides* Svechnikova, fossil sp. nov., map, p. 208].

14101   Svechnikova, I. N. (1964). Predstavitel roda *Cathaya* (Pinaceae) iz Pliotzena Abchazii. Paleontol. Žurn. 2: 125–131. (Russ.).

14110   Swartz, O. P. (1788). Nova genera et species plantarum seu prodromus descriptionum vegetabilium. M. Sweder, Holmiae (Stockholm) & Upsaliae (Uppsala). (x + 152 pp.). [*Pinus occidentalis* sp. nov., p. 103].

14110c   Sweet, R. (1818). Hortus suburbanus Londinensis;... London. [*Podocarpus macrophyllus* (Thunb.) comb. nov., p. 211].

14111   Sweet, R. (1827). Sweet's Hortus britannicus: or, a catalogue of plants cultivated in the gardens of Great Britain; ... London. [*Taxodium distichum* var. *nutans* Sweet = *T. distichum* var. *imbricatum,* p. 372].

14112   Sweet, R. (1830). Sweet's Hortus britannicus: or, a catalogue of plants cultivated in the gardens of Great Britain; ... Ed. 2, London. [*Callitris australis* (Poir) comb. nov. = *C. rhomboidea,* p. 474, *Belis lanceolata* (Lamb.) "comb. nov." [see *Belis lanceolata* (Lamb.) Hoffmans., Verz.: 42 (1824); cf. D. J. Mabberley in Taxon 29 (5–6): 604 (1980)] = *Cunninghamia lanceolata,* p. 475].

14113   Szafer, W. (1913). Przyczynek do znajomości modrzewi eur-azyatyckich ze szczególnem uwzględnieniem modrzewia w Polsce. Kosmos 38: 1281-1322. (Polish, German abstract, ill.). *L.* × *czekanowskii* hybr. nov., p. 1297: *L. gmelinii* × *L. sibirica*].

14120   Szafer, W. (1934). The protection of *Larix polonica* Rac. in Poland. Nadbitka Z "Ochrony Przyrody", Rocznik 14, 1934. (Polish, Eng. summ., maps). [*Larix decidua* var. *polonica*].

14123   Szmidt, A. E., A. Sigurgeirsson, X. R. Wang, J. E. Hällgren & D. Lindgren (1988). Genetic relationships among *Pinus* species based on chloroplast DNA polymorphism. In: J. E. Hällgren, ed. Molecular genetics of forest trees; Proc. Frans Kempe Symp., Rep. No. 9, pp. 33–47. Swedish Univ. Agric. Sci, Dept. Forest Genetics and Plant Physiology, Umeå. [*Pinus,* 20 spp. tested, classified in broad agreement with Little & Critchfield, 1969).

14124 Szmidt, A. E. & X. R. Wang (1993). Molecular systematics and genetic differentiation of *Pinus sylvestris* L. and *P. densiflora* Sieb. & Zucc. Theor. Appl. Genet. 86: 159–165.

14125 Szmidt, A. E., X. R. Wang & S. Changtragoon (1996). Contrasting patterns of genetic diversity in two tropical pines: *Pinus kesiya* Royle ex Gordon and *P. merkusii* Jungh. et De Vriese. Theor. Appl. Genet. 92: 436–441. [*Pinus kesiya*, *P. merkusii*, DNA].

14130 Tahara, M. (1937). Contributions to the morphology of *Sciadopitys verticillata*. Cytologia, Fujii Jub. Vol. 1: 14–19.

14130a Tahara, M. (1940). The gametophytes, fertilization and proembryo of *Sciadopitys verticillata*. Sci. Rep. Tohoku Imp. Univ., Ser. 4, Biol. 15: 28–29.

14130b Taira, H. (2001). A change of *Cryptomeria japonica* distribution from viewpoint of regeneration system and genetic diversity. J. Phytogeogr. Taxon. 49 (2): 111–116. (Japan., Eng. summ., ill.).

14130c Taira, H. & T. Sawada (1977). Occurrence of natural *Cryptomeria japonica* D. Don at the Mt. Tateyama, Toyama Prefecture (2050 m above the sea level). J. Japan. Forest. Soc. 59 (12): 449–452. (Japan.).

14130d Taira, H., Y. Tsumura & K. Ohba (1993). Growing conditions and allozyme analysis of a sugi (*Cryptomeria japonica*) forest at 2050 meters above sea level on Mount Nekomata. J. Japan. Forest. Soc. 75 (6): 541–545. (Japan., Eng. summ.).

14130e Takaso, T. (1981). A developmental study of the integument in gymnosperms 2. *Pinus thunbergii* Parl., *Abies mariesii* Mast. and *A. veitchii* Lindl. J. Japan. Bot. 56: 73–89. [anatomy, ontogeny of ovule and seed in Pinaceae, ill.].

14131 Takaso, T. & J. W. Owens (1995). Ovulate cone morphology and pollination in *Pseudotsuga* and *Cedrus*. Int. J. Plant Sci. 156 (4): 630–639.

14132 Takaso, T. & J. W. Owens (1995). Pollination drop and microdrop secretions in *Cedrus*. Int. J. Plant Sci. 156 (4): 640–649.

14134 Takaso, T. & P. B. Tomlinson (1989). Aspects of cone and ovule ontogeny in *Cryptomeria* (Taxodiaceae). Amer. J. Bot. 76 (5): 692–705. [*Cryptomeria japonica*, ill. LM, SEM].

14135 Takaso, T. & P. B. Tomlinson (1989). Cone and ovule development in *Callitris* (Cupressaceae – Callitroideae). Bot. Gaz. (Crawfordsville) 150: 378–390. (ill. LM, SEM).

14136 Takaso, T. & P. B. Tomlinson (1990). Cone and ovule ontogeny in *Taxodium* and *Glyptostrobus* (Taxodiaceae – Coniferales). Amer. J. Bot. 77 (9): 1209–1221. [*Taxodium distichum*, *Glyptostrobus pensilis*, ill. LM, SEM].

14137 Takaso, T. & P. B. Tomlinson (1991). Cone and ovule development in *Sciadopitys* (Taxodiaceae – Coniferales) Amer. J. Bot. 78 (3): 417–428. [*Sciadopitys verticillata*, ill. LM, SEM].

14138 Takaso, T. & P. B. Tomlinson (1992). Seed cone and ovule ontogeny in *Metasequoia*, *Sequoia* and *Sequoiadendron* (Taxodiaceae – Coniferales). Bot. J. Linn. Soc. 109: 15–37. [*Metasequoia glyptostroboides*, *Sequoia sempervirens*, *Sequoia-dendron giganteum*, ill. LM, SEM].

14139 Takenouchi, M. (1942). A preliminary report of the conifers of Manchukuo. J. Japan. Forest. Soc. 24: 113–129. [*Abies*, *Larix*, *Picea*, *Pinus*].

14140 Takenouchi, M. & J. J. Chien (1957). On *Abies sibirica* and a new hybrid of the genus *Abies* in Heilungkiang, China. Acta Phytotax. Sin. 6: 145–158, pl. 37–38. (Chin., Eng. summ.). [Lat. descr. of *A. × sibirico-nephrolepis* hybr. nov., this epithet is invalid under Art. H.10.3 of the Code].

14153 Talhouk, S. N., R. Zurayk & S. Khuri (2001). Conservation of the coniferous forests of Lebanon: past, present and future prospects. Oryx 35 (3): 206–215. [*Abies*, *Cedrus*, *Cupressus*, *Juniperus*, *Pinus*].

14155 Tan, K., G. Sfikas & G. Vold (1999). *Juniperus drupacea* (Cupressaceae) in the southern Peloponnese. Acta Bot. Fennica 162: 133–135. (maps).

14158 Tarbaeva, V. M. (1991). (Structure of seeds of *Juniperus* species, growing in the Komi SSR). Proc. XI Komi Sci. Conf., Syktyvkar, 1991: 108–114. (Russ.). [*Juniperus communis*, *J. sibirica*].

14160 Tatewaki, M. (1943). Sociological studies on the *Picea glehni* forests. Res. Bull. Coll. Exp. Forests Coll. Agric. Hokkaido Imp. Univ. 13 (2): 1–81, pl. 1–20, f. 1–4. (Japan.). [*Picea glehnii*, *P. jezoensis*, ecol. Hokkaido & southern Kuriles].

14161 Tatewaki, M. (1958). Forest ecology of the islands of the North Pacific Ocean. J. Fac. Agric. Hokkaido Univ. 50 (4): 371–486, fig. 1–15. (Eng.). [*Abies sachalinensis*, *Picea glehnii*, *P. jezoensis*, *Pinus* spp., *Larix gmelinii*, ecology of Hokkaido to Aleutian Islands].

14170 Tatewaki, M., K. Ito & M. Tohyama (1963). Phytosociological study on the forests of Japanese hemlock (*Tsuga diversifolia*). Res. Bull. Coll. Exp. Forests Coll. Agric. Hokkaido Imp. Univ. 23: 83–146.

14180 Tatewaki, M., K. Ito & M. Tohyama (1965). Phytosociological study on the forests of Japanese larch (*Larix leptolepis* Gordon). Res. Bull. Coll. Exp. Forests Coll. Agric. Hokkaido Imp. Univ. 24: 1–176. [= *L. kaempferi*].

14190   Taylor, R. J. (1972). The relationship and origin of *Tsuga heterophylla* and *Tsuga mertensiana* based on phytochemical and morphological interpretations. Amer. J. Bot. 59 (2): 149–157.

14191   Taylor, R. J. (1972). *Tsuga heterophylla* (Raf.) Sarg., Pacific or Western hemlock. Davidsonia 3: 49–54.

14200   Taylor, R. J. & T. F. Patterson (1980). Biosystematics of Mexican Spruce Species and populations. Taxon 29 (4): 421–469. [*Picea chihuahuana, P. engelmannii* ssp. *mexicana*].

14205   Taylor, R. J., T. F. Patterson & R. J. Harrod (1994). Systematics of Mexican Spruce – Revisited. Syst. Bot. 19: 47–59. [*Picea chihuahuana, P. engelmannii, P. engelmannii* var. *mexicana, P. martinezii*].

14210   Taylor, R. J., S. Williams & R. Daubenmire (1975). Interspecific relationships and the questions of introgression between *Picea engelmannii* and *Picea pungens*. Canad. J. Bot. 53: 2547–2555.

14220   Taylor, R. L. & S. Taylor (1980). *Tsuga mertensiana* in British Columbia. Davidsonia 11 (4): 78–84.

14230   Taylor, S. & D. Sziklai (1976). *Chamaecyparis nootkatensis* yellow-cedar member of the family Cupressaceae. Davidsonia 7 (4): 56–62. (ill.). [occurrence in British Columbia].

14240   Taylor, T. M. C. (1959). The taxonomic relationship between *Picea glauca* (Moench) Voss and *P. engelmannii* Parry. Madroño 15: 111–115. [*P. glauca* ssp. *engelmannii* (Engelm.) comb. et stat. nov.].

14244   Taylor, T. N. & K. L. Alvin (1984). Ultrastructure and development of Mesozoic pollen: *Classopollis*. Amer. J. Bot. 71 (4): 575–587. [Cheirolepidiaceae, palynology, palaeobotany, SEM ill.].

14250   Templado, J. (1975). El araar, *Tetraclinis articulata* (Vahl) Masters en las sierras de Cartagena (Espagna). Bol. Estac. Centr. Ecol. 3 (5): 43–56.

14280   Teng, S. C. (1948). Notes on the genus *Metasequoia*. Bot. Bull. Acad. Sin. 2: 204–206. [nomenclature, taxonomy].

14290   Tenore, M. (1811–15). Flora napolitana, ossia descrizione delle piante indigene del regno di Napoli, ... Vol. I. Prodromus florae neapolitanae. [*Pinus brutia* sp. nov.: suppl. I, p. lix, 1811?].

14300   Tenore, M. (1853). Index seminum quae anno 1853 in Horto Regio Neapolitano offerentur. Ann. Sci. Nat. Bot., sér. 3, 19: 355–356. [*Taxodium mucronatum* sp. nov.].

14301   Tenore, M. (1855). Sopra alcune specie di Cupressi. Mem. Soc. Ital. Sci. Modena 25 (2): 187–204. [*Cupressus tournefortii* sp. nov., p. 194 = *C. lusitanica* Mill.].

14310   Teplouchov, T. (Teplouchoff) (1868). Ein Beitrag zur Kenntnis der Siberischen Fichte *Picea obovata* Ledeb. Bull. Soc. Imp. Naturalistes Moscou 41 (2): 244–252, f. 1–4. (Germ.). [*P. obovata, P. vulgaris* var. *altaica* et var. *uralensis* = *P.* aff. *abies* according to the author].

14313   Terry, R. G., R. S. Nowak & R. J. Tausch (2000). Genetic variation in chloroplast and nuclear ribosomal DNA in Utah Juniper (*Juniperus osteosperma*, Cupressaceae): evidence for interspecific gene flow. Amer. J. Bot. 87 (2): 250–258. [*Juniperus occidentalis, J. osteosperma*, introgression].

14320   Teuscher, H. (1920). Die Unterscheidungsmerkmale der häufigsten *Larix*-Arten. Mitt. Deutsch. Dendrol. Ges. 29: 69–72, 1 fig.

14330   Teuscher, H. (1921). Bestimmungstabelle für die im Deutschen Klima Kultivierbaren *Pinus*-Arten. Mitt. Deutsch. Dendrol. Ges. 30: 68–114.

14337   Thiel, W. (1989). Bei den ältesten Bäumen der Erde, den Grannenkiefern. Mitt. Deutsch. Dendrol. Ges. 79: 149–157. [*Pinus longaeva*].

14338   Thielges, B. A. (1969). A chromatographic investigation of interspecific relationships in *Pinus* (subsection *Sylvestres*). Amer. J. Bot. 56: 406–409. [=*Pinus* subsect. *Pinus, P.* sect. *Pinea*].

14340   Thiselton-Dyer, W. T. (1905). *Sciadopitys verticillata*. Bot. Mag. 131: pl. 8050.

14341a   Thomas, J. C. & W. J. Bond (1997). Genetic variation in an endangered cedar (*Widdringtonia cedarbergensis*) versus two congeneric species. S. African J. Bot. 63 (3): 133–140. [*W. cedarbergensis, W. nodiflora, W. schwarzii*, population genetics in relation to fragmentation].

14341b   Thomas, P. A. & A. Polwart (2003). *Taxus baccata* L. J. Ecol. 91 (3): 489–524. [anatomy, morphology, cytology/karyology, embryology, reproductive biology, ill., maps].

14341e   Thompson, J. (1961). Cupressaceae. In: R. H. Anderson. Flora of New South Wales. Contr. New South Wales Natl. Herb., Flora Ser. 1–18 (5): 46–55. [*Callitris*, 9 spp. in N.S.W.].

14342   Thompson, J. & L. A. S. Johnson (1986). *Callitris glaucophylla*, Australia's "White Cypress Pine" a new name for an old species. Telopea 2 (6): 731–736. [*C. columellaris, C. intratropica, C. glaucophylla* nom. nov., p. 731; all are one species: *C. columellaris*, perhaps with an inland, predominantly glaucous ecotype].

14350   Thomson, R. B. (1914). The spur shoot of the pines. Bot. Gaz. (Crawfordsville) 56: 362–385, pl. 20–23. [*Pinus*].

14360   Thor, E. (1974). Taxonomy of *Abies* in southern Appalachians; variation in balsam monoterpenes and wood properties. Forest Sci. 20: 32–40. [*A. fraseri*].

14361 Thorne, R. F. (1978). New subspecific combinations for southern California plants. Aliso 9 (2): 189–196. [*Cupressus guadalupensis* Wats. ssp. *forbesii* (Jepson) Beauchamp, comb. nov., p. 191].

14362 Thunberg, C. P. (1783). Kaempferus illustratus seu explicatio plantarum Japonicarum, ... sectio secunda. Nova Acta Regiae Soc. Sci. Upsal. (ser. 2) 4: 31–40. (repr. in: C. P. Thunberg. Miscellaneous Papers Regarding Japanese Plants. Tokyo, 1935). [*Cupressus pendula* sp. nov. = cultivar of *Platycladus orientalis*, p. 40].

14375 Tidwell, W. D., L. R. Parker & V. K. Folkman (1986). *Pinuxylon woolardii* sp. nov., a new petrified taxon of Pinaceae from the Miocene basalts of eastern Oregon. Amer. J. Bot. 73: 1517–1524.

14390 Tieghem, P. van (1891). Sur la structure primaire et les affinités des pins. J. Bot. (Morot) 5: 265–271, 281–288. [*Pinus*].

14395 Tieghem, P. van (1891). Structure et affinités des *Stachycarpus*, genre nouveau de la famille des Conifères. Bull. Soc. Bot. France 38: 162–176. [*Stachycarpus* gen. nov. = *Prumnopitys*, p. 163, *S. andinus* (Poepp. ex Endl.) comb. nov. = *P. andina*, p. 173, *S. ferrugineus* (G. Benn ex D. Don) comb. nov. = *P. ferruginea*, p. 173, *S. spicatus* (R. Br.) comb. nov. = *Prumnopitys taxifolia*, p. 173, *S. taxifolius* (Kunth) comb. nov. = *P. montana*, p. 173].

14400 Tieghem, P. van (1891). Structures et affinités des *Abies* et des genres les plus voisins. Bull. Soc. Bot. France 38: 406–416. [*Abies chensiensis* sp. nov., p. 413; contains the earliest (informal) bipartite division of the Pinaceae as presently circumscribed].

14410 Tikhomirov, B. A. (1949). The dwarf cedar, its biology and utilization. Mater. Pozn. Fauny Fl. SSSR, Otd. Bot., n. s., 6: 1–106, f. 1–26. (Russ.). [*Pinus pumila*, monographic paper].

14415 Tiong, S. K. K. (1984). *Podocarpus laubenfelsii*, a new species from Borneo (Podocarpaceae). Blumea 29 (2): 523–524.

14420 Tjaden, W. L. (1980). The Chinese Golden Larch. Taxon 29 (3): 314–315. [*Pseudolarix amabilis*].

14430 Tolmachev, A. I. (1954). Notae de Abietibus sachalinensibus. Bot. Mater. Gerb. Bot. Inst. Komarova Akad. Nauk. SSSR 16: 29–38, f. 1–2. (Russ.). [*Abies veitchii*, *A. sachalinensis*, *A. mayriana* = *A. sachalinensis* var. *mayriana*].

14431 Tolmachev, A. I. (1954). K istorii vozniknovenija i razvitija temnokhvojnoj taigi. Isd. Akad. Nauk SSR, Sakhalin filial. (Russ.). [*Abies*, *Pinus*, maps].

14432 Tomaru, N., Y. Tsumura & K. Ohba (1994). Genetic variation and population differentiation in natural populations of *Cryptomeria japonica*. Plant Sp. Biol. 9: 191–199.

14432a Tomback, D. F. (1982). Dispersal of whitebark pine seeds by Clark's nutcracker: a mutualism hypothesis. J. Animal Ecol. 51: 451–467. [*Pinus albicaulis, P. edulis*, coevolution].

14433 Tomback, D. F. & Y. B. Linhart (1990). The evolution of bird-dispersed pines. Evol. Ecol. 4: 185–219. [*Pinus* sect. *Strobus, P.* sect. *Parrya*, Nutcrackers & Jays].

14433c Tomlinson, P. B. & T. Takaso (1990). Transition to the uni-ovulate condition in *Juniperus* (Cupressaceae). Amer. J. Bot. 77: 28.

14434 Tomlinson, P. B., T. Takaso & E. K. Cameron (1993). Cone development in *Libocedrus* (Cupressaceae) – phenological and morphological aspects. Amer. J. Bot. 80 (6): 649–659. [*Libocedrus plumosa*, ontogeny, ill. LM, SEM].

14435 Tomlinson, P. B. & E. H. Zacharias (2001). Phyllotaxis, phenology and architecture in *Cephalotaxus*, *Torreya* and *Amentotaxus* (Coniferales).Bot. J. Linn. Soc. 135 (3): 215–228. (SEM ill.).

14440 Torrey, J. (1853). Plantae Frémontianae; or descriptions of plants collected by Col. J. C. Frémont in California. Smithsonian Contr. Knowl. 6 (2): 3–24. [*Libocedrus decurrens* sp. nov. = *Calocedrus decurrens*, pp. 7–8, pl. 3].

14440a Torrey, J. (1854). Notice of the California nutmeg. New York J. Pharm. 3: 49–51. [*Torreya californica* sp. nov.; IK-IPNI cite as year of publ. 1852, Bradley Bibliography & A. Rehder's Bibl. (No. 11950) cite 1854, original n.v.].

14441 Torrey, J. (1857). Report on the botany of the expedition. in: War Dept. Explorations and surveys for a railroad route from the Mississippi River to the Pacific Ocean. Washington, D.C. (pp. 59–161). [Coniferales, pp. 140–142, *Juniperus tetragona* Schltdl. var. *osteosperma* var. nov. = *J. osteosperma*, p. 141, *J. pachyphlaea* sp. nov. = *J. deppeana* var. *pachyphlaea*, p. 142].

14450 Torrey, J. & J. C. Frémont (1845). Descriptions of some genera and species of plants, collected in Captain J. C. Frémont's exploring expedition to Oregon and North California, in the years 1843–1844. in: J. C. Frémont. Report of the Exploring Expedition to the Rocky Mountains in the year 1842 and to Oregon and North California in the years 1843–'44. Washington, D.C. [*Pinus monophylla* sp. nov., p. 319, t. 4].

14460a Townrow, J. A. (1965). Notes on Tasmanian pines II. *Athrotaxis* from the Lower Tertiary. Pap. Proc. Roy. Soc. Tasmania 99: 109–113.

14460b Townrow, J. A. (1967). On *Rissikia* and *Mataia* podocarpaceous conifers from the Lower Mesozoic of southern lands. Pap. & Proc. Roy. Soc. Tasmania 101: 103–136. [fossil Podocarpaceae, ill.].

14460c Townrow, J. A. (1967). The *Brachyphyllum crassum* complex of fossil conifers. Pap. Proc. Roy. Soc. Tasmania 101: 137–147.

14460d Townrow, J. A. (1967). On *Voltziopsis*, a southern conifer of Lower Triassic Age. Pap. Proc. Roy. Soc. Tasmania 101: 173–188. [*Voltziopsis wolganensis* Townrow].

14461 Trabut, L. (1906). Sur la présence d'un *Abies* nouveau au Maroc (*Abies marocana*). Bull. Soc. Bot. France 53: 154–155, t. III. [*A. marocana* sp. nov. = *A. pinsapo* var. *marocana*].

14463 Trautvetter, E. R. von (1844). Plantarum imagines et descriptiones floram russicam illustrantes... Vol. I, Fasc. 1–2: 12, t. 7. München. [*Cupressus americana* sp. nov. = *Chamaecyparis nootkatensis*].

14464 Trautvetter, E. R. von (1846). Plantarum imagines et descriptiones floram russicam illustrantes... Vol. III, Fasc. 7: [47]–54, t. 31–35. München. [*Larix dahurica* Turcz. ex Trautv. = *L. gmelinii*, p. 48, t. 32].

14465 Trautvetter, E. R. von (1884). Incrementa florae phaenogamae rossicae... Trudy Imp. S.-Petersburgsk. Bot. Sada 9 (1): [1]–220. [*Larix russica* (Endl.) Sabine ex Trautv., comb. nov. (as "*Rossica*"), p. 212].

14466 Trautvetter, E. R. von & C. A. A. von Meyer (1856). Vol. I (2) Lief. 3: Botanik. Florula Ochotensis phaenogama. (133 pp., ill.). in: A. T. von Middendorf. Reise in den äussersten Norden und Osten Siberiens. St. Petersburg. [*Picea ajanensis* Fischer ex Trautv. et Meyer, sp. nov. = *P. jezoensis*, p. 87, *Larix dahurica* = *L. gmelinii*, p. 88].

14468 Tredici, P. del (1998). Lignotubers in *Sequoia sempervirens*: development and ecological significance. Madroño 45 (3): 255–260.

14469 Tredici, P. del (1999). Redwood burls: immortality underground. Arnoldia 59 (3): 14–22. [*Sequoia sempervirens*, morphology, physiology, ecology, ill.].

14470 Trew, C. J. (1757). Cedrorum libani historia ... cum illo Laricis, Abietis Pinique comparatus. Nürnberg. [*Cedrus* Trew, nom. cons.].

14471 Trew, C. J. (1767). Apologia et mantissa observationis de Cedro Libani et Cedrorum libani historiae. App. in N. Acta Leopold. 3: 446–496; also as "pars altera" to Trew, 1757, pp. 1–50 (Nürnberg). [see TL-2, 15.133].

14474 Tripp, K. E. (1995). *Cephalotaxus*: The Plum Yews. Arnoldia 55 (1): 24–39.

14480 Trombulak, S. C. & M. L. Cody (1980). Elevational distribution of *Pinus edulis* and *P. monophylla* (Pinaceae) in the New York Mountains. Madroño 27: 61–67.

14500 Tsiang, Y. (1948). A new pine from south China. Sunyatsenia 7: 111–114. [*Pinus kwangtungensis* sp. nov.].

14510 Tsukada, M. (1982). *Cryptomeria japonica*: glacial refugia and late-glacial and postglacial migration. Ecology 63 (4): 1091–1105.

14512a Tsumura, Y., N. Tomaru, Y. Suyama & S. Bacchus (1999). Genetic diversity and differentiation of *Taxodium* in the southeastern United States using cleaved amplified polymorphic sequences. Heredity 83: 229–238. [*Taxodium distichum* var. *distichum*, *T. distichum* var. *imbricatum*, DNA analysis supports close relationship and recognition at varietal rank of two taxa].

14530 Tucker, J. M. (1960). A range extension for the Chihuahua pine in New Mexico. Southw. Naturalist 5 (4): 226. [*Pinus chihuahuana* = *P. leiophylla* var. *chihuahuana*].

14540 Turra, A. (1765). Dei vegetabili di Monte Baldo. Giorn. Italia Sci. Nat. 1: 152. [*Pinus mugo* sp. nov.].

14545 Turrill, W. B. (1955). *Abies pinsapo* var. *vel hybrida* (Coniferae). Curtis's Bot. Mag. 170, t. 242. [suggesting that *A. numidica* could be considered conspecific with *A. pinsapo*].

14560 Twisselman, E. C. (1962). The Piute Cypress. Leafl. W. Bot. 9 (15): 248–253. [*Cupressus arizonica* var. *nevadensis*].

14568 Uehara, K. & N. Sahashi (2000). Pollen wall development in *Cryptomeria japonica* (Taxodiaceae). Grana 39 (6): 267–274. (ill.).

14570 Ueno, J. (1951). Morphology of pollen of *Metasequoia*, *Sciadopitys* and *Taiwania*. J. Inst. Polytechn. Osaka City Univ., ser. D., Biol. 2: 22–26, pl. 1–2, f. 1.

14580 Ueno, J. (1957). Relationships of genus *Tsuga* from pollen morphology. J. Inst. Polytechn. Osaka City Univ., ser. D., Biol. 8: 191–196.

14600 Unger, A. & L. Beissner (1900). *Juniperus sanderi*. Mitt. Deutsch. Dendrol. Ges. 9: 213–217. [syn. *Retinispora sanderi* hort. = cultivar !].

14600a Uotila, P. (1984). *Abies nebrodensis*. Sorbifolia 15: 35–39.

14601 Urban, I. (1913). Symbolae Antillanae seu fundamenta florae Indiae occidentalis. Vol. 7. Leipzig. [see R. Pilger, 1913].

14601a Urban, I. (1924). Sertum antillanum XIX. Fedde's Repert. Sp. Nov. Regni Veg. 19 (16–21): 298–308. [*Podocarpus buchii* sp. nov. = *P. aristulatus*, p. 298].

14602 Urban, I. (1926). Plantae Haitienses novae vel rariores III a cl. E. L. Ekman 1924–26 lectae. Ark. Bot. 20 (4) A15. [see R. Pilger, 1926].

14620 Uyeki, H. (1925). On the species and their distribution of wild pines in Chosen and Manchuria. J. Chosen Nat. Hist. Soc. 3: 35–47. (Japan., Eng. descr.). [*Pinus tabuliformis* var. *mukdensis* (Nakai) comb. nov. (basion. : *P. mukdensis* Nakai), *P. tabuliformis* var. *rubescens* var. nov.].

14630 Uyeki, H. (1926). Corean timber trees. Vol. 1. Ginkgoales and Coniferae. Bull. Forest Exp. Sta. (Seoul) 4: 1–16, 1–154. (ill.). [*Picea pungsanensis*, nom. nud., pp. 98, 99; see also T. Nakai, 1941; = *P. koraiensis* Nakai var. *pungsanensis* (Nakai) Farjon].

14640 Uyeki, H. (1927). The seeds of the genus *Pinus*, as an aid to the identification of species. Bull. Agric. Coll. Suwon 2: 1–129, pl. 1–18. [key to all known spp.].

14650 Uyeki, H. (1929). Four new ligneous plants from Corea and Manchuria. J. Chosen Nat. Hist. Soc. 9: 20–21. [*Pinus yamazutai* sp. nov. from Manchuria].

14660 Uyeki, H. (1950). Novae varietae Chamaecyparis obtusae. J. Japan. Forest. Soc. 32: 274–276, f. 1–2. [Japan., Lat descr. of *Chamaecyparis obtusa* var. *fastigiato-ovata* var. nov.].

14661 Vahl, M. (1791). Symbolae botanicae, sive plantarum... Pars 2. Hauniae (Kjøbenhavn). [*Thuja articulata* sp. nov. = *Tetraclinis articulata*, p. 96].

14670 Vall, W. B. de (1941). The taxonomic status of *Pinus caribaea* Mor. Proc. Florida Acad. Sci. 5: 121–132. [*P. palustris*, *P. elliottii*, *P. caribaea* auct., non Morelet].

14680 Vasek, F. C. (1966). The distribution and taxonomy of three western junipers. Brittonia 18: 350–372. [*Juniperus californica*, *J. occidentalis*, *J. osteosperma*, *J. occidentalis* ssp. *australis* ssp. nov., p. 352, ill., maps].

14681 Vasey, G. (1876). *Abies macrocarpa* A new coniferous tree. Gard. Monthly & Hort. 18: 21 (publ. Jan. 1876). [*Abies macrocarpa* sp. nov. = *Pseudotsuga macrocarpa*].

14688 Vasil, V. & R. K. Sahni (1964). Morphology and embryology of *Taxodium mucronatum* Tenore. Phytomorphology 14: 369–384.

14690 Vasiljev, V. N. (1950). Far eastern spruces of the section *Omorica* Willk. Bot. Žurn. (Moscow & Leningrad) 35: 498–511, f. 1–7. (Russ.). [*Picea komarovii* sp. nov. = *P. jezoensis* ssp. *jezoensis* var. *komarovii*, p. 504].

14691 Vasiljeva, G. V. (1972). Materials on the comparative anatomy of leaf [in] species of *Agathis* Salisb. (Araucariaceae). Bot. Žurn. 57: 108–118. (Russ., ill., key). [*Agathis* spp., *Nageia nagi* (*Podocarpus nagi*)].

14699 Veblen, T. T. (1982). Regeneration patterns in *Araucaria araucana* forests in Chile. J. Biogeogr. 9: 11–28.

14700 Veblen, T. T. & D. H. Ashton (1982). The regeneration status of *Fitzroya cupressoides* in the Cordilleran Pelada, Chile. Biol. Conservation (Barking) 23 (2): 141–161. [forest profiles, map].

14701 Veblen, T. T., R. J. Delmastro & J. E. Schlatter (1976). The conservation of *Fitzroya cupressoides* and its environment in southern Chile. Environ. Conservation 3: 291–301.

14702 Veblen, T. T. & G. H. Stewart (1982). On the conifer regeneration gap in New Zealand: The dynamics of *Libocedrus bidwillii* stands on South Island. J. Ecol. 70: 413–436.

14707 Veillon, J.-M. (1980). Architecture des espèces néo-calédoniennes du genre *Araucaria*. Candollea 35 (2): 609–640. [*Araucaria* spp., key, ill.].

14710 Veitch & Sons (1881). A Manual of the Coniferae; containing a general view of the order; ... pp. 1–350, pl. 1–20, f. 1–63. London. (2nd. ed. enlarged by A. H. Kent; pp. 1–562, f. 1–141; 1900). [*Laricopsis* Kent, gen. nov. = *Pseudolarix* nom. cons., p. 403; see also A. H. Kent, 1900].

14710a Veldkamp, J. F. & D. J. de Laubenfels (1984). (745) Proposal to reject *Pinus dammara* (Araucariaceae). Taxon 33 (2): 337–347. [*Agathis* spp., *A. celebica* ssp. *flavescens* (Ridl.) Veldkamp & Whitmore, p. 346, nomenclature].

14710c Vellozo, J. M. da Conceição (1831). Florae fluminensis, seu descriptionum plantarum praefectura fluminensis sponte nascentium... Icones Vol. 1–11. Rio de Janeiro. [*Pinus dioica* Vell., Vol. 10 [1]: t. 55, 56 = *Araucaria angustifolia*].

14710e Vendramin, G. G., M. Michelozzi, R. Tognetti & F. Vicario (1997). *Abies nebrodensis* (Lojac.) Mattei, a relevant example of a relic and highly endangered species. Bocconea 7: 383–388. [*Abies alba*, *A. nebrodensis*, allozyme, DNA markers, monoterpenes, species delimitation].

14710f Venema, H. J. (1942). Enkele monstrueuse kegels van coniferen. Gedenkboek J. Valckenier Suringar pp. 237–259 (reprint). (ill.). [*Abies*, *Cunninghamia*, *Sciadopitys*, teratology, morphology, homology; conifer cones investigated are compound structures or 'inflorescences'].

14711 Venning, J. (1979). Character Variation in Australian Species of *Callitris* Vent. (Cupressaceae). Ph. D. diss., Univ. of Adelaide, Australia.

14712 Venning, J. (1979). Seed protein patterns of Australian species of *Callitris* Vent. (Cupressaceae). [mscr. of paper presented at the ANZAAS conference held in Auckland, New Zealand, 1979; classifications evaluated].

14720 Ventenat, E. P. (1808). Decas generum novorum, aut parum cognitorum, ... Paris. (repr. + annot. by W. T. Stearn in: P. Smit & R. J. C. V. ter Laage, eds. (1970). Essays in biohistory. Regnum Veg. Vol. 71: 342–352; see also W. T. Stearn, 1970). [*Callitris* gen. nov., no species named or described; for lectotype see A. A. Bullock, 1957].

14728a   Vieillard, E. (1862). Plantes utiles de la Nouvelle Calédonie. Ann. Sci. Nat. Bot., sér. 4, 16: 26–76. [*Araucaria intermedia* sp. nov. = *A. columnaris*, p. 54, *A. subulata* sp. nov., p. 55, *Dammara lanceolata* sp. nov. = *Agathis moorei*, p. 56, *D. ovata* C. Moore ex Vieill. = *Agathis ovata*, p. 56, *Dacrydium ustum* sp. nov. = *Parasitaxus usta*, p. 56, *Podocarpus novae-caledoniae* sp. nov., p. 56].

14729   Vieira, R. M. S. (1997). Uma nova forma de Cedro-da-Madeira *Juniperus cedrus* Webb & Berth. for. *fastigiata* R. Vieira. Bol. Mus. Mun. Funchal 49 (280): 143–148.

14770   Viguié, M. T. & H. Gaussen (1929). Révision du genre *Abies*. Trav. Lab. Forest. Toulouse T. 2, 2 (1), (1-bis): 1–66, 67–386. [*A. forrestii* var. *smithii* var. nov., earlier publ. in Bull. Soc. Hist. Nat. Toulouse 58: 355, 1929].

14770e   Vining, T. F. & C. S. Campbell (1997). Phylogenetic signal in sequence repeats within nuclear ribosomal DNA internal transcribed spacer 1 in *Tsuga*. Amer. J. Bot. 84 Suppl.: 241. (abstract). [includes *Nothotsuga* in *Tsuga*; abstract gives no voucher evidence].

14771   Visscher, G. E. & R. Jagels (2003). Separation of *Metasequoia* and *Glyptostrobus* (Cupressaceae) based on wood anatomy. I.A.W.A. Journal 24 (4): 439–450.

14820   Vriese, W. H. de (1845). Plantae novae et minus cognitae Indiae batavae orientalis. Nouvelles recherches sur la flore... II. Sur une nouvelle espèce de Pin de l'île de Sumatra, pp. 5–8, pl. 2. (with F. W. Junghuhn). [*Pinus merkusii* Jungh. et de Vriese sp. nov., p. 5].

14830   Vriese, W. H. de (1855). De hiba-boom van Japan, *Thujopsis dolabrata* Sieb. et Zucc. Tuinb.-Fl. Ned. 2: 1–2, 2 pl. (+ Lat. descr.).

14840   Wagener, W. W. (1960). A comment on the cold susceptibility of Ponderosa and Jeffrey's Pines. Madroño 15: 217–219. [*Pinus ponderosa*, *P. jeffreyi*].

14850   Wagener, W. W. & C. R. Quick (1963). *Cupressus bakeri* – an extension of the known botanical range. Aliso 5 (3): 351–352.

14855   Wagner, D. B., W. L. Nance, C. D. Nelson, T. Li, R. N. Patel & D. R. Govindaraju (1992). Taxonomic patterns and inheritance of chloroplast DNA variation in a survey of *Pinus echinata*, *Pinus elliottii*, *Pinus palustris*, and *Pinus taeda*. Canad. J. Forest. Res. 22 (5): 683–689. (ill.).

14857   Wagner, J. (1992). From Gansu to Kolding: the expedition of J. F. Rock in 1925–1927 and the plants raised by Aksel Olsen. Dansk Dendrol. Årskrift 10: 18–93. [Pinaceae, *Pinus*, *Picea*, *Abies*, *Larix*].

14860   Wahlenberg, W. G. (1946). Longleaf pine. Washington, D.C. [*Pinus palustris*].

14870   Wahlenberg, W. G. (1960). Loblolly pine. Duke Univ. School of Forestry., Durham, N.C. (603 pp.). [*Pinus taeda*].

14882   Wallich, N. (1832). A numerical list of dried specimens of plants, in the East India Company Museum... ("Wallich Catalogue"). [*Pinus smithiana* sp. nov. = *Picea smithiana*, cat. no. 6063 (for date of publ. of cat. nos. see F. A. Stafleu & R. S. Cowan in TL-2, Vol. VII, 1988); also publ. in: Plantae asiaticae rariores. Vol. 3: 24, f. 246, 1832].

14890   Walter, T. (1788). Flora caroliniana, ... London. (263 pp., 1 pl.). [*Pinus glabra* sp. nov., p. 237].

14900   Walther, E. E. (1958). Rare conifers of Mexico. Natl. Hort. Mag. 37: 241–246, 5 figs. [*Taxodium*, *Cupressus*, *Pinus*].

14901   Walther, H. (1989). *Cunninghamia miocenica* Ettingshausen, eine wichtige Taxodiacee im Tertiär Mitteleuropas. Flora 182: 287–311. (ill.).

14930   Wang, D. Y. & H. L. Liu (1982). A new species and a new variety of *Cunninghamia* from Sichuan Province. Acta Phytotax. Sin. 20 (2): 230–232. (Chin., Latin). [*C. unicanaliculata* var. *pyramidalis*, sp. et var. nov.].

14940   Wang, F. (1901–02). Waldbilder aus Japan. Österreich. Forst & Jagd-Zeitung 19 (49): 390–391, (52): 418–419, f. 306–309, 329–331 (1901); 20 (4): 26, f. 19–20 (1902). [ill. of *Cryptomeria japonica* forests in Japan].

14945   Wang, F. H. (1948). The early embryogeny of *Glyptostrobus*. Bot. Bull. Acad. Sin. Shanghai 2: 1–12. (ill.).

14950   Wang, F. H. (1948). Life history of *Keteleeria*. I. Strobili, development of the gametophytes and fertilization in *Keteleeria evelyniana*. Amer. J. Bot. 35: 21–27.

14955   Wang, F. H., ed. (1990). Biology of *Cathaya*. Academia Sinica, Beijing. (Chin., ill.).

14960   Wang, F. H. & Z. K. Chen (1974). The embryogeny of *Cathaya* (Pinaceae). Acta Bot. Sin. 16: 64–69. (Chin., Eng. summ.).

14970   Wang, F. H. & N. F. Chien (1964). The embryogeny of *Metasequoia*. Acta Bot. Sin. 12: 241–262. (Chin., Eng. summ.).

14971   Wang, F. H., S. C. Lee & Z. K. Chen (1980). The embryogeny of *Taiwania* in comparison with that of other genera of Taxodiaceae. Acta Phytotax. Sin. 18: 129–138. (Chin., Eng. summ.).

14971a   Wang, S. J., J. Hilton, B. Tian & J. Galtier (2003). Cordaitalean seed plants from the early Permian of north China. I. Delimitation and reconstruction of the *Shanxioxylon sinense* plant. Int. J. Plant Sci. 164 (1): 89–112. (ill.). [Cordaitales, Coniferales, coniferophytes, *S. sinense* B. Tian & S. J. Wang, reconstructed whole fossil plant].

14971b Wang, W. P., C. Y. Hwang, T. P. Lin & S. Y. Hwang (2003). Historical biogeography and phylogenetic relationships of the genus *Chamaecyparis* (Cupressaceae) inferred from chloroplast DNA polymorphism. Plant Syst. Evol. 241: 13–28. [*Chamaecyparis pisifera, C. obtusa, C. lawsoniana, C. formosensis, C. thyoides, C. nootkatensis* (= *Xanthocyparis nootkatensis*), DNA, cladistics, phytogeography, phylogeny, ill.].

14971c Wang, X., S. Y. Duan & J. Z. Cui (1997). Several species of *Schizolepis* and their significance on the evolution of conifers. Taiwania 42 (2): 73–85. [*Pseudovoltzia, Schizolepis, Tricanolepis,* Pinaceae, palaeobotany].

14971d Wang, X. Q., Y. Han, Z. R. Deng & D. Y. Hong (1997). Phylogeny of the Pinaceae evidenced by molecular biology. Acta Phytotax. Sin. 35 (2): 97–106. [*Abies, Cathaya, Cedrus, Keteleeria, Larix, Picea, Pinus, Pseudotsuga,* cpDNA, cladistic analysis].

14971e Wang, X. Q., Y. Han & D. Y. Hong (1998). A molecular systematic study of *Cathaya,* a relic genus of the Pinaceae in China. Plant Syst. Evol. 213: 165–172. [*rbc*L gene sequence (DNA), cladistic analysis; *Cathaya* appears as sister group to *Pinus* (Fitch tree) or to *Picea* (Neighbour-joining tree)].

14971f Wang, X. Q., Y. Han & D. Y. Hong (1998). PCR-RFLP analysis of the chloroplast gene *trn*K in the Pinaceae, with special reference to the systematic position of *Cathaya.* Israel J. Plant Sci. 46 (4): 265–271.

14972 Wang, X. R. (1992). Genetic diversity and evolution of Eurasian *Pinus* species. Dissertation (thesis based on 7 papers publ. with A. E. Szmidt *et al.* elsewhere). Swedish Univ. Agric. Sci., Faculty of Forestry, Dept. Forest Genetics & Plant Physiol., Umeå, Sweden. [*Pinus* spp., DNA, phylogeny].

14973 Wang, X. R. & A. E. Szmidt (1993). Chloroplast DNA-based phylogeny of Asian *Pinus* species (Pinaceae). Plant Syst. Evol. 188: 197–211. [molecular systematics of 18 *Pinus* species].

14973a Wang, X. R. & A. E. Szmidt (1994). Hybridization and chloroplast DNA variation in a *Pinus* species complex from Asia. Evolution 48 (4): 1020–1031. [*P. densata, P. massoniana, P. tabuliformis, P. yunnanensis*].

14973b Wang, X. R., A. E. Szmidt & H. N. Nguyên (2000). The phylogenetic position of the endemic flat-needle pine *Pinus krempfii* (Pinaceae) from Vietnam, based on PCR-RFLP analysis of chloroplast DNA. Plant Syst. Evol. 220 (1): 21–36.

14973c Wang, X. R., Y. Tsumura, H. Yoshimaru, K. Nagasaka & A. E. Szmidt (1999). Phylogenetic relationships of Eurasian pines (*Pinus,* Pinaceae) based on chloroplast *rbc*L, *mat*K, *rpl*20-*rps*18 spacer, and *trn*V intron sequences. Amer. J. Bot. 86 (12): 1742–1753. [*Pinus* spp., *P. krempfii,* DNA].

14974 Warburg, O. (1900). Monsunia. Beiträge zur Kenntnis der Vegetation des süd- und ostasiatischen Monsungebietes. Band I. Leipzig. [Coniferales pp. 182–194, *Agathis regia* sp. nov., p. 183, *A. labillardieri* sp. nov., p. 183, *A. macrostachys* sp. nov., p. 183, *A. beccari* sp. nov., p. 184, *A. borneensis* sp. nov., p. 184, *A. rhomboidalis* sp. nov., p. 184, pl. 8, *A. celebica* (Koord.) comb. nov., p. 185, *A. philippinensis* sp. nov., p. 185, *A. longifolia* sp. nov., p. 186, *A. ovata* (C. Moore ex Vieill.) comb. nov., p. 186, *A. hypoleuca* (C. Moore ex Henkel & W. Hochst.) comb. nov., p. 186, *Agathis* spp., pl. 10, *Araucaria beccarii* sp. nov., p. 187, *A. schumanniana* sp. nov., p. 187, *Araucaria* spp. (most names in both genera are synonyms); describes *Papuacedrus papuana* (as *Libocedrus papuana*) from the Moluccas, pp. 189–190, *Cephalotaxus celebica* sp. nov. = *Taxus sumatrana,* p. 194].

14980 Ward, D. B. (1963). Contributions to the Flora of Florida 2, *Pinus* (Pinaceae). Castanea 28: 1–10. [*P. clausa* var. *immuginata* var. nov., p. 4].

14990 Ward, D. B. (1963). Southeastern limit of *Chamaecyparis thyoides.* Rhodora 65 (764): 359–362.

15000 Ward, D. B. (1974). On the scientific name of the longleaf pine. Rhodora 76: 20–24. [*Pinus palustris* Mill.].

15010 Wardle, P. (1956). *Picea omorika* in its natural habitat. Forestry 29: 91–117. (ill.).

15011 Wardle, P. (1972). *Podocarpus totara* var. *waihoensis* var. nov.: the result of introgressive hybridization between *P. totara* and *P. acutifolius.* New Zealand J. Bot. 10 (1): 195–201. (maps).

15020 Waring, R. H., W. H. Emmingham & S. W. Running (1975). Environmental limits of an endemic spruce, *Picea breweriana.* Canad. J. Bot. 53: 1599–1613. (ill.).

15020a Waring, R. H. & J. F. Franklin (1979). Evergreen coniferous forests of the Pacific Northwest. Science 204: 1380–1386. [phytogeography and ecology].

15021 Warming, J. E. B. (1884). Haandbog i den systematiske Botanik. Ed. 2. Kjøbenhavn. (German ed. 1890, Eng. ed. 1895). [Taxodiaceae, "fam. nov.", but see Saporta, 1865].

15023 Warren, R. & A. J. Fordham (1978). The fire pines. Arnoldia 38: 1–11. [*Pinus* subsect. *Oocarpae, P.* subsect. *Contortae, P.* subsect. *Australes,* popular paper on pines in these subsections].

15025 Wasscher, J. (1941). The genus *Podocarpus* in the Netherlands Indies. Blumea 4 (3): 359–481. (ill.). [*Podocarpus* s.l., incl. *Dacrycarpus, Nageia, Sundacarpus* (as sections, the latter as *Podocarpus amarus*), *P. imbricatus* var. *kinabaluensis* var. nov. = *D. kinabaluensis,* p.

400, *P. steupii* sp. nov. = *D. steupii*, p. 405, *P. dacrydiifolius* sp. nov. = *D. cinctus*, p. 410, *P. compactus* sp. nov. = *D. compactus*, p. 411, *P. leptophyllus* sp. nov. = *Dacrydium leptophyllum*, p. 414, *P. salomoniensis* sp. nov., p. 430, *P. neriifolius* var. *atjehensis* var. nov. = *P. atjehensis*, p. 450, *P. neriifolius* var. *membranaceus* var. nov., p. 450, var. *timorensis* var. nov., p. 451, var. *linearis* var. nov., p. 452, var. *ridleyi* var. nov., p. 453, var. *teysmannii* (Miq.) comb. nov., p. 453, var. *polyanthus* var. nov., p. 455].

15026 Waters, E. R. & B. A. Schaal (1991). No variation is detected in the chloroplast genome of *Pinus torreyana*. Canad. J. Forest Res. 21: 1832–1835. [cp-DNA].

15027 Waters, T., C. A. Galley, R. Palmer, S. T. Turvey & N. M. Wilkinson (2002). Report of the Oxford University expedition to New Caledonia, December 2000 – January 2001. Published by the authors, University of Oxford, UK. [*Araucaria nemorosa*, ecology, conservation, ill., maps]

15030 Watson, F. D. (1983). A taxonomic study of pondcypress and baldcypress. Ph. D. dissertation, North Carolina State University, Raleigh. [*Taxodium distichum* var. *distichum*, *T. distichum* var. *imbricatum*, syn. : *T. ascendens* Brongn.].

15040 Watson, F. D. (1985). The nomenclature of pondcypress and baldcypress (Taxodiaceae). Taxon 34 (3): 506–509. [*Taxodium distichum* var. *distichum*, *T. distichum* var. *imbricatum* (as "*imbricarium*") (Nuttall) Croom].

15044 Watson, J. (1977). Some Lower Cretaceous conifers of the Cheirolepidiaceae from the U.S.A. and England. Palaeontology 20: 715–749. (ill.).

15045 Watson, J. (1983). A new species of the conifer *Frenelopsis* from the Cretaceous of Sudan. Bot. J. Linn. Soc. 86: 161–167. (ill.). [Cheirolepidiaceae, *Frenelopsis silfloana* sp. nov.].

15047 Watson, J., H. L. Fisher & N. A. Hall (1987). A new species of *Brachyophyllum* from the English Wealden and its probable female cone. Rev. Palaeobot. Palynol. 51: 169–187. (ill.). [palaeobotany].

15050 Watson, S. (1880). Botany of California. Vol. 2. Univ. Press, Cambridge, Mass. [see G. Engelmann, 1880].

15051 Watson, S. (1880). Contributions to American Botany. (containing descriptions of new species of plants). Proc. Amer. Acad. Arts 14: 213–303. [*Cupressus guadalupensis* sp. nov., p. 300].

15060 Watson, S. (1885). Contributions to American botany. 2. Descriptions of some New Species of Plants, chiefly from our Western Territories. Proc. Amer. Acad. Arts 20 (n. s. Vol. 12): (324–)352–378. [*Picea breweriana* sp. nov., p. 378].

15061 Webb, P. B. & S. Berthelot *et al.* (1847). Histoire naturelle des Iles Canaries, ... Tome troisième [Botanique]... Deuxième partie: Phytographia canariensis. Sect. 3, livr. 89: 265–280. Paris. [*Juniperus cedrus* sp. nov., p. 277].

15077 Wehr, W. C. & H. E. Scorn (1992). Current research on Eocene conifers at Republic, Washington. Washington Geol. 20 (2): 20–23. (ill.). [palaeobotany, *Abies*, *Amentotaxus*, *Chamaecyparis*, *Ginkgo*, *Metasequoia*, *Picea*, *Pinus*, *Pseudolarix*, *Sequoia*, *Thuja*, *Tsuga*].

15078 Wei, X. X. & X. Q. Wang (2003). Phylogenetic split of *Larix*: evidence from [the] paternally inherited cpDNA *trn*T-*trn*F region. Plant Syst. Evol. 239: 67–77. [*Larix* spp., molecular phylogeny, biogeography].

15080 Weidmann, R. H. (1939). Evidences of racial variation in a 25-year test of Ponderosa Pine. J. Agric. Res. 59: 855–868. [*Pinus ponderosa*].

15085 Weiss, R. (1999). Distribution of *Pinus monophylla* in the northern Wasatch Range of Utah. Great Basin Naturalist 59 (3): 292–294. (map).

15090 Welch, W. H. (1931). An ecological study of the baldcypress in Indiana. Proc. Indiana Acad. Sci. 41: 207–213. [*Taxodium distichum*].

15097 Welle, B. ter & R. P. Adams (1998). Investigation of the wood anatomy of *Juniperus* (Cupressaceae) for taxonomic utilization. Phytologia 84 (5): 354–362. (ill., table). [23 taxa of North American *Juniperus* investigated: quantitative differences are regarded as not informative for taxonomy].

15100 Wells, O. O. (1964). Geographic variation in ponderosa pine. I. The ecotypes and their distributions. Silvae Genet. 13: 89–103. [*Pinus ponderosa*].

15111 Wells, P. V. (1995). Recognizing the new Single-leaf Pinyon pine (*Pinus californiarum* Bailey) of southern California. The Four Seasons 10 (1): 53–58. [*Pinus californiarum* = *P. monophylla*].

15113 Weng, C. & S. T. Jackson (2000). Species differentiation of North American spruce (*Picea*) based on morphological and anatomical characteristics of needles. Canad. J. Bot. 78: 1367–1383. (ill.). [*Picea breweriana*, *P. chihuahuana**, *P. engelmannii*, *P. glauca*, *P. martinezii**, *P. mariana*, *P. mexicana*, *P. pungens*, *P. rubens*, *P. sitchensis* (** "very closely related")].

15114 Werger, M. J. A. (1978). Biogeography and ecology of southern Africa. Vol. 1–2. The Hague. (botany and vegetation in vol. 1). [*Widdringtonia cedarbergensis*, p. 222, ill., *W. nodiflora*, pp. 493–495, ill.].

15115 Werner, W. L. (1993). *Pinus* in Thailand. Geoecological Research, Band 7. Stuttgart (German). [*Pinus kesiya*, *P. merkusii* = *P. latteri*, phytogeography, ecology, maps].

15118 Whang, S. S. & R. S. Hill (1999). Late Palaeocene Cupressaceae macrofossils at Lake Bungarby, New South Wales. Australian Syst. Bot. 12: 241–254. [fossils assigned to *Libocedrus*: *L. acutifolius* sp. nov., *L. obtusifolius* sp. nov., and organ genera *Bungarbia* and *Monarophyllum*].

15118a Whang, S. S., K. Kim & R. S. Hill (2004). Cuticle micromorphology of leaves of *Pinus* (Pinaceae) from North America. Bot. J. Linn. Soc. 144 (3): 303–320. (SEM ill., tables). [31 taxa studied; some cuticle characters support division into two subgenera and *P. resinosa* is distinct from other North American pines in subgenus *Pinus*].

15119 Whang, S. S., J. H. Pak, R. S. Hill & K. Kim (2001). Cuticle micromorphology of leaves of *Pinus* (Pinaceae) from Mexico and Central America. Bot. J. Linn. Soc. 135 (4): 349–373. (SEM ill., tables). [34 taxa studied; cuticle characters largely confirm classification given in Farjon & Styles (1997) but are at variance with classifications based on chloroplast DNA].

15120 Wheeler, G. M. (1878). Report upon U.S. geographical surveys west of the 100th meridian in charge of G. M. Wheeler. Vol. VI. Botany. Washington, D.C. [see G. Engelmann, 1878].

15127 Wheeler, N. C. & W. B. Critchfield (1985). The distribution and botanical characteristics of Lodgepole Pine: biogeographical and management implications. Pp. 1–13 in: D. M. Baumgartner (ed.). Lodgepole Pine – the species and its management. Washington State University, Pullman. [*Pinus contorta*, 3 ssp., *P. banksiana*].

15130 Wheeler, N. C., R. P. Guries & D. M. O'Malley (1983). Biosystematics of the genus *Pinus*, subsection *Contortae*. Biochem. Syst. Ecol. 11: 333–340. [*P. banksiana, P. contorta, P. clausa, P. virginiana*].

15131 White, C. T. (1923). A new Conifer from southern Queensland. Proc. Linn. Soc. New South Wales 48 (4): 449–450, t. 37. [*Callitris baileyi* sp. nov.]

15132 White, C. T. (1926). Ligneous plants collected in New Caledonia by C. T. White in 1923. J. Arnold Arbor. 7 (2): 74–103. [Gymnospermae pp. 76–85, Araucariaceae, Podocarpaceae, *Podocarpus, Dacrydium, Acmopyle, Agathis, Araucaria*, div. spp., *Araucaria columnaris* f. *luxurians* (Brongn. & Gris) E. H. Wilson].

15133 White, C. T. (1933). Ligneous plants collected for the Arnold Arboretum in North Queensland by S. F. Kajewski in 1929. Contr. Arnold Arbor. 4: 5–115, t. 1–9. [*Podocarpus dispermus* sp. nov., p. 10, t. 1].

15135 White, M. E. (1981). The cones of *Walkomiella australis* (Feist.) Florin. Palaeobotanist 28–29: 75–80. [palaeobotany, Permian conifer].

15136 White, M. E. (1981). Revision of the Talbragar Fish Bed Flora (Jurassic) of New South Wales. Rec. Austral. Mus. 33: 695–721. [*Agathis jurassica* sp. nov. (fossil), palaeobotany].

15140 Whiting, A. F. (1942). Junipers of the Flagstaff region. Plateau 15: 23–31. [*Juniperus* spp.].

15141 Whitmore, T. C. (1977). A first look at *Agathis*. Trop. Forest. Pap. 11. Dept. Forestry, Commonwealth Forestry Inst., Univ. Oxford. (54 pp.). [photographs].

15143 Whitmore, T. C. (1980). A monograph of *Agathis*. Plant Syst. Evol. 135: 41–69. (ill., key, distr. maps). [*A. dammara* ssp. *flavescens* (Ridley) comb. nov., p. 59, *A. robusta* ssp. *nesophila* ssp. nov., p. 64].

15144 Whitmore, T. C. & C. N. Page (1980). Evolutionary implications of the distribution and ecology of the tropical genus *Agathis*. New Phytol. 84: 407–416.

15150 Whyte, A. (1892). Botany of Milanji in Nyassaland. Bull. Misc. Inform. 65: 121–124. [discovery of *Widdringtonia whytei*].

15160 Wiesehuegel, E. G. (1932). Diagnostic characteristics of the xylem of the North American *Abies*. Bot. Gaz. (Crawfordsville) 93: 55–70. (16 figs.).

15161 Wiggins, I. L. (1933). New plants from Baja California. Contr. Dudley Herb. 1 (5): 161–178, pl. 11–17. [*Cupressus montana* sp. nov., p. 161, pl. 11, f. 1].

15170 Wiggins, I. L. (1935). An extension of the known range of the Mexican bald cypress. Torreya 35: 65–67. [*Taxodium mucronatum*].

15170a Wiggins, I. L. (1940). Yellow pines and other conifers observed in Lower California. J. New York Bot. Gard. 41: 267–269. [*Cupressus arizonica* var. *montana, Pinus jeffreyi*].

15171 Wight, W. F. (1908). A new larch from Alaska. Smithsonian Misc. Collect. 1 (50): 174, pl. 17. [*Larix alaskensis* sp. nov. = *L. laricina*].

15172 Wilde, J. J. F. E. de (1961). *Cedrus atlantica* Manetti in Marokko. Med. Bot. Tuinen Belmonte Arbor. Wageningen 5 (4): 93–103. [map, habitus photogr., ecology].

15172a Wilde, M. H. & A. J. Eames (1952). The ovule and 'seed' of *Araucaria bidwillii* with discussion of the taxonomy of the genus. II. Taxonomy. Ann. Bot. (London), n.s. 16: 27–47.

15172c Willard, D. (1994). Giant Sequoia groves of the Sierra Nevada. privately publ., Berkeley, California. [*Sequoiadendron giganteum*].

15173 Willdenow, C. L. (1796). Berlinische Baumzucht, oder Beschreibung der in den Gärten um Berlin, im Freien ausdauernden Bäume und Sträucher, ... Berlin. [*Juniperus nana* sp. nov. = *J. communis*, p. 159].

15174   Willdenow, C. L. (1806). Caroli a Linné Species plantarum exhibente plantas rite cognitas ad genera relatas cum differentiis specificis, ... Vol. 4 (2): 631–1157. Berlin. [*Juniperus excelsa* "sp. nov.", p. 852 (but see F. A. Marschall von Bieberstein, 1800), *J. foetidissima* sp. nov., p. 853, *Taxus montana* Humb. & Bonpl. ex Willd. = *Prumnopitys montana*, p. 857].

15180   Williams, L. (1955). *Pinus caribaea*. Ceiba 4: 299–300. [*P. caribaea, P. oocarpa*, hybridization].

15191   Willkomm, M. (1890). Ueber die Herkunft der "Ceder von Goa" (*Cupressus glauca* Lam.). Wiener Ill. Garten-Zeitung 15 (3): 98–100. [= *C. lusitanica* Mill.].

15210   Wilson, E. H. (1913). A Naturalist in Western China, Vol. 1–2. London. [descr. expedition routes in Sichuan].

15221   Wilson, E. H. (1920). The Liukiu Islands and their ligneous vegetation. J. Arnold Arbor. 1: 171–186. [*Juniperus conferta* Parl. = *J. taxifolia*, see also E. H. Walker, 1976].

15230   Wilson, E. H. (1920). Four new conifers from Korea. J. Arnold Arbor. 1: 186–190. [*Abies koreana* sp. nov., p. 188, formae nov. in *Abies* et *Larix*, (descr. of) *Thuja koraiensis* Nakai].

15240   Wilson, E. H. (1926). A noble Chinese tree. Garden 90: 249–250, 2 figs. [*Platycladus orientalis*].

15260   Wilson, E. H. (1926). *Thuja orientalis* Linnaeus. J. Arnold Arbor. 7: 71–74, pl. 1. [= *Platycladus orientalis*, the plate depicts *Juniperus chinensis* L., see E. H. Wilson, 1930].

15270   Wilson, E. H. (1926). *Taiwania cryptomerioides* Hayata. J. Arnold Arbor. 7: 229–231, pl. 3.

15271   Wilson, E. H. (1927). *Juniperus procera* Hochst. J. Arnold Arbor. 8: 1–2. (ill.).

15280   Wilson, E. H. (1930). *Thuja orientalis* and *Juniperus chinensis*. J. Arnold Arbor. 11: 135–136, pl. 23. (correction to E. H. Wilson, 1926). [*Thuja orientalis = Platycladus orientalis*].

15281   Winslow, C. F. (1854). The "Big Tree". Calif. Farmer 2: 58. [*Washingtonia californica, Taxodium washingtonianum*, nom. prov. sine descr. (see also W. J. Hooker, 1855) = *Sequoiadendron giganteum*].

15290   Wislizenus, F. A. (1848). Memoir of a tour to Northern Mexico, ... Washington. (German transl.: Braunschweig, 1850). [see G. Engelmann, 1848].

15295   Wissman, H. von (1972). Die *Juniperus*-gebirgswälder in Arabien. Ihre Stellung zwischen dem borealen und tropisch-afrikanischen Florenreich. In: C. Troll (ed.). Edrwissenschaftliche Forschung 5: 157–176. (Wiesbaden). [*Juniperus procera*].

15302   Wittlake, E. B. (1975). The androstrobilus of *Glyptostrobus nordenskioldii* (Heer) Brown. Amer. Midl. Naturalist 94: 215–223.

15310   Wolf, C. B. & W. W. Wagener (1948). The New World cypresses. Aliso 1. (444 pp., ill.). [*Cupressus, C. stephensonii* C. B. Wolf, sp. nov., p. 125; *C. abramsiana* C. B. Wolf, sp. nov., p. 215, monograph/revision; new spp. red. by E. L. Little, 1966, 1970 to var. : *C. arizonica* var. *stephensonii* and *C. goveniana* var. *abramsiana*].

15311   Wolf, E. L. (1922). Dendrologische Mitteil-ungen. Mitt. Deutsch. Dendrol. Ges. 1922: 211–217. [*Juniperus niemannii* sp. nov. = *J. communis* var. *communis*; also in: Bot. Mater. Gerb. Glavn. Bot. Sada RSFSR 3: 37, 1922].

15330   Wolf, E. L. (1925). Die dahurische Lärche, *Larix dahurica* Turcz. Mitt. Deutsch. Dendrol. Ges. 35: 328–330. [= *L. gmelinii*].

15350   Wolfe, J. A. (1975). Some aspects of plant geography of the northern hemisphere during the late Cretaceous and Tertiary. Ann. Missouri Bot. Gard. 62: 264–279. [gymnosperms].

15351   Wolfe, J. A. & H. E. Schorn (1990). Taxonomic revision of the Spermatopsida of the Oligocene Creede Flora, southern Colorado. U.S. Geol. Survey Bull. 1923: 1–40, t. 1–13. [palaeobotany, *Juniperus, Abies, Pinus, Picea*, ill.].

15351d   Wolff, R. L., L. G. Deluc, A. M. Marpeau & B. Comps (1997). Chemotaxonomic differentia-tion of conifer families and genera based on the seed oil fatty acid compositions: multivariate analyses. Trees 12: 57–65. [Pinaceae: *Abies, Cedrus, Larix, Picea, Pinus, Pseudotsuga, Tsuga*; Cupressaceae, multivariate analysis of composite data (!), taxonomy].

15351e   Wolff, R. L., B. Comps, A. M. Marpeau & L. G. Deluc (1997). Taxonomy of *Pinus* species based on the seed oil fally acid compositions. Trees 12: 113–118. [*Pinus*, 49 spp. analysed, multivariate analysis on 13 variables (of composite data!), taxonomy, cold-acclimation].

15351w   Woltz, P. (1969). Une nouvelle espèce de *Podocarpus* de Madagascar, *P. gaussenii*. Trav. Lab. Forest. Toulouse T. 1 (8, art. 2): 1–8. (ill.). [*Podocarpus gaussenii* sp. nov. = *Afrocarpus gaussenii*].

15352   Woltz, P. (1983). Lettre Botanique de l'Hémisphère Sud: remarques sur les Gymnospermes du Chili. Rapport de Mission: Ministère Rel. Ext. (Paris); Universidad de Chile (Santiago): 1–49. Univ. d'Aix-Marseille. [Cupressaceae: *Austrocedrus chilensis, Pilgerodendron uviferum, Fitzroya cupressoides; Araucaria araucana*, Podocarpaceae, ill., maps].

15353   Woltz, P. (1985). Place des gymnospermes endémiques des Andes méridionales dans la végétation du Chili. Lazaroa (Madrid) 7 (2): 293–314. [Cupressaceae: *Austrocedrus chilensis, Fitzroya cupressoides, Pilgerodendron uviferum; Araucaria araucana*, Podocarpaceae, map].

15354 Woltz, P. & C. Chatelet (1973). Remarques sur l'évolution vasculaire de quelques plantules d' Austrolibocédrées: *Libocedrus yateensis* Guillaumin, *Libocedrus doniana* Endlicher. Bull. Soc. Bot. France 120: 303–310. (ill.).

15355 Woltz, P. & J. F. Cherrier (1984). A propos du *Neocallitropsis pancheri* (Carrière) Laubenf. Cupressaceae endémique de Nouvelle-Calédonie et d'Evolution vasculaire de la plantule. Bull. Soc. Bot. France 131 (3): 191–199.

15356 Woltz, P., M. L. Rouane & M. Gondran (1993). Apport des cotyledons dans l'évolution des Podocarpineae. Gaussenia 8: 6–15. [Podocarpaceae, seedling anatomy and morphology].

15357 Woltz, P., R. A. Stockey, M. Gondran & J.-F. Cherrier (1995). Interspecific parasitism in the gymnosperms: unpublished data on two endemic New Caledonian Podocarpaceae using scanning electron microscopy. Acta Bot. Gallica 141 (6/7): 731–746. [*Parasitaxus usta, Falcatifolium taxoides*, parasitism linked to mycelian symbiosis].

15360 Woodbury, A. M. (1947). Distribution of pigmy conifers in Utah and northeastern Arizona. Ecology 28: 113–126. [*Juniperus, Pinus*].

15360b Wormald, T. J. (1975). *Pinus patula*. Trop. Forest. Pap. 7: 1–172. (+ appendices & plates). Univ. of Oxford.

15370 Wright, J. W. (1955). Species crossability in spruce in relation to distribution and taxonomy. Forest Sci. 1: 319–349. [*Picea*].

15380 Wright, J. W. (1959). Species hybridization in the white pines. Forest Sci. 5: 210–222. [*Pinus* sect. *Strobus*].

15390 Wu, C. L. (1956). The taxonomic revision and phytogeographical study of Chinese pines. Acta Phytotax. Sin. 5: 131–164, ill. (Chin., Eng. summ.). [*Pinus* spp., *P. massoniana* var. *henryi* (Mast.) comb. et stat. nov., p. 153].

15391 Wu, H. Q. (1987). Materials for the study on the genus *Picea* Dietr. in Northeast China. Bull. Bot. Res. North-East. Forest. Inst. 7 (2): 139–145. (Chin., Eng. summ.). [*P. meyeri* var. *mongolica* var. nov., p. 153].

15400 Wu, M. H. (1976). *Abies beshanzuensis* M. H. Wu, sp. nov. Acta Phytotax. Sin. 14 (2): 16–17, t. 1, f. 1.

15405 Xiang, Q. P. (1997). *Abies fansipanensis* – a new species of the genus *Abies* from Vietnam. Acta Phytotax. Sin. 35 (4): 356–359. (Chin., Lat.) [*Abies fansipanensis* sp. nov., p. 356].

15405a Xiang, Q. P. & A. Farjon (2003). Cuticle morphology of a newly discovered conifer, *Xanthocyparis vietnamensis* (Cupressaceae), and a comparison with some of its nearest relatives. Bot. J. Linn. Soc. 143: 315–322. (ill.). [*X. vietnamensis, X. nootkatensis, Chamaecyparis formosana, C. obtusa, Cupressus arizonica, C. funebris*].

15405b Xiang, Q. P., A. Farjon, Z. Y. Li, L. K. Fu & Z. Y. Liu (2002). *Thuja sutchuenensis*: a rediscovered species of the Cupressaceae. Bot. J. Linn. Soc. 139 (3): 305–310.

15405c Xiang, Q. P., Q. Y. Xiang, A. Liston, L. K. Fu & D. Z. Fu (2000). Length variation of the nuclear ribosomal DNA Internal Transcribed Spacer in the genus *Abies*, with reference to its systematic utility in Pinaceae. (Chin., Eng. summ.). Acta Bot. Sinica 42 (9): 946–951.

Xiang, Q. P., Q. Y. Xiang, A. Liston & X. C. Zhang (2004). Phylogenetic relationships in *Abies* (Pinaceae): evidence from PCR-RFLP of the nuclear ribosomal DNA internal transcribed spacer region. Bot. J. Linn. Soc. 145: 425–435. (ill., cladogram, maps).

15408 Xie, Z. Q., W. L. Chen, M. X. Jiang, H. D. Huang & R. G. Zhu (1995). A preliminary study on the population of *Cathaya argyrophylla* in Bamian-shan mountain. Acta Bot. Sin. 37 (1): 58–65. (Chin., Eng. summ.). [total wild growing trees in China ca. 5000].

15409 Xing, S. P., Q. Zhang, Y. X. Hu, Z. K. Chen & J. X. Lin (1999). The meganism of pollination in *Platycladus orientalis* and *Thuja occidentalis* (Cupressaceae). Acta Bot. Sin. 41: 130–132.

15410 Yamanaka, T. (1975). Ecology of *Pseudotsuga japonica* and other coniferous forests in eastern Shikoku. Mem. Nat. Sci. Mus. 8. Tokyo.

15420 Yamazaki, T. (1935). The natural distribution and association of *Larix dahurica*. Bull. Soc. Forest. Kyoto Imp. Univ. 7: 1–54, 9 pl. (Japan.). [= *Larix gmelinii*, ecology].

15428 Yang, H. & J. H. Jin (2000). Phytogeographic history and evolutionary stasis of *Metasequoia*: geological and genetic information contrasted. Acta Palaeontol. Sin. 39 (suppl.): 288–307.

15430 Yang, Y. C., Y. L. Chou & S. C. Nie (1964). The genus *Larix* in the Hsiaoshinganling-Changpaishan Region. Acta Phytotax. Sin. 9 (2): 168–178, t. 16–22. (Chin., ill.). [*Larix heilingensis* sp. nov. = *L. gmelinii*, p. 173, + new varieties and forms of this species].

15460 Yao, T. Y. (1936). A statement of the hybrid *Pinus tabulaeformis* var. *taihanshanensis* recently discovered at Taihanshan, Honan, China. J. Agric. Assoc. China 144: 67–68. [*P. tabuliformis*].

15461 Yao, X. L., T. N. Taylor & E. L. Taylor (1993). The Triassic seed cone *Telemachus* from Antarctica. Rev. Palaeobot. Palynol. 78: 269–276. [Coniferales, palaeobotany, ill.].

15462 Yao, X. L., T. N. Taylor & E. L. Taylor (1997). A taxodiaceous seed cone from the Triassic of Antarctica. Amer. J. Bot. 84 (3): 343–354. [*Parasciadopitys aequata* gen. et sp. nov., p. 343, Taxodiaceae, Voltziaceae, Sciadopityaceae].

15462a Yao, X. L., Z. Y. Zhou & B. Zhang (1989). On the occurrence of *Sewardiodendron laxum* Florin (Taxodiaceae) in the Middle Jurassic from Yima, Henan. Chin. Sci. Bull. 34, no. 23. (ill.).

15463 Yao, X. L., Z. Y. Zhou & B. Zhang (1998). Reconstruction of the Jurassic conifer *Sewardiodendron laxum* (Taxodiaceae). Amer. J. Bot. 85 (9): 1289–1300. (ill.).

15466 Yao, Z. Q., L. J. Liu, G. W. Rothwell & G. Mapes (2000). *Szecladia* new genus, a Late Permian conifer with multiveined leaves from south China. J. Paleontol. 74 (3): 524–531. [*Szecladia multinervia*, ill.].

15470 Yarie, J. (1983). Forest Community Classification of the Porcupine River Drainage, Interior Alaska, and its application to forest management. U.S. Forest Serv. Gen. Techn. Rep. PNW-154. Pacific Northwest Forest and Range Exp. Stat., Portland, Oregon. [Pinaceae, *Picea glauca*, *P. mariana*, ecology].

15480 Yatsenko-Khmelevsky, A. A. & E. V. Budkevich (1958). On the woody anatomy of *Cathaya argyrophylla* Chun et Kuang (Pinaceae). Bot. Žurn. (Moscow & Leningrad) 43: 477–480. (Russ., Eng. summ.).

15480b Yeaton, R. I. (1982). The altitudinal distribution of the genus *Pinus* in the western United States and Mexico. Bol. Soc. Bot. Mexico 42: 55–71. [*Pinus*, *Haploxylon* (= subgen. *Strobus*), *Diploxylon* (= subgen. *Pinus*].

15480c Yeaton, R. I., R. W. Yeaton & J. P. Waggoner III (1983). Changes in morphological characteristics of *Pinus engelmannii* over an elevational gradient in Durango, Mexico. Madroño 30: 168–175. [*P. engelmannii*, *P. engelmannii* var. *blancoi*, clinal variation].

15481 Ying, S. S. (1974). *Tsuga chinensis* (Franch.) Pritz. var. *daibuensis*, var. nov. in: Bull. Exp. Forest Natl. Taiwan Univ. 114: 150. (Chin., Lat. diagn.). (also publ. in: Quart. J. Chin. Forest. 8 (3): 98, 1975).

15500 Yoshie, F. & A. Sakai (1985). Types of Florin rings, distributional patterns of epicuticular wax, and their relationships in the genus *Pinus*. Canad. J. Bot. 63: 2150–2158. (ill.). [*Pinus* subg. *Pinus*, *Pinus* subg. *Strobus*].

15506 Yu, Y. F. & L. K. Fu (1998). Notes on gymnosperms 2. New taxa and combinations in *Juniperus* (Cupressaceae) and *Ephedra* (Ephedraceae) from China. Novon 7 (4): 443–444. [*Juniperus chengii* sp. nov., p. 443, *J. baimashanensis* sp. nov., p. 443, *J. pingii* W. C. Cheng ex Ferré var. *carinata*, var. nov., p. 443, *J. squamata* Buch.-Ham. ex D. Don var. *parvifolia*, var. nov., p. 444, *J. squamata* Buch.-Ham. ex D. Don var. *hongxiensis*, var. nov., p. 444, *J. sabina* L. var. *yulinensis* (T. C. Chang & L. K. Fu), comb. nov., p. 444, *J. sabina* L. var. *erectopatens* (W. C. Cheng & L. K. Fu), comb. nov., p. 444].

15508 Yuncker, T. G. (1959). Plants of Tonga. Bishop Mus. Bull. 220: 3–283. (ill.). [*Podocarpus pallidus* N. E. Gray, sp. nov., p. 46, f. 6].

15510 Zajkov, G. I. (1968). On the phylogeny and geography of the Siberian spruce (*Picea obovata* Ledeb.). Izv. Omskogo Geogr. Obsc. Sojuza SSSR 9: 134–139. (Russ.).

15513 Zamjatnin, B. (1963). Observationes nonnullae de *Microbiota decussata* Kom. Not. Syst. Herb. Hort. Bot. Petrop. 22: 43–50. (Russ.) [monoecy observed in wild growing plants; comparison with *Platycladus orientalis*].

15514 Zamora, S. C. (1981). Algunos aspectos sobre *Pinus oocarpa* Schiede, en el estado de Chiapas. Revista Ci. Forest. 6 (32): 25–53.

15515 Zamora, S. C. & V. Velasco (1977). *Pinus strobus* var. *chiapensis*, una especie en peligro de extinción en el estado de Chiapas. Revista Ci. Forest. 2 (8): 3–23. [= *Pinus chiapensis*].

15516 Zamora, S. C. & V. Velasco (1978). Contribucion al estudio ecologico de los pinos del estado de Chiapas. SAHR (Forestal & Fauna) Techn. bulletin 56. [*Pinus* spp.].

15520 Zanoni, T. A. (1978). The American junipers of the section *Sabina* (*Juniperus*, Cupressaceae) a century later. Phytologia 38 (6): 433–454. [*J. deppeana* var. *patoniana* (Martínez) comb. et stat. nov., p. 438].

15521 Zanoni, T. A. (1980). Notes on *Cupressus* in Mexico. Bol. Soc. Bot. México 39: 128–133. [*C. benthamii*, nomenclature].

15540 Zanoni, T. A. (1982). Flora de Veracruz: fasciculo 23. Cupressaceae. Xalapa, INIREB. [*Cupressus benthamii*, *Juniperus deppeana*, *J. flaccida*, *J. monticola*].

15545 Zanoni, T. A. (1982). Flora de Veracruz: fasciculo 25. Taxodiaceae. Xalapa, INIREB. [*Taxodium mucronatum*].

15550 Zanoni, T. A. & R. P. Adams (1973). Distribution and synonymy of *Juniperus californica* Carrière (Cupressaceae) in Baja California, Mexico. Bull. Torrey Bot. Club 100 (6): 364–367.

15560 Zanoni, T. A. & R. P. Adams (1975). The genus *Juniperus* (Cupressaceae) in Mexico and Guatemala: numerical and morhological analysis. Bol. Soc. Bot. México 35: 69–92. (fig.). [18 taxa (24 OTUs), numerical taxonomy, 4 similarity groupings].

15565   Zanoni, T. A. & R. P. Adams (1976). The genus *Juniperus* (Cupressaceae) in Mexico and Guatemala: numerical and chemosystematic analysis. Biochem. Syst. Ecol. 4 (3): 147–158. [as in Zanoni & Adams, 1975 but with data from terpenoid compounds].

15570   Zanoni, T. A. & R. P. Adams (1979). The genus *Juniperus* (Cupressaceae) in Mexico and Guatemala: synonymy, key, and distributions of the taxa. Bol. Soc. Bot. México 38: 83–121, figs. 1–7. [16 species recognised, distribution maps].

15588   Zavala C., F. & J. L. Campos D. (1993). Una nueva localidad de *Pinus discolor* Bailey & Hawksworth en el centro de México. Acta Bot. Mex. 25: 21–25. [= *P. cembroides* var. *bicolor*].

15590   Zavarin, E. (1988). Taxonomy of pinyon pines. in: M.-F. Passini, D. Cibrian Tovar & T. Eguiluz Piedra (comp.). II Simposio nacional sobre pinos piñoneros, 6-7-8 de agosto de 1987, pp. 29–40. Chapingo, México, D.F. [*Pinus* subsect. *Cembroides*; new comb. not validly publ.].

15600   Zavarin, E. & K. Snajberk (1965). Chemotaxonomy of the genus *Abies*. I. Survey of the terpenes present in *Abies* balsams. Phytochemistry 4: 141–148.

15608   Zavarin, E. & K. Snajberk (1985). Monoterpenoid and morphological differentiation within *Pinus cembroides*. Biochem. Syst. Ecol. 13 (2): 89–104.

15609   Zavarin, E. & K. Snajberk (1986). Monoterpenoid differentiation in relation to the morphology of *Pinus discolor* and *Pinus johannis*. Biochem. Syst. Ecol. 14 (1): 1–11. [*Pinus cembroides* var. *bicolor*].

15610   Zavarin, E. & K. Snajberk (1987). Monoterpene differentiation in relation to the morphology of *Pinus culminicola*, *Pinus nelsonii*, *Pinus pinceana* and *Pinus maximartinezii*. Biochem. Syst. Ecol. 15: 307–312. (ill.).

15615   Zavarin, E., K. Snajberk & L.G. Cool (1989). Monoterpenoid differentiation in relation to the morphology of *Pinus edulis*. Biochem. Syst. Ecol. 17: 271–282. [*Pinus* subsect. *Cembroides*, *P. monophylla*, *P. discolor*, *P. edulis*, chemistry of monoterpenes].

15617   Zavarin, E., K. Snajberk & R. Debry (1980). Terpenoid and morphological variability of *Pinus quadrifolia* and its natural hybridization with *Pinus monophylla* in northern Baja California and adjoining United States. Biochem. Syst. Ecol. 8: 225–235.

15620   Zavarin, E., K. Snajberk & C. J. Lee (1978). Chemical relationships between firs of Japan and Taiwan. Biochem. Syst. Ecol. 6: 177–184. [*Abies mariesii*, *A. sachalinensis*, *A. veitchii*, *A. firma*, *A. kawakamii*, *A. homolepis*].

15627   Zhang, D., M. A. Dirr & R. A. Price (1999). Classification of cultivated *Cephalotaxus* species based on *rbc*L sequences. In: S. Andrews, A. Leslie & C. Alexander (eds.). Taxonomy of cultivated plants –Third International Symposium, pp. 265–275. Royal Botanic Gardens, Kew. [only 3 species retained: *C. harringtonii* (syn. *C. drupacea*, *C. koreana*, *C. wilsoniana*), *C. fortunei* (syn. *C. sinensis* of possible hybrid origin), *C. oliveri*].

15628   Zhang, S. & S. J. Xu (1997). *Glyptostrobus pensilis* (Staunton) Koch, 1873. Enzyklopädie der Holzgewächse III-1 (7): 1–8. (ill., map).

15629   Zhang, S. J., C. X. Li & X. Y. Yuan (1995). A new variety of *Pinus densiflora* Sieb. & Zucc. Bull. Bot. Res. NE Forest. Univ. 15 (3) 338–341. [*P. densiflora* var. *zhangwuensis* var. nov., p. 338, ill.].

15629a   Zhang, X. K., Z. J. Mao, H. Song & B. Meng (2002). Genetic relationship between 5 species of *Larix* based on allozymes. Bull. Bot. Res. (China) 22 (2): 224–230. (Chin., Eng. summ., maps).

15630   Zhao, N. (1980). Species nova generis Cupressi. Acta Phytotax. Sin. 18 (2): 210. [*Cupressus jiangeensis* sp. nov.].

15633   Zhong, Y. & K. Huang (1990). A new species of *Pinus* L. from Guanxi. Guihaia 10: 287–289. [*Pinus crassicorticea* sp. nov., p. 287, *P. massoniana*].

15635   Zhou, Q. X., S. Ge, Z. J. Gu & Z. S. Yue (1998). Genetic variation and relationships within *Taxus* and between the genus and *Pseudotaxus* in China. Acta Phytotax. Sin. 36 (4): 323–332. (Chin., Eng. summ.).

15636   Zhou, Z. Y. (1987). *Elatides harisii* sp. nov. from the Lower Cretaceous of Liaoning, China. Rev. Palaeobot. Palynol. 51: 189–204.

15640   Zimmermann, W. et al., eds. (1966). Handbuch der Pflanzenanatomie, 2. Aufl. Berlin–Nikolassee. [see K. Napp-Zinn, 1966].

15650   Zobel, B. J. (1951). The natural hybrid between Coulter and Jeffrey pines. Evolution 5: 405–413. (ill.). [*Pinus coulteri* × *P. jeffreyi*].

15651   Zobel, B. J. (1951). Oleoresin composition as a determinant of pine hybridity. Bot. Gaz. (Crawfordsville) 113: 221–227. [*Pinus*].

15655   Zobel, B. J. (1953). Geographic range and intraspecific variation of Coulter pine. Madroño 12 (1): 1–7. [*Pinus coulteri*, morphology, chemistry].

15660   Zobel, B. J. (1969). Factors affecting the distribution of *Pinus pungens*, an Appalachian endemic. Ecol. Monogr. 39: 303–333. (ill.).

15670   Zobel, D. B. (1983). Twig elongation patterns of *Chamaecyparis lawsoniana*. Bot. Gaz. (Crawfordsville) 144: 92–103. [phenology].

15672   Zobel, D. B. & F. Cech (1957). Pines from Nuevo Leon, Mexico. Madroño 14: 133–144. [*Pinus* spp.].

15680   Zodda, G. (1903). Il *Pinus pinea* L. nel Pontico di Messina. Malpighia 17 (11–12): 488–491.

15690c   Zou, H. Y. & F. C. Zhang (1996). A study on numerical taxonomy of longbracted hemlock. J. Nanjing Forest. Univ. 20 (1): 43–47. (Chin., Eng. summ.). [*Nothotsuga longibracteata, Tsuga* sect. *Canadensis* sect. nov., p. 46].

15691   Zuccarini, J. G. (1832). Plantarum novarum vel minus cognitarum, quae in Horto Botanico Herbarioque Regio Monacensi servantur, descriptio. Abh. Math.-Phys. Cl. Königl. Bayer. Akad. Wiss. 1: 287–396. (publ. date of Abh. 1: 1832). [*Pinus cembroides* sp. nov., p. 392; also publ. in Flora 15 (2), Beibl. 93, 1832].

INDEX OF BOTANICAL NAMES, WITH REFERENCES TO PUBLICATIONS

The index of botanical names lists alphabetically all the names of taxa (living and fossil) mentioned in this bibliography, both within the titles and in the accompanying annotations. The index only enumerates names of gymnosperms, even when on a few occasions names of angiosperms appear in the bibliography. The single exception is *Thuja aphylla* L., a misidentification of *Tamarix articulata*, on which some authors have based new combinations in Cupressaceae. The emphasis on the families Araucariaceae, Cupressaceae (including Taxodiaceae) and Pinaceae in the first edition (1990) has been reduced by the inclusion of many additional taxa, including names of fossils. Despite this, the list is less complete for these newly added taxa, especially species in Podocarpaceae (see the Introduction to this edition) than for the originally included families.

It has been one of the aims of this bibliography to enumerate the publications where protologues appear. Accordingly, new names and/or combinations were annotated with priority over names cited in a different context. However, the (often numerous) new names which appeared in a few publications of doubtful scientific quality, or names apparently based on horticultural varieties in case they were not (later) attributed to botanical taxa, were deliberately left out. It must be expressly stated that in a selected bibliography of this kind no claim of completeness can or will be made. Since publication of the first edition of A Bibliography of Conifers, I have published a World Checklist and Bibliography of Conifers (Farjon, 1998, 2nd edition 2001) in which I have attempted to include all validly published names for all conifer species and their infraspecific taxa. The bibliography in that work is only a very selective choice of important publications.

Only the correct spellings of names are given in the index. This means, that they refer equally to the spelling as originally given by the author of that name and to the orthographically corrected form of that name. Both spellings (if appropriate) are given in the bibliography. Orthographic correction (if not carried out earlier) has been employed very restrictively in this bibliography: only grammatical errors have been amended.

With the names of the taxa the authorities are cited in the abbreviated form as given in Authors of Plant Names (Brummitt & Powell, 1992); names of authors not occurring in that book are cited as full as the relevant publication permits and is considered necessary. This applies in particular to authors of names of fossil conifers. (Later) homonyms are cited with their 'authors' following the name with the correct authority.

The publications are cited with their numbers as given in the bibliography. Numbers in bold print refer to the protologues or to the publications where new combinations were made. Reference is usually limited to entries where the name of a particular taxon is made explicit in this bibliography, but exceptions are made when it was deemed appropriate to do so. Especially in floras, manuals and monographs of large groups, too many names are cited to be indexed for this bibliography. Such publications may be indexed with the names of genera; more commonly they have been listed in the bibliography under the appropriate category headings (see the Introduction and below) and, when of a general nature in relation to conifers, are not indexed. This implies that it may be worthwhile to consult them in addition to the references given with a taxon at family rank or lower. The reorganisation of more general publications on conifers in this edition has relocated many publications from the main list to their respective categories. In order to find these references from the Index of botanical names; abbreviated references to the categories have been added in parentheses to the numbers. These abbreviations refer to the following category headings, appearing in that order in the Bibliography.

(Biblio)	BIBLIOGRAPHIES	pp. 5–6
(Floras)	FLORAS	pp. 6–14
(Manuals)	MANUALS	pp. 14–19
(Gymnos)	GYMNOSPERMS (general titles)	pp. 19–24
(Conif)	CONIFERALES (general titles on conifers, incl. "taxads")	pp. 24–42
(Arauca)	ARAUCARIACEAE (general titles)	pp. 42–43
(Cupres)	CUPRESSACEAE (general titles)	pp. 43–45
(Phyllo)	PHYLLOCLADACEAE (general titles)	p. 45
(Pina)	PINACEAE (general titles)	pp. 45–49
(Podocarp)	PODOCARPACEAE (general titles)	pp. 49–51
(Taxa)	TAXACEAE	p. 51
(Taxod)	TAXODIACEAE	pp. 51–52

Aachenia debeyi **07315**
Abies Mill. 00156 (Pina), 00157, 00870, 00940, 01390, 01495 (Manuals), 01550 (Pina), 01870, 02120a, 02332 (Conif), 02360 (Floras), 02501, 02550, 02580, 02731, 03141, 03390 (Conif), 03520, 03654, 03690 (Pina), 03710, 03720 (Biblio), 03765, 03855 (Pina), 04390, 04450, 04520, 04770, 04915, 05030 (Floras), 05049 (Floras), 05840, 05900, 05910, 06421, 06430, 06460, 06510b, 06660, 06770a, 06870, 07190, 07543, 07732, 07780, 07840, 08227d (Pina), 08229 (Pina), 08260, 08360 (Conif), 08410, 08710, 08880, 09180, 09191 (Conif), 09340 (Conif), 09650, 09660, 09680, 09690, 09700, 09710, 09750 (Pina), 09801, 09810, 09850, 09860

Index

The compilation of this index of botanical names was completed on 1 August 2004. Botanical literature up to that date may have been incorporated in the preceding bibliography; if not, such names as were (first) published in them are consequently not indexed. The author would be greatly indebted to workers in the field of conifer taxonomy and related disciplines for any indications of omissions of literature which might be added to a future edition of this bibliography.

Kew, August 2004

Address of the author:
Aljos Farjon,
Herbarium, Royal Botanic Gardens, Kew
Richmond, Surrey, TW9 3AB England, UK
a.farjon@rbgkew.org.uk